The Ethical Frontier of AI and Data Analysis

Rajeev Kumar
Moradabad Institute of Technology, India

Ankush Joshi
COER University, Roorkee, India

Hari Om Sharan
Rama University, Kanpur, India

Sheng-Lung Peng
College of Innovative Design and Management, National Taipei University of Business, Taiwan

Chetan R. Dudhagara
Anand Agricultural University, India

A volume in the Advances in Computational Intelligence and Robotics (ACIR) Book Series

Published in the United States of America by
IGI Global
Engineering Science Reference (an imprint of IGI Global)
701 E. Chocolate Avenue
Hershey PA, USA 17033
Tel: 717-533-8845
Fax: 717-533-8661
E-mail: cust@igi-global.com
Web site: http://www.igi-global.com

Library of Congress Cataloging-in-Publication Data

CIP Data in progress

This book is published in the IGI Global book series Advances in Computational Intelligence and Robotics (ACIR) (ISSN: 2327-0411; eISSN: 2327-042X)

British Cataloguing in Publication Data
A Cataloguing in Publication record for this book is available from the British Library.

All work contributed to this book is new, previously-unpublished material. The views expressed in this book are those of the authors, but not necessarily of the publisher.

For electronic access to this publication, please contact: eresources@igi-global.com.

Advances in Computational Intelligence and Robotics (ACIR) Book Series

Ivan Giannoccaro
University of Salento, Italy

ISSN:2327-0411
EISSN:2327-042X

MISSION

While intelligence is traditionally a term applied to humans and human cognition, technology has progressed in such a way to allow for the development of intelligent systems able to simulate many human traits. With this new era of simulated and artificial intelligence, much research is needed in order to continue to advance the field and also to evaluate the ethical and societal concerns of the existence of artificial life and machine learning.

The **Advances in Computational Intelligence and Robotics (ACIR) Book Series** encourages scholarly discourse on all topics pertaining to evolutionary computing, artificial life, computational intelligence, machine learning, and robotics. ACIR presents the latest research being conducted on diverse topics in intelligence technologies with the goal of advancing knowledge and applications in this rapidly evolving field.

COVERAGE

- Brain Simulation
- Natural Language Processing
- Synthetic Emotions
- Cyborgs
- Computer Vision
- Evolutionary Computing
- Computational Logic
- Pattern Recognition
- Automated Reasoning
- Artificial Intelligence

IGI Global is currently accepting manuscripts for publication within this series. To submit a proposal for a volume in this series, please contact our Acquisition Editors at Acquisitions@igi-global.com or visit: http://www.igi-global.com/publish/.

Titles in this Series

For a list of additional titles in this series, please visit:
http://www.igi-global.com/book-series/advances-computational-intelligence-robotics/73674

Predicting Natural Disasters With AI and Machine Learning
D. Satishkumar (Nehru Institute of Technology, India) and M. Sivaraja (Nehru Institute of Technology, India)
Engineering Science Reference • © 2024 • 340pp • H/C (ISBN: 9798369322802) • US $315.00

Impact of AI on Advancing Women's Safety
Sivaram Ponnusamy (Sandip University, Nashik, India) Vibha Bora (G.H. Raisoni College of Engineering, Nagpur, India) Prema M. Daigavane (G.H. Raisoni College of Engineering, Nagpur, India) and Sampada S. Wazalwar (G.H. Raisoni College of Engineering, Nagpur, India)
Engineering Science Reference • © 2024 • 315pp • H/C (ISBN: 9798369326794) • US $315.00

Empowering Low-Resource Languages With NLP Solutions
Partha Pakray (National Institute of Technology, Silchar, India) Pankaj Dadure (University of Petroleum and Energy Studies, India) and Sivaji Bandyopadhyay (Jadavpur University, India)
Engineering Science Reference • © 2024 • 330pp • H/C (ISBN: 9798369307281) • US $300.00

AIoT and Smart Sensing Technologies for Smart Devices
Fadi Al-Turjman (AI and Robotics Institute, Near East University, Nicosia, Turkey & Faculty of Engineering, University of Kyrenia, Kyrenia, Turkey)
Engineering Science Reference • © 2024 • 250pp • H/C (ISBN: 9798369307861) • US $300.00

Industrial Applications of Big Data, AI, and Blockchain
Mahmoud El Samad (Lebanese International University, Lebanon) Ghalia Nassreddine (Rafik Hariri University, Lebanon) Hani El-Chaarani (Beirut Arab University, Lebanon) and Sam El Nemar (AZM University, Lebanon)
Engineering Science Reference • © 2024 • 348pp • H/C (ISBN: 9798369310465) • US $275.00

Principles and Applications of Adaptive Artificial Intelligence
Zhihan Lv (Uppsala University, Sweden)
Engineering Science Reference • © 2024 • 316pp • H/C (ISBN: 9798369302309) • US $325.00

AI Tools and Applications for Women's Safety
Sivaram Ponnusamy (Sandip University, Nashik, India) Vibha Bora (G.H. Raisoni College of Engineering, Nagpur, India) Prema M. Daigavane (G.H. Raisoni College of Engineering, Nagpur, India) and Sampada S. Wazalwar (G.H. Raisoni College of Engineering, Nagpur, India)
Engineering Science Reference • © 2024 • 362pp • H/C (ISBN: 9798369314357) • US $300.00

701 East Chocolate Avenue, Hershey, PA 17033, USA
Tel: 717-533-8845 x100 • Fax: 717-533-8661
E-Mail: cust@igi-global.com • www.igi-global.com

Table of Contents

Preface...xviii

Chapter 1
Advanced Technologies in Clinical Research and Drug Development 1
    N. L. Swathi, Jawaharlal Nehru Technological University, Anantapur, India
    Achukutla Kumar, Indian Council of Medical Research, India

Chapter 2
AI in Education: Ethical Challenges and Opportunities 18
    Gurwinder Singh, Department of AIT-CSE, Chandigarh University, Punjab, India
    Anshika Thakur, Department of AIT-CSE, Chandigarh University, Punjab, India

Chapter 3
AI in Education: Ethical Challenges and Opportunities 39
    Ashok Singh Gaur, Noida Institute of Engineering and Technology, India
    Hari Om Sharan, Rama University, India
    Rajeev Kumar, Moradabad Institute of Technology, India

Chapter 4
An Inquiry Into the Obstacles Hindering the Widespread Use of Artificial Intelligence in
Environmental, Social, and Governance Practices 55
    Sabyasachi Pramanik, Haldia Institute of Technology, India

Chapter 5
Artificial Intelligence at the Helm: Steering the Modern Business Landscape Toward Progress............72
    Geetha Manoharan, SR University, India
    Abdul Razak, Entrepreneurship Development Institute of India, India
    C. V. Guru Rao, SR University, India
    Sunitha Purushottam Ashtikar, SR University, India
    M. Nivedha, Robert Gordon University, UK

Chapter 6

Artificial Intelligence-Powered Political Advertising: Harnessing Data-Driven Insights for
Campaign Strategies 100
Roop Kamal, Chandigarh University, India
Manpreet Kaur, Chandigarh University, India
Jaspreet Kaur, Chandigarh University, India
Shivani Malhan, Chitkara University, India

Chapter 7

Brain Tumor Detection From MRI Images Using Deep Learning Techniques 110
Anu Sharma, Moradabad Institute of Technology, Moradabad, India

Chapter 8

Data Privacy and E-Consent in the Public Sector 118
Abhay Bhatia, Roorkee Institute of Technology, India
Anil Kumar, Ajay Kumar Garg Engineering College, India
Pankhuri Bhatia, GRD IMT Dehradun, India

Chapter 9

Ensuring Privacy and Security in Machine Learning: A Novel Approach to Efficient Data
Removal 138
Velammal, Anna University, Chennai, India
N. Aarthy, Anna University, Chennai, India

Chapter 10

Ethical Considerations in AI Development 156
Bhanu Pratap Singh, COER University, Roorkee, India
Ankush Joshi, COER University, Roorkee, India

Chapter 11

Exploring the Ethical Implications of Generative AI in Healthcare 180
Dinesh Kumar, Lovely Professional University, India
Rohit Dhalwal, Lovely Professional University, India
Ayushi Chaudhary, Lovely Professional University, India

Chapter 12

Guardians of the Algorithm: Human Oversight in the Ethical Evolution of AI and Data Analysis 196
Dwijendra Nath Dwivedi, Krakow University of Economics, Poland
Ghanashyama Mahanty, Utkal University, India

Chapter 13
Investigation Into the Use of IoT Technology and Machine Learning for the Identification of Crop Diseases..211
    K. Manikandan, Vellore Institute of Technology, India
    Vivek Veeraiah, Sri Siddharth Institute of Technology, Sri Siddhartha Academy of Higher
       Education, India
    Dharmesh Dhabliya, Vishwakarma Institute of Information Technology, India
    Sanjiv Kumar Jain, Medi-Caps University, India
    Sukhvinder Singh Dari, Symbiosis Law School, Symbiosis International University, India
    Ankur Gupta, Vaish College of Engineering, India
    Sabyasachi Pramanik, Haldia Institute of Technology, India

Chapter 14
Narrating Spatial Data With Responsibility: Balancing Ethics and Decision Making........................225
    Munir Ahmad, Survey of Pakistan, Pakistan
    Saadia Ureeb, Zarai Taraqiati Bank Ltd., Pakistan

Chapter 15
Revolution Ethics of Data Science and AI...245
    Anil Meher, Sri Sri University, India

Chapter 16
Security Analysis of the Cyber Crime 257
    Ratnesh Kumar Shukla, Shambhunath Institute of Engineering and Technology, India
    Arvind Kumar Tiwari, Kamla Nehru Institute of Technology, Sultanpur, India

Chapter 17
Structural Systems Powered by AI and Machine Learning Technologies..272
    Sabyasachi Pramanik, Haldia Institute of Technology, India

Chapter 18
The Evolution of AI and Data Science 295
    A. S. Anurag, Central University of Kerala, India

Chapter 19
The Impact of 5G on the Future Development of the Healthcare Industry 313
    Saurabh Srivastava, Department of Computer Science and Engineering (Data Science),
       Moradabad Institute of Technology, India
    Harish Chandra Verma, ICAR-Central Institute for Subtropical Horticulture, Lucknow, India
    Syed Adnan Ahaq, Advanced Computing Research Laboratory, Department of Computer
       Application, Integral University, Lucknow, India
    Mohammad Faisal, Advanced Computing Research Laboratory, Department of Computer
       Application, Integral University, Lucknow, India
    Tasneem Ahmed, Advanced Computing Research Laboratory, Department of Computer
       Application, Integral University, Lucknow, India

Chapter 20

The Impact of Data Science and Participated Geographic Metadata on Improving Government
Service Deliveries: Prospects and Obstacles 320
> *Vivek Veeraiah, Sri Siddharth Institute of Technology, Sri Siddhartha Academy of Higher
> Education, India*
> *Dharmesh Dhabliya, Vishwakarma Institute of Information Technology, India*
> *Sukhvinder Singh Dari, Symbiosis Law School, Symbiosis International University, India*
> *Jambi Ratna Raja Kumar, Genba Sopanrao Moze College of Engineering, India*
> *Ritika Dhabliya, ResearcherConnect, India*
> *Sabyasachi Pramanik, Haldia Institute of Technology, India*
> *Ankur Gupta, Vaish College of Engineering, India*

Chapter 21

The Proposed Framework of View-Dependent Data Integration Architecture 343
> *Pradeep Kumar, Teerthanker Mahaveer University, India*
> *Madhurendra Kumar, Ministry of Electronics and Information Technology, CDAC, India*
> *Rajeev Kumar, Moradabad Institute of Technology, India*

Chapter 22

Toward a More Ethical Future of Artificial Intelligence and Data Science 362
> *Wasswa Shafik, School of Digital Science, Universiti Brunei Darussalam, Brunei & Dig
> Connectivity Research Laboratory (DCRLab), Kampala, Uganda*

Compilation of References 389

About the Contributors 448

Index 454

Detailed Table of Contents

Preface...xviii

Chapter 1

Advanced Technologies in Clinical Research and Drug Development 1

    N. L. Swathi, Jawaharlal Nehru Technological University, Anantapur, India
    Achukutla Kumar, Indian Council of Medical Research, India

This chapter explores the synergistic potential of decentralized trials, gene editing (e.g., CRISPR-Cas9), and the integration of artificial intelligence (AI) and machine learning (ML) in clinical trials and drug development. Decentralized trials enhance diversity and expedite timelines, while gene editing ensures precision in treating genetic diseases, necessitating robust ethical guidelines. AI and ML streamline processes, improving efficiency from patient recruitment to data analysis. Digital biomarkers and real-time monitoring systems provide rich data streams. This confluence marks a transformative era, promoting patient-centric research, accelerating innovation, and optimizing trial design. Ethical and regulatory challenges require careful navigation. Integrating digital biomarkers and continuous monitoring will enhance data quality. This synergy holds promise for personalized medicine and improved outcomes, emphasizing the need for stakeholders to balance innovation with ethical responsibility for optimal healthcare advancement.

Chapter 2

AI in Education: Ethical Challenges and Opportunities... 18

    Gurwinder Singh, Department of AIT-CSE, Chandigarh University, Punjab, India
    Anshika Thakur, Department of AIT-CSE, Chandigarh University, Punjab, India

Artificial intelligence (AI) offers transformative opportunities in education, promising personalized learning and streamlined tasks. However, its integration raises ethical concerns like algorithmic bias and data privacy. AI can revolutionize education by tailoring learning experiences and optimizing administrative tasks, yet it risks eroding student autonomy and perpetuating biases. To mitigate these challenges, transparent algorithms, robust data governance, and educator training are essential. Successful AI implementations demonstrate improved engagement and outcomes. Moving forward, collaborative efforts are crucial to navigate ethical complexities and ensure AI enhances education responsibly, fostering equity and inclusivity.

Chapter 3

AI in Education: Ethical Challenges and Opportunities 39

Ashok Singh Gaur, Noida Institute of Engineering and Technology, India
Hari Om Sharan, Rama University, India
Rajeev Kumar, Moradabad Institute of Technology, India

As artificial intelligence (AI) continues to advance, its integration into the field of education presents both promising opportunities and ethical challenges. This chapter explores the multifaceted landscape of AI in education, examining the ethical considerations associated with its implementation. The opportunities encompass personalized learning experiences, adaptive assessment tools, and efficient administrative processes. However, ethical concerns arise regarding data privacy, algorithmic bias, accountability, and the potential exacerbation of educational inequalities. Artificial intelligence is a field of study that combines the applications of machine learning, algorithm production, and natural language processing. Applications of AI transform the tools of education. AI has a variety of educational applications, such as personalized learning platforms to promote students' learning, automated assessment systems to aid teachers, and facial recognition systems to generate insights about learners' behaviors.

Chapter 4

An Inquiry Into the Obstacles Hindering the Widespread Use of Artificial Intelligence in
Environmental, Social, and Governance Practices 55

Sabyasachi Pramanik, Haldia Institute of Technology, India

This chapter examines the correlation between artificial intelligence (AI) and the integration of environmental, social, and governance (ESG) factors in contemporary business and technology. The chapter explores how AI might provide practical answers to current urgent social issues regarding sustainability. The chapter focuses on the ethical dimensions of sustainable technologies that address ecological issues, as well as AI-driven forecasts for ESG indicators, ethical supply chains, and forthcoming legislation. The growing integration of AI and ESG presents opportunities for sustainable-oriented organizations to expand their market share via the adoption of environmentally friendly strategies. This convergence allows for the alignment of sustainability goals with advanced technology, resulting in a truly transformative partnership.

Chapter 5

Artificial Intelligence at the Helm: Steering the Modern Business Landscape Toward Progress 72

Geetha Manoharan, SR University, India
Abdul Razak, Entrepreneurship Development Institute of India, India
C. V. Guru Rao, SR University, India
Sunitha Purushottam Ashtikar, SR University, India
M. Nivedha, Robert Gordon University, UK

Computers are part of our everyday life. Advances in technology allow these machines to duplicate human skills with remarkable accuracy. AI managed all global business functions. Progress is mostly due to AI. Using AI, computers can automate human work. Many organizations employ AI to simplify. Business uses it extensively. Companies use AI to automate, analyze, and interact with consumers and workers. Companies want market domination and industry growth. Business insights are gained by many successful global firms using AI, automation, big data analytics, and NLP. AI is transforming business by automating processes, analyzing data, improving decision-making, and connecting customers. AI may improve corporate chatbots, inventory management, fraud protection, and predictive maintenance. AI can raise productivity, cut expenses, and increase profits by speeding up procedures and enhancing customer service. However, businesses must use AI responsibly and protect data. This chapter discusses AI's business impacts.

Chapter 6
Artificial Intelligence-Powered Political Advertising: Harnessing Data-Driven Insights for
Campaign Strategies 100
 Roop Kamal, Chandigarh University, India
 Manpreet Kaur, Chandigarh University, India
 Jaspreet Kaur, Chandigarh University, India
 Shivani Malhan, Chitkara University, India

This study examines how political advertising is changing in the age of artificial intelligence (AI) technologies. The objective of this chapter is to investigate the complex relationship between political advertising and AI, stressing both its potential advantages and moral dilemmas. This chapter explores the complex interaction between political advertising and machine learning, revealing how algorithms and data-driven insights are transforming political campaigns. In this chapter, the ethical ramifications of AI in political advertising are also covered. The chapter underlines the moral issues related to data security, privacy, and manipulation risk. It also examines the moral conundrums raised by deep fake technologies, microtargeting, and the potential bias present in AI systems. The study also looks into the necessity for accountability and openness in AI-powered political advertising to protect the integrity of democratic processes. In the research, the impact of political strategies is explored, and 150 respondents participated in the primary study.

Chapter 7
Brain Tumor Detection From MRI Images Using Deep Learning Techniques 110
 Anu Sharma, Moradabad Institute of Technology, Moradabad, India

Machine learning and deep learning algorithms are utilized to identify brain tumors in a number of research papers. When these algorithms are applied to MRI images, it takes exceedingly slight time to expect a brain tumor, and the increased accuracy makes it easier to treat patients. The performance of the hybrid Convolution Neural Network (CNN) used in the proposed work to detect the existence of brain tumours is examined. In this study, we suggested a hybrid convolutional neural network followed by deep learning techniques using 2D magnetic resonance brain pictures, segment brain tumors (MRI). In our research, hybrid CNN achieved an accuracy of 98.73%, outperforming the results so far.

Chapter 8
Data Privacy and E-Consent in the Public Sector 118
 Abhay Bhatia, Roorkee Institute of Technology, India
 Anil Kumar, Ajay Kumar Garg Engineering College, India
 Pankhuri Bhatia, GRD IMT Dehradun, India

In the era of the internet, all face administrative and legal responsibilities obtaining informed consent and safeguarding personal information, with the public growing mistrust to data collection. Moral consent management takes place in account of person's views, subjective norms, and sense of control. When obtaining consent, this chapter aims to combat this cynicism. It accomplishes this by creating a novel conceptual model of online informed consent that combines the TPB with the autonomous authorisation model of informed consent. It is argued logically and is bolstered. As a result, it develops a model for online informed consent that is based on the ethic of autonomy and makes use of theory based on behaviour to enable a method of eliciting agreement that can put interest of users first and then promotes moral the information management and the marketing techniques. This approach also presents an innovative idea, the informed attitude for the validity of informed consent. It also indicates that informed permission may be given against.

Chapter 9
Ensuring Privacy and Security in Machine Learning: A Novel Approach to Efficient Data Removal 138
Velammal, Anna University, Chennai, India
N. Aarthy, Anna University, Chennai, India

Modern systems generate vast amounts of data, creating complex data networks. Users prioritize the safety, security, and privacy of their data. This project focuses on efficiently removing or erasing data from the machine learning model upon user request, addressing privacy concerns. Under GDPR, users can request the deletion of sensitive data from both user records and the machine learning model that has processed the data. Additionally, the project employs the SISA approach to address errors and attacks by dividing the dataset into shards and implementing a slice-based ensemble learning technique. Each shard functions as an independent model, and after training, a majority voting approach aggregates these models into a final model. Experimental results demonstrate reduced retraining costs, as only the remaining slices are retrained instead of the entire model.

Chapter 10
Ethical Considerations in AI Development 156
Bhanu Pratap Singh, COER University, Roorkee, India
Ankush Joshi, COER University, Roorkee, India

This chapter serves not only as a repository of ethical insights, but as a clarion call for a multidisciplinary and collaborative approach. Ethical considerations in AI development demand a collective effort, where technologists, ethicists, policymakers, and society engage in an ongoing dialogue. It is through such collaboration that we can envision and craft a future where AI technologies not only excel in technical prowess but also uphold the principles of fairness, accountability, and societal benefits. In essence, this chapter invites readers to embark on a reflective journey, encouraging a deeper understanding of the ethical intricacies that underlie AI development. It is a call to action, challenging us to wield the power of AI responsibly and ethically, shaping a future where technology becomes a force for positive societal transformation.

Chapter 11
Exploring the Ethical Implications of Generative AI in Healthcare 180
Dinesh Kumar, Lovely Professional University, India
Rohit Dhalwal, Lovely Professional University, India
Ayushi Chaudhary, Lovely Professional University, India

This chapter critically evaluates the ethical challenges posed by the advent of generative artificial intelligence (GenAI) in healthcare. It investigates how GenAI's potential to revolutionize patient care and medical research is counterbalanced by significant ethical concerns, including privacy, security, and equity. An extensive literature review supports a deep dive into these issues, comparing GenAI's impact on traditional healthcare ethics. Through case studies and theoretical analysis, the chapter seeks to understand GenAI's ethical implications thoroughly, aiming to contribute to the development of nuanced ethical frameworks in this rapidly advancing area.

Chapter 12
Guardians of the Algorithm: Human Oversight in the Ethical Evolution of AI and Data Analysis 196
    Dwijendra Nath Dwivedi, Krakow University of Economics, Poland
    Ghanashyama Mahanty, Utkal University, India

The emergence of artificial intelligence (AI) and data enquiry priciples uncovered immese technological possibilities, but it has also presented a range of ethical concerns that require careful supervision and moderation to avoid unintended consequences. This chapter is a thorough examination that emphasizes the crucial importance of human intervention in upholding the ethical integrity of AI systems and data-driven processes. It emphasizes the importance of human supervision not only as a regulatory structure, but also as an essential element in the development and execution of AI systems. The study examines many approaches to human oversight, including both direct intervention and advanced monitoring techniques, that can be incorporated at every stage of the AI lifecycle, from original creation to post-deployment. The study showcases many case studies and real-world situations to illustrate instances when the lack of human supervision resulted in ethical violations, and conversely, where its presence effectively reduced dangers.

Chapter 13
Investigation Into the Use of IoT Technology and Machine Learning for the Identification of Crop Diseases.. 211
    K. Manikandan, Vellore Institute of Technology, India
    Vivek Veeraiah, Sri Siddharth Institute of Technology, Sri Siddhartha Academy of Higher Education, India
    Dharmesh Dhabliya, Vishwakarma Institute of Information Technology, India
    Sanjiv Kumar Jain, Medi-Caps University, India
    Sukhvinder Singh Dari, Symbiosis Law School, Symbiosis International University, India
    Ankur Gupta, Vaish College of Engineering, India
    Sabyasachi Pramanik, Haldia Institute of Technology, India

The control and management of crop diseases has always been a focal point of study in the agricultural domain. The growth of agricultural planting areas has posed several obstacles in monitoring, identifying, and managing large-scale illnesses. Insufficient disease identification capacity in relation to the expanding planting area results in heightened disease intensity, leading to decreased crop production and reduced yield per unit area. Evidence indicates that the reduction in crop productivity resulting from illnesses often surpasses 40%, leading to both financial setbacks for farmers and a certain degree of impact on local economic growth. A total of 1406 photos were gathered from 50 image sensor nodes. These images consist of 433 healthy images, 354 images showing big spot disease, 187 images showing tiny spot disease, and 432 images showing rust disease. This chapter examines the cultivation of maize fields in open-air environments and integrates internet of things (IoT) technologies.

Chapter 14

Narrating Spatial Data With Responsibility: Balancing Ethics and Decision Making.......................225
Munir Ahmad, Survey of Pakistan, Pakistan
Saadia Ureeb, Zarai Taraqiati Bank Ltd., Pakistan

This chapter discussed the responsible use of spatial data in decision-making and the need to balance ethics with decision-making. The role of spatial data in decision-making and storytelling is discussed. Ethical issues in spatial data storytelling such as privacy and confidentiality, bias and discrimination, accuracy and misinformation, accessibility and inclusivity, environmental and social impacts, data ownership and credit, and transparency are described, along with the strategies to overcome these issues. The different roles of storytelling in decision-making are elaborated on, along with the challenges and risks associated with using storytelling in decision-making. The chapter also discussed the challenges of striking a balance between ethics and decision-making, such as conflicting interests, cognitive biases, time constraints, political and legal constraints, and many more. Strategies to overcome these challenges, including developing clear ethical guidelines and engaging diverse stakeholders, are underpinned in this chapter as well.

Chapter 15

Revolution Ethics of Data Science and AI...245
Anil Meher, Sri Sri University, India

Artificial intelligence is becoming more and more widespread in our increasingly connected world. Artificial intelligence is slowly but surely modifying the way we live and work, from self-driving cars to automated customer service agents. As artificial intelligence becomes more sophisticated, the ethical implications of its use become more complex. There are several key issues to consider regarding the ethics of artificial intelligence, such as data privacy, algorithmic bias, and socioeconomic inequality. The rapid development of AI brings with it several ethical issues. However, we must remain vigilant in protecting our fundamental rights and freedoms. We must ensure that artificial intelligence is not used to discriminate against vulnerable groups or invade our privacy. We must also be careful that AI does not become a tool for the powerful to control and manipulate the masses. But while there are risks, the author believes the potential benefits of AI are too great to ignore.

Chapter 16

Security Analysis of the Cyber Crime 257
Ratnesh Kumar Shukla, Shambhunath Institute of Engineering and Technology, India
Arvind Kumar Tiwari, Kamla Nehru Institute of Technology, Sultanpur, India

The primary driver of this expansion is the internet user, who is expected to connect 64 billion devices worldwide by 2026. Nearly $20 trillion will be spent on IoT devices, services, and infrastructure, according to Business Insider. Many cybercrimes and vulnerabilities related to cybercrime are committed with the use of data. Asset management, fitness tracking, and smart cities and homes are examples of internet security applications. The average person will most likely own two to six connected internet security devices by the end of the year, a significant increase over the total number of cell phones, desktop computers, and tablets. Although data provides a plethora of opportunities for its users, some have taken advantage of these advantages for illegal purposes. In particular, a great deal of cybercrime is made possible by the gathering, storing, analyzing, and sharing of data as well as the widespread gathering, storing, and distribution of data without the users' knowledge or consent and without the required security and legal protections. Furthermore, because data gathering, analysis, and transfer happen at scales that governments and organisations are unprepared for, there are a plethora of cybersecurity threats. Protection, privacy, and system and network security are all related.

Chapter 17
Structural Systems Powered by AI and Machine Learning Technologies 272
    Sabyasachi Pramanik, Haldia Institute of Technology, India

This chapter investigates the use of artificial intelligence (AI) and machine learning (ML) technologies in structural engineering, with an emphasis on their applications in automating design processes, optimizing structural configurations, and evaluating performance measures. It demonstrates the effectiveness of AI-powered algorithms in creating design alternatives, anticipating structural behavior, and improving sustainability. The chapter also includes a framework for comparing the performance of various structural designs, taking into account safety, cost-effectiveness, and environmental impact. It provides case studies and practical examples that show how AI/ML-driven autonomous design may achieve greater structural performance while using fewer resources. The chapter stresses the potential of AI and ML to revolutionize structural engineering by allowing engineers to design more sustainable and high-performing buildings, so contributing to a more ecologically aware and economically viable built environment.

Chapter 18
The Evolution of AI and Data Science 295
    A. S. Anurag, Central University of Kerala, India

The history of artificial intelligence (AI) and data science has their origins in the 1940s and 1950s respectively. However, it has been through many changes throughout its history. AI is a vast and fascinating subject. There are many more elements to discover and understand. This chapter aims to outline the history of AI and data science, from its origin to its current developments. It will also explore the ethical considerations within AI and data science, such as bias and fairness, transparency, data privacy, etc. In the end, the chapter sheds light on the ethical concerns regarding the implementation of AI and the security concerns that data science poses. The chapter also provides insights into the role of individuals, government, and society in mitigating these issues. This chapter aims to furnish the reader with the scientific foundation and essential understanding required for embarking on the journey to comprehend the realm of artificial intelligence and data science.

Chapter 19
The Impact of 5G on the Future Development of the Healthcare Industry 313
    Saurabh Srivastava, Department of Computer Science and Engineering (Data Science),
        Moradabad Institute of Technology, India
    Harish Chandra Verma, ICAR-Central Institute for Subtropical Horticulture, Lucknow, India
    Syed Adnan Ahaq, Advanced Computing Research Laboratory, Department of Computer
        Application, Integral University, Lucknow, India
    Mohammad Faisal, Advanced Computing Research Laboratory, Department of Computer
        Application, Integral University, Lucknow, India
    Tasneem Ahmed, Advanced Computing Research Laboratory, Department of Computer
        Application, Integral University, Lucknow, India

A 5G network can enable services like real-time remote patient monitoring and the distribution of huge files, including medical data for e-health systems. The internet of things (IoT), sensors, and other cutting-edge technologies will be used in the future to identify patients' illnesses and offer advice on how to treat them. The popularity of electric health care is increasing day by day, there are many applications available that can be used by the patient for routine checkups from the smartphone. Patients' private information is taken at the time of application downloads such as name, gender, and age is used by the application to increase its accuracy, as well as, the results of routine checkups are stored on the application's server (storage). The stored data can be used in different kinds of promotions. Hence, hackers are trying to steal information from users for their benefit and the IoT-based applications are not so reliable in terms of security.

Chapter 20

The Impact of Data Science and Participated Geographic Metadata on Improving Government Service Deliveries: Prospects and Obstacles 320

    Vivek Veeraiah, Sri Siddharth Institute of Technology, Sri Siddhartha Academy of Higher Education, India
    Dharmesh Dhabliya, Vishwakarma Institute of Information Technology, India
    Sukhvinder Singh Dari, Symbiosis Law School, Symbiosis International University, India
    Jambi Ratna Raja Kumar, Genba Sopanrao Moze College of Engineering, India
    Ritika Dhabliya, ResearcherConnect, India
    Sabyasachi Pramanik, Haldia Institute of Technology, India
    Ankur Gupta, Vaish College of Engineering, India

This chapter examined the profound influence of data science and volunteered geographic information (VGI) on the delivery of public services. Volunteered geographic information, being material created by users, has had a substantial impact on making geographic information accessible to everybody, enabling people to actively engage in the creation and management of data. The incorporation of VGI into government operations has introduced novel prospects for enhancing service provision in diverse sectors such as education, health, transportation, and waste management. In addition, data science has enhanced VGI by using sophisticated methodologies like artificial intelligence (AI), internet of things (IoT), big data, and blockchain, thereby transforming the whole framework of government service provision. Nevertheless, in order to effectively use VGI in public sector services, it is essential to tackle significant obstacles such as data accuracy, safeguarding, inclusiveness, technical framework, and specialized expertise.

Chapter 21

The Proposed Framework of View-Dependent Data Integration Architecture 343

    Pradeep Kumar, Teerthanker Mahaveer University, India
    Madhurendra Kumar, Ministry of Electronics and Information Technology, CDAC, India
    Rajeev Kumar, Moradabad Institute of Technology, India

In this chapter the authors have proposed framework that have we are using various techniques to overcome the above mentioned challenges and proving the exact identification and extraction of WQI by classifying them according to their domain. The framework has been represented in system level design as a high level view of the view dependent data integration system and the architectural framework of the system design. Being a multi-database-oriented system, it is scalable to structure as well as unstructured data source. The wrapper and mediator module of the system is used to map the web-query-interface to the global schema of the integrated web query interfaces. In this the authors have implemented high level view of the system modeling from the end users' point of view; and the operational framework design also been represented.

Chapter 22
Toward a More Ethical Future of Artificial Intelligence and Data Science..362
Wasswa Shafik, School of Digital Science, Universiti Brunei Darussalam, Brunei & Dig
Connectivity Research Laboratory (DCRLab), Kampala, Uganda

Examining the ethical aspects of artificial intelligence (AI) and data science (DS) recognizes their impressive progress in innovation while emphasizing the pressing necessity to tackle intricate ethical dilemmas. The chapter provides a detailed framework for navigating the changing environment, beginning with an examination of the increasing ethical challenges. The study highlights transparency, fairness, and responsibility as crucial for cultivating confidence in AI systems. The chapter emphasizes the urgent requirement to address problems such as algorithmic bias and privacy breaches with strong mitigation techniques. Furthermore, it promotes flexible policies that strike a balance between innovation and ethical safeguards. The examination of societal effects, particularly on various socioeconomic groups, economies, and cultures, is conducted thoroughly, with a focus on equity and the protection of individual rights. Finally, to proactively tackle future ethical challenges in technology, it is advisable to employ proactive solutions such as implementing AI ethics by design.

Compilation of References 389

About the Contributors 448

Index..454

Preface

In the ever-evolving landscape of technology, the fusion of artificial intelligence (AI) and data science has emerged as a powerful force, revolutionizing industries and reshaping the very fabric of our society. As these transformative technologies continue to advance, it becomes imperative for individuals, organizations, and policymakers to navigate the ethical frontiers that accompany their development and deployment. It is within this context that this edited volume, titled *The Ethical Frontier of AI and Data Analysis*, seeks to serve as a comprehensive guide and indispensable resource.

The world today is witnessing an unprecedented reliance on AI and Data Science, and their integration into various domains raises profound ethical challenges. This book is crafted to be a beacon for students, researchers, technocrats, and policymakers, shedding light on the ethical dimensions inherent in the realm of AI and data analysis. Its key features are designed to provide a holistic understanding of the subject, making it a valuable asset for those navigating the complexities of ethical decision-making in the rapidly advancing technological landscape.

Comprehensive Exploration: The book embarks on an in-depth exploration of the ethical dimensions of AI and data science, traversing a wide spectrum of critical topics. From algorithmic bias to privacy concerns, transparency, fairness, accountability, and the responsible use of data, the book leaves no stone unturned in its examination of the ethical challenges that accompany these technologies.

Real-World Relevance: Through real-world case studies and examples, readers gain a practical understanding of the ethical dilemmas that can manifest when AI and data science technologies are deployed across diverse sectors, including finance, healthcare, criminal justice, and marketing.

Ethical Frameworks: A solid grounding in ethical frameworks and principles relevant to AI and data analysis is provided, equipping readers with the tools to critically assess and address ethical challenges. The book explores utilitarianism, deontology, virtue ethics, and their application to technology.

Practical Guidance: The focus on actionable insights ensures that the book offers practical guidance for individuals and organizations involved in AI and data science projects. Strategies for ensuring fairness, transparency, and accountability are explored, along with recommendations for fostering ethical decision-making throughout the development lifecycle.

Legal and Regulatory Considerations: An in-depth overview of existing and emerging regulations, such as GDPR, HIPAA, and AI ethics guidelines, is provided. The book elucidates their impact on technology development and compliance.

Societal Implications: Beyond technical aspects, the book explores the broader societal implications of AI and data science, examining their impact on employment, social justice, and human rights. This well-rounded perspective offers a nuanced understanding of the ethical challenges posed by these technologies.

Interdisciplinary Approach: Authored by experts in AI, data science, ethics, and law, this book bridges the gap between technical and ethical domains, making it accessible and relevant to a diverse audience, from data scientists and engineers to ethicists and legal professionals.

Roadmap for Responsible Innovation: The book culminates with a forward-looking roadmap for responsible innovation in AI and data science. It outlines actionable steps for promoting ethical technology development, responsible data handling, and transparent decision-making processes.

"Ethics in AI and Data Science: Navigating the Moral Landscape of Artificial Intelligence and Data Analysis" is not just a compilation of insights; it is an essential resource for anyone seeking to understand, address, and uphold the ethical principles that should underpin the transformative technologies shaping our future. As we stand at the intersection of innovation and responsibility, this book invites readers to embark on a journey of ethical exploration, ensuring that the ethical frontier of AI and data analysis is charted with thoughtfulness, integrity, and a commitment to the greater good.

Rajeev Kumar
Moradabad Institute of Technology, India

Ankush Joshi
COER University, Roorkee, India

Hari Om Sharan
Rama University, Kanpur, India

Sheng-Lung Peng
College of Innovative Design and Management, National Taipei University of Business, Taiwan

Chetan R. Dudhagara
Anand Agricultural University, India

Chapter 1
Advanced Technologies in Clinical Research and Drug Development

N. L. Swathi

https://orcid.org/0000-0002-3695-0732

Jawaharlal Nehru Technological University, Anantapur, India

Achukutla Kumar

Indian Council of Medical Research, India

ABSTRACT

This chapter explores the synergistic potential of decentralized trials, gene editing (e.g., CRISPR-Cas9), and the integration of artificial intelligence (AI) and machine learning (ML) in clinical trials and drug development. Decentralized trials enhance diversity and expedite timelines, while gene editing ensures precision in treating genetic diseases, necessitating robust ethical guidelines. AI and ML streamline processes, improving efficiency from patient recruitment to data analysis. Digital biomarkers and real-time monitoring systems provide rich data streams. This confluence marks a transformative era, promoting patient-centric research, accelerating innovation, and optimizing trial design. Ethical and regulatory challenges require careful navigation. Integrating digital biomarkers and continuous monitoring will enhance data quality. This synergy holds promise for personalized medicine and improved outcomes, emphasizing the need for stakeholders to balance innovation with ethical responsibility for optimal healthcare advancement.

INTRODUCTION

Clinical trials are essential for advancing medical understanding and enhancing patient care. They are crucial for assessing the security and effectiveness of novel therapies, interventions, or regulations (Kiley et al., 2017). However carrying out clinical trials can be difficult for a variety of reasons, including difficulties with recruiting participants, organizational obstacles, and methodological problems (Schmitt,

DOI: 10.4018/979-8-3693-2964-1.ch001

2002). Clinical trials are increasingly using qualitative research to address more general research questions that quantitative methods by themselves cannot address (Elliott et al., 2017). Researchers can better understand the challenges of recruitment and informed consent in clinical trials with the aid of qualitative research, which sheds light on participants' perspectives and behaviors (Elliott et al., 2017).

The investigation of obstacles to minority recruitment in cancer clinical trials is also made possible (Jackson, 2020). According to one study (Briel et al., 2021), randomized controlled trials (RCTs) are frequently halted or revised as a result of poor recruitment. Narrow eligibility criteria, investigator sponsorship, increased workload for patients and recruiters, dated control interventions, and the launch of RCTs later than those with good recruitment are all factors that contribute to poor recruitment (Briel et al., 2021). Clinical trials can only be carried out successfully with the help of research coordinators. They are in charge of several duties, including patient registration and randomization, recruitment follow-up, case report form completion, collaboration with other stakeholders like Clinical Research Associates (CRAs), reporting of serious adverse events, handling investigator files, and preparing sites for audits (Rico-Villademoros et al., 2004).

Thus, technology now plays a role in reducing these concerns by combining technology with clinical research. Dentistry is one industry where cutting-edge technologies have had a significant impact. A narrative review of three-dimensional printed complete dentures was done by Anadioti et al. in 2020 (Anadioti et al., 2020). The successful fabrication of removable dental prostheses using CAD/CAM technologies was discovered to be a result of recent advancements in digital dentistry. However, they pointed out that in the available literature, milled dentures have received more attention than 3D-printed ones. Garbayo et al. (2020) reviewed developments in cancer treatment and regenerative medicine using nanomedicine and drug delivery technologies (Garbayo et al., 2020) in the field of drug delivery systems and nanomedicine. They emphasized how chemotherapeutic drugs can have their biodistribution and target site accumulation altered by nanomedicine, which lowers their toxicity.

The ability of drug delivery systems to deliver therapeutic proteins and peptides in a controlled manner while preventing their deterioration was also covered. The authors gave illustrations of anticancer drug-loaded nanoparticles that had demonstrated effectiveness in preclinical settings. Another area where advanced technologies have been utilized is in the field of RNA interference (RNAi) therapeutics.(Weng et al., 2019) provided an overview of the cutting-edge biotechnological development of RNAi therapeutics (Weng et al., 2019). They discussed the approval of Alnylam Pharmaceuticals' RNA interference drug ONPATTROTM as a treatment for hereditary forms of transthyretin-mediated amyloidosis. They emphasized how this approval has created new opportunities for the creation and application of RNAi therapeutics in several diseases.

Clinical trials use a variety of cutting-edge technologies, not just biomaterials and drug delivery systems. As part of the BRAIN Initiative, Litvina et al. (2019) talked about cutting-edge resources and tools for neuroscience research (Litvina et al., 2019). They emphasized how technological developments like deep learning and spectral flow cytometry are opening up new perspectives on Brain circuit function and allowing researchers to create ground-breaking treatments for neurological diseases. The difficulties in conventional clinical trials have also been addressed by the emergence of innovative trial designs. Innovative human trial designs that have produced notable therapeutic compounds were reviewed by Chen et al. (2020) (Chen & Qi, 2020). They discussed expedited clinical trial modes that aim to improve efficiency and reduce costs while maintaining rigorous scientific standards. They comprise master protocols, platform trials, basket trials, and adaptive trial designs.

Clinical research has made significant strides due to the incorporation of cutting-edge technologies. Precision medicine, genomics, big data analytics, artificial intelligence (AI), and machine learning (ML) are just a few of the cutting-edge technologies that are transforming clinical trials and drug development (Rosenberg et al., 1990). Researchers may increase the effectiveness, speed, and efficiency of clinical research operations by utilizing these cutting-edge technologies, resulting in better patient outcomes and a shortened regulatory approval process (Obermeyer & Emanuel, 2016).

Artificial intelligence and machine learning advancements have had a significant positive impact on clinical research and drug development. Medical imaging, genetic data, and electronic health records (EHRs) are just a few examples of the various patient data types that AI algorithms can thoroughly examine (Mayo et al., 2017). By seeing patterns and predicting the progression of illnesses, artificial intelligence (AI) can enhance treatment strategies for particular individuals. Additionally, AI algorithms have the potential to significantly reduce the time and costs associated with traditional trial-and-error procedures and alter the drug discovery process by forecasting the efficacy and safety of novel pharmaceutical candidates(Hargrove et al., 2013) .

Randomized clinical trials (RCTs) are often regarded as the gold standard for demonstrating a connection between a health intervention and its intended outcomes. The highest grade of evidence in medical research and innovation, according to (Sen et al., 2017)and (Mayo et al., 2017), is provided by these studies. Nevertheless, the industry is worried about the rising costs associated with conducting clinical trials and the overall decline in medical discoveries (Ngayua et al., 2021). Drug launch failures have increased significantly during the past ten years, reaching previously unheard-of proportions (Ngayua et al., 2021). Furthermore, due to difficulties with the strict eligibility and inclusion requirements, almost one in five clinical trials does not finish participant enrollment. To solve problems, investigate opportunities, and track development, a system that can use the data generated during clinical trial design and execution is urgently required. A system like this would improve adaptability to the constantly changing clinical procedures and regulatory standards related to RCTs.

Big data analytics is another field where cutting-edge technology has a huge impact. Researchers now have more options thanks to abundant healthcare data, including data from clinical trials and empirical evidence. Researchers can get complete insights into disease processes, treatment responses, and adverse events by integrating and evaluating various data sources. Big data analytics makes it possible to identify patient subpopulations, classify patients according to their genetic profiles, and forecast treatment outcomes, which makes it easier to deploy customized medicine techniques. Big data analytics can also improve trial design, recruitment procedures, and patient selection, resulting in clinical trials that are more effective and productive (Steinhubl et al., 2013).

Drug development has been significantly changed by genetics. Now that the whole human genome can be sequenced affordably, researchers can better understand the hereditary causes of illnesses, find the biomarkers linked with certain illnesses, and create customized treatments. Genomic profiling enables individualized medical treatments in which a patient's genetic profile can be considered when making treatment decisions. This method, known as precision medicine, chooses therapies based on the patient's genomic traits to optimize therapeutic success while reducing side effects. Clinical trial reform and patient outcomes can both be significantly improved by precision medicine (Collins & Varmus, 2015).

In clinical research, wearable and remote monitoring technologies have become essential resources. Outside the conventional clinical setting, these technologies allow for the collection and monitoring of real-time patient data. Wearable technologies provide a complete picture of a person's health status by continuously recording vital signs, activity levels, and patient-reported outcomes (Steinhubl et al., 2013).

Figure 1. The role of technologies
(Author)

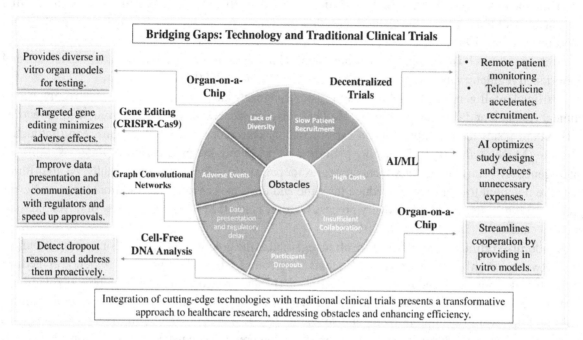

Clinical trials' data accuracy is improved via remote monitoring, which also boosts patient participation and allows for the early discovery of negative occurrences

Blockchain technology, which is renowned for being safe and decentralized, can transform the clinical research and medication development processes completely. Blockchain can improve information sharing, teamwork, and trust between participants in clinical trials by ensuring the transparency, integrity, and immutability of data. Blockchain-based smart contracts can safeguard data privacy, offer real-time trial data monitoring, and automate consent management. In addition, blockchain can make it easier for medicinal supply chains to be authenticated and tracked, reducing the risk of patients receiving substandard medications (Agbo et al., 2019)

Decentralized Trails

DTMs use a variety of techniques and technology to make remote data collection, monitoring, and patient contact possible. These could include electronic health records (EHRs), wearable technology, mobile health (mHealth) apps, secure data management systems, and telemedicine platforms. By including these digital components, DTMs streamline the trial process, improve patient access, and offer many benefits (Khozin & Coravos, 2019). Diversity and improved patient recruitment are important since traditional trials frequently have trouble enrolling a patient population that reflects the population being studied, which could bias the results of the study. To increase diversity and inclusivity in clinical research, DTMs can get beyond geographical restrictions and allow a wider spectrum of people to participate (Van Norman, 2021).

- **Increased patient comfort and involvement:** DTMs cut down on the number of site visits required, lowering participant travel and associated costs. Since patients can participate from the comfort of their homes, convenience is improved, and engagement and retention rates may rise.
- **Real-time data monitoring and collection:** Digital tools incorporated into DTMs enable continuous data collection of things like vital signs, patient-reported outcomes, and medication adherence. This real-time monitoring improves the timeliness and accuracy of data, allowing researchers to assess results more quickly.
- **Rapid and adaptive trial design:** DTMs make it possible to construct adaptive trial designs, which let changes be made to the trial protocol in reaction to new information. Due to this flexibility, decisions, and changes can be made more quickly, thereby hastening the drug development process.
- **Cost and time effectiveness:** DTMs have the potential to lower overall trial costs and duration by obviating the requirement for extensive site infrastructure and lessening patient stress. This benefit is especially beneficial in light of the intricate and drawn-out clinical trial processes (Van Norman, 2021)

RBM and RBQM

Clinical trials employ the techniques of risk-based monitoring (RBM) and risk-based quality management (RBQM) to increase the effectiveness and efficiency of the monitoring process. In contrast to RBM, which concentrates on monitoring tasks, RBQM adopts a more comprehensive approach by integrating quality management ideas across the entire clinical trial process. Since it goes beyond merely monitoring activities, RBQM might be considered an improvement over RBM (Adams et al., 2023). With an emphasis on crucial information and procedures, RBM is a monitoring strategy that tries to identify and rank risks in clinical trials. On-site monitoring has historically been used in clinical trials to guarantee protocol adherence and data quality. This entailed repeated visits to experimental sites. Nevertheless, this strategy can require a lot of resources and might not fully handle all potential problems. RBM developed the idea of risk assessment to pinpoint crucial information and procedures that are crucial to the trial's goals and patient security. Concentrating on high-risk locations enables sponsors to more effectively devote monitoring resources. This strategy can use remote monitoring, centralized monitoring, or a mix of the two, which eliminates the need for regular on-site visits. Building on the fundamental principles of RBM, RBQM develops the concepts of risk assessment and management throughout the clinical trial process. To ensure the overall quality and integrity of the trial, RBQM applies quality management principles. It adopts a comprehensive approach by considering methods for risk assessment and reduction concerning study design, site selection, data collection, and analysis (Stansbury et al., 2022). The focus of RBQM is on proactive and ongoing risk management. To identify potential dangers and immediately take mitigation actions, it promotes the use of technology and data analytics. RBQM makes it possible to analyze risks more thoroughly and quickly by utilizing data from a variety of sources, including electronic health records, patient-reported outcomes, and electronic data capture tools. Additionally, RBQM encourages cooperation among trial participants, such as sponsors, clinical research companies (CROs), researchers, and regulatory authorities. The early detection and avoidance of potential threats are made easier by the encouragement of effective communication and information sharing (Adams et al., 2023; Stansbury et al., 2022). RBM to RBQM signifies a shift in perspective from a clinical trial quality management approach that is primarily monitoring-oriented to one that is more broadly focused on quality management. Risk assessment and management are continual activities that ought to be incorporated into the entire trial lifecycle, according to RBQM. Sponsors and stakeholders can improve trial efficiency, data quality, patient safety, and regulatory compliance by implementing RBQM concepts (Adams et al., 2023).

AI and Clinical Trials

Clinical trials are one area where AI (Artificial Intelligence) is already making notable strides. Before new medications, therapies, and medical technologies are deemed safe and effective enough to be made available to the general public, they must first undergo clinical trials. However, conventional clinical trials can be time-consuming, costly, and occasionally ineffective. Numerous advantages and chances for enhancing the effectiveness and efficiency of clinical trials are provided by AI

Patient Recruitment and Selection

Finding participants who are qualified and meet certain requirements is one of the difficulties in clinical studies. AI can swiftly find possible participants who fit the requirements by analyzing vast amounts of patient data from numerous sources, including electronic health records, genetic data, and medical literature. This can considerably hasten the recruitment process and enhance the research population's representation (Harrer et al., 2019).

Data Analysis and Prediction

AI can examine complicated data sets produced during clinical trials, such as patient information, laboratory findings, imaging information, and patient-reported outcomes. AI uses machine learning algorithms to find patterns, forecast outcomes, and unearth insights that may not be visible to human researchers. This can make it easier for researchers to spot potential hazards, choose the best dosages, and more accurately gauge treatment response (Harrer et al., 2019).

Adverse Event Monitoring

To guarantee patient safety throughout clinical trials, monitoring and detecting adverse events are essential. By real-time evaluating patient data, AI can assist in automating the detection and tracking of adverse occurrences. To collect and analyze data from clinical notes, patient diaries, and social media posts to identify potential adverse events early and enable timely intervention, natural language processing (NLP) techniques can be used (Kolluri et al., 2022).

Precision Medicine and Individualized Care

By analyzing extensive genetic and proteomic data, AI can help advance precision medicine. AI can help identify patient subgroups that are more likely to respond favorably to a certain treatment by discovering genetic markers and molecular signatures. This makes it possible for researchers to develop more specialized, focused clinical studies that result in more individualized treatment methods (Vamathevan et al., 2019).

DRUG REPURPOSING

AI can be used to find current medications that might be used for new purposes. AI algorithms can find patterns and links in the analysis of massive amounts of biomedical data that point to a drug's efficacy for a particular ailment. Repurposing medications that have already undergone safety testing can save time and costs and speed up the clinical trial procedure for the treatments (Pushpakom et al., 2019).

TRIAL DESIGN OPTIMIZATION

By modeling various situations and outcomes, AI can help with trial design optimization for clinical trials. To determine the most successful and economical trial design, researchers can utilize AI models to simulate the impact of many aspects, including sample size, treatment regimens, and endpoints. This can save expenses, decrease the number of patients required for studies, and hasten the medication development process.

Despite the many advantages of AI in clinical trials, it's vital to remember that these technologies should be utilized to support rather than replace human judgment. When incorporating AI into the clinical trial process, it is essential to carefully consider ethical issues, data protection, and regulatory compliance (Harrer et al., 2019).

GENE EDITING AND GENE THERAPY

In molecular biology, the topics of gene editing and gene therapy are connected, and both focus on modifying genetic material to treat or prevent disease. There are significant differences between the two, even though they have comparable objectives.

Gene Editing: The exact alteration of an organism's DNA at a particular point within the genome is referred to as gene editing. It entails modifying the DNA sequence by adding, removing, or substituting particular nucleotides (DNA's building blocks). CRISPR-Cas9, one of the most effective and popular gene-editing methods, uses a guide RNA molecule to target particular DNA sequences and a Cas9 enzyme to cut the DNA at that site. This causes a break in the DNA strand, which the cell's natural repair mechanisms can repair by changing the DNA sequence (Singh et al., 2017)

The use of gene editing in science and medicine has enormous potential. The introduction of targeted mutations and the observation of the outcomes enables researchers to investigate the function of particular genes. Gene editing can be used to fix or eliminate malfunctioning genes that cause genetic illnesses in a therapeutic setting. For instance, it might be used to correct a mutation that causes a disease by swapping it out with a healthy version of the gene or by turning off a gene that isn't working properly. Gene editing, albeit still in its infancy, has promise for treating a variety of genetic illnesses, including some forms of cancer, blood abnormalities, and hereditary ailments (Prakash et al., 2016)

Gene therapy: To treat or prevent illness, new genetic material is inserted into a patient's cells using this procedure. Gene therapy tries to add functional genes or change gene expression patterns to address a specific medical disease, in contrast to gene editing, which concentrates on changing existing DNA sequences. Gene therapy can be divided into two categories (Rosenberg et al., 1990)

Somatic Gene Therapy

This strategy focuses on non-reproductive body cells, such as those in the liver, muscles, or lungs. These cells are given the therapeutic genes to have a therapeutic effect on the particular patient. Clinical trials using somatic gene therapy have been conducted for a variety of illnesses, including acquired problems like some types of cancer and hereditary conditions like cystic fibrosis (Tachibana et al., 2013).

Germline Gene Therapy

In germline gene therapy, genetic alterations are made to early embryos or sperm that can be passed on to the next generation. Due to the possibility of modifying the human germline and the potential long-term effects on subsequent generations, this technique poses ethical and safety concerns. Germline gene therapy is not commonly used and is now subject to strict regulation.

Gene therapy can treat genetic illnesses permanently or for a very long time. By replacing the damaged gene or adding a functional version, it can send therapeutic genes right to the affected cells. Gene therapy techniques can also be used to modify how genes are expressed, for example, by turning off overactive genes or adding regulatory components to manage gene activity (Tachibana et al., 2013).

ORGAN-ON-A-CHIP

Organ-on-a-chip (OOC) is a cutting-edge technology that aspires to duplicate the composition and operation of human organs on a tiny, chip-sized substrate. To replicate the intricate physiology of organs in a controlled laboratory environment, it includes integrating living cells, biomaterials, and microfluidic technologies.

The desire for more realistic and dependable models for disease research, toxicity testing, and medication development gave rise to the idea of OOC. Traditional in vitro cell culture and animal models have limitations in their capacity to mimic the complexity of human organs and forecast how they will react to treatments or illnesses. OOC gadgets offer a more realistic and human-relevant alternative (Bhatia & Ingber, 2014; Osório et al., 2021)

An OOC device's essential elements are as follows:

Microfluidic channels: OOC devices have a network of tiny channels that resemble the blood veins or other fluid-transporting systems found inside an organ. By simulating the circulation and transport mechanisms within the organ, these channels enable the controlled flow of culture media, nutrients, and other substances (Bhatia & Ingber, 2014).

Biomaterial scaffolds: To give the cells a 3D habitat, biomaterial scaffolds are frequently used to fill or coat the microfluidic channels. To provide structural support and cues for cell architecture and function, these scaffolds can mimic the extracellular matrix (Osório et al., 2021)

Cell cultures: OOC devices are filled with relevant human cell types that make up the particular organ being mimicked. Epithelial, endothelial, immunological, and other cell types that are important to the organ's operation can be included in these cells. On biomaterial scaffolds inside microfluidic channels, the cells are normally nurtured.

OOC technology is still developing quickly, and efforts are still being made to increase the intricacy, functionality, and scalability of these devices. To construct "body-on-a-chip" systems that simulate the interactions between various organs and organ systems within the human body, researchers are merging diverse OOC platforms (Zhang et al., 2018)

BIOMARKERS AND PRECISION MEDICINE

Biomarkers are quantifiable biological traits that can be used to detect numerous aspects of health and disease. They can be discovered in a variety of physiological fluids, including blood, urine, tissue samples, and even genetic material. The presence, development, or severity of a disease, as well as the response to treatment, are all determined by the biomarkers.

Biomarkers are essential in the context of precision medicine for identifying patient populations that are more likely to benefit from a certain treatment. Healthcare practitioners can identify which individuals are more likely to respond favorably to a given therapy by examining particular biomarkers. Instead of using a one-size-fits-all strategy, this technique enables a more individualized and focused treatment plan (Yang et al., 2010)

Traditional medicine, which frequently includes a method of trial and error, falls short of precision medicine. Instead, it makes use of advanced analytics, biomarkers, and genomic data to learn more about the distinctive traits of each patient. Precision medicine strives to personalize treatments for each patient's unique needs by considering their genetic makeup, molecular profile, and other pertinent information.

Precision medicine's potential to increase therapeutic efficacy while reducing adverse effects is one of its key benefits. Healthcare practitioners can choose the best therapy for each patient by finding biomarkers linked to an illness or its response to treatment. With this focused strategy, there are more chances for positive results and fewer chances for negative side effects or ineffective therapies.

Biomarkers can potentially be useful for early disease monitoring and detection. Clinicians can intervene earlier by locating specific biomarkers linked to a disease's early stages by potentially improving treatment outcomes. Biomarkers can also be used to track the development of a disease over time or the effectiveness of a treatment, allowing for any necessary modifications to the course of action.(Yang et al., 2010)

HIGH THROUGHPUT SCREENING TECHNOLOGIES

Clinical studies do not often directly use high-throughput screening (HTS) technologies. Instead, HTS is mostly used in the early stages of drug discovery and preclinical research to find new therapeutic candidates. However, through assisting in target discovery, lead optimization, and compound profiling, several HTS technologies indirectly support the clinical trial process (A. P. Li et al., 2004)

Target Identification

HTS can be used to test sizable compound libraries against particular targets or disease-related biological pathways. Using HTS, researchers can find prospective therapeutic candidates by discovering substances that affect the target of interest.

Lead Optimization

After the initial HTS hits have been discovered, more optimization is required to enhance the potency, selectivity, and other pharmacological characteristics of the leads. To improve the desired attributes of the lead compounds, HTS techniques, including structure-activity relationship (SAR) investigations, can be utilized to screen sizable compound libraries with structural alterations.

ADME-Tox Profiling

Absorption, distribution, metabolism, excretion, and toxicity (ADME-Tox) profiling is an important stage in drug development. HTS can be used to evaluate the solubility, stability, metabolic profiles, and potentially hazardous consequences of prospective medication candidates. By using this data, researchers can more efficiently produce drugs with good ADME-Tox profiles (Fox et al., 2006)

Biomarker Discovery

HTS technologies, such as genomics, proteomics, and metabolomics, can help with the identification and validation of biomarkers. The use of biomarkers helps in patient stratification, illness progression prediction, and therapy response prediction. Large sample sets can be screened using HTS to find possible biomarkers that can then be further examined in clinical trials.

Patient Stratification

Using HTS technology, patients can be categorized into subgroups based on their molecular profiles, genetic variants, or other traits. This stratification enables more individualized treatment approaches by enabling the identification and enrollment in clinical trials of specific patient populations that are likely to respond to a given treatment, resulting in more focused and effective trials (Isgut et al., 2018)

Companion Diagnostics

HTS technologies play a critical role in the creation of companion diagnostics, or tests that are used to pinpoint patients who will most likely benefit from certain medications. These tests may be used to select patients for clinical trials and to direct clinical practice therapy choices.

CELL-FREE DNA IN DRUG DISCOVERY AND DEVELOPMENT

Due to its ability to offer insightful data on disease states, therapeutic response, and toxicity, cell-free DNA (cfDNA) has become a viable biomarker in drug discovery and development. When a cell dies or decomposes, it releases fragments of DNA called CfDNA into the blood or other physiological fluids. This genetic material, which can come from both healthy cells and cancerous cells, can be utilized to track different elements of the development of the illness and the effectiveness of various treatments. (Agostini et al., 2011)

Disease Biomarker

Cancer is one of many disorders for which CfDNA can be used as a non-invasive biomarker. Tumor-derived cfDNA (ctDNA) contains genetic changes, including mutations, copy number variations, and rearrangements, that are unique to the tumor. Researchers can find malignancies, track the development of diseases, and find prospective therapeutic targets by studying ctDNA.

Early Detection and Diagnosis

CfDNA analysis may make it possible to identify and diagnose diseases before symptoms appear. For instance, ctDNA can be found in circulation in cases of cancer even before the development of clinical signs or the appearance of tumors on imaging. Monitoring ctDNA levels and certain mutations can help identify cancer recurrence early and enable prompt treatment.

Treatment Response Monitoring

By examining alterations in ctDNA levels and genetic changes, scientists can assess the efficacy of a certain medication or treatment plan. Clinicians can make educated decisions about therapy modifications based on the presence of certain mutations or alterations in ctDNA that may suggest drug resistance or disease progression.

Personalized Medicine

CfDNA analysis makes personalized treatment techniques possible by revealing details about a person's genetic makeup. This knowledge can inform treatment choices, such as choosing targeted medicines that are most likely to be successful in light of the precise ctDNA mutations found.(Agostini et al., 2011)

Pharmacodynamics and Pharmacokinetics

CfDNA may be used to evaluate the pharmacodynamics and pharmacokinetics of a certain medication. The effect of a medicine on tumor burden, cell death, and therapeutic response can be assessed by monitoring cfDNA alterations. Additionally, cfDNA can offer information on medication metabolism and clearance rates, which can be used to improve drug dosage protocols.

Safety and Toxicity Monitoring

Monitoring cfDNA for safety and toxicity can help in the early identification of drug-related toxicities. Changes in cfDNA levels or specific genetic modifications can signal potential side effects, enabling quick intervention and readjusting treatment strategies to reduce patient harm.

Drug Development and Clinical Trials

CfDNA analysis can be useful in the creation of new medications and clinical studies. Researchers can assess drug efficacy, find possible biomarkers for patient stratification, and track treatment response across various patient populations by examining cfDNA from participants in clinical trials (Agostini et al., 2011)

CONDITIONAL DENOVO DRUG GENERATIVE MODELING GRAPH CONVOLUTIONAL NETWORKS

In the world of drug development, conditional de novo drug creation is a difficult task. Deep learning methods, such as Graph Convolutional Networks (GCNs), have recently demonstrated promising results in several molecule production problems. We can create a potent model for producing innovative medications with certain features by using GCNs in clinical trials (Landrum et al., 2020)

The fundamental goal of a conditional de novo drug-generating model is to produce molecules with desired properties or that satisfy predetermined requirements. Clinical trial eligibility criteria could include things like effectiveness against a specific illness target, desired pharmacokinetic features, safety profiles, and more. Clinical trials can make use of the GCN model for:

Data Representation

The first step is to create an appropriate format for GCN-based models that represent the drug molecules and their attributes. One typical illustration is a graph, where the atoms are the nodes, and the connections between them are the edges. As node and edge properties, additional details like atom kinds, bond types, and other chemical aspects can also be included (Mallick & Bhadra, 2023)

Training Data

A dataset with details on drug compounds and the desired attributes is required to train the model. This dataset can be retrieved from a variety of sources, including open databases, data from clinical trials, and confidential datasets. The dataset's molecules need to have the desired features or qualities marked on them.

Graph Convolutional Networks

Deep learning models called "graph convolutional networks" (GCNs) use graph-structured data to operate on it. They have demonstrated remarkable success in understanding the structural characteristics of molecules and learning from molecular graphs. Message transfer between nodes is carried out via the GCN layers, which also update node representations based on local neighborhood data (Y. Li et al., 2018).

Conditional Generation

Following training, the GCN model can be used to create new molecules with specific attributes. The model is fed the required attributes as input throughout the conditional generation process, and it acts as a guide for generating molecules that satisfy those properties. To encourage the model to produce molecules with the required properties, this can be accomplished by adding the desired attributes as extra input features or by altering the loss function during training (Mallick & Bhadra, 2023)

Evaluation and Optimization

The created molecules can be improved via iterative refinement, genetic algorithms, or reinforcement learning. Numerous optimization goals can be taken into account, such as maximizing desired properties, limiting undesirable features, or balancing several goals at once. To evaluate the potential efficacy and safety of the created compounds, molecular docking, scoring functions, or other molecular modeling methods should be used (Tong et al., 2021)

Iterative Refinement

By incorporating feedback from experimental data or the findings of clinical trials, the model can be refined repeatedly. The model can be retrained using this feedback, and new molecules can be created using the improved understanding (Landrum et al., 2020; Mallick & Bhadra, 2023)

CONCLUSION

In conclusion, a transformational trajectory for clinical trials and drug development is presented by the convergence of decentralized trials, gene editing, artificial intelligence (AI), and machine learning (ML). Adopting decentralized trials speeds up the drug development process by utilizing patient-centricity and real-world data. The ability to precisely target therapeutic interventions through gene editing could revolutionise the way that hereditary illnesses are treated. Data-driven insights made possible by AI and ML speed up trial design and patient enrollment while enhancing medication discovery. The symbiotic link between these developments holds great promise for improving the effectiveness of clinical trials, patient outcomes, and the trajectory of medical advancement in general. A collaborative and interdisciplinary approach will be crucial in the future for maximising the potential of these technologies for the improvement of healthcare globally.

REFERENCES

Adams, A., Adelfio, A., Barnes, B., Berlien, R., Branco, D., Coogan, A., Garson, L., Ramirez, N., Stansbury, N., Stewart, J., Worman, G., Butler, P. J., & Brown, D. (2023). Risk-Based Monitoring in Clinical Trials: 2021 Update. *Therapeutic Innovation & Regulatory Science*, 57(3), 529–537. doi:10.1007/s43441-022-00496-9 PMID:36622566

Agbo, C., Mahmoud, Q., & Eklund, J. (2019). Blockchain Technology in Healthcare: A Systematic Review. *Health Care*, 7(2), 56. doi:10.3390/healthcare7020056 PMID:30987333

Agostini, M., Pucciarelli, S., Enzo, M. V., Del Bianco, P., Briarava, M., Bedin, C., Maretto, I., Friso, M. L., Lonardi, S., Mescoli, C., Toppan, P., Urso, E., & Nitti, D. (2011). Circulating Cell-Free DNA: A Promising Marker of Pathologic Tumor Response in Rectal Cancer Patients Receiving Preoperative Chemoradiotherapy. *Annals of Surgical Oncology*, 18(9), 2461–2468. doi:10.1245/s10434-011-1638-y PMID:21416156

Anadioti, E., Musharbash, L., Blatz, M. B., Papavasiliou, G., & Kamposiora, P. (2020). 3D printed complete removable dental prostheses: A narrative review. *BMC Oral Health*, 20(1), 343. doi:10.1186/s12903-020-01328-8 PMID:33246466

Bhatia, S. N., & Ingber, D. E. (2014). Microfluidic organs-on-chips. *Nature Biotechnology*, 32(8), 760–772. doi:10.1038/nbt.2989 PMID:25093883

Briel, M., Elger, B. S., McLennan, S., Schandelmaier, S., von Elm, E., & Satalkar, P. (2021). Exploring reasons for recruitment failure in clinical trials: A qualitative study with clinical trial stakeholders in Switzerland, Germany, and Canada. *Trials*, 22(1), 844. doi:10.1186/s13063-021-05818-0 PMID:34823582

Chen, D., & Qi, E. Y. (2020). Innovative highlights of clinical drug trial design. *Translational Research; the Journal of Laboratory and Clinical Medicine*, 224, 71–77. doi:10.1016/j.trsl.2020.05.007 PMID:32504825

Collins, F. S., & Varmus, H. (2015). A new initiative on precision medicine. *The New England Journal of Medicine*, 372(9), 793–795. doi:10.1056/NEJMp1500523 PMID:25635347

Elliott, D., Husbands, S., Hamdy, F. C., Holmberg, L., & Donovan, J. L. (2017). Understanding and Improving Recruitment to Randomised Controlled Trials: Qualitative Research Approaches. *European Urology*, 72(5), 789–798. doi:10.1016/j.eururo.2017.04.036 PMID:28578829

Fox, S., Farr-Jones, S., Sopchak, L., Boggs, A., Nicely, H. W., Khoury, R., & Biros, M. (2006). High-Throughput Screening: Update on Practices and Success. *SLAS Discovery*, 11(7), 864–869. doi:10.1177/1087057106292473 PMID:16973922

Garbayo, E., Pascual-Gil, S., Rodríguez-Nogales, C., Saludas, L., Estella-Hermoso de Mendoza, A., & Blanco-Prieto, M. J. (2020). Nanomedicine and drug delivery systems in cancer and regenerative medicine. *Wiley Interdisciplinary Reviews. Nanomedicine and Nanobiotechnology*, 12(5), e1637. doi:10.1002/wnan.1637 PMID:32351045

Harrer, S., Shah, P., Antony, B., & Hu, J. (2019). Artificial Intelligence for Clinical Trial Design. *Trends in Pharmacological Sciences*, 40(8), 577–591. doi:10.1016/j.tips.2019.05.005 PMID:31326235

Isgut, M., Rao, M., Yang, C., Subrahmanyam, V., Rida, P. C. G., & Aneja, R. (2018). Application of Combination High-Throughput Phenotypic Screening and Target Identification Methods for the Discovery of Natural Product-Based Combination Drugs. *Medicinal Research Reviews*, *38*(2), 504–524. doi:10.1002/med.21444 PMID:28510271

Jackson, M. (2020). Good Financial Practice and Clinical Research Coordinator Responsibilities. *Seminars in Oncology Nursing*, *36*(2), 150999. doi:10.1016/j.soncn.2020.150999 PMID:32253048

Khozin, S., & Coravos, A. (2019). Decentralized Trials in the Age of Real-World Evidence and Inclusivity in Clinical Investigations. *Clinical Pharmacology and Therapeutics*, *106*(1), 25–27. doi:10.1002/cpt.1441 PMID:31013350

Kiley, R., Peatfield, T., Hansen, J., & Reddington, F. (2017). Data Sharing from Clinical Trials—A Research Funder's Perspective. *The New England Journal of Medicine*, *377*(20), 1990–1992. doi:10.1056/NEJMsb1708278 PMID:29141170

Kolluri, S., Lin, J., Liu, R., Zhang, Y., & Zhang, W. (2022). Machine Learning and Artificial Intelligence in Pharmaceutical Research and Development: A Review. *The AAPS Journal*, *24*(1), 19. doi:10.1208/s12248-021-00644-3 PMID:34984579

Landrum, G., Tosco, P., & Kelley, B. Sriniker, Gedeck, NadineSchneider, Vianello, R., Ric, Dalke, A., Cole, B., AlexanderSavelyev, Swain, M., Turk, S., N, D., Vaucher, A., Kawashima, E., Wójcikowski, M., Probst, D., Godin, G., & Doliath, G. (2020). *rdkit/rdkit: 2020_03_1 (Q1 2020) Release* (Release_2020_03_1) [Computer software]. Zenodo. doi:10.5281/ZENODO.3732262

Li, A. P., Bode, C., & Sakai, Y. (2004). A novel in vitro system, the integrated discrete multiple organ cell culture (IdMOC) system, for the evaluation of human drug toxicity: Comparative cytotoxicity of tamoxifen towards normal human cells from five major organs and MCF-7 adenocarcinoma breast cancer cells. *Chemico-Biological Interactions*, *150*(1), 129–136. doi:10.1016/j.cbi.2004.09.010 PMID:15522266

Li, Y., Zhang, L., & Liu, Z. (2018). Multi-objective de novo drug design with conditional graph generative model. *Journal of Cheminformatics*, *10*(1), 33. doi:10.1186/s13321-018-0287-6 PMID:30043127

Litvina, E., Adams, A., Barth, A., Bruchez, M., Carson, J., Chung, J. E., Dupre, K. B., Frank, L. M., Gates, K. M., Harris, K. M., Joo, H., William Lichtman, J., Ramos, K. M., Sejnowski, T., Trimmer, J. S., White, S., & Koroshetz, W. (2019). BRAIN Initiative: Cutting-Edge Tools and Resources for the Community. *The Journal of Neuroscience : The Official Journal of the Society for Neuroscience*, *39*(42), 8275–8284. doi:10.1523/JNEUROSCI.1169-19.2019 PMID:31619497

Mallick, S., & Bhadra, S. (2023). CDGCN: Conditional de novo Drug Generative Model Using Graph Convolution Networks. In H. Tang (Ed.), *Research in Computational Molecular Biology* (Vol. 13976, pp. 104–119). Springer Nature Switzerland. doi:10.1007/978-3-031-29119-7_7

Mayo, C. S., Matuszak, M. M., Schipper, M. J., Jolly, S., Hayman, J. A., & Ten Haken, R. K. (2017). Big Data in Designing Clinical Trials: Opportunities and Challenges. *Frontiers in Oncology*, *7*, 187. doi:10.3389/fonc.2017.00187 PMID:28913177

Ngayua, E. N., He, J., & Agyei-Boahene, K. (2021). Applying advanced technologies to improve clinical trials: A systematic mapping study. *Scientometrics*, *126*(2), 1217–1238. doi:10.1007/s11192-020-03774-1 PMID:33250544

Obermeyer, Z., & Emanuel, E. J. (2016). Predicting the Future—Big Data, Machine Learning, and Clinical Medicine. *The New England Journal of Medicine*, *375*(13), 1216–1219. doi:10.1056/NEJMp1606181 PMID:27682033

Osório, L. A., Silva, E., & Mackay, R. E. (2021). A Review of Biomaterials and Scaffold Fabrication for Organ-on-a-Chip (OOAC) Systems. *Bioengineering (Basel, Switzerland)*, *8*(8), 113. doi:10.3390/bioengineering8080113 PMID:34436116

Prakash, V., Moore, M., & Yáñez-Muñoz, R. J. (2016). Current Progress in Therapeutic Gene Editing for Monogenic Diseases. *Molecular Therapy*, *24*(3), 465–474. doi:10.1038/mt.2016.5 PMID:26765770

Pushpakom, S., Iorio, F., Eyers, P. A., Escott, K. J., Hopper, S., Wells, A., Doig, A., Guilliams, T., Latimer, J., McNamee, C., Norris, A., Sanseau, P., Cavalla, D., & Pirmohamed, M. (2019). Drug repurposing: Progress, challenges and recommendations. *Nature Reviews. Drug Discovery*, *18*(1), 41–58. doi:10.1038/nrd.2018.168 PMID:30310233

Rico-Villademoros, F., Hernando, T., Sanz, J.-L., López-Alonso, A., Salamanca, O., Camps, C., & Rosell, R. (2004). The role of the clinical research coordinator—Data manager—In oncology clinical trials. *BMC Medical Research Methodology*, *4*(1), 6. doi:10.1186/1471-2288-4-6 PMID:15043760

Rosenberg, S. A., Aebersold, P., Cornetta, K., Kasid, A., Morgan, R. A., Moen, R., Karson, E. M., Lotze, M. T., Yang, J. C., Topalian, S. L., Merino, M. J., Culver, K., Miller, A. D., Blaese, R. M., & Anderson, W. F. (1990). Gene Transfer into Humans—Immunotherapy of Patients with Advanced Melanoma, Using Tumor-Infiltrating Lymphocytes Modified by Retroviral Gene Transduction. *The New England Journal of Medicine*, *323*(9), 570–578. doi:10.1056/NEJM199008303230904 PMID:2381442

Schmitt, C. M. (2002). Clinical research. *Gastrointestinal Endoscopy Clinics of North America*, *12*(2), 395–419. doi:10.1016/S1052-5157(01)00018-6 PMID:12180169

Sen, A., Ryan, P. B., Goldstein, A., Chakrabarti, S., Wang, S., Koski, E., & Weng, C. (2017). Correlating eligibility criteria generalizability and adverse events using Big Data for patients and clinical trials. *Annals of the New York Academy of Sciences*, *1387*(1), 34–43. doi:10.1111/nyas.13195 PMID:27598694

Singh, V., Braddick, D., & Dhar, P. K. (2017). Exploring the potential of genome editing CRISPR-Cas9 technology. *Gene*, *599*, 1–18. doi:10.1016/j.gene.2016.11.008 PMID:27836667

Stansbury, N., Barnes, B., Adams, A., Berlien, R., Branco, D., Brown, D., Butler, P., Garson, L., Jendrasek, D., Manasco, G., Ramirez, N., Sanjuan, N., Worman, G., & Adelfio, A. (2022). Risk-Based Monitoring in Clinical Trials: Increased Adoption Throughout 2020. *Therapeutic Innovation & Regulatory Science*, *56*(3), 415–422. doi:10.1007/s43441-022-00387-z PMID:35235192

Steinhubl, S. R., Muse, E. D., & Topol, E. J. (2013). Can Mobile Health Technologies Transform Health Care? *Journal of the American Medical Association*, *310*(22), 2395. doi:10.1001/jama.2013.281078 PMID:24158428

Tachibana, M., Amato, P., Sparman, M., Woodward, J., Sanchis, D. M., Ma, H., Gutierrez, N. M., Tippner-Hedges, R., Kang, E., Lee, H.-S., Ramsey, C., Masterson, K., Battaglia, D., Lee, D., Wu, D., Jensen, J., Patton, P., Gokhale, S., Stouffer, R., & Mitalipov, S. (2013). Towards germline gene therapy of inherited mitochondrial diseases. *Nature*, *493*(7434), 627–631. doi:10.1038/nature11647 PMID:23103867

Tong, X., Liu, X., Tan, X., Li, X., Jiang, J., Xiong, Z., Xu, T., Jiang, H., Qiao, N., & Zheng, M. (2021). Generative Models for De Novo Drug Design. *Journal of Medicinal Chemistry*, *64*(19), 14011–14027. doi:10.1021/acs.jmedchem.1c00927 PMID:34533311

Vamathevan, J., Clark, D., Czodrowski, P., Dunham, I., Ferran, E., Lee, G., Li, B., Madabhushi, A., Shah, P., Spitzer, M., & Zhao, S. (2019). Applications of machine learning in drug discovery and development. *Nature Reviews. Drug Discovery*, *18*(6), 463–477. doi:10.1038/s41573-019-0024-5 PMID:30976107

Van Norman, G. A. (2021). Decentralized Clinical Trials. *JACC. Basic to Translational Science*, *6*(4), 384–387. doi:10.1016/j.jacbts.2021.01.011 PMID:33997523

Weng, Y., Xiao, H., Zhang, J., Liang, X.-J., & Huang, Y. (2019). RNAi therapeutic and its innovative biotechnological evolution. *Biotechnology Advances*, *37*(5), 801–825. doi:10.1016/j.biotechadv.2019.04.012 PMID:31034960

Yang, D., Liu, H., Goga, A., Kim, S., Yuneva, M., & Bishop, J. M. (2010). Therapeutic potential of a synthetic lethal interaction between the *MYC* proto-oncogene and inhibition of aurora-B kinase. *Proceedings of the National Academy of Sciences of the United States of America*, *107*(31), 13836–13841. doi:10.1073/pnas.1008366107 PMID:20643922

Zhang, B., Korolj, A., Lai, B. F. L., & Radisic, M. (2018). Advances in organ-on-a-chip engineering. *Nature Reviews. Materials*, *3*(8), 257–278. doi:10.1038/s41578-018-0034-7

Chapter 2
AI in Education:
Ethical Challenges and Opportunities

Gurwinder Singh

Department of AIT-CSE, Chandigarh University, Punjab, India

Anshika Thakur

Department of AIT-CSE, Chandigarh University, Punjab, India

ABSTRACT

Artificial intelligence (AI) offers transformative opportunities in education, promising personalized learning and streamlined tasks. However, its integration raises ethical concerns like algorithmic bias and data privacy. AI can revolutionize education by tailoring learning experiences and optimizing administrative tasks, yet it risks eroding student autonomy and perpetuating biases. To mitigate these challenges, transparent algorithms, robust data governance, and educator training are essential. Successful AI implementations demonstrate improved engagement and outcomes. Moving forward, collaborative efforts are crucial to navigate ethical complexities and ensure AI enhances education responsibly, fostering equity and inclusivity.

INTRODUCTION

The classrooms of the future are undergoing a transformative shift, where the integration of artificial intelligence (AI) is shaping a dynamic and personalized learning environment (Black & Wiliam, 1998; Luxton-Earl, 2020). In this futuristic educational landscape, intelligent assistants take center stage, revolutionizing traditional teaching methodologies. These assistants are equipped with the capability to discern and adapt to individual learning styles, paces, and strengths. Through sophisticated analysis of student performance and interests, AI crafts personalized curricula and activities, ushering in an era where educational experiences are tailored to the unique needs of each learner. This departure from the conventional one-size-fits-all approach marks a significant advancement in educational practices, as supported by research highlighting the benefits of personalized learning for student engagement and achievement (Hattie, 2008; Shernoff et al., 2020).

DOI: 10.4018/979-8-3693-2964-1.ch002

One of the notable contributions of AI in education lies in automation. Mundane and timeconsuming tasks, such as grading assessments, are now streamlined through AI-powered systems, leading to the elimination of the arduous process of handling mountains of papers (Weller, 2017). Automated grading not only saves time but also facilitates instant feedback, enabling students to promptly grasp their strengths and areas for improvement, as research from Black and Wiliam (1998) has shown. Consequently, teachers are liberated from the shackles of administrative burdens, allowing them to redirect their focus towards providing more profound guidance and personalized support. This shift transforms educators into facilitators and mentors, fostering a more engaging and interactive teacher-student dynamic, in line with the principles of humanized learning advocated by Hargreaves and Avelino (2022) and Honey and Osborne (2017).

Furthermore, AI's impact extends beyond administrative efficiency, as it plays a crucial role in predicting and addressing student difficulties before they escalate. By analyzing vast amounts of data, AI can identify patterns indicative of potential learning challenges. Adaptive learning platforms, driven by AI algorithms, dynamically adjust difficulty levels to match the skill progression of individual students (Shute, 2008). This proactive intervention ensures that learners are consistently challenged at an appropriate level, contributing to improved learning outcomes for all students, as studies by Kettering et al. (2020) have demonstrated. The application of AI in education, therefore, transcends the mere integration of advanced technologies; it becomes a powerful tool for enhancing the overall quality and effectiveness of the educational experience (Luxton-Earl, 2020).

In essence, the incorporation of AI in education represents a paradigm shift, aiming not just to introduce innovative gadgets but to reimagine the entire learning experience. The overarching goal is to create an educational ecosystem that is not only more engaging but also efficient and effective. The personalized nature of AI-driven learning endeavors to unlock the latent potential within each student, fostering a truly individualized educational journey. In this way, AI in education emerges as a catalyst for transformation, propelling education into a future characterized by adaptability, engagement, and the fulfillment of every student's unique learning needs.

As exciting as AI in education appears, it's crucial to acknowledge the looming ethical challenges that come with its integration. Here are some key concerns:

1. Algorithmic Bias: AI systems learn from the data they're fed, and if that data reflects existing biases, it can perpetuate those biases in the classroom. Imagine an AI recommending certain career paths based on a student's gender or socioeconomic background, replicating and reinforcing inequalities.
2. Data Privacy and Security: Collecting and storing student data for AI analysis raises privacy concerns. Who owns this data? How is it secured? What happens in case of breaches? These questions require robust data governance frameworks and transparent policies to empower students and parents.
3. Surveillance and Student Autonomy: Overreliance on AI-powered monitoring and assessment risks creating a "Big Brother" atmosphere. Are students constantly under surveillance, with their every click and keystroke tracked? This can undermine student autonomy and create an unhealthy learning environment.
4. Job Displacement: While AI can automate administrative tasks, concerns linger about the potential for educator job displacement. How can we ensure AI complements, rather than replaces, teachers, focusing on tasks AI does best while preserving the irreplaceable human element of education?

5.   The Human Element and Social-Emotional Learning: AI excels at certain tasks, but it cannot replace the human connection and warmth necessary for fostering social-emotional learning and holistic development. How can we ensure AI doesn't dehumanize education and that students still receive the emotional support and guidance they need to thrive?

Existing Works

The Table 1 encompasses a range of perspectives on artificial intelligence (AI) in education, emphasizing its multifaceted nature. It delves into ethical challenges and opportunities presented by AI integration, including concerns about algorithmic bias, data privacy, and surveillance. Pedagogical impacts and societal challenges of AI in educational settings are explored, alongside the need for ethical principles and governance frameworks. Practical considerations, such as the transformation of education through AI and the role of EdTech companies, are highlighted. Moreover, there is a focus on responsible AI integration and the importance of teacher preparation. Collectively, these references contribute to a comprehensive understanding of AI's potential in education, while also addressing the ethical and practical complexities that arise from its implementation.

This chapter navigates the exciting possibilities and pressing ethical concerns surrounding AI's integration into education. It has explored AI's potential to personalize learning, automate tasks, and empower educators, while critically examining challenges like algorithmic bias, data privacy, surveillance, and the future of teacher roles. Through real-world case studies and ethical strategies, it has charted a roadmap for harnessing AI's potential for a more personalized, equitable, and responsible future of education, where technology complements, rather than replaces, the irreplaceable human element in learning.

OPPORTUNITIES AND BENEFITS

AI's Potential Revolution in Education: From Personalized Learning to Enhanced Efficiency

Artificial intelligence (AI) is poised to shake the dusty shelves of the traditional classroom and bring about a transformative learning experience. Its vast capabilities touch every corner of education, offering exciting possibilities to:

1.   Personalize Learning: Imagine textbooks morphing into dynamic pathways that adapt to your pace and understanding. AI can analyze your strengths, weaknesses, and learning styles, crafting unique curriculums, recommending resources, and adjusting difficulty levels on the fly. This customized approach ensures everyone stays engaged and challenged, from gifted minds to those needing extra support.

2.   Automate and Amplify Educators: Goodbye mountains of papers, hello instant feedback and automated grading! AI can handle tedious tasks like marking quizzes and generating detailed reports, freeing up teachers to focus on what they do best - guiding, mentoring, and providing personalized support. This efficiency boost allows educators to delve deeper into individual needs, offering focused attention and nuanced feedback.

Table 1. References on AI in education

Label	Authors	Title	Findings
(Akgun & Greenhow, 2022)	S. Akgun and C. Greenhow	Addressing ethical challenges in K-12 settings	Ethical challenges in integrating AI in K-12 education
(Pedro et al., 2019)	F. Pedro, M. Subosa, A. Rivas, and P. Valverde	Challenges and opportunities for sustainable development	Exploration of AI's potential for sustainable education
(Han et al., 2023)	B. Han, S. Nawaz, G. Buchanan, and D. McKay	Ethical and Pedagogical Impacts of AI in Education	Ethical and pedagogical implications of AI integration in education
(Slimi & Villarejo Carballido, 2023)	Z. Slimi and B. Villarejo Carballido	Navigating the Ethical Challenges of Artificial Intelligence in Higher Education: An Analysis of Seven Global AI Ethics Policies	Analysis of global AI ethics policies in higher education
(Vincent-Lancrin & Van der Vlies, 2020)	S. VincentLancrin and R. Van der Vlies	Trustworthy artificial intelligence (AI) in education: Promises and challenges	Discussion on building trust in AI applications for education
(Nguyen et al., 2023)	A. Nguyen, H. Ngo, Y. Hong, B. Dang, and B. Nguyen	Ethical principles for artificial intelligence in education	Proposal of ethical principles for AI integration in education
(Cath, 2018)	C. Cath	Governing artificial intelligence: ethical, legal and technical opportunities and challenges	Examination of governance challenges in AI adoption in education
(Castello-Sirvent et al., 2023)	F. Castello-´ Sirvent, V. E. Garc´ıa Felix, and´ L. Canos-Dar´ os´	AI IN HIGHER EDUCATION: NEW ETHICAL CHAL-LENGES FOR STUDENTS AND TEACHERS	Identification of emerging ethical challenges in higher education due to AI integration
(Zaman, 2023)	B. U. Zaman	Transforming Education Through AI, Benefits, Risks, and Ethical Considerations	Examination of benefits, risks, and ethical considerations of AI in education
(Kousa & Niemi, 2023)	P. Kousa and H. Niemi	AI ethics and learning: EdTech companies' challenges and solutions	Discussion on challenges faced by EdTech companies in adhering to AI ethics
(Remian, 2019)	D. Remian	Augmenting education: ethical considerations for incorporating artificial intelligence in education	Ethical considerations for integrating AI into educational practices
(Aler Tubella et al., 2024)	A. Aler Tubella, M. Mora-Cantallops, and J. C. Nieves	How to teach responsible AI in Higher Education: challenges and opportunities	Strategies for teaching responsible AI in higher education
(Holmes, 2021)	W. Holmes et al.	Ethics of AI in education: Towards a community-wide framework	Proposal of a community-wide framework for addressing AI ethics in education
(Akgun & Greenhow, 2022)	S. Akgun and C. Greenhow	Addressing Societal and Ethical Challenges in K-12 Settings	Examination of societal and ethical challenges in K-12 education due to AI integration
(Sijing & Lan, 2018)	L. Sijing and W. Lan	Artificial intelligence education ethical problems and solutions	Identification and solutions for ethical problems in AI education
(Kassymova, 2023)	G. K. Kassymova et al.	Ethical Problems of Digitalization and Artificial Intelligence in Education: A Global Perspective	Global perspective on ethical problems arising from digitalization and AI integration in education

3. Create Adaptive Learning Environments: Imagine a virtual learning assistant whispering hints when you stumble and celebrating your triumphs. AI can personalize the learning environment, offering real-time support, suggesting relevant study materials, and providing targeted interventions for students struggling with specific concepts. This dynamic environment constantly adapts, ensuring everyone receives the help they need, exactly when they need it.

4. Bridge the Accessibility Gap: AI can be a powerful tool for inclusivity. Text-to-speech tools can assist students with reading difficulties, while speech recognition aids those with writing challenges. Customized interfaces and learning pathways can cater to diverse learning styles and disabilities, ensuring everyone has equal access to a quality education.

5. Enhance Engagement and Motivation: Learning doesn't have to be a dreary slog. AI can infuse classrooms with interactive games, simulations, and personalized learning adventures. Imagine exploring historical events through virtual reality or mastering math concepts through gamified challenges. This engaging approach sparks curiosity, fuels motivation, and makes learning an active, rewarding experience.

6. Empower Educators with Data-Driven Insights: AI isn't meant to replace teachers, but to empower them. Powerful analytics tools can provide educators with real-time data on student performance, identifying areas of difficulty, tracking progress, and informing individualized teaching strategies. This data-driven approach allows educators to tailor their instruction to each student's unique needs, ensuring everyone reaches their full potential.

Figure 1. The potential benefits of AI in education (Getting Smart, 2023)

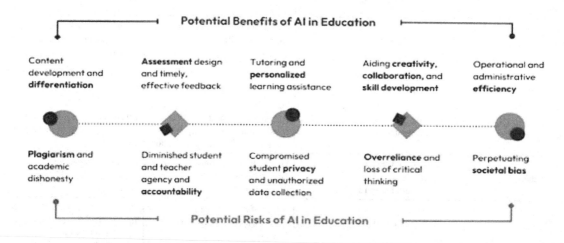

ETHICAL CHALLENGES AND CONCERNS

Ethical Challenges in the Integration of AI in Education

While AI's potential to revolutionize education is undeniable, its integration raises significant ethical concerns that we must address head-on. Here are some major challenges:

1. Algorithmic Bias: AI algorithms learn from the data they're fed, and if that data reflects existing biases, it can perpetuate those biases in the classroom. Imagine an AI recommending certain career paths based on a student's gender or socioeconomic background, replicating and reinforcing inequalities.
2. Data Privacy and Security: Collecting and storing student data for AI analysis raises privacy concerns. Who owns this data? How is it secured? What happens in case of breaches? These questions require robust data governance frameworks and transparent policies to empower students and parents.
3. Surveillance and Student Autonomy: Overreliance on AI-powered monitoring and assessment risks creating a "Big Brother" atmosphere. Are students constantly under surveillance, with their every click and keystroke tracked? This can undermine student autonomy and create an unhealthy learning environment.
4. Job Displacement and Teacher Roles: While AI can automate administrative tasks, concerns linger about the potential for educator job displacement. How can we ensure AI complements, rather than replaces, teachers, focusing on tasks AI does best while preserving the irreplaceable human element of education?
5. The Human Element and Social-Emotional Learning: AI excels at certain tasks, but it cannot replace the human connection and warmth necessary for fostering social-emotional learning and holistic development. How can we ensure AI doesn't dehumanize education and that students still receive the emotional support and guidance they need to thrive?
6. Unintended Consequences and Lack of Transparency: AI algorithms are complex, and their long-term impacts can be difficult to predict. What unintended consequences might arise from integrating AI into education? How can we ensure transparency and accountability in AI development and implementation?
7. Equitable Access and Digital Divide: The benefits of AI in education shouldn't be limited to those with access to technology and resources. How can we ensure equitable access to AIpowered learning tools and bridge the digital divide between underprivileged communities and wealthier ones?
8. Lack of Teacher Preparation and Support: Successfully integrating AI into education requires teachers to develop new skills and adapt their teaching styles. How can we prepare educators for this shift and provide ongoing support to ensure they thrive in an AI-infused classroom?

In the landscape of inclusive education, assistive technologies (AT) serve as crucial allies, empowering students with disabilities to navigate academic hurdles and unlock their full potential. A recent infographic aptly captures this diverse ecosystem of support, presenting a comprehensive visual of the various AT categories readily available. At the heart of the Figure 3 lies a central circle, aptly titled "Assistive Technology for Student Use." This encompassing sphere branches out into eight distinct sections, each representing a key category of AT: Text-to-Speech, Seeing AI, Summarizing & Composing, Readability & Focus, Closed Caption, Voice Recognition, Translators, and Others. Within each section, a multitude of specific tools and services are showcased, providing a glimpse into the vast and ever-evolving landscape of AT solutions. For instance, students struggling with reading comprehension can find solace in Text-to-Speech tools like Beeline Reader or Read Aloud, which transform written text into spoken words, easing the decoding process. Visual impairments are addressed through Seeing AI applications like Seeing Aloud, which narrates the surrounding environment, while tools like Google Dictionary and Summarizer within the Summarizing & Composing category empower students with writing difficulties to express their ideas more effectively. This visual emphasizes the synergy between

Figure 2. The potential benefits of AI in education (Akgun & Greenhow, 2021)

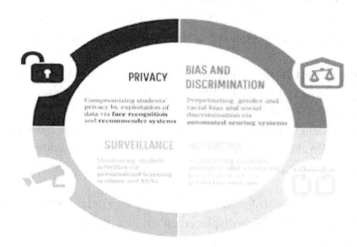

different tools, encouraging educators and students to explore combined approaches that cater to a range of needs. A student using a Text-to-Speech tool, for example, may find their productivity further enhanced by employing a note-taking application or a summarizer to consolidate the newly vocalized information. This multifaceted portrayal of AT underscores the profound impact these technologies can have on the educational journeys of students with disabilities. From tackling learning roadblocks to fostering greater independence and self-confidence, AT acts as a bridge, enabling students to actively participate in and excel within the learning environment.

However, it is crucial to acknowledge that the specific tools and services listed in the infographic represent merely a snapshot of the ever-expanding realm of AT possibilities. The most effective approach to harnessing the power of AT lies in personalized assessments that delve into the unique needs and preferences of each student. With the guidance of qualified professionals, educators can curate a tailored AT toolkit that empowers students to navigate academic challenges and achieve their full potential.

Figure 3. The potential benefits of AI in education (Kerr, n.d.)

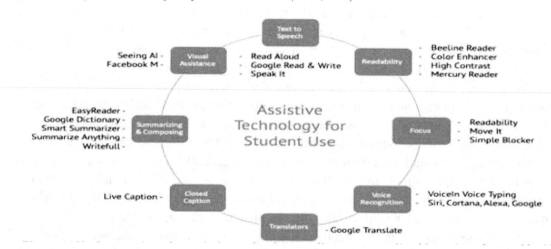

STRATEGIES FOR MITIGATING CHALLENGES

The Figure 4 visualizes the complex interplay between four key AI-driven privacy concerns: ownership of information, non-discrimination, surveillance, and autonomy. Each concern occupies a distinct, yet partially overlapping, sphere, highlighting their intrinsic interconnectedness. For instance, data ownership directly impacts its potential for discriminatory utilization within surveillance practices, consequently impinging upon individual autonomy. This interconnectedness is crucial in comprehending the multifaceted privacy challenges posed by AI. The data-hungry nature of AI algorithms raises ownership concerns, as personal information gathered across various domains, from financial transactions to facial recognition, fuels their operation. This collected data can then be leveraged for profiling and decision-making processes, raising potential risks of discriminatory outcomes, particularly where biases exist within the training data. Furthermore, the extensive data collection inherent in AI systems inevitably fuels concerns about pervasive surveillance, potentially infringing upon individuals' right to privacy and limiting their freedom of movement and expression. Ultimately, the intertwined nature of these concerns necessitates a holistic approach to safeguard privacy in the AI age. Balancing the undeniable benefits of AI with robust privacy protections demands careful consideration of the ethical implications embedded within each overlapping sphere. Only then can we harness the power of AI while ensuring individual rights and freedoms remain uncompromised.

Proactive Strategies for Ethical AI in Education

The Figure 5 depicts the adaptability levels of students across various demographic factors. These factors include education level, gender, age, institution type, IT background, location, financial condition, and internet access. The data reveals variations in adaptability across all categories, with higher levels generally observed among university students, boys, older age groups, nongovernment institutions, IT students, urban residents, wealthier backgrounds, and Wi-Fi users. This suggests the potential of AI in education to personalize learning experiences based on individual student characteristics, although potential challenges like bias and teacher replacement require careful consideration.

Figure 4. Six privacy concerns (Kerr, n.d.)

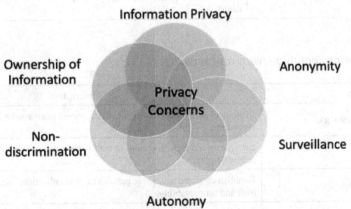

Figure 5. Comparison of adaptability at different levels

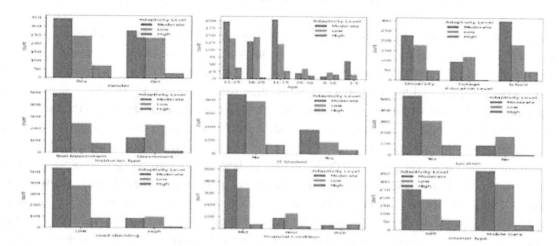

Navigating the ethical minefield of AI in education requires proactive strategies to address concerns without stifling potential. Here are some potential approaches: The Tables 2, 3, 4 and 5 present comprehensive strategies for navigating the integration of AI in education, addressing critical concerns such as bias, privacy, surveillance, and inclusive access. Strategies to address bias emphasize diversifying datasets, conducting algorithmic audits, ensuring transparency, and maintaining human oversight. For

Table 2. Strategies for addressing bias

Strategy	Description
Data diversification	Use diverse datasets in AI development and training to prevent perpetuation of existing biases.
Algorithmic auditing	Regularly analyze algorithms for bias and implement bias mitigation techniques like counterfactual analysis.
Transparency and explainability	Make AI algorithms and decision-making processes transparent, allowing for scrutiny and public engagement.
Human oversight and accountability	Ensure human control and oversight over AI systems, preventing automated biased decisions.

Table 3. Strategies for protecting privacy and security

Strategy	Description
Robust data governance frameworks	Implement strict data security measures, secure storage solutions, and clear consent procedures.
Student agency and control	Empower students with options to control their data, including access, correction, and deletion rights.
Transparency and privacy policies	Establish clear and accessible policies on data collection, use, and sharing, fostering trust and accountability.
Independent oversight and auditing	Implement independent oversight bodies to audit data practices and ensure compliance with regulations.

Table 4. Strategies for balancing surveillance and autonomy

Strategy	Description
Minimize student monitoring	Limit data collection to essential purposes and avoid pervasive surveillance practices.
Transparency and student awareness	Inform students about data collection practices and provide mechanisms for opting out or raising concerns.
Focus on formative assessment	Utilize AI for personalized learning and feedback, rather than for constant monitoring and grading.
Promote student agency and critical thinking	Empower students to make informed choices about their online presence and develop critical thinking skills for navigating digital environments.

Table 5. Strategies for ensuring inclusive access and bridging the divide

Strategy	Description
Universal access to technology and resources	Bridge the digital divide by providing affordable technology and internet access to underprivileged communities.
Culturally sensitive and inclusive design	Design AI tools and learning platforms that are culturally sensitive and cater to diverse learners' needs and abilities.
Community engagement and co-creation	Involve local communities in AI development and implementation to ensure solutions address their specific challenges and needs.
Prioritize human connection and support	Ensure AI complements, rather than replaces, human interaction and personalized support for students from diverse backgrounds.

privacy and security, institutions should implement robust data governance, empower students with data control, establish transparent policies, and enforce independent oversight. Balancing surveillance and autonomy involves minimizing monitoring, promoting transparency, emphasizing formative assessment, and empowering student decision-making. Ensuring inclusive access entails bridging the digital divide, designing culturally sensitive tools, engaging communities, and prioritizing human interaction over automated processes. These strategies collectively aim to foster ethical, inclusive, and effective integration of AI in educational settings.

Table 6 presents a comprehensive overview of student-centric methods aimed at addressing discrimination in AI implementation within educational settings. Under the subsection "Integrating Cognitive Diversity," strategies include inclusive algorithmic design, user-centered prototyping, and adaptive learning paths, all tailored to accommodate diverse learning styles. "Bias Response Teams" highlight the importance of assembling diverse expert groups, conducting ongoing training sessions, and establishing anonymous reporting mechanisms to address biases effectively. In the context of "Community Audits," initiatives such as co-creation workshops, transparency measures, and training programs for citizen data scientists are outlined, emphasizing community involvement in detecting and mitigating potential biases. Overall, these strategies underscore the importance of collaborative efforts and proactive measures to foster inclusivity and fairness in AI-driven educational environments.

Table 7 summarizes the key strategies for prioritizing privacy in AI frameworks and addresses ethical concerns in the classroom. It outlines various privacy-enhancing techniques such as decentralized model training and differential privacy techniques, along with solutions like blockchain for data owner-

Table 6. Comprehensive student-centric methods: Handling discrimination

Subsection	Strategies
Integrating Cognitive Diversity	• Inclusive Algorithmic Design: Make sure algorithms take into account a wide range of learning waysby actively involving students with varied cognitive styles in the design phase. • User-Centered Prototyping: Conduct prototyping sessions with students of all backgrounds and cog-nitive capacities to get immediate input and improve AI models to be inclusive. • Adaptive Learning Paths: Create AI systems that can easily accommodate different learning methodsby dynamically adjusting learning paths based on individual cognitive preferences.
Bias Response Teams	• Diverse Expert Inclusion: Assemble diverse expert teams of practitioners in education, experts froma range of fields, and members of underrepresented populations. • Ongoing Training Programs: Hold frequent training sessions for reaction teams to keep them in-formed about new ethical issues and growing prejudices, ensuring prompt and efficient responses. • Establish Anonymous Reporting Mechanisms: Foster open dialogue and prompt remedial action byproviding students, teachers, and community members with anonymous routes to report observed biases.
Community Audits	• Community Co-Creation Workshops: Conduct workshops where participants in the community areincluded in the co-creation of AI tools. This way, the community may detect potential biases and contribute to the development process. • Transparency Initiatives: Put into practice measures to promote transparency, such as opening up AIalgorithms and decision-making procedures to public inspection in order to foster mutual respect and cooperative problem-solving. • Citizen Data Scientists: Through training programs, enable community people to become citizen datascientists so they may actively audit AI systems and make ethical advancements.

Table 7. Tech solutions that put privacy first

Protecting Privacy with AI Models	Blockchain Technology for Data Ownership
Decentralized Model Training: Put into practice decentralized AI model training frameworks that disperse learning activities among devices and aggregate insights collectively while guaranteeing individual privacy protection.	Self-Sovereign Identity Platforms: Students can take control of their educational data by using self-sovereign identity platforms on blockchain, which provide them the ability to share it with certain educational institutions or other appropriate parties.
Differential Privacy Techniques: Use techniques for differential privacy to protect sensitive data while maintaining overall trends. These techniques involve adding noise to particular data points during model training.	Immutable Educational Records: Students will have ownership over their academic history and data integrity will be improved by implementing blockchain to produce immutable educational records.
Edge Computing for Inference: Make use of edge computing to reduce privacy issues by processing data locally on devices and eliminating the requirement for centralized data storage.	Smart Contracts for Consent: Students can specify precise conditions under which their data can be accessed or used by utilizing smart contracts on blockchain to manage consent for data management.
Safe Multi-Party Computation	Sustaining Autonomy while Encouraging Ethical AI Classroom Practices
Homomorphic Encryption: Include homomorphic encryption into data processing to ensure privacy is preserved throughout analysis and to enable computation on encrypted material without first decrypting it.	Student-Developed AI Manual: Collaborative Code of Conduct: To foster a feeling of group accountability for moral AI practices, encourage students to work together to develop a code of conduct for the use of AI in the classroom.
Privacy-Preserving Data Aggregation: To obtain insights without disclosing raw data, employ privacy-preserving data aggregation strategies such as federated learning or safe enclaves.	Making Ethical Decisions Through Gamification: EthicsBased Gamified Modules: Provide students with gamified modules that revolve around making ethical decisions in AI scenarios. This will enable them to explore and comprehend the consequences of different decisions made in a virtual world.
Dynamic Role-Based Access: Make sure that only authorized people or entities can access particular sections of the educational material by implementing dynamic role-based access controls through secure multiparty computation.	Scenario-Based Role Plays: Assist students in making ethical decisions and honing their critical thinking abilities by facilitating scenario-based role plays that mimic moral conundrums pertaining to AI.

ship. Additionally, it presents methods for sustaining autonomy while encouraging ethical AI classroom practices, including student-led initiatives and gamification for making ethical decisions.

Communications With Stakeholders

We look at ways to have a continuous conversation on AI in education in this part. Informed public conversation and inclusive decision-making are promoted by accessible AI impact studies, community workshops, and stakeholder roundtables.

Roundtables With Stakeholders

- Diverse Panel Representation: Prioritize diverse panels for stakeholder roundtables, ensur-ing cultural, socioeconomic, and academic variation. Rotation systems and easily available documen-tation preserve openness and encourage continuous learning of many perspectives.

- Themed Discussion Sessions: Organize roundtables with specific issues like privacy or cur-riculum enhancement. talks are facilitated by knowledgeable moderators, and stakeholders can review earlier talks for future reference thanks to open access documentation.

- Open Access Documentation: Ensure transparency through accessible documentation, pro-viding stakeholders with summaries and anonymized transcripts. Online platforms facilitate ongoing ac-cess, keeping stakeholders well-informed about key points discussed during roundtable sessions.

Engaging Community Workshops

- Customized Workshop Content: Provide examples from real-world situations and contentthat is appropriate for a range of audiences. Q/A sessions encourage participation and deepen knowledge of the ramifications of AI by creating an interactive atmosphere.

- Interactive Learning Activities: Experiment and go beyond theory in workshops by incorpo-rating interactive activities and simulations. Questions and answers sessions give participants the chance to ask questions and improve teamwork and critical thinking.

- Community Ambassador Program: Empower individuals through an ambassador programfor lo-calized sessions. Supporting ambassadors on an ongoing basis builds a network of advocates who are actively sharing knowledge, starting with seminars.

AI Impact Reports Available to the Public

- Visual Representation of Impact: Enhance accessibility in impact reports using visual el-ements like infographics and data visualizations. Make reports accessible by employing narrative ap-proaches and simplifying complex findings into measures that are easy to understand.

- Periodic Impact Assessments: Keep an eye on the changing scene by conducting periodicassess-ments and updates on the impact of AI in education. Accuracy and credibility are ensured through stakeholder feedback and expert collaboration.

- Interactive Online Platform: Develop an intuitive user interface, multimedia content, andgraphs for AI impact reports on an interactive online platform. To promote a communitydriven knowl-edge of the implications of AI, invite public views and conversations.

Stakeholder-inclusive communication techniques must be put into practice if artificial intelligence is to be responsibly included into education. The ethical trajectory of AI in learning environments is shaped by the voices of diverse stakeholders, who are empowered through accessible impact reports, community workshops, and regular roundtable conversations. By valuing cooperation and openness, these tactics support a group-based, knowledgeable strategy that harmonizes technology development with the principles and requirements of the larger educational community.

AI IN INDIAN EDUCATION: SUCCESS STORIES AND OPPORTUNITIES WITH AN ETHICAL LENS

India's adoption of AI in education is characterized by uplifting success stories that take ethical issues into account. These programs highlight responsible use, putting individualized learning first, promoting diversity in language learning, early childhood education, filling the teacher shortage, and democratizing STEM education.

Personalized Learning for Rural Students

A lighthouse of individualized learning for rural kids, the Pratham platform from Akshara Foundation is highlighted in the landscape of AI deployment in Indian education (Akshara Foundation, n.d.). Pratham uses artificial intelligence to power its adaptive reading tools, supports a variety of learning methods, and maintains strict data protection policies. Examining the platform's dedication to customized curriculum, multimodal interaction, and language inclusion, this investigation digs into the subtle aspects of its success and serves as a monument to the responsible and inclusive integration of AI in educational settings (Achar, 2017).

Tailored Educational Journeys

- Dynamic Curriculum Adaptation: Every rural student has a customized learning experience thanks to Pratham's AI, which dynamically adjusts the curriculum based on individual reading levels (Achar, 2017).
- Learning Style Analysis: By applying AI to analyze a variety of learning styles, the platform can make recommendations for content that best suits each user's requirements and preferences (Achar, 2017).
- Continuous Improvement: To improve its comprehension of students' learning preferences and styles, the adaptive system incorporates feedback and continues to progress (Achar, 2017).

Interactive Narration and Exercises

- Captivating Storytelling: By using AI to create captivating and culturally relevant stories, the platform piques students' interest and encourages a love of reading (Achar, 2017).
- Multimodal Learning: To accommodate varying learning preferences, this approach incorporates multiple learning modalities, including multimedia information and interactive exercises (Achar, 2017).

- Integration of User Feedback: Proactively solicits user input to improve and grow its collection of tales and activities, guaranteeing continued interest and relevance (Achar, 2017).

Empowerment of Regional Languages

- Native Language Emphasis: Places an emphasis on local languages to preserve linguisticdiversity, improve accessibility to learning, and establish a link between education and students' cultural origins (Achar, 2017).
- Parental Involvement: Promotes parental involvement through establishing a collaborativelearning environment, obtaining agreement for the use of data, and cultivating trust (Tacho, 2019).
- LocalizationStrategies:Consistentlyallocatesresourcestowardslocalizationendeavors,customizing interface and content to accommodate the linguistic subtleties of various places, guaranteeing inclusivity and significance (World Innovation Summit for Education, 2015).

Therefore, the Pratham platform developed by the Akshara Foundation is a prime example of how artificial intelligence may significantly improve the educational outcomes of rural kids. Pratham is a role model for responsible AI adoption because it combines rigorous data protection controls, cultural inclusion, and tailored learning in a seamless manner (UNESCO Mahatma Gandhi Institute of Education for Peace and Sustainable Development, 2020). The success of such initiatives establishes the groundwork for equal access, customized learning experiences, and a dedication to fostering the full spectrum of students across varied communities as India continues its journey toward an inclusive and ethical educational future (Tacho, 2019).

Playful AI in Early Childhood Education

With its novel blend of augmented reality and interactive activities, Kinderlabs' "Boondi" app is a shining example of innovative learning in the early childhood education space, providing a dynamic learning environment for young learners. This section explores the essential elements of "Boondi," including its individualized learning paths, compelling features, and steadfast dedication to ethical principles.

Immersive Educational Experiences

- Augmented Reality Integration: "Boondi" harnesses augmented reality to create immersive learning experiences, making educational information come to life for young learners (Collins & Halverson, 2019).
- Narrative-driven Interactions: In order to improve comprehension and engagement, artificial intelligence (AI) creates narrative-driven interactions that skillfully integrate instructional content into engrossing stories (Collins & Halverson, 2019).
- Sensory Stimulation: Assists in a comprehensive early learning experience by including sensory aspects into AR activities to activate many senses (Baroody, 2004).

Flexible Skill Development

- AI-Driven Skill Assessment: Boondi's AI evaluates each student's talents and customizes learning routes so that each learns at their own speed (UNESCO, 2021).

- Progress Tracking: This feature tracks progress in real time, enabling parents and teachers to keep an eye on students' skill growth and pinpoint areas that need more help (UNESCO, 2021).
- Skill Diversity Embrace: Acknowledges and meets the various skill levels of children while encouraging inclusivity and attending to their unique needs (Grant & Sonesh, 2019).

Technology and Human Connection in Balance

- Interactive Learning Partners: Artificial Intelligence is positioned as a learning partner that encourages human-technology collaboration for a comprehensive educational experience (Firth, 2023).
- Integration of Parental Guidance: Promotes meaningful conversations and cooperative learning opportunities by giving parents insights into their children's app usage (Epstein, 2008).
- Child Safety procedures: Gives top priority to strict safety procedures that guarantee children's online safety and parents' and educators' peace of mind (NSPCC, 2023).

By the time we finish up our investigation of Kinderlabs' "Boondi" app, it's clear that there is a lot of potential at the nexus of early childhood education and technology. Our youngest scholars can be inspired to love studying by the app's dedication to interactive learning, tailored skill development, and a healthy balance between technology and human connection. "Boondi" sets the stage for a future in which technology elevates rather than undermines the formative years of education with its moral design and focus on safety.

Overcoming Linguistic Divide With AI-Powered Translation

Language hurdles have frequently impeded access to high-quality education in India's diversified educational landscape. Through the use of cutting-edge AI, the TranslatorsIndia program breaks down linguistic barriers and becomes a revolutionary force. In order to ensure that instructional content reaches diverse groups in their original languages, this section explores the initiative's varied approach to overcoming language gaps.

Translation with Cultural Richness

- Cultural Sensitivity: To ensure that the instructional content relates to local subtleties and sensitivities, the TranslatorsIndia initiative uses AI for both cultural adaptation and translation.
- Integration of Regional Context: The effort improves the cultural relevance of educational materials and promotes a more inclusive and enjoyable learning experience by incorporating regional context into translations.
- Dialect Recognition: Artificial intelligence algorithms have been refined to identify various dialects, hence enhancing the translations to correspond with the linguistic variations in each area.

Translation Platforms for Collaboration

- Community Involvement: Acknowledging the value of community involvement in the creation of pertinent educational resources, the program actively incorporates local communities in the translation process.

- Training Initiatives: Provides community members with the necessary training to enable them to translate effectively using AI, fostering sustainability and local control over educational materials.
- Feedback Loops: Creates channels for end users to provide feedback in order to get insights that guarantee ongoing development and adaptability to the community's changing language needs.

Foundations of Ethics in Translation

- Data Privacy Assurance: The TranslatorsIndia project prioritizes data privacy, putting strong security measures in place to safeguard private data and win over consumers.
- Transparent Translation Procedures: Preserves transparency in translation procedures so that people can comprehend the use of AI, hence bolstering the moral application of technology.
- User Education Initiatives: Promotes transparency and informed interaction with the translated educational materials by running awareness campaigns to enlighten users about the AI-powered translation process.

The TranslatorsIndia project is a shining example of inclusive education that shows how powerful AI can be in bridging language gaps. It not only changes educational access but also establishes a standard for responsible and culturally aware AI use by utilizing technology in an ethical manner, empowering local communities, and adopting transparent procedures. These initiatives provide a hopeful window into how technology might be used to build a more varied and inclusive learning environment as we navigate the future of education.

Using AI Assistants to Help Teachers Shortage Situation "Vidya"

Artificial Intelligence (AI) has brought about revolutionary changes in the Indian educational scene. Notably, ethical use of AI in education has produced remarkable success stories that demonstrate the country's dedication to inclusive, egalitarian, and morally sound learning environments. Let's explore some concrete examples of how AI is benefiting education, with a focus on individualized learning, early childhood education, language inclusion, and cooperative assistance for teachers.

Cooperative Administrative Support

- Teacher Task Optimization: Vidya makes administrative work more efficient so teachers can spend more time interacting directly with students.
- Dynamic effort Management: By adjusting to changing workloads, the AI assistant makes sure that tasks are assigned effectively and lessens the effort for instructors.
- Feedback-Driven Improvements: Takes into account instructor input on administrative procedures, continuously enhancing its capacities to meet the changing demands of educators.

Customized Instructional Assistance

- Customized Feedback Mechanisms: Vidya's customized feedback system offers helpful insights for advancement while adapting to each student's particular learning requirements.

- Adaptive Learning Plans: By addressing each person's strengths and shortcomings, AI is used to develop adaptive learning plans that promote a more individualized educational experience.
- Ongoing Student Progress Monitoring: The assistant makes sure that student progress is tracked in real time, which allows for prompt support and interventions.

AI-Human Synergy

- Collaborative Teaching Models: These models place a strong emphasis on collaborative teaching, in which artificial intelligence (AI) enhances rather than replaces teachers' ability to create a human-centered learning environment.
- Integration with Professional Development: Vidya enhances teachers' skills in tandem with AI integration by integrating training in AI literacy into their professional development programs.
- Community Engagement: Promotes community involvement in determining AI's place in education, guaranteeing that local needs and viewpoints are taken into account when implementing the technology.

In the pursuit of an AI-powered educational future, India must continue to prioritize resolving ethical issues. Striking a balance between technical innovation and ethical considerations, such as data privacy, unbiased algorithms, teacher training, and digital accessibility, will be crucial. India has the ability to create a future in which all students, from all backgrounds, receive individualized, inclusive, and morally sound education through continued cooperation, research, and dedication to ethical AI practices.

Increasing Accessibility to STEM Education

Embibe's dedication to utilizing AI-driven solutions to make STEM education accessible in India is evidence of the innovative nature of the education sector. Embibe uses artificial intelligence to try to break down barriers to STEM education, encourage critical thinking, and personalize learning experiences.

AI-Powered Learning Routes

- Adaptive Practice Algorithms: Embibe's STEM platform uses artificial intelligence (AI) algorithms to modify learning routes in real time, giving students practice problems that are tailored to their skill levels.
- Simulated Learning Environments: Combines simulations with real-world activities to improve comprehension and memory of STEM subjects.
- Real-Time Progress Tracking: Students may monitor their progress in real-time thanks to AI-driven analytics, which increases motivation and a sense of accomplishment.

Addressing these challenges through open dialogue, continuous research, and collaborative efforts will allow India to maximize the potential of AI while upholding ethical principles. By following the path of responsible AI use seen in these success stories, India can pave the way for a more personalized, inclusive, and ethical future of education for all its students.

Various Learning Approaches

- Interactive Simulations: By utilizing AI, interactive simulations may be created that accommodate different learning styles and encourage a deeper comprehension of STEM concepts.
- Collaborative Problem Solving: Promotes cooperative problem-solving exercises that develop critical thinking and teamwork.
- Tailored Feedback: AI-generated feedback helps with ongoing learning by offering insights into a person's areas of strength and growth.

Attainable STEM Education

- Cost-effective Learning Solutions: Embibe places a high priority on cost-effectiveness, opening up STEM education to a larger range of people.
- Culturally Relevant Content: Content is adapted to speak to a range of cultural backgrounds while maintaining inclusion and relevancy.
- Community Engagement Initiatives: Involves communities in an active manner to comprehend their unique demands and issues, thereby helping to shape STEM programs that meet local needs.

To sum up, Embibe's efforts highlight how AI has the ability to drastically alter STEM education in the future. Embibe's ethical considerations, data privacy precautions, and dedication to affordability offer a template for responsible AI integration in the pursuit of fair STEM learning possibilities, as India moves toward a more technologically advanced and inclusive educational landscape.

CONCLUSION

The chapter outlines the vast potential of Artificial Intelligence (AI) in revolutionizing education, emphasizing personalized learning, enhanced efficiency, adaptive learning environments, bridging accessibility gaps, and increasing engagement and motivation among students. However, it also highlights significant ethical challenges associated with AI integration in education, including algorithmic bias, data privacy and security concerns, surveillance issues, job displacement fears, and the risk of dehumanizing the learning experience. Strategies for mitigating these challenges are proposed, such as diversifying datasets, implementing transparent policies, empowering student agency, minimizing surveillance, and prioritizing inclusive access. Moreover, it discusses the importance of proactive measures to address discrimination, protect privacy, and sustain autonomy in AI-driven educational environments. Various tables and figures illustrate comprehensive strategies, including data diversification, algorithmic auditing, robust data governance, user-centered prototyping, and community engagement initiatives, underscoring the multifaceted approach required to ensure ethical AI integration in education.

The integration of AI in education offers personalized learning opportunities while also raising ethical considerations. Variations in adaptability levels across demographic factors highlight the potential of AI to tailor educational experiences. Proactive strategies outlined, including data diversification and transparent policies, aim to mitigate biases and safeguard student privacy. Effective stakeholder engagement through roundtable discussions and community workshops fosters transparency and accountability in AI implementation. Success stories like the Pratham platform and Kinderlabs' "Boondi" app demonstrate

AI's potential in addressing educational challenges while promoting inclusivity and adhering to ethical principles. India's commitment to responsible AI adoption, prioritizing data privacy and cultural sensitivity, underscores its dedication to fostering an ethical trajectory in AI-driven education, ensuring equitable learning environments for all students. Additionally, Artificial Intelligence (AI) is reshaping education in India with a focus on inclusive and ethically sound learning environments. Highlighted examples include the Vidya AI assistant, which optimizes administrative tasks, provides customized instructional assistance, and fosters AI-human synergy in teaching. The pursuit of an AI-powered educational future emphasizes the resolution of ethical issues, such as data privacy and unbiased algorithms, while prioritizing accessibility and inclusivity. Furthermore, Embibe's AI-driven solutions aim to increase accessibility to STEM education by offering adaptive practice algorithms, interactive simulations, and cost-effective learning solutions. These initiatives underscore the transformative potential of AI in education while advocating for responsible AI integration and collaborative efforts to address diverse learning needs.

REFERENCES

Achar, A. (2017). *Artificial intelligence for early literacy in India: A case study of Pratham.* Retrieved from https://files.eric.ed.gov/fulltext/ED496345.pdf

Akgun, S., & Greenhow, C. (2021). Artificial Intelligence in Education: Addressing Ethical Challenges in K-12 Settings. *AI and Ethics*, 1–10. PMID:34790956

Akgun, S., & Greenhow, C. (2022). Artificial intelligence in education: Addressing ethical challenges in K-12 settings. *AI and Ethics*, 2(3), 431–440. doi:10.1007/s43681-021-00096-7 PMID:34790956

Akgun, S., & Greenhow, C. (2022). Artificial Intelligence (AI) in Education: Addressing Societal and Ethical Challenges in K-12 Settings. *Proceedings of the 16th International Conference of the Learning Sciences-ICLS 2022*, (pp. 1373-1376). IEEE.

Akshara Foundation. (n.d.). *Home.* Akshara Foundation. https://akshara.org.in/

Aler Tubella, A., Mora-Cantallops, M., & Nieves, J. C. (2024). How to teach responsible AI in Higher Education: Challenges and opportunities. *Ethics and Information Technology*, 26(1), 3. doi:10.1007/s10676-023-09733-7

Baroody, R. A. (2004). *Sensory experiences and the development of mathematical reasoning.* RoutledgeFalmer. https://link.springer.com/article/10.1007/s10763-004-3224-2

Black, P., & Wiliam, D. (1998). *Assessment and classroom learning.* Guilford Publications. doi:10.1080/0969595980050102

Castello-Sirvent, F., Garcıa Felix, V., & Canos-Daros, L. (2023). AI In Higher Education: New Ethical Challenges For Students And Teachers. *EDULEARN23 Proceedings* (pp. 4463-4470). Research Gate.

Cath, C. (2018). Governing artificial intelligence: Ethical, legal and technical opportunities and challenges. *Philosophical Transactions. Series A, Mathematical, Physical, and Engineering Sciences*, 376(2133), 20180080. doi:10.1098/rsta.2018.0080 PMID:30322996

Collins, A., & Halverson, J. (Eds.). (2019). *Rethinking the Future of Learning: Augmented Reality and Mixed Reality in Education*. Vanderbilt Center for Teaching. https://cft. vanderbilt.edu/

Epstein, A. (2008). *What can schools do to promote the involvement of parents?* American Education Consulting Firm. https://www.aecf.org/resources/ parental-involvement-in-education

Firth, K. (2023). The future of the workforce: How human-AI collaboration will redefine the industry. *Forbes*. https://www.forbes.com/ sites/forbestechcouncil/2023/05/04/the-future-of-the-workforce-\ \ how-human-ai-collaboration-will-redefine-the-industry

Gee, J. (2013). *What video games have to teach us about learning and literacy*. Routledge. https:// blog.ufes.br/kyriafinardi/files/2017/10/ What-Video-Games-Have-to-Teach-us-About-Learning-and-Literacy-2003. -ilovepdf-compressed.pdf

Getting Smart. (2023). *Home*. Getting Smart. https://www.gettingsmart.com/2023/10/25/ one-year-into-the-ai-revolution-and-most-schools-are-still\ \-seeking-direction

Grant, C., & Sonesh, L. (2019). Skillset diversity: Why it matters and how to harness it. Skill Up Powered by Sertifier. *Medium*. https://medium.com/skill-up-powered-by-sertifier/the-importance-of-skillset-diversity-in-the-modern-workplace-4a9d581dd72

Han, B., Nawaz, S., Buchanan, G., & McKay, D. (2023). Ethical and Pedagogical Impacts of AI in Education. *International Conference on Artificial Intelligence in Education*, (pp. 667-673). Springer. 10.1007/978-3-031-36272-9_54

Hargreaves, A., & Avelino, A. (2022). *Humanized learning: Rethinking education in a globalized age*. Routledge.

Hattie, J. (2008). *Visible learning: The key to improving education*. Routledge. doi:10.4324/9780203887332

Holmes, W. (2021). Ethics of AI in education: Towards a community-wide framework. *International Journal of Artificial Intelligence in Education*, 1–23.

Honey, M., & Osborne, A. (2017). *The future of jobs*. Oxford University Press.

Kapp, K. (2014). *Learning in the age of disruption: How personalized learning and microlearning are changing education*. John Wiley & Sons. https://mitpress.mit.edu/ 9780262542210/the-next-age-of-disruption/

Kassymova, G. K. (2023). Ethical Problems of Digitalization and Artificial Intelligence in Education: A Global Perspective. *Journal of Pharmaceutical Negative Results*, 21502161.

Kerr, K. (2020). *Ethical Considerations When Using Artificial Intelligence-Based Assistive Technologies in Education*. Open Educational Berta. https:// openeducationalberta.ca/educationaltechnologyethics/chapter/ ethical-considerations-when-using-artificial-intelligence-based-\ \assistive-technologies-in-education

Kettering, A. H., Baker, R. S., & Mathews, M. M. (2020). *Personalized learning with adaptive systems: From theory to practice*. Cambridge University Press.

Kousa, P., & Niemi, H. (2023). AI ethics and learning: EdTech companies' challenges and solutions. *Interactive Learning Environments*, *31*(10), 6735–6746. doi:10.1080/10494820.2022.2043908

Luxton-Earl, C. (2020). *The future of artificial intelligence in education: Promises and perils*. John Wiley & Sons.

Nguyen, A., Ngo, H., Hong, Y., Dang, B., & Nguyen, B. (2023). Ethical principles for artificial intelligence in education. *Education and Information Technologies, 28*(4), 42214241. doi:10.1007/s10639-022-11316-w PMID:36254344

NSPCC. (2023). *Keeping children safe online*. NSPCC. https://learning.nspcc.org.uk/safeguardingchild-protection/

Pedro, F., Subosa, M., Rivas, A., & Valverde, P. (2019). *Artificial intelligence in education: Challenges and opportunities for sustainable development*. Unesco.

Remian, D. (2019). *Augmenting education: ethical considerations for incorporating artificial intelligence in education*. Academic Press.

Shernoff, D. J., Cukier, K. N., & Anderson, M. (2020). *Learning by doing: A new approach to education*. W. W. Norton & Company.

Shute, V. J. (2008). Unfair and unproductive: Grading and reporting practices undermining public confidence. *Educational Researcher, 37*(4), 19–33.

Sijing, L., & Lan, W. (2018). Artificial intelligence education ethical problems and solutions. *2018 13th International Conference on Computer Science & Education (ICCSE)*, (pp. 1-5). IEEE. 10.1109/ICCSE.2018.8468773

Slimi, Z., & Villarejo Carballido, B. (2023). Navigating the Ethical Challenges of Artificial Intelligence in Higher Education: An Analysis of Seven Global AI Ethics Policies. TEM Journal, 12(2).

Tacho, A. (2019). AI in Indian education: Opportunities and challenges. In: *International Journal of Education and Development, 54*(3). https://www.researchgate.net/publication359046086_Education_and_the_ Use_of_Artificial_Intelligence

UNESCO. (2021). *Artificial intelligence in education: Opportunities and challenges for schools*. UNESCO. https://unesdoc.unesco.org/ark:/48223/pf0000366994

UNESCO Mahatma Gandhi Institute of Education for Peace and Sustainable Development. (2020). *Ethical frameworks for AI in education in India*. UNESCO. https://mgiep.unesco.org/

Vincent-Lancrin, S., & Van der Vlies, R. (2020). *Trustworthy artificial intelligence (AI) in education: Promises and challenges*. OECD.

Weller, M. (2017). *The machine classroom: Artificial intelligence and education*. John Wiley & Sons.

World Innovation Summit for Education. (2015). *The Pratham Education Foundation: Transforming learning in India*. WISE. https://www.wise-qatar.org/2015-summiteducation-invest-impact/

ZamanB. U. (2023). Transforming Education Through AI, Benefits, Risks, and Ethical Considerations. Authorea Preprints.

Chapter 3
AI in Education:
Ethical Challenges and Opportunities

Ashok Singh Gaur

ⓘ https://orcid.org/0009-0002-3307-4674

Noida Institute of Engineering and Technology, India

Hari Om Sharan

ⓘ https://orcid.org/0000-0001-8866-1089

Rama University, India

Rajeev Kumar

ⓘ https://orcid.org/0000-0002-4141-1282

Moradabad Institute of Technology, India

ABSTRACT

As artificial intelligence (AI) continues to advance, its integration into the field of education presents both promising opportunities and ethical challenges. This chapter explores the multifaceted landscape of AI in education, examining the ethical considerations associated with its implementation. The opportunities encompass personalized learning experiences, adaptive assessment tools, and efficient administrative processes. However, ethical concerns arise regarding data privacy, algorithmic bias, accountability, and the potential exacerbation of educational inequalities. Artificial intelligence is a field of study that combines the applications of machine learning, algorithm production, and natural language processing. Applications of AI transform the tools of education. AI has a variety of educational applications, such as personalized learning platforms to promote students' learning, automated assessment systems to aid teachers, and facial recognition systems to generate insights about learners' behaviors.

INTRODUCTION

Artificial intelligence (AI) began to be applied to education about 50 years ago and only a decade after AI itself was established as a research field in 1956 in a workshop at Dartmouth College in Hanover, New Hampshire, USA (Moor, 2006).In 1970, Carbonell's article "AI en CAI: An Artificial-Intelligence

DOI: 10.4018/979-8-3693-2964-1.ch003

Approach to Computer-Assisted Instruction" described a semantic web-based tutor and creation system called SCHULARO for geography (Carbonell, 1970). This "Information Structure Oriented (ISO)" instructor differentiated his teaching strategy from knowledge of South American geography essentially by applying the layout of the other world geography and applying a teaching strategy or another teaching strategy. to the geography of South America. Furthermore, because its geographic knowledge was explicitly represented through semantic networks, the system could reason about its knowledge to make inferences that were not explicitly encoded, and also to answer questions about what it knew. Thus, its "mixed initiative" teaching strategy can include both a system that challenges the student using context and the meaning of questions, and a system that the student asks, both in very limited English. The system tracked which pieces of the geographic region the student understood, marking the significant parts of the semantic network, creating an evolving model of the student's knowledge. This adaptation to the student was one of the factors that distinguished this system from the computer-assisted instruction (CAI) systems that preceded it. The system also demonstrated what became the standard conceptual architecture for Artificial Intelligence in a Learner (AIED) systems..

The Early Days of AI in Education

An early collection of AIEd papers showed what could be achieved as early as a decade later (Sleeman and Brown, 1979). Among other things, this collection included articles on computer-assisted instructional systems in a game setting (Burton and Brown, 1979) and added expert system instructional rules to explain and teach expert system rules (Clancey, 1979)., a knowledge representation to capture the student's evolving understanding (Goldstein, 1979), an entry-level programming tutor (Miller, 1979), and a quadratic equation teaching system that administered tests to assess its teaching effectiveness and then updated itself as a result of teaching tactics (O'Shea, 1979).

Those early publications essentially mapped out what is now often called "learning tools," a conceptual architecture, namely an explicit model of what is being taught, an explicit model of how it should be taught, an evolving learning model. understanding and skills and the user interface through which learner and system interactions communicate. Hartley (1973) gave an early definition of this architecture as follows, where (3) and (4) together are explicit instruction models, and the user interface was not mentioned because of its limited scope:

1. A representation of the task
2. A representation of the student and his performance
3. A vocabulary of (teaching) operations
4. A pay-off matrix or set of means-ends guidance rules (Hartley, 1973, p. 424)

The standalone nature of these early systems, their unsophisticated interfaces, and their lack of interest in collecting large amounts of learner data meant that many of the contemporary ethical issues around the use of AIEd were not in evidence.

From the start, the general field of AI has had intertwined scientific and engineering aspects (Buchanan, 1988). The scientific aspect of AI in education has concerned itself with questions around the nature of human learning and teaching, often with the goal of understanding and then duplicating human expert teaching performance. This aspect has focused largely on learner-facing tools but more recently has expanded into teacher-facing tools. The science has been pursued as a kind of computational psychology for its own sake or as a way to improve educational practice and opportunity in the world. The engineering aspect of applying AIEd has exploited a wide range of computational technologies such as Carbonell's semantic networks, mentioned above, and more recently machine learning techniques of various kinds. This aspect of the work has pursued even wider goals that also include the development of educational administrator-facing tools.

CONTEMPORARY AI IN EDUCATION

Today, the AIEd industry is divided into three broad overlapping companies. The first continues the development of student-centered learning tools, implementing different pedagogical roles, such as guiding certain skills (Koedinger and Aleven, 2016) or helping the acquisition of concepts (Biswas, Segedy, and Bunchongchit, 2016) or supporting metacognitive awareness and regulation . (Azevedo and Aleven, 2013) e.g. Another company develops tools for teachers ("AI and Teacher-Facing Tools"), and a third company develops tools designed to help educational administrators (see the "AI and Teacher-Facing Tools" section). A useful summary of AIEd applications for the reader working in ODDE can be found in Kose and Koc (2015).

Figure 1. Potential ethical and societal risks of AI applications in education

AI and Learner-Facing Tools

An example of a learner-centered tool is Betty's Brain, a system designed to help students develop an understanding of ecological concepts (Biswas et al., 2016). In this system, the user interface is one of the most important parts of the system. Using the interface, the student draws a conceptual map consisting of nodes and arrows that describe some processes in the river ecosystem, such as oxygen consumption and carbon dioxide production. The system also provides reading material from which the student is expected to create a concept map. At any time, the student can ask the system to check and test their concept map for accuracy and completeness, and the system provides feedback that helps them create a better concept map. The system is presented as a story where the student creates a concept map for the artificial student Betty, or Betty's brain. The review and testing is sort of assigned and marked by Mr. Davis, the artificial teacher. Mr. Davis also provides metacognitive cues to a student when it appears that they are not paying enough attention to their learning, such as not using enough of the available reading material.

One of the directions in the development of AIEd since its early days has been to focus on learners as people with feelings and aspirations as well as knowledge and skills. This broader focus on student nature and learning has emerged as a result of our increased understanding of student motivation (Schunk, Pintrich, & Meece, 2008), mindset (Dweck, 2002), and academic emotions (Pekrun, 2014). name only three aspects of human learning. While such developments help to humanize interactions between systems and learners, they also open up more opportunities for privacy and ethical issues related to the data collected and stored. This expanded focus included the development of techniques that attempt to assess the transient emotional and motivational states of learners in order to enhance positive states of mind, such as focus, and combat negative states of mind, such as frustration or boredom.

An example of the previous application can be found in the school's math teacher. Arroyo et al. (2014) used the work of Dweck (2002) and others to augment an existing math instruction system by grouping students' learning behaviors into a small number of profiles based on how they use cues and how much time they spent solving problems. and the number of mistakes they made. Each of these profiles was defined based on both cognitive and affective/motivational factors. Each profile included cognitive and affective/motivational activities and feedback from the instructor, such as presenting a more difficult problem (cognitive), giving praise, or de-emphasizing immediate success (affective/motivational).

The suite of language learning tools, Enskill, provides another example of a contemporary interface for learner-facing systems (Johnson, 2019). This is a suite of tools for learning a language, the contextualized use of the correct language register, and for learning how to speak effectively, e.g., making a forceful case for some course of action. The tools use game-based technology to set up an on-screen scenario containing one or more characters with whom the learner speaks and who can reply in speech. The analysis and feedback of the learner's language can be at different levels depending on the context, e.g., pronunciation, grammar, and appropriateness. Moreover, the tools log all interactions with learners, and these link into a mechanism to improve the systems' performance when mistakes or glitches occur ("data-driven development (D3) of learning environments").

A particular result of the learner-analytic aspect of AI in education has been the growth of "dashboards" (Schwendimann et al., 2017). They can be directed at students to help them reflect on their progress either now or after a lesson or session, or even reflect on the effectiveness of reflective tools (Jivet, Wong, Scheffel, Specht, & Drachsler, 2021). Learner-centered dashboards evolved from an earlier learner-centered technology called "Open Learner Models" (see, e.g., Bull and Kay, 2016).

Are Learner-Facing Tools Effective and Being Used?

At least seven meta-studies and meta-analyses have been conducted on the effectiveness of student-directed tools in comparison to either a teacher working with a whole class of students or an experienced teacher working with a single student (see summary in Boulay, 2016). The general message of the 182 comparative studies is that student-directed tools outperform a human teacher working with a whole class (effect size = 0.47), but slightly worse than a professional teacher working with a single student (effect size = -0.19). In addition to the seven meta-studies, a 2-year, large-scale, multisite evaluation of the Algebra cognitive tutor was conducted in matched pairs of schools across the United States (Pane, Griffin, McCaffrey, & Karam, 2014). Each pair consisted of one school that continued to teach algebra in its own way and another where the school also used the Algebra-Cognitive Tutor (although not necessarily using it as instructed by the designers of the teaching system). In the second year of the study, when teachers used to use the teaching system effectively, there was a small comparative learning gain in favor of schools using the teaching system (effect = 0.21).

Despite the positive results for learner-facing tools above, the penetration of artificial intelligence tools of all kinds into schools and colleges has been slow, but with some notable exceptions, such as the Cognitive Tutors in the USA, mentioned above, and now trading under the name Carnegie Learning (Koedinger & Aleven, 2016). More positively, Baker, Smith, and Anissa (2019) say:

Despite minimal attention, AIEd tools are already being used in schools and colleges in the UK and around the world – today.

We find learner-facing tools, such as adaptive learning platforms that 'personalise' content based on a child's strengths and weaknesses. We find teacher-facing tools, such as those which automate marking and administration (one government-backed pilot in China sees children in around 60,000 schools having their homework marked by a computer). We find system-facing tools, such as those which analyse data from across multiple schools and colleges to predict which are likely to perform less well in inspections. (p. 5)

According to a systematic review of research on artificial intelligence applications in higher education, the penetration into universities is still patchy with few papers referring either to the ethical dimensions or to learning theory (Zawacki-Richter, Marín, Bond, & Gouverneur, 2019):

Descriptive results show that most majors involved in AIEd writings are from computer science and STEM, and quantitative methods were the most used in empirical research. A synthesis of the results presents four application areas of AIEs in academic support services and institutional and administrative services: 1. profiling and forecasting, 2. assessment and evaluation, 3. adaptive systems and personalization, and 4. intelligent guidance systems. The conclusions reflect an almost complete lack of critical reflection on AIED challenges and risks, a weak link with theoretical pedagogical perspectives, and the need to further explore ethical and educational approaches to AIED implementation in higher education. (Zawacki-Richter et al., 2019).In their editorial to a special issue on AI in university education, that included the paper mentioned above, the editors noted that "there is little evidence at the moment of a major breakthrough in the application of 'modern' AI specifically to teaching and learning, in higher education, with the exception of perhaps learning analytics" (Bates, Cobo, Mariño, & Wheeler, 2020).

AI and Teacher-Facing Tools

Recently, teaching tools have been developed that focus on teachers to help them either guide the use of technology in the classroom or think about its organization. They also help teachers (i) allocate their valuable time efficiently to students who need it most, and (ii) analyze student work to identify common classroom problems. We can see this as an evolution of the learner model that includes the individuals in the group as well as the group itself. For example, the Lumilo system gave the teacher glasses that provided an augmented reality view of her class of students, each working alone with an AIEd system (Holstein, McLaren, & Aleven, 2018). There were two kinds of augmentation in this view. The first involved an augmented reality symbol, apparently hovering above each student's head, that indicated their current learning state. These symbols included those for designating the following learner states: idle, (too) rapid attempts, hint abuse or gaming the system, high local error after hints, or unproductive persistence. These symbols were designed to give the teacher information on which to base her decision about which student she should go and help in person. The second augmentation involved an analysis of how the students were doing as a whole to provide a synopsis of problems common to the class. This synopsis was designed to give the teacher information on what might be the focus of her whole class interventions.

AI and Administrator-Facing Tools

A third broad area of AIEd has been the dissemination of analyzes applied to data generated at the class or cohort level in the educational environment, and is directed to administrative tools. Such analyzes are used, for example, to examine the relationship of student engagement with overall success in massive open online courses (Rienties et al., 2016), different models of engagement (Rizvi, Rienties, Rogaten, & Kizilcec, 2020). and identify individual and class-wide difficulties from the course material and ways to quickly identify and correct potential problems and gaps in the system's interaction with students (Johnson, 2019).

For example, Peach, Yaliraki, Lefevre, and Barahona (2019) analyzed the time behavior of students in online courses at the Imperial College Business School and the UK Open University. This information included the frequency, timing and interaction with the learning system of tasks. They mapped individuals' task times to the average of all students and used clustering techniques to create groups that included early birders, attenders, low-committers, occasional dropouts, and dropouts. They found that poor performance (based on outcome measures) tended to be associated with clogging behavior (not surprisingly), but high performance was found in all time-related groups, including low engagement and clogging.

In their wide-ranging systematic review, Zawacki-Richter et al. (2019) found a number of papers related to the application of AI in admissions decisions. For example, Acikkar and Akay (2009) used machine learning techniques to generate a predictive model of whether students would be admitted to university to study physical education and sports based on their "performance in the physical ability test as well as [their] scores in the National Selection and Placement Examination and graduation grade point average (GPA) at high school" (p. 7228). These analyses were undertaken retrospectively and were very accurate (e.g., >90%). The ethical dimension of such predictions comes into sharp focus if such predications are made prospectively when students apply, either as advice to admission tutors or more worryingly as actual decisions with no human in the loop.

ETHICAL ISSUES

Artificial intelligence was feared long before the emergence of this field (see for example Golemon (Meyrink, 1915) - a retelling of an ancient story about a clay figure that animates a living being). However, the early creators of tools for students using artificial intelligence had few ethical issues in mind. For them, the questions were mostly technological and pedagogical, such as how to build such systems in the first place and how effective they were in educational institutions. Today, ethical issues have become much more urgent, as educational technology (including AI-based technology) has penetrated education at all levels, data collection in educational situations is much greater, and surveillance companies have entered the market. capitalism to the educational ecosystem (Williamson, 2018).

Ethical Issues Around Education in General

In most countries, human teachers already operate within an ethical framework. In Scotland, for example, this covers a number of areas, including doing one's best for one's students, e.g., by keeping up to date with changes in the curriculum, and treating students equitably. It also includes respecting students' confidentiality (see, for example, General Teaching Council Scotland, 2012).

The rise of educational technology of all kinds, whether involving AI or not, and its creation of logs of interactions, has produced a huge amount of student data at all levels in education from primary (elementary) schools to universities. Teachers' ethical guidelines, such as those above, need to encompass these extra sources of data. There are many unanswered questions about who owns this data, who has access to it, how long it will be kept, and so on. The European Framework on General Data Protection Regulation (GDPR) provides guidance on managing all kinds of personal data (Li, Yu, & He, 2019). However, there are still issues for students around understanding what data about them counts as "personal" (Marković, Debeljak, & Kadoić, 2019), as well as around their degree of ownership and rights over educational log data.

Ethical Issues of AI in Education

Involving AI into educational technology must also be required to do its best and treat students equitably. For learner-facing tools, one should expect that designers of the educational technology will ensure that the technology will do the best that is possible in the circumstances, whether it is teaching, tutoring, mentoring, or counseling students. One should also expect that the technology treats students in an equitable fashion and does not favor one student over another either inadvertently or deliberately.

Learner-Facing Tools

We might ask, how can technology treat students unfairly? Many learner-centered systems choose the next most useful learning task, such as the next problem to solve for a student from a special educational background. Incorrectly grouping students based on, for example, gender, prior achievement, motivation, or self-regulated learning ability can lead to students being given inappropriate or much more difficult assignments than they can handle on their own. easier tasks as they learn. Of course, this bias can also occur with human teachers, where their low expectations for some students can become self-fulfilling prophecies. But just because teachers can sometimes be biased doesn't mean we should close our eyes to the potential biases of AI-based educational technology.

Teacher-Facing Tools

Similar considerations apply to teacher-facing tools. The Lumilo orchestration system we described above flagged up students who were doing OK or who were experiencing different kinds of difficulty. This aimed to enable the teacher to make choices about who to help. Clearly, this is an ethically charged decision. Should the teacher prioritize those who are in most difficulty or spread her effort more evenly across the whole class? That is a human dilemma. But the orchestration system had better get its diagnostics correct about who it thinks is doing OK and who it thinks needs help. Even without the use of AI, systems for helping the teacher manage a classroom can have unexpected negative effects on the students. In a study of ClassDojo, used by teachers to record student behaviors, Lu, Marcu, Ackerman, and Dillahunt (2021) noted that:

In particular, the use of ClassDojo runs the risk of measuring, codifying, and simplifying the nuanced psycho-social factors that drive children's behavior and performance, thereby serving as a "Band-Aid" for deeper issues. We discuss how this process could perpetuate existing inequality and bias in education. (Lu et al., 2021)

Administrator-Facing Tools

See Zeide (2019) for a discussion of the widespread use of artificial intelligence in universities. Administrators are sometimes used to predict which students are likely to do well and have evidence of failing or dropping out of a course. Such assessments are often based on learning from analysis using artificial intelligence methods. The problem here is the consequences of false negatives and false positives due to incomplete data analysis. For example, missing signs that a student really has a problem can mean that no one is notified to help. Labeling a struggling student who is doing well can also have a negative impact on credit scores, not the wrong label. For an interesting example of artifacts that can occur when analyzing cohort data, see Alexandron, Yoo, Ruipérez-Valiente, Lee, and Pritchard (2019). They showed that sometimes students using MOOCs created two accounts to game the system. With one account (with a fake name), the system would get a lot of help in finding the right answers, while with another account (with the student's real name), all questions would be answered quickly and correctly.

Using predictions to drive admissions of students to schools or colleges (Acikkar & Akay, 2009) or to predict grades when exams could not be taken because of COVID are fraught with ethical issues. The recent creation and then abandonment of an algorithm to predict UK student grades for entry to university is a salutary reminder about both potential AI biases and the potential human teacher biases the algorithm was intended to mitigate (see, for example, Hao, 2020).

Dealing With Ethical Issues in the Design, Implementation, and Deployment of AIEd Systems

Many different systems are now designed that include AI elements, from smartphone apps to big bank data systems. There has been increasing concern about the ethical questions that arise in the design, implementation, and deployment of such systems, with the EU proposing legislation to manage the situation (European Commission, 2020). Many different frameworks have been proposed to manage the development of such systems. A useful summary of such frameworks can be found in Floridi and Cowls (2019). They developed a framework from bioethics, under the general headings of beneficence, non-maleficence, autonomy, and justice, to also include "explicability." Most systems should work under the control of (or at least in tandem with) humans, so it is important that the system employing AI is able to offer an explanation or justification for exactly why it is suggesting a decision, a course of action, an outcome, or whatever, in order that the human can weigh up the degree to which he or she should agree with the machine. Particularly in education, autonomy and explicability must play a central role.

The issue of collecting, analyzing, and managing learner data has become more pressing for many reasons, including (i) greater general awareness of data privacy issues, (ii) the sheer quantity of learner data being collected, (iii) the increased use of AI and other methodologies for finding patterns in that data and drawing inferences from them, and (iv) the use of learner (and thus user) data for commercial purposes which have nothing to do with education (Williamson, 2018).

For example, Williamson (2018) warns against the movement of "Big Tech" companies into education, typically through learner-oriented tools, to harvest learner data for commercial purposes. Startup schools are analyzed as prototypes of educational institutions that emerge from the culture, discourse, and ideals of the venture capital and start-up culture of Silicon Valley, and which plan to transfer its practices throughout the social, technical, political, and economic infrastructure of the school. These new schools will be designed as scalable technology platforms; funded by commercial "corporate philanthropy" sources; and is staffed and managed by executives and engineers from Silicon Valley's most successful startups and Internet companies. Together, they form a powerful joint "algorithmic fantasy" that aims to "disrupt" public education with the technocratic expertise of a Silicon Valley venture philanthropist. (Williamson, 2018, p. 218).

Researchers within AI in education are starting to be aware of these ethical issues, even though Zawacki-Richter et al. (2019) found only two papers in their systematic review of the applications of AI in universities that dealt with ethical issues. So, for example, we see both the emergence of general design frameworks, such as that of Floridi and Cowls (2019) above, for including AI in software products, and those aimed specifically at the development of AI applications in education (see, for example, Drachsler & Greller, 2016) and, most notably, the creation of an Institute of Ethical AI in Education which has set out guidelines particularly for teachers in their use of applications of AI (Seldon, Lakhani, & Luckin, 2021).

OPPORTUNITIES AND SOLUTIONS

Personalised Learning

AI-driven personalised learning offers a powerful opportunity to revolutionise education. By harnessing the power of data, AI tools can quickly identify a student's individual learning needs, allowing teachers to develop customised plans tailored specifically to them. Students will be able to learn at their own pace and in formats that suit their individual preferences, leading to greater engagement and, ultimately, improved educational outcomes. Personalized learning in AI in education refers to the use of artificial intelligence technologies to tailor educational experiences to individual students' needs, preferences, and learning styles. This approach aims to move away from the traditional one-size-fits-all model of education and create a more adaptive and responsive learning environment.

Implement adaptive learning platforms that use AI algorithms to adjust content delivery, assessments, and resources based on each student's progress and needs.

Automation of Repetitive Tasks

With advances in AI technology, the opportunities for educators are endless! Teaching is no longer confined to the traditional model of just delivering a lecture - automation of repetitive tasks can open up possibilities of more efficient use of time and resources, ultimately allowing teachers to do more meaningful work. By taking mundane tasks like grading assignments off their hands, they can focus on activities that garner greater academic outcomes and higher levels of student engagement.

Use AI for grading, administrative tasks, and resource organization to free up teachers' time for more impactful interactions with students.

Access to Information

AI-powered tools like ChatGPT and Jasper offer a revolutionary new way to access information in an incredibly convenient manner. Instead of having to manually search through articles or learning materials, these technologies allow us to ask questions and receive back the answers we need directly, all with natural language! This opens up incredible opportunities – students now have the capacity to learn more efficiently and quickly. In this constantly changing digitally connected world, AI technology is making it easier, faster, and more enjoyable for everyone to expand their knowledge base.

Integrate AI-driven tools, such as speech-to-text or text-to-speech applications, to assist students with disabilities and ensure inclusivity in the learning environment.

Figure 2. Visual representation of the opportunities and corresponding solutions of AI in education

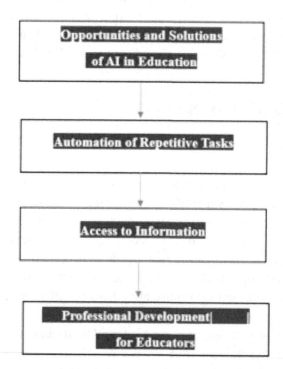

OPEN QUESTIONS AND DIRECTIONS FOR FUTURE RESEARCH

From an ethical perspective, the big question is how we can ensure that learners have more control over the data generated by interacting with educational technology and that they are protected from misuse of their data. This section outlines some open questions and research directions for the science and design of AIEd applications for each of the aforementioned system categories, namely learner, teacher, and system administrator.

Given the growing interest in collecting and using learner affect data to improve the adaptability of learner-centered tools, two scientific questions are (i) what might be the most useful affect categories for the development of affect pedagogy and (ii) what different pedagogical. rules should be used due to the current cognitive and affective states of the student, in order to maximize the possibility of fruitful learning. For example, should hope, respect and pride be important alongside confusion, frustration and focus, and how should they be 'managed'?.

For systems aimed at teachers, an engineering question is: How best to manage and support the division of labor between the human teacher(s) and the system, given the manifest complexity and dynamic nature of most classrooms full of learners? For example, how should a tool used for dynamic management differ from one used for reflective practice?

For systems aimed at understanding cohorts, an engineering question is: How best can learning management systems be developed to measure and potentially answer academic questions about learning rather than administrative ones? For example, did students on this course show strong evidence of improvements in their self-regulated learning capability?

IMPLICATIONS FOR OPEN, DISTANCE, AND DIGITAL EDUCATION (ODDE)

There are three main implications for ODDE. The first is that one of the oldest technologies for distance learning, the textbook, has been enhanced by the application of AI, either through adapting the content or the route through that content to the reader (see, for example, Thaker, Huang, Brusilovsky, & He, 2018). The second implication is that online, distance, and digital systems have increasingly incorporated elements of AI in order to make such systems smarter and more responsive to the needs of learners and teachers (see, for example, Kose, 2015; UNESCO, 2021). The third implication is that the developers and deployers of ODDE systems are already taking an ethical stance on how the systems are designed and built, how they are used in practice, and how their data is collected, stored, and analyzed (Prinsloo & Slade, 2016). For example, with particular respect to ODDE, Sharma, Kawachi, and Bozkurt (2019) state:

First, there should be some control mechanisms that should be put into place to ensure transparency in collection, use and dissemination of the AI data. Second, we need to develop ethical codes and standards proactively so that we truly benefit from AI in education without harming anything; not only humans but any entity. Third, we should ensure learners' privacy and protect them for any potential harm. Next, we must raise awareness about the AI so that individuals can protect themselves and take a critical position when needed. (p. 2)

CONCLUSION

To provide a longitudinal overview of AIEs and ethics, this chapter describes the early days of the development of learner-oriented systems in the 1970s, as well as examples of much more recent systems. While early systems were primarily aimed at learners, modern AI applications now include tools for teachers and administrators and are used both on-site and through online, remote and digital technologies.

The interface is one area where there have been big changes. One of the earliest systems had an interface that involved the learner typing in answers (and indeed questions) in stilted English, whereas contemporary learner-facing tools can show lifelike pedagogical agents with whom learners have a spoken dialogue in everyday English. Moreover, tools for other kinds of user make use of complex interactive dashboards.

In the early days, learner-oriented tools were mostly designed to act as tutors with one learner. Today, some tools are designed to work with a single learner, although they can now be adapted to the mood and motivational state of the learner, as well as what the learner knows and understands. Other tools can work with more than one student (see, e.g., Walker, Rummel, & Koedinger, 2009), while others work with teachers rather than students to help them accomplish the complex task of managing a class full of students. allocate your limited time in the most efficient way.

The creation of log data from educational systems and the use of data mining and other analytic techniques have given rise to the thriving field of learner analytics. This in turn has enabled the creation of dashboards for learners, teachers, and administrators to interrogate data at varying levels of granularity.

Ethics has played a strong role in education for many years, most obviously via the codes of professional practice that teachers are expected to act within. In the early days of AI, ethical issues around education tool design and deployment were not uppermost in the minds of designers: simply making the systems work effectively was the main goal. Nowadays, ethics is very much in people's minds whether they be system designers, teachers, parents, administrators, or indeed learners, but there is still a long way to go to make educational technology a place of trust and safety.

Artificial intelligence has a mixed reputation. On the other hand, it is so common that we hardly notice it, for example when interacting with a chat on a website or optimizing a photo with a camera. On the other hand, there are scary stories about artificial intelligence taking over the world, or equally scary reports about biased decisions that can affect well-being (like denying a mortgage or a job) or life (like a system that generates a false positive). or a false negative for a tumor). In education, there are problems about how analysis can produce biased results, or companies using artificial intelligence participate in education not for the benefit of the learner, but to collect their data for commercial purposes. To solve these problems, various codes of ethics have been developed, covering all aspects of the design and implementation of educational technology based on artificial intelligence, both internationally and locally. As artificial intelligence continues to shape the educational landscape, there is a need to strike a balance between harnessing its opportunities and the ethical challenges that come with it. By actively participating in conversations about privacy, bias, transparency, and accessibility, stakeholders can work together to ensure that AI in education promotes equity, inclusion, and ethical behavior, ultimately enriching the educational experience for all.

REFERENCES

Acikkar, M., & Akay, M. F. (2009). Support vector machines for predicting the admission decision of a candidate to the School of Physical Education and Sports at Cukurova University. *Expert Systems with Applications*, *36*(3), 7228–7233. doi:10.1016/j.eswa.2008.09.007

Alexandron, G., Yoo, L. Y., Ruipérez-Valiente, J. A., Lee, S., & Pritchard, D. E. (2019). Are MOOC learning analytics results trustworthy? With fake learners, they might not be! *International Journal of Artificial Intelligence in Education*, *29*(4), 484–506. doi:10.1007/s40593-019-00183-1

Arroyo, I., Woolf, B. P., Burleson, W., Muldner, K., Rai, D., & Tai, M. (2014). A multimedia adaptive tutoring system for mathematics that addresses cognition, metacognition and affect. *International Journal of Artificial Intelligence in Education*, 24(4), 387–426. doi:10.1007/s40593-014-0023-y

Azevedo, R., & Aleven, V. (Eds.). (2013). *International handbook of metacognition and learning technologies*. Springer. doi:10.1007/978-1-4419-5546-3

Baker, T., Smith, L., & Anissa, N. (2019). *Educ-AI-tion rebooted? Exploring the future of artificial intelligence in schools and colleges*. NESTA. https://media.nesta.org.uk/documents/Future_of_AI_and_education_v5_WEB.pdf

Bates, T., Cobo, C., Mariño, O., & Wheeler, S. (2020). Can artificial intelligence transform higher education? *International Journal of Educational Technology in Higher Education*, 17(1), 42. doi:10.1186/s41239-020-00218-x

Biswas, G., Segedy, J. R., & Bunchongchit, K. (2016). From design to implementation to practice a learning by teaching system: Betty's Brain. *International Journal of Artificial Intelligence in Education*, 26(1), 350–364. doi:10.1007/s40593-015-0057-9

Buchanan, B. G. (1988). Artificial intelligence as an experimental science. In J. H. Fetzer (Ed.), *Aspects of artificial intelligence* (pp. 209–250). Kluwer Academic Publishers. doi:10.1007/978-94-009-2699-8_8

Bull, S., & Kay, J. (2016). SMILI☺: A framework for interfaces to learning data in open learner models, learning analytics and related fields. *International Journal of Artificial Intelligence in Education*, 26(1), 293–331. doi:10.1007/s40593-015-0090-8

Burton, R. R., & Brown, J. S. (1979). An investigation of computer coaching for informal learning activities. *International Journal of Man-Machine Studies*, 11(1), 5–24. doi:10.1016/S0020-7373(79)80003-6

Carbonell, J. R. (1970). AI in CAI: An artificial-intelligence approach to computer-assisted instruction. *IEEE Transactions on Man-Machine Systems*, 11(4), 190–202. doi:10.1109/TMMS.1970.299942

Clancey, W. J. (1979). Tutoring rules for guiding a case method dialogue. *International Journal of Man-Machine Studies*, 11(1), 25–50. doi:10.1016/S0020-7373(79)80004-8

Drachsler, H., & Greller, W. (2016). Privacy and analytics: It's a DELICATE issue a checklist for trusted learning analytics. In S. Dawson, H. Drachsler, & C. P. Rosé (Eds.), *Enhancing impact: Convergence of communities for grounding, implementation, and validation* (pp. 89–98). ACM. doi:10.1145/2883851.2883893

du Boulay, B. (2016). Artificial intelligence as an effective classroom assistant. *IEEE Intelligent Systems*, 31(6), 76–81. doi:10.1109/MIS.2016.93

Dweck, C. S. (2002). Beliefs that make smart people dumb. In R. J. Sternberg (Ed.), *Why smart people can be so stupid* (pp. 24–41). Yale University Press.

European Commission. (2020). *White Paper on Artificial Intelligence: A European approach to excellence and trust* (COM(2020) 65 final). EC. https://ec.europa.eu/info/publications/white-paper-artificial-intelligence-european-approach-excellence-and-trust_en

Floridi, L., & Cowls, J. (2019). A unified framework of five principles for AI in society. *Harvard Data Science Review*, *1*(1), 1–13. doi:10.1162/99608f92.8cd550d1

General Teaching Council Scotland. (2012). *Code of Professionalism and Conduct*. GTCS. https://www.gtcs.org.uk/regulation/copac.aspx

Goldstein, I. P. (1979). The genetic graph: A representation for the evolution of procedural knowledge. *International Journal of Man-Machine Studies*, *11*(1), 51–78. doi:10.1016/S0020-7373(79)80005-X

Hao, K. (2020). The UK exam debacle reminds us that algorithms can't fix broken systems. *The MIT Technology Review*. https://www.technologyreview.com/2020/08/20/1007502/uk-exam-algorithm-cant-fix-broken-system/

Hartley, J. R. (1973). The design and evaluation of an adaptive teaching system. *International Journal of Man-Machine Studies*, *5*(3), 421–436. doi:10.1016/S0020-7373(73)80029-X

Holstein, K., McLaren, B. M., & Aleven, V. (2018). Student learning benefits of a mixed-reality teacher awareness tool in ai-enhanced classrooms. In C. P. Rosé, R. Martínez-Maldonado, H. U. Hoppe, R. Luckin, M. Mavrikis, K. Porayska-Pomsta, B. McLaren, & B. du Boulay (Eds.), *Artificial intelligence in education: 19th international conference,* (pp. 154–168). Cham: Springer. 10.1007/978-3-319-93843-1_12

Jivet, I., Wong, J., Scheffel, M., Specht, M., & Drachsler, H. (2021). Quantum of choice: How learners' feedback monitoring decisions, goals and self-regulated learning skills are related. In M. Scheffel, N. Dowell, S. Joksimovic, & G. Siemens (Eds.), *The impact we make: The contributions of learning analytics to learning* (pp. 416–427). ACM. doi:10.1145/3448139.3448179

Johnson, W. L. (2019). Data-driven development and evaluation of Enskill English. *International Journal of Artificial Intelligence in Education*, *29*(3), 425–457. doi:10.1007/s40593-019-00182-2

Koedinger, K. R., & Aleven, V. (2016). An interview reflection on "Intelligent Tutoring Goes to School in the Big City". *International Journal of Artificial Intelligence in Education*, *16*(1), 13–24. doi:10.1007/s40593-015-0082-8

Kose, U. (2015). On the intersection of artificial intelligence and distance education. In U. Kose & D. Koc (Eds.), *Artificial intelligence applications in distance education* (pp. 1–11). IGI Global. doi:10.4018/978-1-4666-6276-6.ch001

Kose, U., & Koc, D. (2015). *Artificial intelligence applications in distance education*. IGI Global. doi:10.4018/978-1-4666-6276-6

Li, H., Yu, L., & He, W. (2019). The impact of GDPR on global technology development. *Journal of Global Information Technology Management*, *22*(1), 1–6. doi:10.1080/1097198X.2019.1569186

Lu, A. J., Marcu, G., Ackerman, M. S., & Dillahunt, T. R. (2021). *Coding bias in the use of behavior management technologies: Uncovering socio-technical consequences of data-driven surveillance in classrooms*. Paper presented at the DIS'21: Conference on Designing Interactive Systems, Virtual Event, USA. https://static1.squarespace.com/static/5ebb1d874617b44f913c6d4b/t/609afa7f6ca7b40f39e55106/1620769435258/lu_dis21.pdf

Marković, M. G., Debeljak, S., & Kadoić, N. (2019). Preparing students for the era of the General Data Protection Regulation (GDPR). TEM Journal: Technology, Education, Management. *Informatics (MDPI)*, *8*, 150–156. doi:10.18421/TEM81-21

Meyrink, G. (1915). *Dr Golem*. Kurt Wolff.

Miller, M. L. (1979). A structured planning and debugging environment for elementary programming. *International Journal of Man-Machine Studies*, *11*(1), 79–95. doi:10.1016/S0020-7373(79)80006-1

Moor, J. (2006). The Dartmouth College Artificial Intelligence conference: The next fifty years. *AI Magazine*, *27*(4), 87–91. doi:10.1609/aimag.v27i4.1911

O'Shea, T. (1979). A self-improving quadratic tutor. *International Journal of Man-Machine Studies*, *11*(1), 97–124. doi:10.1016/S0020-7373(79)80007-3

Pane, J. F., Griffin, B. A., McCaffrey, D. F., & Karam, R. (2014). Effectiveness of Cognitive Tutor Algebra I at scale. *Educational Evaluation and Policy Analysis*, *36*(2), 127–144. doi:10.3102/0162373713507480

Peach, R. L., Yaliraki, S. N., Lefevre, D., & Barahona, M. (2019). Data-driven unsupervised clustering of online learner behaviour. *npj Science of Learning*, *4*, 14. doi:10.1038/s41539-019-0054-0

Pekrun, R. (2014). Emotions and learning. Retrieved from https://www.ibe.unesco.org/en/document/emotions-and-learning-educational-practices-24

Prinsloo, P., & Slade, S. (2016). Big data, higher education and learning analytics: Beyond justice, towards an ethics of care. In K. D. Ben (Ed.), *Big data and learning analytics in higher education: Current theory and practice* (pp. 109–124). Springer.

Rienties, B., Boroowa, A., Cross, S., Farrington-Flint, L., Herodotou, C., Prescott, L., & Woodthorpe, J. (2016). Reviewing three case-studies of learning analytics interventions at the Open University UK. In S. Dawson, H. Drachsler, & C. P. Rosé (Eds.), *Enhancing impact: Convergence of communities for grounding, implementation, and validation* (pp. 534–535). ACM. doi:10.1145/2883851.2883886

Rizvi, S., Rienties, B., Rogaten, J., & Kizilcec, R. F. (2020). Investigating variation in learning processes in a FutureLearn MOOC. *Journal of Computing in Higher Education*, *32*(1), 162–181. doi:10.1007/s12528-019-09231-0

Schunk, D. H., Pintrich, P. R., & Meece, J. L. (2008). *Motivation in education: Theory, research and applications* (3rd ed.). Pearson/Merrill Prentice Hall.

Schwendimann, B. A., Rodriguez-Triana, M. J., Vozniuk, A., Prieto, L. P., Boroujeni, M. S., Holzer, A., Gillet, D., & Dillenbourg, P. (2017). Perceiving learning at a glance: A systematic literature review of learning dashboard research. *IEEE Transactions on Learning Technologies*, *10*(1), 30–41. doi:10.1109/TLT.2016.2599522

Seldon, A., Lakhani, P., & Luckin, R. (2021). *The ethical framework for AI in education*. Buckingham. https://www.buckingham.ac.uk/wp-content/uploads/2021/03/The-Institute-for-Ethical-AI-in-Education-The-Ethical-Framework-for-AI-in-Education.pdf

Sharma, R. C., Kawachi, P., & Bozkurt, A. (2019). The landscape of artificial intelligence in open, online and distance education: Promises and concerns. *Asian Journal of Distance Education*, *14*, 1–2. http://www.asianjde.com/ojs/index.php/AsianJDE/article/view/432

Sleeman, D. H., & Brown, J. S. (1979). Editorial: Intelligent tutoring systems. *International Journal of Man-Machine Studies*, *11*(1), 1–3. doi:10.1016/S0020-7373(79)80002-4

Thaker, K., Huang, Y., Brusilovsky, P., & He, D. (2018*). Dynamic knowledge modeling with heterogeneous activities for adaptive textbooks.* Paper presented at the 11th International conference Educational Data Mining (EDM 2018), Buffalo.

UNESCO. (2021). *The Open University of China awarded UNESCO Prize for its use of AI to empower rural learners.* UNESCO. https://en.unesco.org/news/open-university-china-awarded-unesco-prize-its-use-ai-empower-rural-learners

Walker, E., Rummel, N., & Koedinger, K. R. (2009). Integrating collaboration and intelligent tutoring data in the evaluation of a reciprocal peer tutoring environment. *Research and Practice in Technology Enhanced Learning*, *4*(3), 221–251. doi:10.1142/S179320680900074X

Williamson, B. (2018). Silicon startup schools: Technocracy, algorithmic imaginaries and venture philanthropy in corporate education reform. *Critical Studies in Education*, *59*(2), 218–236. doi:10.1080/17508487.2016.1186710

Zawacki-Richter, O., Marín, V. I., Bond, M., & Gouverneur, F. (2019). Systematic review of research on artificial intelligence applications in higher education – Where are the educators? *International Journal of Educational Technology in Higher Education*, *16*(1), 39. doi:10.1186/s41239-019-0171-0

Zeide, E. (2019). Artificial intelligence in higher education: Applications, promise and perils, and ethical questions. *EDUCAUSE Review*, 31–39. https://er.educause.edu/-/media/files/articles/2019/8/er193104.pdf

Chapter 4
An Inquiry Into the Obstacles Hindering the Widespread Use of Artificial Intelligence in Environmental, Social, and Governance Practices

Sabyasachi Pramanik

ⓘD https://orcid.org/0000-0002-9431-8751

Haldia Institute of Technology, India

ABSTRACT

This chapter examines the correlation between artificial intelligence (AI) and the integration of environmental, social, and governance (ESG) factors in contemporary business and technology. The chapter explores how AI might provide practical answers to current urgent social issues regarding sustainability. The chapter focuses on the ethical dimensions of sustainable technologies that address ecological issues, as well as AI-driven forecasts for ESG indicators, ethical supply chains, and forthcoming legislation. The growing integration of AI and ESG presents opportunities for sustainable-oriented organizations to expand their market share via the adoption of environmentally friendly strategies. This convergence allows for the alignment of sustainability goals with advanced technology, resulting in a truly transformative partnership.

INTRODUCTION

An overview of how the integration of artificial intelligence (AI) with environmental, social, and governance (ESG) principles contribute to the promotion of sustainability.

Artificial intelligence (AI) and environmental, social, and governance (ESG) are closely linked areas that have a prominent role in modern business and technology (Makridakis, 2017; Saetra, 2021). Artificial intelligence (AI) refers to a broad spectrum of technical advancements that enable computers to

DOI: 10.4018/979-8-3693-2964-1.ch004

imitate human intellect. The term "intelligence of machines" is sometimes used to describe it (Trujillo, 2021). These technologies include robotics, computer vision, machine learning, and natural language processing. Artificial intelligence revolutionizes business practices and sectors, offering the capacity to address significant social challenges, such as the urgent need for sustainable solutions to contemporary difficulties (Nishant et al., 2020; Sestino and De Mauro, 2022).

ESG, as described by Johnson Jr. et al. (2020), is a complete taxonomy that considers an organization's non-financial requirements related to the environment, society, and governance. ESG may be attributed to two primary factors. Businesses worldwide are subject to regulations and legislation that emphasize the need of adhering to specified standards and demonstrating knowledge in areas unrelated to finance (Krishnamoorthy, 2021). Environmental considerations include subjects such as carbon emissions, climate change, and the responsible use of limited resources such as air, water, and waste. Social concerns such as human trafficking, child labor, health and safety, inclusion and diversity, racial and social justice, data privacy, employment, and the general well-being of workers and humanity are considered in the ongoing battle. Johnson Jr. et al. (2020) identify the following as governance variables: organizational objective; legislative and social effect; concerns related to pay and corruption; and the independence, control, and assessment of the board and management. The heightened global awareness and urgency around the ecological disaster have led to the development of technological solutions that seek to tackle or mitigate the underlying issues at its heart (Falk & van Wynsberghe, 2023).

Sustainable development refers to the practice of fulfilling the requirements of the current generation without jeopardizing the capacity of future generations to fulfill their own requirements (UN, 1987). The United Nations formulated Agenda 2030 along with the 17 sustainable development objectives (Schrijver, 2008). Achieving sustainable development requires a firm commitment to corporate social responsibility, abbreviated as CSR. Corporate Social Responsibility (CSR) encompasses socially responsible investment, safeguarding the interests of stakeholders, promoting sustainable development, and enhancing corporate governance (Zhao and Fariñas, 2023). By enhancing their efficacy and efficiency, artificial intelligence (AI) has the capacity to enhance corporate social responsibility (CSR) initiatives (Naqvi, 2021). With the increasing awareness of the importance of environmental and social responsibility, sustainability has become a paramount concern for companies. Unilivers, a firm that recognized the need of sustainability, actively endeavored to encourage other companies to adopt similar practices. The initiation of Unilever's Sustainable Living Plan in 2010 serves as a prime example of an ESG case study. The Unilever Sustainability Living Plan is characterized as a cutting-edge strategy, referred to as a "virtual circle of growth" (Lawrence et al., 2018). The strategy was underpinned by three key principles of sustainability: enhancing the living standards of individuals inside the company's value chain, mitigating the adverse environmental effects of Unilever goods, and enhancing the well-being and health of one billion people. Unilever's commitment to integrating social, governance, and environmental aspects into its core business strategy is seen in their extensive sustainability initiative. Unilever promoted transparency and proper hiring procedures, ensuring that board membership and leadership aligned with the company's commitment to sustainability, as outlined by its governance principles. This project demonstrates the potential of a major multinational consumer goods corporation to effectively achieve a harmonious equilibrium between profitability and an unwavering commitment to social responsibility, environmental conservation, and transparent governance procedures. Carney (2021) identifies Unilever as the leading corporation in terms of embodying ESG efforts.

Robots equipped with artificial intelligence (AI) have the ability to acquire knowledge via experience, adjust to novel information, and perform activities that resemble human skills (Duan et al., 2019, p. 63).

Consequently, AI emerges as a very auspicious field. Currently, artificial intelligence (AI) is generally recognized as a crucial component of the current digital transformation of several industries. AI has grown widespread, even in organizations and households (Chawla & Goyal, 2022; Singh & Chouhan, 2023). It aids organizations in automating processes, making decisions based on data, and extracting valuable information from extensive databases. However, as a growing number of organizations worldwide recognize the advantages of adopting sustainable practices, the assessment of an organization's impact on environmental, social, and governance (ESG) concerns using the ESG framework gains significance (Nishant et al., 2020; Kar et al., 2022). Corporate sustainability is crucial in the current fast-paced world, since companies are seeing a decrease in their average lifetime, making long-term survival an urgent issue (Dhanda & Shrotryia, 2021).

The increasing focus on AI in the fields of finance and sustainability, as well as its connection to ESG issues, is noteworthy. The growing incorporation of environmental, social, and governance (ESG) concerns in enterprises and the finance sector has resulted in the development of new financial tools and technology. These tools aim to provide financial returns while simultaneously considering environmental and social impacts, so achieving a balanced approach. In addition, throughout the process of integration, AI promotes a culture that encourages the implementation of ethical principles and environmental preservation, while simultaneously mitigating poverty, air pollution, and resource depletion.

Artificial intelligence has significant value in addressing poverty, pollution, resource depletion, and actively facilitating environmental and social governance (Nishant et al., 2020). ESG principles assist organizations in making strategic choices that will enhance their financial gains and foster a fair and robust global society. These rules outline the expected ethical and responsible behavior for enterprises. Dandha and Shrotrya (2021) examine the motivations for corporations choosing to prioritize sustainability for gaining a competitive edge, rather than just making charitable donations or completely replacing conventional business practices with sustainable ones.

The period of using sophisticated methodologies to address pressing issues related to management, society, and environment. By collecting and examining extensive amounts of data on environmental, social, and governance (ESG) factors, artificial intelligence (AI) aids firms in swiftly assessing and enhancing their performances (Hoa and Demir, 2023). Studies have shown that firms worldwide increasingly use sustainability reports to effectively convey information about social and environmental problems, mostly driven by significant demand from stakeholders.

Artificial intelligence will greatly contribute to the pursuit of sustainability. This offers several benefits, including improving corporate governance, reducing environmental impact, and fostering ethical behavior in the workplace. Artificial intelligence (AI) has the capability to enhance the accountability and transparency of environmental, social, and governance (ESG) processes. This, in turn, would enhance stakeholders' confidence in enterprises' commitment to socially and ecologically responsible objectives (Chong et al., 2022). The combination of ecological social responsibility (ESG) and intelligent technology (AI) has the capacity to revolutionize business practices and provide a future where ethics, social responsibility, and sustainable development are imperative.

Entities overseeing several aspects of the enduring viability of human endeavors have proliferated among governing bodies, large corporations, and the public sector (Walker et al., 2019). Significant advancements have been achieved in recent years in the incorporation of environmentally sustainable principles into corporate strategic planning. In the present day, it is essential for large corporations to elucidate and document significant actions, while also being cognizant of their influence on society and the environment (Saetra, 2021). This process involves the progressive substitution of a range of ESG (environmental, social, and governance) activities, frameworks, and indicators for the outdated phrase "corporate social responsibility" (Verbin, 2020).

The primary objective of this chapter is to address social, governance, and environmental issues and explore the potential of AI in resolving them. Companies must assess their social, administrative, and environmental accomplishments due to the growing significance of eco-friendly business practices in the global market. Numerous research papers have been undertaken on this issue area, including those by Yoon et al. (2018), Lam et al. (2016), and Tammuruji et al. (2016). According to the study conducted by Sood et al. in 2023, the concept of "governance" ranks second in importance, following closely behind the "environmental" component. The third component, "social," seems to have the least impact on an investigator's decision-making process. In a study conducted by Ellili (2022), it was discovered that the incorporation of ESG data resulted in improved financial statement reporting and investment efficiency for a sample of 30 publicly traded firms in the United Arab Emirates between 2010 and 2019.

According to Lourenço et al., (2012), contemporary companies should engage in sustainable business practices. By implementing environmentally sustainable practices, corporations may decrease their carbon emissions and gain a competitive advantage over other enterprises in their own sector. Furthermore, according to Albuquerque et al. (2019), the integration of environmental, social, and governance (ESG) aspects is a commendable strategy that enables enterprises to improve overall operational efficiency and effectiveness, reduce systemic risk, and increase returns. The techniques have shown potential. According to Lee and Zhang (2019), the use of ESG standards and data created by artificial intelligence may aid in the identification of abnormalities and catastrophic disruptions. Essentially, this entails forecasting or anticipating future events, which is especially crucial in the present global context where markets and corporate financial outcomes are heavily impacted by social, political, and environmental factors. In their study, Lee and Zhang (2019) identified several similarities between sustainability metrics and data produced by intelligent devices. The results may indicate that in order to attain long-term success, certain ESG indicators must be used.

Nevertheless, incorporating all AI into ESG is not a straightforward task due to many problems that arise throughout the process. Prior to integrating AI into ESG operations, it is essential to address certain challenges, as highlighted by Hao and Demir (2023). Several obstacles to consider include the responsible and ethical use of AI, accurate assessment of morality and performance, data quality, data security and privacy concerns, technical challenges, legal uncertainties, and the perpetuation of stereotypes against certain populations. Another set of challenges is developing a separate communication channel between human supervisors and autonomous robotics AI systems.

This chapter will explore the notion of artificial intelligence (AI) and its correlation with environmental, social, and governance (ESG) concepts. This entails addressing the difficulties associated with incorporating AI into environmental, social, and governance (ESG) practices, as well as examining the specific functions of AI in promoting favorable societal and environmental results. This publication aims to provide corporate managers and executives with clear direction on how to harness the creative capabilities of AI to promote development in different ESG goals. It does this by explaining the potential benefits and dangers associated with combining AI with ESG.

LITERATURE REVIEW

The body of research on AI for sustainability tackles environmental issues and spans across several fields and disciplines. Transportation, energy, water, and biodiversity are all fertile ground for AI research.

Scholars have used machine learning (ML) models in their investigations of environmental, economic, and, to a lesser extent, social issues. These include climate change and smart cities, which are especially conducive to AI study. For example, water quality metrics are examined and stream flow is predicted using ML models. Linear regression, genetic algorithms, artificial neural networks, support vector machines, autoregressive moving average models, adaptive neuro-fuzzy inference systems, and network-based fuzzy inference systems (ANFIS) have been the main methodologies used in water systems research to develop and test machine learning (ML) models. In a similar vein, a large body of research on biodiversity has modeled habitat suitability and availability for biodiversity protection, produced machine learning (ML) or natural language processing (NLP) methods to anticipate ecosystem services, and analyzed social media data to predict public opinion. Fuzzy logic models, expert systems, ANNs, and pattern recognition are the key areas of interest in energy research. The primary areas of study in this arena have been the energy usage of buildings, energy production, energy distribution, operations and maintenance, and public opinion about renewable energy.

Up until now, the majority of research has been limited to engineering fields that are technical and design-focused, with an emphasis on finding technological answers to problems. There hasn't been any theory development on using AI for sustainability, which presents a chance for IS researchers. The next natural step for the sub-domain of Green IS (IS for sustainability research) is to concentrate on AI for sustainability, which will increase our understanding of Green IS solutions. Our review of the literature, which was compiled using the Web of Science and Google Scholar databases, is then presented. Journal papers published between 2015 and 2023, when research on AI grew dramatically, were given priority. We also used terms and ideas from past research.

THE USE OF ARTIFICIAL INTELLIGENCE IN THE INTEGRATION OF ENVIRONMENTAL, SOCIAL, AND GOVERNANCE FACTORS

This chapter examines the use of artificial intelligence in tackling environmental, social, and governance (ESG) challenges. The integration of AI with environmental, social and governance factors in the context of growing commerce is a significant achievement (EY, 2023). We must understand the intricate connections that hinder the rapid use of AI in sustainable enterprises. Jaber (2022) states that artificial intelligence systems are designed to analyze extensive information, uncover specific patterns, and ultimately provide hypotheses or even predictions. This chapter presents a range of tools and approaches that corporations and organizations may use to directly confront and overcome difficulties. Through collaboration, corporations may establish the groundwork for a future AI-driven society that would significantly enhance the current state of affairs in the community.

In a recent study, Hao and Demir (2023) thoroughly examined the factors that facilitate or impede the integration of artificial intelligence in ESG domains. However, their investigation transcends surface-level analysis and delves into intricate aspects such as stimuli and technical challenges. This chapter examines their work because it is innovative in discovering triggers that are closely related to environmental, socio-economic, or governance challenges. These technical triggers and restrictions, as stated by them, provide informative insights that serve as our starting point for examining the many barriers impeding AI inside the ESG. Businesses that are sustainable and adhere to environmental, social and governance (ESG) standards have superior financial performance compared to those that do not (Moro–Visconti, 2022).

Nevertheless, drawing upon the research conducted by Hao and Demir (2023), this study aims to contribute to the existing body of knowledge by examining the many obstacles that impede the complete incorporation of artificial intelligence in the environmental, social, and governance (ESG) context. By conducting a comprehensive analysis of the intricate interplay between technical, regulatory, ethical, and data-related issues, our objective is to make a valuable contribution to the expanding knowledge base. This will empower companies to effectively navigate this demanding landscape and go forward with confidence. The primary objective is to enable companies to effectively utilize AI, converting it from a fascinating technological tool into a powerful enabler of sustainable practices. This will establish the foundation for a future where operational success not only embraces but also relies on ESG factors.

Per Hirsch's (2021) findings, corporations possess little comprehension of the carbon footprint associated with their IT system. Through the use of cleaner technologies, energy-efficient procedures, and sustainable behaviors, both people and corporations may reduce their environmental impact. This is accomplished by developing an understanding of carbon footprints and actively striving to reduce them. Nevertheless, some organizations possess a well-defined approach to quantifying their carbon footprint. Zhao & Fariñas (2023) argue that a collaborative effort between governments and companies is necessary to address the environmental challenges and health hazards associated with AI.

The objectives of this chapter are to outline the goals and research methodology of the study. This chapter explores the use of artificial intelligence (AI) and sustainable initiatives to tackle social, governance, and environmental issues. The objective of this chapter is to identify significant obstacles associated with the integration process and provide many recommendations for enterprises to navigate these hurdles. Examine case studies that demonstrate effective practices, ethical dilemmas, legal considerations, and potential consequences for future operations. This chapter examines existing literature on the challenges related to AI-ESG and provides guidance for those aiming to achieve sustainability and fairness in the results produced by AI. Thus, this chapter refers to the relevant extant literature on the integration of AI and ESG. It is important to acknowledge that implementing realistic strategies may effectively overcome the barriers that hinder widespread use of artificial intelligence for environmental, social, and governance purposes.

OBSTACLES TO THE USE OF ARTIFICIAL INTELLIGENCE IN THE INTEGRATION OF ENVIRONMENTAL, SOCIAL, AND GOVERNANCE FACTORS: A COMPLEX AND DIVERSE CHALLENGE

Integrating AI with ESG poses several hurdles, with each difficulty highlighting an intricate facet of this topic. In order to have a deeper understanding of the challenges that companies encounter, we shall thoroughly examine these obstacles (as seen in Figure 1).

The quality and availability of data: The current age of oil production requires a highly effective gathering and processing of diverse and multi-user data for artificial intelligence-driven ESG (Environmental, Social, and Governance) initiatives. Nevertheless, it is a challenging endeavor. This chapter examines the intricate landscape of data availability, quality, and dependability. It emphasizes the challenges that companies have in obtaining verified and dependable data. Organizations are now experiencing digital transformation as a result of the technology (Andrushia, A. D. et al. 2023) revolution, particularly in the areas of data analytics (Ray, A. M. et al. 2023), artificial intelligence (Khanh, P. T. et al. 2024), automation, and cloud computing (Dhamodaran S. 2023). The importance of unbiased, high-quality data becomes apparent in this procedure since it is crucial for generating long-lasting good results. Data serves as more than simply a resource; it works as the vital fluid for AI-powered ESG initiatives, and its caliber directly influences their effectiveness and precision. ProPublica, an investigative journalism group, found that in a controversial case involving algorithmic decision-making in the US for predicting recidivism (the probability of a convicted person committing another crime), the COMPAS system demonstrated a higher accuracy in identifying black defendants as being at high risk. Furthermore, it was erroneous to presume that white offenders would see the events in a dissimilar manner (Angwin et al., 2016). Currently, fintech (Jain, V. et al. 2023) and AI-driven ESG screening and analysis solutions are acknowledged as "strategic facilitators" capable of mitigating disparities in ESG ratings and rectifying biases in ESG data (Macpherson et al., 2021).

Ethical and Privacy Considerations: Key ethical concerns that emerge in the context of using AI for sustainable decision-making are questions of equity, clarity, and potential inadvertent adverse impacts on human beings. This part provides an in-depth analysis of these ethical challenges, emphasizing AI-driven sustainability initiatives that are based on logical reasoning and ethically defensible principles. ESG encompasses concerns around the collection, storage, and use of data, including those related to privacy. Ensuring security and secrecy are crucial while handling sensitive ESG-specific information. Data privacy (Pramanik, S. 2023) and ethics are crucial for the attainment of long-term objectives. This is evident when addressing the ethical and privacy considerations that arise from effectively incorporating AI within the framework of environmental social governance (ESG). A good AI is characterized by its ability to provide beneficial social and environmental outcomes. This involves actively promoting corporate social responsibility via both official and informal means, engaging in campaigns that support these values, and adhering to established rules and regulations. The sustainable expansion of AI is guaranteed by the effective governance of its domain and the adoption of an ethical framework. Utilizing accurate, moral, and lawful data may provide firms with vital knowledge and choices for decision-making, granting them a clear "data advantage" over rivals (Stuck & Grunes, 2016).

Limitations on expenses and available resources: Hence, it is important for firms to thoroughly analyze the many cost-related obstacles they encounter. These charges include the costs associated with installing the AI, as well as the ongoing costs of maintaining the systems and providing specialized training for staff. Resolving issues related to the pooling of resources is crucial for companies. Therefore, it is crucial to guarantee the optimal use of all available resources. Therefore, it is imperative that AI programs be aligned with overarching objectives. Achieving a harmonious equilibrium between environmental goals, ambitions, and cost-oriented strategies may pose significant challenges for a corporation. An equitable allocation of the resources and a systematic strategy would undoubtedly be beneficial.

Insufficient Expertise and Awareness: The development and implementation of AI-enabled ESG integration solutions need a certain set of competencies. This section addresses the fundamental elements of navigating a challenging rocky terrain with success. This demonstrates the equal importance of implementing programs that educate, train, and engage stakeholders in understanding the advantages and disadvantages of AI. An obstacle in addressing this knowledge gap is the willingness of prospective responders to actively engage in public education and information sharing initiatives. (Enholm et al. 2022) assert that there are inconsistencies about the primary activities that create value and the misunderstanding around the incorporation of AI into a business strategy. Hence, it is imperative to investigate the premise that autonomous learning always engenders unconscious discrimination against certain demographic groups (Larsson et al., 2019). Moreover, there were concerns over the potential formality of this approach.

Regulatory and Legal Complexities: It is necessary to establish a comprehensive legal and regulatory framework to effectively incorporate AI and ESG practices into businesses. The research will include a comprehensive examination of the legal frameworks that regulate the activities of artificial intelligence in relation to EGS. Hence, it is crucial to comply with both existing and emerging regulations to ensure the long-term viability of AI systems. In order to navigate complicated surroundings, firms must thoroughly analyze and address current compliance-related challenges and concerns, and proactively implement measures to mitigate these intricate circumstances. The dynamic nature and possible legal ramifications of artificial intelligence need the implementation of rigorous compliance processes. Given the unexpected nature of AI, it is imperative to develop sustainable and regulated AI systems, which presents a very persuasive case. Nevertheless, in order for AI to be completely included into ESG, a number of crucial obstacles must be carefully and thoroughly tackled. First and foremost, these problems must be addressed in order for AI to achieve a comprehensive degree of disruption and to ensure that organizations do not become immoral or unethical.

Table 1. Barriers to AI adoption in ESG integration

Barriers to AI Adoption in ESG Integration	Data Quality and Availability	Ethical and Privacy Concerns	Cost and Resource Constraints	Lack of Expertise and Awareness	Regulatory and Legal Complexities

OVERCOMING BARRIERS: FACILITATING THE INTEGRATION OF AI IN ESG

AI has several obstacles when trying to expand into the environmental, social, and governance domains. Nevertheless, innovative strategies might overcome these obstacles to facilitate the integration of AI into environmentally friendly efforts. Larsson et al., 2019 conducted further study aimed at enhancing confidence in the use of artificial intelligence and machine learning. Accountability and transparency are closely linked. One of the accountability difficulties that arise is the lack of transparency or "black boxes" that develop when decisions are made using algorithms (Roy, A. et al. 2023). Another crucial need for sustainable AI is to establish multidisciplinary cooperation, foster confidence in AI applications, and continuously monitor legal and governance concerns. Section Two examines potential remedies (refer to Table 1) focused on eliminating a hindrance and enabling the use of ESG in AI.

Strategy for Overcoming Barriers:

- Data Solutions
- Ethical AI Frameworks
- Cost-Effective Approaches
- Capacity Building
- Advocacy and Regulation

Data Solutions: The integration of AI-driven ESG centers on data as its fundamental component. This section covers numerous ways for managing data, including open data projects, data standards, and data cleansing. Implementing these techniques guarantees that the data required for seamless integration of AI with ESG rules is of superior quality and readily available. This chapter provides many instances of fruitful cooperation or alliances between corporations, governments, and data providers to emphasize the essentiality of collective efforts in data research. (Smith et al. 2020) provide actual examples demonstrating the collective ability to address data issues, hence ensuring the sustainability of AI systems.

Ethical AI Frameworks: This section focuses on the development of AI systems that are based on principles of transparency, fairness, and accountability, specifically for the integration of environmental, social, and governance (ESG) considerations. Consequently, organizations will need to set ethical principles and standards for artificial intelligence that align with environmental, social, and governance (ESG) criteria. The demonstration should illustrate how adherence to these principles may guide the development of AI systems with ethical awareness. Readers will get genuine case studies on the use of AI in environmental protection and social responsibility. A method that uses illustrative instances of such applications to structure the implementation of AI for ESG objectives entails knowledge-based and principled techniques.

Economical Methods: Monetary constraints and limits sometimes impede the incorporation of artificial intelligence into environmental, social, and governance (ESG) practices. This section specifically examines strategies for reducing costs, with a particular emphasis on estimating return on investment (ROI) and evaluating expense-benefit ratios. Consequently, it offers decision-makers a structured approach to incorporating AI, which may result in cost reductions and facilitate intelligent financial decision-making. This portion of the book provides a detailed explanation of how to effectively and efficiently implement resource restrictions and economies for AI deployment, specifically focusing on scalability and cloud integration. The study illustrates that the integration of AI into businesses may increase their economic viability by enabling low-cost and scalable use of AI throughout the whole organization (Kamble et.al., 2020).

Enhancing Organizational Development: Nevertheless there are several methods by which individuals and businesses may use the present technology to improve their future performance. Hence, it is essential for organizations to prioritize continuous education throughout one's career, with guidance on enhancing skills for artificial intelligence and environmental, social, and governance (ESG) management. By using ESG-focused training and education, one may effectively address knowledge gaps and cultivate a specialized mentality in sustainable AI. Hence, this section is crucial for developing a skilled workforce capable of addressing the future of artificial intelligence, as impacted by environmental, social, and governance factors. This segment provides people and organizations with the necessary resources to effectively apply sustainable AI (Brauer et al., 2019).

Promotion and Governance: The significance of lobbying methods in influencing legislative and regulatory changes pertaining to the integration of AI and ESG. Legislators are consistently engaged in effecting substantial changes. This chapter provides examples of cooperatives that successfully advanced morally acceptable AI and adapted to evolving legal requirements. These conversations demonstrate the perspectives on how regulators and interested parties work to advance sustainability. The provisions in this chapter promote moral and legal reform as means to enable regulatory reforms that may support the deployment of AI-based ESG (Environmental, Social, and Governance) practices (Bietti et al., 2021). After taking into account all of these elements, we propose a comprehensive plan for any firm that wants to overcome the limitations and obstacles to using AI in ESG integration. The text provides practical answers and concrete illustrations to help readers enhance their understanding and proficiency in AI-driven sustainability, so contributing to a more equitable and environmentally sustainable future.

Case Studies: Successful Instances of Overcoming Obstacles to the Adoption of Artificial Intelligence in ESG Integration

This chapter contains examples of actual cases in which corporations have effectively tackled the obstacles of adopting AI in relation to environmental, social, and governance (ESG) factors. The papers are remarkable as they address the challenges that arose throughout the implementation of AI for sustainability, providing answers and valuable insights.

IBM's Watson for Sustainability: One example of a solution to environmental, social, and governance (ESG) challenges is IBM's Watson for Sustainability. IBM's cognitive technology, Watson, extensively evaluated a substantial amount of ESG-related data. The overview above illustrates how IBM addressed data quality challenges by using innovative techniques to merge and cleanse corrupted data. As a consequence, IBM's ESG performance was improved by using data-driven decision making and integrating verified data from various sources. This underscores the need of maintaining data accuracy and consistency in AI-powered ESG initiatives.

Microsoft's Ethical AI Framework: Microsoft's ethical commitment to environmental, social and governance (ESG) is shown via a case study that demonstrates their dedication to ESG through the use of ethical artificial intelligence (AI) concepts. In pursuit of this objective, the technology behemoth established an AI framework centered on ethics, which is founded on principles of accountability, justice, and openness. Microsoft guaranteed that their AI-driven sustainability initiative adhered to these criteria. This instance exemplifies the need of considering ethics while using AI in ESG and offers a method for responsibly leveraging AI inside enterprises.

Amazon's Cloud-Based AI Solutions: Amazon's use of economic methods is evident in their integration of EGS via their cloud-based AI capabilities. Amazon decreased the initial cost of artificial intelligence by using scalable cloud services such as QPoint. This is a demonstration of the cost-effectiveness and adaptability of a cloud-based artificial intelligence system over a prolonged period of time.

Google's Educational and Capacity Building Programs: For example, Google demonstrated this when they took steps to raise public understanding about how businesses may address gaps in perception and knowledge, and improve their capabilities. Financial assistance has been provided via initiatives focused on generating public awareness, training programs, and educational resources to address current gaps in AI understanding. This support aims to foster a culture that promotes the long-term viability of technology. Furthermore, this scenario exemplifies the need of employing adequately skilled and educated personnel, such as stakeholders, who can thoroughly evaluate the advantages and disadvantages of artificial intelligence within the framework of environmental, social, and governance factors.

Regulatory Framework for Artificial Intelligence in the European Union: The European Union's efforts to establish a legal framework for the integration of artificial intelligence (AI) with environmental, social, and governance (ESG) practices provide a thought-provoking case study on lobbying and regulation. The European Union has fostered an environment conducive to the ethical and sustainable use of AI via engagement with stakeholders and the implementation of diverse legislations. This case study exemplifies the efficacy of collaborative efforts between business and government, as well as the implementation of proactive laws, in fostering sustainable behavior.

This chapter examines the significant insights and standards of excellence derived from these case studies by analyzing real-life scenarios. Readers may get valuable insights on several topics, including resolving data challenges, creating cost-effective solutions, upholding ethical standards in AI, improving AI skills, and advocating for regulatory changes. These case studies provide as inspiring examples for other firms who face similar challenges in their business models, yet use AI to achieve their long-term goals. These examples serve as a source of inspiration that encourages other firms facing similar obstacles to use AI in order to tackle long-term objectives. They exemplify the transformative potential of AI in revolutionizing the integration of environmental, social, and governance (ESG) factors, provided that challenges are addressed by strategic deliberation, unwavering dedication to sustainability, and appropriate use of technology.

Prospects for the Future and Final Remarks: This chapter presents take-home messages that are substantiated by a high standard benchmark derived from real-life situations, to be taken into account. The themes include using suitable methodologies to address data issues economically, advocating for ethical AI models, enhancing the capabilities of AIs, and aiding in the enforcement of regulatory frameworks.

This chapter explores the projected challenges and modifications in this field that are likely to have an influence on the future. There are viable solutions for ethical managers concerning the supply chain, and the societal impact of artificial intelligence is on the rise. Artificial intelligence forecasting for ESG indicators is a technology that is being increasingly used to enhance the pace of adoption. Furthermore, these issues will include difficulties pertaining to the securitization of data, the establishment of regulatory frameworks, and the establishment of ethical criteria for AI architecture.

Anticipating Future Trends and Challenges: Given the increasing prevalence of artificial intelligence in the integration of environmental, social, and governance (ESG) factors, it is crucial for sustainability-focused firms to adjust accordingly. The forthcoming matters in this area include anticipated problems and advancements. Developing ethical management strategies for supply chains and using AI-based tools to quantify social impact. Artificial intelligence-based forecasting for ESG indicators is a popular technology for the futures market. Businesses should be prepared for extra concerns such as data security, evolving regulatory systems, and the need for ethical AI frameworks.

Future Prospects: Companies that use sustainable development practices and integrate artificial intelligence will get significant benefits. The following section offers comprehensive analysis on potential applications of artificial intelligence in advancing sustainability initiatives, such as the development of eco-friendly products, enhancing stakeholder engagement in ESG programs, and generating additional revenue opportunities through AI-driven ESG services. The nascent green AI ecosystem offers organizations an opportunity to enhance their market position via the use of an environmentally conscious strategy.

An Overview of the Developing Discipline: The combination of AI and ESG combines the pressing issue of sustainability with cutting-edge technology. Transparent, open, and data-driven practices have the potential to revolutionize how firms use ESG. This enables artificial intelligence to address critical concerns, while enhancing the environment and societal well-being.

CONCLUSION: OVERCOMING BARRIERS TO ACHIEVE LONG-TERM GOALS

The key take away from this chapter is the need of eliminating obstacles that hinder the use of artificial intelligence. Organizations and stakeholders must address data concerns, adopt ethical AI initiatives, promote affordability, enhance capacity development, and lobby for improved regulation in order to achieve a sustainable future. This will enable them to recognize the disruptive power of AI and take a leading role in the area of sustainability. Numerous advancements are taking place at the intersection of technology and sustainability. Businesses may maintain their leading position in terms of ethics, collaboration, and innovation by using AI and addressing environmental and social concerns. AI and ESG, as a confluence of technologies, provide a set of ideas aimed at fostering a fairer and ecologically responsible future. We may strive towards a future when AI and ESG collaborate to establish a global community that reaps the advantages of their combined efforts.

REFERENCES

Albuquerque, R., Koskinen, Y., & Zhang, C. (2019). Corporate social responsibility and firm risk: Theory and empirical evidence. *Management Science*, *65*(10), 4451–4469. doi:10.1287/mnsc.2018.3043

Andrushia, A. D., Neebha, T. M., Patricia, A. T., Sagayam, K. M., & Pramanik, S. (2023). Capsule Network based Disease Classification for VitisVinifera Leaves. *Neural Computing & Applications*. doi:10.1007/s00521-023-09058-y

Angwin, J., Larson, J., Mattu, S., & Kirchner, L. (2016). *Machine Bias*. ProPublica. https://www.pro-publica.org/article/machine-bias-risk-assessments-in-criminal-sentencing.

Buallay, A., Fadel, S. M., Al-Ajmi, J. Y., & Saudagaran, S. (2020). Sustainability reporting and performance of MENA banks: Is there a trade-off? *Measuring Business Excellence*, *24*(2), 197–221. doi:10.1108/MBE-09-2018-0078

Carney, M. (2021). *Value(s): Building a Better World for All*. Public Affairs.

Chawla, R. N., & Goyal, P. (2022). Emerging trends in digital transformation: A bibliometric analysis. *Benchmarking*, *29*(4), 1069–1112. doi:10.1108/BIJ-01-2021-0009

Chong, S., Rahman, A., & Narayan, A. K. (2022). Guest editorial: Accounting in transition: influence of technology, sustainability and diversity. *Pacific Accounting Review*, *34*(4), 517–525. doi:10.1108/PAR-07-2022-210

Datta, A., Tschantz, M. C., & Datta, A. (2015). Automated Experiments on Ad Privacy Settings – A Tale of Opacity, Choice, and Discrimination. *Proceedings on Privacy Enhancing Technologies. Privacy Enhancing Technologies Symposium*, *1*(1), 92–112. doi:10.1515/popets-2015-0007

Dhamodaran, S., Ahamad, S., Ramesh, J. V. N., Sathappan, S., Namdev, A., Kanse, R. R., & Pramanik, S. (2023). *Fire Detection System Utilizing an Aggregate Technique in UAV and Cloud Computing, Thrust Technologies' Effect on Image Processing*. IGI Global.

Dhanda, U., & Shrotryia, V. K. (2021). Corporate sustainability: The new organizational reality. *Qualitative Research in Organizations and Management, 16*(3/4), 464–487. doi:10.1108/QROM-01-2020-1886

Duan, Y., Edwards, J. S., & Dwivedi, Y. K. (2019). Artificial Intelligence for Decision Making in the Era of Big Data–Evolution, Challenges, and Research Agenda. *International Journal of Information Management, 48*, 63–71. doi:10.1016/j.ijinfomgt.2019.01.021

Ellili, N. O. D. (2022). Impact of ESG disclosure and financial reporting quality on investment efficiency. *Corporate Governance: An International Journal of Business in Society*.

Enholm, I. M., Papagiannidis, E., Mikalef, P., & Krogstie, J. (2022). Artificial Intelligence and business value: A literature review. *Information Systems Frontiers, 24*(8), 1709–1734. doi:10.1007/s10796-021-10186-w

EY. (2023). *Artificial intelligence ESG stakes, Discussion paper*. Assets. https://assets.ey.com/content/dam/ey-sites/ey-com/en_ca/topics/ai/ey-artificial-intelligence-esg-stakes-discussion-paper.pdf

Falk, S., & van Wynsberghe, A. (2023). Challenging AI for Sustainability: What ought it mean? *AI and Ethics*. doi:10.1007/s43681-023-00323-3

Hao, X., & Demir, E. (2023). Artificial intelligence in supply chain decision-making: an environmental, social, and governance triggering and technological inhibiting protocol. *Journal of Modelling in Management*. doi:10.1108/JM2-01-2023-0009

Hirsch, P. B. (2021). Footprints in the cloud: The hidden cost of IT infrastructure. *The Journal of Business Strategy, 43*(1), 65–68. doi:10.1108/JBS-11-2021-0175

Jaber, T. A. (2022). Artificial intelligence in computer networks. *Periodicals of Engineering and Natural Sciences, 10*(1), 309–322. doi:10.21533/pen.v10i1.2616

Jain, V., Rastogi, M., Ramesh, J. V. N., Chauhan, A., Agarwal, P., Pramanik, S., & Gupta, A. (2023). FinTech and Artificial Intelligence in Relationship Banking and Computer Technology. In K. Saini, A. Mummoorthy, R. Chandrika, N. S. Gowri Ganesh, & I. G. I. Global (Eds.), *AI, IoT, and Blockchain Breakthroughs in E-Governance*. doi:10.4018/978-1-6684-7697-0.ch011

Johnson, C. E. Jr, Stout, J. H., & Walter, A. C. (2020). Profound Change: The Evolution of ESG. *Business Lawyer, 75*, 2567–2608.

Kar, A. K., Choudhary, S. K., & Singh, V. K. (2022). How can artificial intelligence impact sustainability: A systematic literature review. *Journal of Cleaner Production, 376*, 134120. doi:10.1016/j.jclepro.2022.134120

Khanh, P. T., Ngoc, T. T. H., & Pramanik, S. (2024). AI-Decision Support System: Engineering, Geology, Climate, and Socioeconomic Aspects' Implications on Machine Learning. Using Traditional Design Methods to Enhance AI-Driven Decision Making. IGI Global.

Kitsios, F., & Kamariotou, M. (2021). Artificial Intelligence and Business Strategy towards Digital Transformation: A Research Agenda. *Sustainability (Basel)*, *13*(4), 2025. doi:10.3390/su13042025

Krishnamoorthy, R. (2021). Environmental, Social, and Governance (ESG) Investing: Doing Good to Do Well. *Open Journal of Social Sciences*, *9*(7), 189–197. doi:10.4236/jss.2021.97013

La Torre, M., Sabelfeld, S., Blomkvist, M., Tarquinio, L., & Dumay, J. (2028). Harmonising non-financial reporting regulation in Europe: Practical forces and projections for future research. *Meditari Accountancy Research, 26*(4), 598-621.

Larsson, S. (2019). Artificial Intelligence as a Normative Societal Challenge: Bias, Responsibility, and Transparency. Banakar, Dahlstrand & Ryberg-Welander (Eds.), Festschrift for Håkan Hydén. Lund: Juristförlaget.

Lawrence, J., Rasche, A., & Kenny, K. (2018). Sustainability as Opportunity: Unilever's Sustainable Living Plan. In G. Lenssen & N. Smith (Eds.), *Managing Sustainable Business* (pp. 435–455). Springer. doi:10.1007/978-94-024-1144-7_21

Lee, Y.-J., & Zhang, X. T. (2019). AI-Generated Corporate Environmental Data: An Event Study with Predictive Power. In J. J. Choi & B. Ozkan (Eds.), Disruptive Innovation in Business and Finance in the Digital World. Emerald Publishing Limited. doi:10.1108/S1569-376720190000020009

Lourenço, I., Branco, M., Curto, J., & Eugénio, T. (2012). How does the market value corporate sustainability performance? *Journal of Business Ethics*, *108*(4), 417–428. doi:10.1007/s10551-011-1102-8

MacphersonM.GasperiniA.BoscoM. (2021). Artificial Intelligence and FinTech Technologies for ESG Data and Analysis. SSRN. https://ssrn.com/abstract=3790774 or doi:10.2139/ssrn.3790774

Makridakis, S. (2017). The forthcoming Artificial Intelligence (AI) revolution: Its impact on society and firms. *Futures*, *90*, 46–60. doi:10.1016/j.futures.2017.03.006

McDonnell, M.-H., & Cobb, J. (2020). Take a Stand or Keep Your Seat: Board Turnover after Social Movement Boycotts. *Academy of Management Journal*, *63*(4), 1028–1053. doi:10.5465/amj.2017.0890

Miller, T. (2019). Explanation in artificial intelligence: Insights from the social sciences. *Artificial Intelligence*, *267*, 1–38. doi:10.1016/j.artint.2018.07.007

Moro-Visconti, R. (2022). *Augmented Corporate Valuation: From Digital Networking to ESG Compliance*. Palgrave Macmillan. doi:10.1007/978-3-030-97117-5

Musleh Al-Sartawi, A. M., Hussainey, K., & Razzaque, A. (2022). The role of artificial intelligence in sustainable finance. *Journal of Sustainable Finance & Investment*, 1–6. doi:10.1080/20430795.2022.2057405

Musleh Al-Sartawi, A. M., Razzaque, A., & Kamal, M. M. (Eds.). (2021). *Artificial Intelligence Systems and the Internet of Things in the Digital Era. EAMMIS 2021. Lecture Notes in Networks and Systems* (Vol. 239). Springer.

Naqvi, A. (Ed.). (2021). *Artificial intelligence for asset management and investment: a strategic perspective*. Wiley. doi:10.1002/9781119601838

Nishant, R., Kennedy, M., & Corbett, J. (2020). Artificial Intelligence for Sustainability: Challenges, Opportunities, and a Research Agenda. *International Journal of Information Management*, *53*, 102104. doi:10.1016/j.ijinfomgt.2020.102104

Porro, C. & Bierce, K. (2018). *AI for good: what CSR professionals should know*. CECP.

Pramanik, S. (2023). A Novel Data Hiding Locating Approach in Image Steganography. *Multimedia Tools and Applications*. doi:10.1007/s11042-023-16762-3

Ray, A. M., Pramanik, S., Das, B., Khanna, A. (2023). Hybrid Cryptography and Steganography Method to Provide Safe Data Transmission in IoT. *ICDAM 2023*. Research Gate.

Roy, A., & Pramanik, S. (2023). A Review of the Hydrogen Fuel Path to Emission Reduction in the Surface Transport Industry. *International Journal of Hydrogen Energy*. doi:10.1016/j.ijhydene.2023.07.010

Sætra, H. S. (2021). A Framework for Evaluating and Disclosing the ESG Related Impacts of AI with the SDGs. *Sustainability (Basel)*, *13*(15), 8503. doi:10.3390/su13158503

Sestino, A., & De Mauro, A. (2022). Leveraging artificial intelligence in business: Implications, applications, and methods. *Technology Analysis and Strategic Management*, *34*(1), 16–29. doi:10.1080/0 9537325.2021.1883583

Singh, A., & Chouhan, T. (2023). Artificial Intelligence in HRM: Role of Emotional–Social Intelligence and Future Work Skill. In P. Tyagi, N. Chilamkurti, S. Grima, K. Sood, & B. Balusamy (Eds.), *The Adoption and Effect of Artificial Intelligence on Human Resources Management, Part A* (pp. 175–196). Emerald Studies in Finance, Insurance, and Risk Management. doi:10.1108/978-1-80382-027-920231009

Sood, K., Pathak, P., Jain, J., & Gupta, S. (2023). How does an investor prioritize ESG factors in India? An assessment based on fuzzy AHP. *Managerial Finance*, *49*(1), 66–87. doi:10.1108/MF-04-2022-0162

Stuck, M., & Grunes, A. (2016). *Big data and competition policy*. Oxford University Press.

Tarmuji, I., Maelah, R., & Tarmuji, N. H. (2016). The impact of environmental, social and governance practices (ESG) on economic performance: Evidence from ESG score. International Journal of Trade. *Economics and Finance*, *7*(3), 67–74.

Trujillo, J. (2021). The Intelligence of Machines. Filosofija. *Sociologija, 32*(1), 84–92. doi:10.6001/ fil-soc.v32i1.4383

United Nations. (1987). *Report of the World Commission on Environment and Development: Our Common Future ('Brundtland Report')*. Oxford University Press.

Verbin, I. (2020). *Corporate Responsibility in the Digital Age: A Practitioner's Roadmap for Corporate Responsibility in the Digital Age*. Routledge. doi:10.4324/9781003054795

Vinuesa, R., Azizpour, H., Leite, I., Balaam, M., Dignum, V., Domisch, S., Felländer, A., Daniela Langhans, S., Tegmark, M., & Fuso Nerini, F. (2020). The role of artificial intelligence in achieving the Sustainable Development Goals. *Nature Communications*, *11*(1), 1–10. doi:10.1038/s41467-019-14108-y PMID:31932590

Walker, J., Pekmezovic, A., & Walker, G. (2019). *Sustainable Development Goals: Harnessing Business to Achieve the SDGs through Finance, Technology and Law Reform.* John Wiley & Sons. doi:10.1002/9781119541851

Xie, J., Nozawa, W., Yagi, M., Fujii, H., & Managi, S. (2019). Do environmental, social, and governance activities improve corporate financial performance? *Business Strategy and the Environment, 28*(2), 286–300. doi:10.1002/bse.2224

Yoon, B., Lee, J. H., & Byun, R. (2018). Does ESG performance enhance firm value? Evidence from Korea. [DOI]. *Sustainability (Basel), 10*(10), 3635. doi:10.3390/su10103635

Zhao, J., & Fariñas, B. G. (2023). Artificial Intelligence and Sustainable Decisions. *European Business Organization Law Review, 24*(1), 1–39. doi:10.1007/s40804-022-00262-2

KEY TERMS AND DEFINITIONS

AI Frameworks: These should be governed by ethical principles and standards to ensure the integration of sustainability and social responsibility. The practical implementation of ethical AI in conjunction with an Environmental, Social, and Governance (ESG) framework is shown via case studies drawn from daily scenarios.

AI Solutions: Provided by Amazon That Are Hosted in the Cloud: Scalability may decrease the initial costs linked to the implementation of AI, hence enhancing the financial viability of sustainability objectives. This was shown by Amazon's use of cloud-based solutions.

Collaboration Between Industry Stakeholders and Regulators: It is essential to advance ethical AI and maintain compliance with legal obligations. Promoting ethical and regulatory changes fosters a favorable environment for the integration of AI in ESG.

Data Integration: This involves the merging and analysis of data from both environmental, social, and governance (ESG) sources and artificial intelligence (AI) systems. This process requires thorough examination and combination of the data. These include a variety of information, including environmental statistics, governance indicators, and other measures of development outcomes. In order to enhance the accuracy of predictions and evaluations in AI-powered ESG activities, it is crucial to use data of superior quality.

Data Quality and Availability: For AI-powered ESG integrations to be efficient, it is necessary to have easily accessible qualitative and quantitative data that is readily available. Open data efforts, data standards, and data cleansing all contribute to enhancing the accessibility and quality of data.

Data Solutions: Resolving data problems necessitates cooperative data initiatives including enterprises, government organizations, and data source firms. One advantage of effective collaborations is that they enhance both the quality and dependability of data.

Difficulties and Barriers: The integration of artificial intelligence (AI) into environmental, social, and governance (ESG) practices faces several challenges. In order to get major advantages from AI in their ESD activities, entities must overcome many key obstacles, including technical challenges, regulatory frameworks, data security and privacy concerns, ethical considerations, biases, and data quality issues.

Environmental, Social, and Governance (ESG): This refers to a structured approach for evaluating the potential negative impact an organization may have on environmental, social, or governance matters.

Sustainability continues to be a popular term in several domains. In the realm of business, enterprises have recognized the need of aligning their operations with environmental protocols. Companies are guided by ESG principles to behave ethically, so promoting environmental sustainability and fostering a more equitable contemporary world.

Ethical and Privacy Concerns: These arise while using AI, since it must adhere to ethical standards for the integration of environmental, social, and governance (ESG) principles. Hence, it is important to thoroughly contemplate judgments about equality, transparency, and the potential for unexpected consequences. Furthermore, it is essential to implement robust security protocols and privacy controls to safeguard ESD sensitive data.

Forecasting Future Trends and Anticipating Challenges: Anticipated developments include the implementation of ethical supply chain management systems, the use of artificial intelligence for social impact evaluations, and the use of predictive analytics to evaluate environmental, social, and governance (ESG) aspects. Obstacles arise due to concerns over data privacy and the ever-evolving legal frameworks.

IBM's Watson: for Sustainability: This used advanced data cleaning and integration approaches to address data quality issues, highlighting the crucial need of data integrity in AI-driven ESG initiatives.

Inadequate Knowledge and Understanding: Attaining success in an AI-driven ESG approach requires specialized knowledge and abilities. It is crucial to bridge the gaps in understanding among the relevant stakeholders, and implement essential initiatives such as public awareness, education, and training to demonstrate the potential benefits and limitations of incorporating AI into ESG practices.

Integrating Artificial Intelligence (AI) With Environmental, Social, and Governance (ESG) Factors: It is a combination of sustainable technological progress and inventions. The integration of artificial intelligence into ESG guarantees objectivity, traceability, and reliability for stakeholders. This is a novel approach to addressing environmental and social issues, providing organizations with a transformational outlook on their environmental, social, and governance (ESG) challenges.

Limitations on Expenses and Available Resources: ESG-integrated AI-based financial usage is hindered by many hurdles. Efficiency in resource distribution and cost allocation for both initial expenditures and ongoing maintenance is also essential. It involves a thorough and careful planning process with the goal of minimizing costs while achieving sustainability goals.

Microsoft's Ethical AI Framework: This illustrates the company's dedication to ethical ideals by showcasing the ethical concerns involved in adopting AI for environmental, social, and governance (ESG) purposes.

Predictive Analytics: The ability to forecast future outcomes is of utmost importance in the context of Environmental, Social, and Governance (ESG) factors and Artificial Intelligence (AI). Organizations will use AI-driven data to predict and prevent environmental, social, and governance risks in order to enhance their sustainability and overall performance.

Chapter 5
Artificial Intelligence at the Helm:
Steering the Modern Business Landscape Toward Progress

Geetha Manoharan
ⓘ https://orcid.org/0000-0002-8644-8871
SR University, India

C. V. Guru Rao
ⓘ https://orcid.org/0000-0002-9210-6122
SR University, India

Abdul Razak
ⓘ https://orcid.org/0000-0003-2553-4992
Entrepreneurship Development Institute of India, India

Sunitha Purushottam Ashtikar
SR University, India

M. Nivedha
Robert Gordon University, UK

ABSTRACT

Computers are part of our everyday life. Advances in technology allow these machines to duplicate human skills with remarkable accuracy. AI managed all global business functions. Progress is mostly due to AI. Using AI, computers can automate human work. Many organizations employ AI to simplify. Business uses it extensively. Companies use AI to automate, analyze, and interact with consumers and workers. Companies want market domination and industry growth. Business insights are gained by many successful global firms using AI, automation, big data analytics, and NLP. AI is transforming business by automating processes, analyzing data, improving decision-making, and connecting customers. AI may improve corporate chatbots, inventory management, fraud protection, and predictive maintenance. AI can raise productivity, cut expenses, and increase profits by speeding up procedures and enhancing customer service. However, businesses must use AI responsibly and protect data. This chapter discusses AI's business impacts.

DOI: 10.4018/979-8-3693-2964-1.ch005

INTRODUCTION

Business procedures in the 21st century are complex, involving tasks that are difficult and unproductive for humans. The digital age dominates modern business. Data may tell companies about efforts that might boost growth. Thus, in today's competitive business climate, enterprises must understand customers' needs and preferences to prosper and stay relevant. Artificial intelligence may help companies understand their customers, simplify processes, boost productivity and revenue, and save expenses. Over time, AI has become more common. It improves human capacities and changes corporate practices. West found that 37% of organizations employ AI in some capacity. In the previous four years, AI use by enterprises has grown 270%. It is increasingly important for anticipating customer behavior and eliminating human data entry.

Beyond contemporary technology, smart technologies are entering our everyday lives in work, commerce, and housing. AI is restructuring library management and industrial processes globally and reshaping our economy (Abdulwahid, A. H., Pattnaik, M., et al., 2023) Smart home gadgets, virtual assistants, chatbots, financial forecasting models, social media, e-commerce, and manufacturing employ AI. As useful, reachable, and relevant, AI systems may have a promising future. When used appropriately and under supervision and control, AI in the workplace is important and likely to develop. Our operations depend on our ability to use deep-learning technologies. The future of AI seems bright since it makes decisions quicker and more accurately than before (Shameem, A., Ramachandran, K. K., et al., 2023) This technology is versatile despite its recent application. Top business AI apps will be reviewed in this blog.

ARTIFICIAL INTELLIGENCE IN BUSINESS

Both large and small organizations quickly see the need of using AI to fulfil their current and long-term goals. The integration of AI in business has the capacity to fundamentally transform infrastructure.

- AI has the ability to automate various operational tasks, thereby allowing corporate executives to focus on more intricate business challenges and decision-making processes.
- AI technology is capable of performing tasks like data analysis, which would typically take human workers several hours to complete, in a matter of seconds. This not only helps organizations save significantly on salaries but also leads to increased revenue.
- It is a reality that humans are prone to making mistakes. Individuals must consider the surrounding conditions and understand complex situations, but data science thrives on reduced inaccuracies, leading to more precise estimations and data analysis.

Despite increased interest, Teoh, T.T., & Goh, Y.J. (2023) say integrating AI into organizations remains difficult. Recent studies show that up to 85% of AI systems fail to achieve their aims. Effective AI application research is scarce, which might help organizations starting AI projects. This study aims to improve corporate management by understanding AI technology, individuals, and processes. This study examines Taobao's e-commerce fulfillment center's resource orchestration approach and AI application for business management (AL-MAHAIRAH, M. S., MANOHARAN, G., et al., 2022). Results indicate that data, AI algorithms, and robotics are AI's major assets. Use and apply these resources and ensure their coordination with other relevant resources, such as storage facilities and current information systems, to build major AI capabilities. Knowledge acquisition, strategic decision-making, and anticipatory analysis are AI's key functions. AI's ability to improve effectiveness (e.g., labor productivity and space utilization), efficacy (e.g., advanced data analytics), and learning and training efficiency determines its worth to enterprises. The talk also covers how AI's social impacts affect corporate management.

REVIEW OF LITERATURE

The study by Alblooshi, M.A., Mohamed, A.M., and Yusr, M.M. had an objective. This research examines how artificial intelligence moderates leadership characteristics and company continuity in 2023. Combination theory underpins the research's theoretical basis. Artificial intelligence as a moderator is the study's main theoretical contribution. The instrument was analyzed for reliability and validity. A basic random sample of 384 people was used to attain this goal. The data is from 2021. The research found that AI and leadership skills affect business continuity. The findings also confirmed the importance of artificial intelligence in regulating leadership and business continuity. The study validated artificial intelligence's moderating effect on combination theory, advancing research, practice, and society. The findings demonstrate the relevance of AI for leaders and businesses today. Companies need AI to survive.

Benabed (2023) explored the use of artificial intelligence in energy optimization organizations and worldwide expansion. Digital and globalization's tremendous combination affects company internationalization. Research involved literature review and quantitative qualitative data analysis. Scopus data was used for a case study. VOSviewer enhanced the research. The research asks: "To what extent is artificial intelligence relevant, interconnected, and beneficial for business internationalization, companies, and energy optimization?" Textless. Studies show that artificial intelligence may speed up business processes. However, as the world moves toward Industry 5 and the future economy, artificial intelligence has become essential in many economic areas, including energy. Since energy is related to business possibilities, globalization, and operations, it is a supply-and-demand industry that requires ongoing improvement. AI is useful in smart grids, smart cities, smart residences, renewable energy (such as solar and wind), industrial applications, and electric cars. Digitalization simplifies organizational operations and connections, making worldwide expansion possible. AI aids business internationalization, operations, and energy efficiency.

The research conducted by Widayanti, R., and Meria, L. (2023) indicated that there is now a rapid emergence of firms that are using artificial intelligence as an integral part of their business strategies. The integration of AI technology in business has been ongoing for a considerable duration, despite recent research indicating the implementation of novel or alternative business models. It can be argued that AI technology has long been employed in business models, raising doubts about the distinctiveness of these models. This research aims to compare the business models of AI firms with those of conventional IT companies to get a deeper understanding of their differences. The first step is to create a taxonomy for AI business models using 162 global startups. The four main business models are deep technology researcher, data analytics supplier, AI product and service provider, and AI development facilitator. The subsequent analysis focuses on three key components of startup business models in the AI industry: (1) novel value propositions enabled by AI, (2) diverse utilization of data for value creation, and (3) the impact of AI technology on overall business rationale. This research proposes stimulating avenues for further exploration in the domain of entrepreneurship. It is designed to promote entrepreneurial activities. Taxonomies and models serve as tools or instruments.

Technological advancements in recent times have had a significant influence on business operations, as discussed by Rama Krishna, S., Rathor, K., Ranga, J., Soni, A., D, S., & N, A.K. (2023). Compared to other emerging technologies, artificial intelligence (AI) can change marketing strategy the most. AI can benefit modern businesses in several ways. Intellectuals and professionals believe artificial intelligence will determine human civilization's fate. Information and communication technologies have transformed the world into a vast network of nodes. Technology has increased investments in AI to extract valuable insights from large amounts of data to improve commercial decision-making. In addition to popular belief, healthcare, e-commerce, education, government, and business use AI technology. AI-using companies are increasing. Advertising professionals worldwide are seeking the best AI solutions. However, a thorough analysis of the study's findings may demonstrate the importance of artificial intelligence and large-scale data analysis in marketing and suggest future research topics.

The research was conducted by Chankoson, T., Chen, F., Wang, Z., Wang, M., and Sukpasjaroen, K. The objective of (2023) is to provide a comprehensive and organized knowledge framework for scholars engaged in studying the use of artificial intelligence in the area of education. Furthermore, it aims to facilitate researchers in promptly comprehending author cooperation patterns, institutional collaboration patterns, emerging research themes, evolutionary patterns, and cutting-edge areas of study from the standpoint of library informatics. This research used a bibliometric technique to statistically examine the literature using the bibliometric analysis program CiteSpace. This document presents the analytical findings in the form of tables and visual graphics. This study found that academic collaboration needs improvement and that research institution collaboration is fragmented. This study also highlights its limitations: Many key aspects of AI and education integration and advancement are still unclear to Chinese educational researchers and practitioners. This study uses quantitative research methods, such as bibliometrics and visualisation images, to methodically and intuitively uncover advancements and patterns in education's use of artificial intelligence based on published literature. This study informs future research in this field.

Chandgude, V., and Kawade, B. (2023) found that need drives innovation. Artificial Intelligence influences corporate development by influencing decision-making. Artificial Intelligence and Machine Learning affect corporate decision-making. AI and Machine Learning have a major impact on businesses and the global economy due to rapid technological advancement. Companies are using AI and ML to improve industry operations. Artificial Intelligence and Machine Learning improve their products and services, boosting their business. A research article on how AI and ML affect company growth and improvement was reviewed. The text discusses AI and ML's many applications in various fields. AI and ML have improved human life and business growth, according to the literature review. Decision makers must understand AI and ML approaches and technologies to use them for their businesses. Preventing common issues requires thorough risk research and early investment assessment. It will boost business value.

The resource-based perspective was used by Abrokwah-Larbi and Y. Awuku-Larbi (2023) to study AI in marketing (AIM) and company success. Surveyed were 225 Eastern Region Ghana Enterprise Agency-listed SMEs. Durai, Krishnaveni, & Manoharan (2022) examined AIM's impact on SMEs using structural equation modeling-route analysis. Financial, customer, internal business process, and learning and growth performance of Ghanaian SMEs increase significantly with AIM. According to study, AIM improves financial, customer, internal corporate process, and learning and development. Personalization, CDMS, VAR, and IoT do this. Limited yet useful research. SMEs may be included (Geetha M., Pavan K. B., & Sunitha P. A., 2022). non-sampled sources. To increase customer engagement, future research should use AIM to evaluate consumer communications and information like social media posts.

Its practical impacts are two. According to this study, SME owners and management should prioritize AIM to boost performance (Manoharan, G., Durai, S., & Rajesh, G. A., 2022). SME managers should also adopt this study's four AIM determinants: IOT, CDMS, VAR, and customization. They might obtain AIM resources to boost performance.

Ahmad, H.A., Hanandeh, R., Alazzawi, F.R., Al-Daradkah, A., ElDmrat, A.T., Ghaith, Y.M., and Darawsheh, S.R. assessed Jordanian telecommunications enterprises' e-learning and commercial performance in 2023. After sampling, 269 were collected. No inquiry data was analyzed without PLS. Results demonstrate connecting improves e-learning and corporate processes. Using big data, AI, and BI may enhance e-learning infrastructure and organizational productivity (García-Tadeo et al., 2022). Corporate performance improved immediately using Big Data External and Internal, Innovative Usage, Indexing, and Sources Accuracy. AI promotes company performance by improving data accuracy, transparency, speed, innovation, and learning. Enterprise performance is improved by business intelligence, including Data Warehouse, Data Mining, BPM, and CI. The findings show that e-learning system, content, and self-confidence boost firm performance. Previously, no relationship was identified between the parameters and e-learning and corporate performance. This research suggests e-learning to boost company performance. To boost Jordanian e-learning and telecommunications, this study emphasizes big data, AI, and business intelligence.

Hassan, W.M., Aldoseri, D.T., Saeed, M.M., Khder, M.A., and Ali, B. (2022) explained artificial intelligence and robotics and their differences. This study also explains robot strategies and challenges. Cognitive processes are studied in robotics. AI must be used to improve connectivity. AI investigates the data needed for each cognitive process, the best ways to represent that knowledge, and the best ways to use it. Future advances in AI and robots will greatly impact humanity. Artificial intelligence is used to describe any artificial computer system that performs complex actions to achieve goals.

Kaushik, D.P. (2022) claimed that AI in business analytics tools across industries is widely used due to the expanding volume and complexity of company data. Business intelligence businesses use AI and ML to analyze huge, complicated datasets. That allows them provide people across sectors clear, practical business advice. IT's business analytics employs computing to get data insights. These data might come from the firm's ERP system, warehouse and mart data, third-party data suppliers, or other public sources. Artificial intelligence in business analytics was assessed by surveying 198 executives from diverse industries. AI affects business analytics greatly. This research examines literature on value-creating approaches and AI technologies for companies. This research explores how companies employ AI. AI's direct and indirect consequences and uses are examined. AI acceptance and utilization are influenced by important aspects, according to the report. We streamlined the research.

Enterprise Cognitive Computing is highlighted by Tarafdar, Beath, and Ross (2019), who shed light on the revolutionary effects of AI on company operations (ECC). ECC applications automate processes by incorporating algorithms, which improves the speed of data analysis and the reliability of output. The authors showcase ECC's excellence in call centres by accomplishing quick response times, high rates of issue resolution, and smooth interactions with customers. In sum, ECC apps prove to be potent instruments for enhancing operational excellence, raising customer happiness, and bettering the employee experience.

Using trends and opportunities as examples, Bharadiya (2023) investigates how AI and machine learning are revolutionising business intelligence. Predictive analytics, highlighted in the article, help firms make smart decisions by seeing trends in past data. Chatbots and virtual assistants powered by artificial intelligence have the ability to optimise processes while reducing costs. In sum, the article stresses how these technologies, when used to business intelligence, can help organisations thrive in the digital age by giving them a leg up in the competition, boosting innovation, and gaining an advantage over their rivals.

Focusing on the business value gained from AI-based transformation projects, Wamba-Taguimdje et al. (2020) examine how AI affects firm performance. Analyzing AI concepts, exploring case studies across industrial sectors, collecting data from AI solution providers, and reviewing relevant literature make up the study's four-step sequential approach. Organizations can adapt, optimise processes, and gain strategic advantages with the help of AI's diverse technologies, like chatbots and machine translation, according to the findings. This study highlights the potential of AI to improve financial and operational performance of organisations through process optimization and operational reconfiguration.

In their 2018 article, Chui, Manyika, and Miremadi offered a sophisticated view of where AI is at the moment and how it can be used in the corporate world. The authors warn of the dangers and restrictions of deep learning and machine learning while highlighting the pervasiveness of AI in many parts of our lives. Insights into promising developments that could overcome these limitations and open new opportunities are offered in this article, which aims to assist executives in understanding potential obstacles that could hinder AI efforts. An invaluable resource for company executives, the writers present AI's accomplishments and concerns in a fair and balanced manner.

The extensive research by Wamba-Taguimdje et al. (2020) examines how AI affects company performance, with a focus on the commercial value created by AI-powered transformation initiatives. This research takes a four-pronged approach to its findings by reviewing the literature, collecting data from AI solution providers, analysing AI, and exploring case studies. Applying the theory of IT capabilities, the study examines 500 case studies to determine how AI affects process and organisational performance. The results show how AI can optimise processes, improve automation, and boost organisational performance through its diverse technologies. For academics and professionals alike, this study is a gold mine of information about the many ways AI can improve business operations.

The authors Mishra and Tripathi (2021) looked at how AI is being used in business models, highlighting how it is a cognitive engine in the digital ecosystem. The authors present a model for an AI business platform that is similar to the SaaS model in the cloud. In this model, AI solutions work together with other digital systems like ERP and CRM (ERP). A key finding of this model is that it relies on subscriptions. It offers strategic implications and analytics-based innovations with an emphasis on AI and ML technology's preventative features in enterprise digital platforms. The ever-changing terrain of AI-driven business transformations gains valuable insights from this contribution.

In a 2020 essay, Reim, Åström, and Eriksson discuss how AI might transform business models, while acknowledging the challenges managers have in adapting to this revolutionary transition. The authors provide a literature review-based roadmap for a company to successfully integrate AI into its operations: (1) understanding AI and the capabilities needed by the organization; (2) evaluating the roles and current business models in the ecosystem; (3) creating and improving key capabilities; and (4) encouraging organizational acceptance by developing internal competencies. This report provides guidelines for managers who wish to utilize AI to reinvent their business models to bridge the academic support gap and mitigate AI implementation risks. AI integration's revolutionary implications on firms were examined by Allioui and Mourdi (2023) in supply chain optimization and digital offering reliability. AI can improve processing speeds, consumer insights, customer service, and customization, according to the report. A comprehensive literature evaluation and methodology synthesis fills a knowledge vacuum by revealing prospective advantages, drawbacks, and untapped potential. The report also covers future prospects, enabling firms to maximize AI in today's competitive business climate.

APPLICATIONS OF AI IN BUSINESS FUNCTIONING

Those that embrace the reality of building technology as business tools have benefited greatly from AI integration. AI capabilities have a wide and rising influence, showing a promising and continuous development that will provide organizations new and exciting benefits.

Artificial Intelligence for Business Strategy

ML (Tripathi, M. A., Tripathi, R., et al., 2023) and AI have a profound and ongoing impact on the corporate environment, particularly in terms of strategic planning. Instead of just depending on last year's sales to forecast for this year, companies are influenced by several factors such as regulations, seasonal needs, supply chain modifications, unforeseen weather fluctuations, trends, prospective staffing challenges, and other significant variables. Artificial Intelligence may be used in business to design strategic methods, taking into consideration certain elements.

Artificial Intelligence for Business Operations

AI systems provide enterprises with several benefits for optimizing various processes, including the automation of repetitive work, the elimination of human mistake, and the enhancement of human productivity in employment. An advantageous feature is the decreasing cost of technology. If you have not yet begun incorporating artificial intelligence (AI) into your business, it is advisable to do so immediately.

Artificial Intelligence (AI) for the Implementation of Advanced Technology in the Field of Finance and Financial Technology (FinTech)

The use of Artificial Intelligence in finance and fintech has empowered financial institutions to make informed decisions by analyzing extensive quantities of data obtained in real-time from financial markets. This approach is very dependable as it enables real-time data collection, processing, and analysis. Insurance organizations enhance the customer experience by automating regular insurance administration and underwriting procedures using machine learning. The integration of data science into the decision-making process has revolutionized the way financial businesses make decisions, thanks to the use of predictive analysis. Artificial intelligence and machine learning have the potential to aid organizations in evaluating diverse consumer and market data, hence accelerating corporate development and streamlining management operations.

Artificial Intelligence (AI) for Marketing and E-Commerce Applications

AI-driven marketing leverages consumer data and Artificial Intelligence to forecast the user's purchasing behavior and provide tailored concepts. Simultaneously, AI has reduced the time and workload of marketers in research and development by delegating tasks to robots.

Artificial Intelligence (AI) and its Implications in the Supply Chain

Supply chain professionals are using Artificial Intelligence to address significant challenges and enhance global operations. AI-powered solutions are used throughout supply chains to boost efficiency, address the challenges posed by a worldwide scarcity of personnel, and discover superior and safer approaches to transporting goods.

The use of Artificial Intelligence in commercial applications may be seen throughout several stages, ranging from the manufacturing process to the final delivery at the customer's doorstep. Shipping organizations use Internet of Things (IoT) devices (Thavamani, S., Mahesh, D., Sinthuja, U., & Manoharan, G., 2022) to gather and analyze data about items during transportation. These devices also monitor the mechanical condition and real-time location of valuable vehicles and other transportation equipment.

EXAMPLES OF AI APPLICATIONS IN BUSINESS

Businesses can't do without AI apps that help with decision-making, customer service, and efficiency. Here are a few noteworthy instances:

Intelligent Virtual Assistants

Certainly, this is a domain in which you possess expertise in the field of Artificial Intelligence. Siri, Cortana, Alexa, and Google Assistant are the leading smart assistants available in the current market. They provide a range of functionalities and services that enable users to control and operate smart home devices and technologies using voice commands.

- Conduct online research.
- Initiate phone calls and send SMS messages
- Generate reminders — and perform several more functions.

Chatbots for Assistance in Technical Support

A chatbot is a kind of computer software designed to replicate human speech patterns and behavior. It is crucial to highlight that the chatbot's operation replicates a genuine conversation. Users interact using a chat interface or voice, and chatbots interpret the words and provide a pre-programmed response.

Chatbots may be categorized into three distinct types:

- Governed by a set of rules
- Possessing the ability to understand and reason
- Powered by artificial intelligence

Technology for Identifying and Verifying Individuals Based on Their Facial Features

The application utilizes facial recognition technology to identify certain facial landmarks, resulting in the creation of a mathematical representation of your unique facial features. This representation is then compared to a database of pre-existing facial profiles. Facial recognition technology is widely used by many entities, such as the United States government at airports, law enforcement agencies, marketers, advertisers, and social media corporations.

Customized Suggestions

User behavior is used to produce tailored recommendations. These are items that have often been seen, investigated, or bought in relation to the one the client is now contemplating. For instance, several online retailers provide "frequently bought together" recommendations, as well as suggestions based on customers' browsing and buying history, including items seen by other customers who looked at the same product.

Anticipatory Maintenance

Artificial Intelligence aids firms in pre-emptively repairing or replacing components or machinery prior to their obsolescence. Predictive maintenance integrates data from several sources, including historical maintenance schedules, device sensor data, and weather data, to forecast the timing for device repairs. Operators enhance their decision-making on device repairs by using real-time asset data and analysing past data.

Both the company's internal processes and the customer's experience are improved by these AI applications. With the rapid advancement of technology, AI is set to become more integrated into different industries, opening up exciting new possibilities for optimization and innovation.

USAGE OF ARTIFICIAL INTELLIGENCE IN BUSINESS

Lu (2022) advises exploring AI-based business administration innovation to boost retail organization efficiency. Assess retail firms' value network, company growth, and benefits using a questionnaire (Lourens, M., Raman, R., et al., 2022). Run regression. Testing reveals internal value network innovation has 0.413 and external 0.258 before business model innovation. By including business model innovation, the coefficients of the two variables drop to 0.208 ($p = 0.012 < 0.05$) and 0.113 ($p = 0.005 < 0.01$). The two business model innovation coefficients are significant at 1%. Conclusion. Both internal and external value network innovation boost retail enterprises' functional and AI business models. Functional and artificial intelligence business model innovation improve retail enterprises' profits. Innovation in internal and external value networks benefits retail enterprises financially and commercially. Retailers get financial and commercial benefits from business model innovation and internal and external value network innovation. Companies can no longer use traditional commercial methods to expand due to rising data, changing customer preferences, and complexity. With AI deriving practical insights from consumer data, fundamental developments have produced expanding business prospects. Revenue, customer experience, productivity, efficiency, and corporate growth and transformation are improved by AI.

AI is becoming increasingly important in business. AI technology like machine learning, natural language processing, and robotics may automate tedious tasks, enhance business processes, and improve decision-making. Manufacturing, automotive, healthcare, and finance employ AI (Lourens, M., Sharma, S., et al., 2023). Companies are partnering with AI companies and developing their own AI solutions to improve operations, customer experience, and competitiveness. AI may help businesses grow

Figure 1. AI in business fact
(Forrester/IDC/Narrative Science)

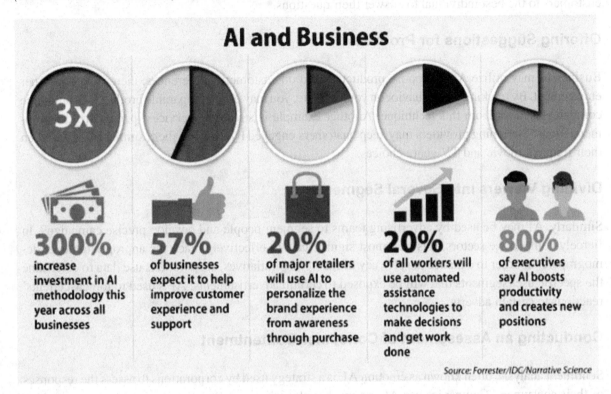

in numerous ways (Satpathy, A., Samal, A., et al., 2023). AI and machine learning are improving firm performance in new ways (Krishna, S. H., & Manoharan, G., 2022; Gopinathan, R., & Manoharan, G., 2022). AI may boost productivity by automating procedures, improving service speed and consistency, leveraging customer insights to make decisions, and identifying new products and services. AI can fit into most company plans. Understand data collection and analysis before diving into AI. By studying AI methodologies, you may better predict their uses in your industry.

EXAMPLES OF ARTIFICIAL INTELLIGENCE IN BUSINESS

Artificial intelligence is used by corporations with more frequency than one may fully comprehend. The applications of AI span across several domains, including marketing (Manoharan, G., & Narayanan, S., 2021), operations, and customer service, exhibiting a near limitless potential. Here are some instances of artificial intelligence applications in business.

Enhancing the Quality of Client Assistance

Has a website chatbot ever caught your eye? Clients often interact with chatbots. Chatbots help businesses streamline customer assistance and free up staff time for personalized tasks. Natural language processing, machine learning, and AI help chatbots understand customer queries. Chatbots may advise customers to the best individual to answer their questions.

Offering Suggestions for Products

Businesses may utilize AI to propose products based on customers' preferences, increasing consumer engagement. By tracking user behavior on your website, you may offer comparable products. Ecommerce companies benefit from this technique. Another example is streaming services' personalized recommendations. Streaming providers may keep customers engaged by showing them similar titles based on their frequent movie and TV genre choices.

Dividing Viewers Into Several Segments

Similarly, AI may be used by advertising teams to segment people and develop precise campaigns. In fiercely competitive sectors, it is of utmost significance to effectively reach the appropriate target demographic. In order to enhance the efficacy of marketing initiatives, corporations use data to determine the specific user segments that will be exposed to certain advertisements. AI is used to forecast clients' reactions to certain adverts.

Conducting an Assessment of Consumer Contentment

Sentiment analysis, often known as emotion AI, is a strategy used by corporations to assess the responses of their consumers. Companies use AI and machine learning to collect data on consumer perceptions (Manoharan, G., Ashtikar, S. P., et al., 2023) of their brand. This might include using artificial intelligence algorithms to analyze social media postings, reviews, and ratings that specifically reference the brand. The study provides valuable insights that enable firms to discover areas where improvements may be made.

Fraud Detection

Artificial intelligence may also be used to aid firms in identifying and promptly addressing instances of fraudulent activities. Financial technologies use machine learning algorithms to detect and flag suspicious activity. The programme stops the transaction and notifies parties if fraud is suspected.

Enhancing the Efficiency of Supply Chain Operations

AI might help your company deliver items on schedule. AI can estimate material and transportation costs and supply chain speed (Gulati, N., Sethi, A., et al., 2022). These insights help supply chain experts choose the best delivery option. Small-scale AI may help delivery trucks find better routes.

AI IN BUSINESS

The substitution of human workers with artificial intelligence in customer service is an additional contentious instance of the integration of artificial intelligence in the corporate sector (Shaikh, I. A. K., Kumar,

C. N. S., et al., 2023). Thanks to the emergence of chat-bots, consumers now have the ability to engage with firms instantly in order to address grievances, make purchases, get information, and do almost any task that would typically involve conversing with a human customer service agent. Gartner predicts that by 2020, 85% of consumer contacts will be automated, eliminating the need for human involvement. AI would reduce human involvement in customer engagement, lowering company costs and improving customer accessibility and conversational experience.

Business Intelligence

Data mining and advanced analytics are used in business intelligence to acquire, analyze, and interpret huge volumes of data for insights and choices. Growing corporate data may make judgments harder. AI is often used in business intelligence to get insights from data. These insights boost marketing, consumer understanding, segmentation to personalize experiences, business strategies to assist corporate decision-making, and AI use in management. Business intelligence tools often employ AI. Popular technologies Microsoft Power BI helps firms make KPI-impacting choices using essential facts. Power BI may assist organizations construct machine learning models for data analysis and business automation (Deviprasad, S., Madhumithaa, N., et al., 2023).

Customized and Focused Marketing

Understanding client preferences and targeting items to each consumer may boost corporation revenue. Buyers have little time in modern digital environment, where numerous companies are trying to sell their products. Increasingly, companies must market goods that appeal to each consumer and know which customers to target for each product. By leveraging online consumer data, companies may utilize AI to predict and target particular people for product marketing. This method boosts sales and optimizes marketing while lowering marketing costs.

RECOMMENDATION OF PRODUCTS AND ANALYSIS OF FUTURE OUTCOMES USING PREDICTIVE ANALYTICS

Companies must provide items that satisfy consumers to boost marketing and consumer engagement. Amazon, Spotify, Netflix, and others utilize AI to track and promote products. Spotify utilizes AI to predict individual music tastes based on their listening patterns and song selections. AI recommendation systems from Netflix and others select movies based on users' likes and past interactions to engage and amaze. Netflix's AI recommends 75% of programming and saves $1 billion annually. Artificial intelligence helps banks and Fintech businesses detect fraud and assess loan applicants' creditworthiness.

AI IN MANUFACTURING

Natural Language Processing

At what point will computers be able to read, write, and comprehend human languages? Here is the answer! Companies now provide customers with intelligent digital assistants to aid them with mundane chores, thanks to advancements in natural language processing. Modern businesses rely on AI to automate the creation of reports for management and to analyse customer sentiment towards their brand in social media and other online platforms. Using sentiment analysis, businesses can track how customers feel about their offerings. Service and personalised product offerings are enhanced as a result.

Benefits of Artificial Intelligence in Business/ Workplace

There are virtually endless advantages of using AI in the workplace. Have a look at some of the most essential perks below. The benefits businesses stand to gain from the use of AI is endless and includes:

- Robotic control of operations
- Better outcomes from marketing efforts and higher income
- A deeper familiarity with client needs and a more satisfying service experience
- Identifying fraudulent activities;
- Customer service that is both better and more reliable

The above are the fundamental ones but nevertheless below are the important general elements.

Reducing Human Mistakes

Humans err. Unfortunately, errors occur despite caution and quality control. The finest attempts cannot prevent this. UK retailer Argos understands that even a little error may have serious effects. A £349.99 TV offered for 49p for 7 hours in 2005. After rejecting hundreds of requests, Argos had to wage a protracted legal battle to avoid giving away thousands of TVs for a bar of chocolate. Successfully removes such mistakes using AI. Companies save millions by detecting errors before launch. Producing damage, equipment failure, and revenue estimate mistakes are reduced.

Reducing Human Workloads

As AI can handle a lot of the "heavy lifting" for firms, human workers consequently have more time to concentrate on other activities. Labour-intensive jobs like as customer service and sales may be done with chatbots, who deliver rapid and precise replies to consumers. This decreases the amount of contact center and technical support people that are required by firms, enabling them to concentrate on management, development and expansion. AI may assist firms with repetitive and mundane jobs, leaving staff free to focus on essential parts of the company.

Increased Efficiency and Accuracy

By improving task speed and accuracy, AI helps organizations work more efficiently. AI technologies reduce human error, speeding up and improving processes. Data-intensive fields like finance and engineering need more time to analyze than AI-powered tools. Hyper-efficiency saves money and minimizes embarrassing mistakes that might damage a company's image.

Increased Productivity

Pauses are needed because humans become tired, procrastinate, and lose focus. AI never tires. It does not need a coffee break or complain about overtime. It works flawlessly 24/7/365. This increases productivity and revenue. AI can do tedious tasks perfectly every time. A human-supervised AI workforce is a big benefit to global businesses.

So, Artificial intelligence is here and it's not going away. It's taken a few hits since its debut in the previous several decades, with many forecasting the mass layoff of the human labor and even going so far as to anticipate some type of Skynet-terminator tragedy to emerge as a consequence.

CHALLENGES

Artificial intelligence in business is growing. This rise is hindered by three main factors that limit business AI adoption.

Data Scarcity

Organizations have a lot of data, but AI adoption is tricky. Machine learning, which underpins most corporate AI applications, needs enormous data sets to train the model. AI use in new industries without data is limited. Business adoption of AI is problematic since most AI applications need supervised training on labelled data yet our enormous data collection is generally unstructured and unlabeled.

Algorithm Bias

Due to its ethnic, racial, and gender biases, Microsoft and Amazon prevented law enforcement from purchasing their AI face recognition software. This illustrates how biased data may hurt AI. AI systems will ultimately address such biases, but they currently threaten AI acceptance in several industries. Because AI collects sensitive personal data, computer constraints and data security and privacy hazards are also issues.

Businesses That Have Altered Operations Using AI

Most of the top firms in the world have significantly embraced the usage of AI. Here are some of the top firms leaning on the potential of AI.

AI IN CUSTOMER SERVICE

AI helps Alibaba predict what customers want to buy and automatically write product descriptions. Uber is another AI-powered business disruptor. Uber predicts demand using AI, reducing estimated arrival time and efficiently matching riders and drivers, lowering church rate. Uber drivers can automatically respond to riders' messages with a single click using AI one-click chat.

Figure 2. AI in customer service
(Author)

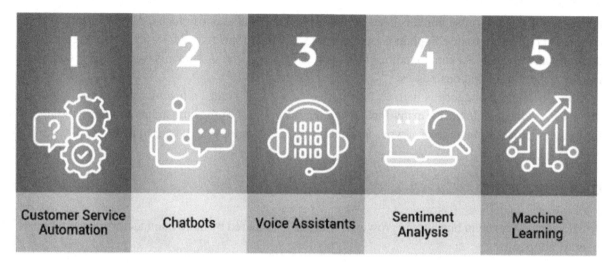

- AI-recommended products, automated factories, and the Amazon Alexa digital assistant are other great AI-powered businesses.
- Autonomous Tesla vehicles
- Azure machine learning and power BI from Microsoft

AI IN TRAVEL

AI has transformed the travel industry, improving recommendations, booking, forecasting, flying, and itineraries. Let's examine each area in detail:

Figure 3. AI in travel
(Author)

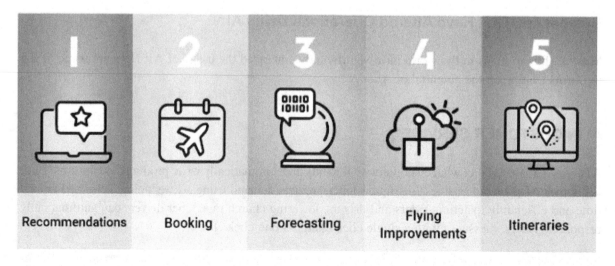

Recommendations

Personalized Travel Suggestions: AI algorithms recommend destinations, accommodations, activities, and more based on user preferences, travel history, and behaviour.

Contextual Suggestions: AI takes into account factors like weather, local events, and user preferences to provide context-aware recommendations, ensuring a more relevant and enjoyable travel experience.

Booking

Chatbots and Virtual Assistants: AI-powered chatbots assist users in real-time, answering queries, providing information, and facilitating the booking process. Virtual assistants can handle complex interactions, helping users find and book flights, hotels, and activities seamlessly.

Dynamic Pricing: AI algorithms analyze various data points, such as demand, seasonal trends, and competitor pricing, to optimize pricing strategies dynamically. This ensures that prices are competitive and reflective of market conditions.

Forecasting

Demand Forecasting: Travel demand is predicted by AI models using historical data, current trends, and external factors. Businesses can anticipate peak periods, optimise pricing, and efficiently allocate resources.

Weather and Disruption Forecasting: AI utilizes weather data and other relevant information to forecast potential disruptions to travel plans, enabling proactive measures and better customer communication in the case of delays or cancellations.

Flying Improvements

Predictive Maintenance: AI monitors the condition of aircraft components in real time, predicting maintenance needs before issues arise. This helps airlines schedule maintenance proactively, reducing the risk of unexpected breakdowns and improving overall safety.

Route Optimization: AI algorithms optimize flight routes based on various factors, including weather conditions, fuel efficiency, and air traffic. This not only saves fuel but also reduces emissions and enhances on-time performance.

Itineraries:

Smart Itinerary Management: AI assists travellers in creating well-organized itineraries by considering factors like travel preferences, time constraints, and local events. It can provide real-time updates, suggest alternative plans in case of disruptions, and optimize the overall travel experience.

Automated Trip Planning: AI can analyze user preferences and constraints to automatically generate personalized travel itineraries, considering factors like budget, preferred activities, and available time. AI has revolutionized the travel industry by offering personalized recommendations, streamlining booking processes, improving forecasting accuracy, enhancing flying operations, and optimizing travel itineraries. These innovations improve business efficiency and customer satisfaction and make travel easier and more enjoyable.

Figure 4. AI in retail
(Author)

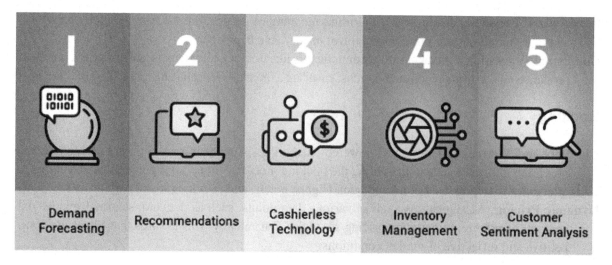

AI IN RETAIL

Artificial Intelligence (AI) has been increasingly integrated into various aspects of the retail industry, transforming operations, enhancing customer experiences, and optimizing business processes. Here's an overview of how AI is applied in different areas of retail, as you've mentioned:

Demand Forecasting

Machine Learning Models: AI algorithms predict future demand using historical sales data, seasonal patterns, loan prediction (Meenaakumari, M., Jayasuriya, P., et al., 2022) and other factors.
Dynamic Pricing: AI helps retailers adjust pricing dynamically based on demand fluctuations, competitor pricing, and other market conditions.

Recommendation Systems

Personalized Recommendations: AI algorithms make personalised product recommendations based on customer behaviour, preferences, and purchase history, improving cross-selling and upselling.
Content Recommendations: AI-driven content recommendations help retailers optimize website and app content to enhance the overall shopping experience.

Cashier-less Technology

Computer Vision: AI-powered computer vision systems enable cashierless checkout experiences by tracking items selected by customers and automatically charging them.
Sensor Fusion: Combining data from various sensors, such as cameras and weight sensors, allows accurate tracking of products and customers within the store.

Inventory Management

Predictive Analytics: AI helps retailers optimize inventory levels by predicting which products will be in demand, reducing overstock and stockouts.

Supply Chain Optimization: AI is used to enhance supply chain efficiency, ensuring the timely delivery of products while minimizing costs.

Customer Sentiment Analysis

Natural Language Processing (NLP): AI analyzes customer reviews, social media mentions, and feedback to understand customer sentiment and gather insights.

Chatbots and Virtual Assistants: AI-driven chatbots provide real-time customer support, addressing queries and concerns, and gathering valuable information about customer satisfaction. These apps boost operational efficiency and make shopping more convenient and personalised. As technology advances, retail AI may offer more advanced solutions and capabilities.

Figure 5. AI in finance
(Author)

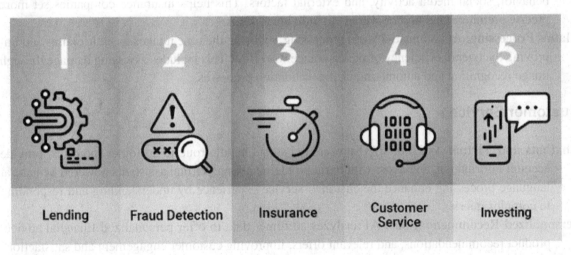

AI IN FINANCE

Artificial Intelligence (AI) has had a profound impact on the finance industry, transforming various aspects including lending, fraud detection, insurance, customer service, and investing. Let's explore each of these areas in detail:

Lending

Credit Scoring: AI algorithms analyze vast amounts of data, including traditional credit history and alternative data sources, to assess an individual's creditworthiness more accurately. This enables lenders to make better-informed decisions and expand access to credit for individuals with limited credit histories.

Automated Underwriting: AI streamlines the underwriting process by automating the analysis of borrower information, reducing the time and resources required for loan approvals. This results in faster decision-making and increased efficiency in lending operations.

Fraud Detection

Behavioral Analysis: AI systems analyze patterns of user behavior and transactions to identify anomalies that may indicate fraudulent activity. Machine learning models continuously adapt to new fraud patterns, improving detection accuracy over time.

Biometric Authentication: AI-powered biometric systems, such as facial recognition and fingerprint scanning, enhance security measures for identity verification, reducing the risk of identity theft and fraudulent transactions.

Insurance

Risk Assessment: AI assesses risk more accurately by analyzing diverse data sets, including customer behavior, social media activity, and external factors. This helps insurance companies set more precise premiums and tailor coverage to individual needs.

Claims Processing: AI automates claims processing, reducing the time it takes to settle claims and improving the overall efficiency of the insurance workflow. This includes assessing damage through image recognition and automating claims validation processes.

Customer Service

Chatbots and Virtual Assistants: AI-powered chatbots handle routine customer inquiries, provide account information, and assist with basic problem-solving. Virtual assistants powered by natural language processing enhance the customer service experience by understanding and responding to complex queries.

Personalized Recommendations: AI analyzes customer data to offer personalized financial advice, product recommendations, and relevant offers, improving customer engagement and satisfaction.

Investing

Algorithmic Trading: Using massive volumes of market data, news, and sentiment analysis from social media, algorithms powered by artificial intelligence carry out trades according to predetermined criteria. Quicker and more informed investment choices are made possible by this.

Portfolio Management: AI assists in optimizing investment portfolios by considering risk tolerance, market conditions, and individual goals. It can automatically rebalance portfolios and suggest adjustments based on changing market dynamics.

AI has transformed the banking business, improving productivity, accuracy, and customer service. AI is transforming financial services by speeding lending, fraud detection, insurance underwriting, customer service, and investment methods. These innovations increase financial institutions' operational efficiency, customer services, and decision-making.

EXAMPLES OF AI IN TOP COMPANIES

Alibaba

Chinese firm Amazon and eBay lose worldwide e-commerce sales to Alibaba. Alibaba predicts customer purchases via AI. A automobile website is automated by the corporation (Ramalingam, Manoharan, & Puviarasi, 2021). Natural language descriptions. AI-powered Alibaba City Brain produces smart cities. City car movements are tracked by AI systems to decrease traffic. Artificial intelligence from Alibaba Cloud helps farmers boost output and decrease expenses.

Alphabet - Google

Google is owned by Alphabet. Google founded Waymo, its autonomous driving technology division. Waymo aims to globally implement autonomous driving technology to improve transportation and reduce accidents. Self-driving taxis from the company are now moving passengers in California. In the trial programme, a human driver is in the vehicle and the corporation cannot charge. It was clear that Google was serious about deep learning when it acquired DeepMind. AlphaGo learnt to play 49 Atari games and beat a professional Go player. Duplex is another one of Google's AI breakthroughs. An AI voice interface that uses natural language processing can do things like set up appointments and make phone calls. Investigate Google's use of AI and ML.

Amazon

Alexa is one of several Amazon AI applications. Artificial intelligence helps Amazon deliver items before customers purchase. Using customer purchasing data, the company claims it can propose products. They use predictive analytics to forecast customer needs. As traditional retailers suffer, Amazon Go is a new convenience store model. Not everyone must use the checkout, unlike other businesses. Your Amazon Go phone app monitors and charges your purchases using artificial intelligence. There is no conventional checkout, therefore consumers must provide their own bags. On-site security cameras monitor and identify every bag. Data is used to charge customers correctly.

Apple

Apple, a multinational technology company, makes iPhones, Apple Watches, computer software, and online services. Apple uses artificial intelligence and machine learning in its iPhones to enable FaceID. Other Apple products like the AirPods, Apple Watch, and HomePod smart speakers use these technologies to power Siri. Apple is adding services and using AI to suggest music on Apple Music, find photos in iCloud, and navigate meetings on Maps.

Baidu is a Chinese Technology Company

Baidu, the Chinese counterpart of Google, extensively employs artificial intelligence across several domains. Their product, Deep Voice, utilizes artificial intelligence and deep learning techniques to replicate a voice using only 3.7 seconds of audio. Utilizing the identical technology, a tool has been developed to automatically read books to individuals in the voice of the author, eliminating the need for a recording studio.

Facebook is a Social Networking Platform

To a large extent, Facebook relies on AI and deep learning to structure its chaotic data. Their text comprehension engine, DeepText, can automatically decipher and analyse the meaning and tone of the millions of posts made by users in a variety of languages every single second.DeepFace lets the popular social media company instantly recognise you in a photo posted on their site. This technology outperforms humans in face recognition. The startup uses AI to automatically identify and remove revenge pornography from its platform.

International Business Machines Corporation (IBM)

IBM has long led artificial intelligence. Over 20 years have gone since IBM's Deep Blue computer defeated a world chess champion. The company held other tournaments putting people against robots, including one when its Watson computer won Jeopardy. IBM's latest AI project is Project Debater. This cognitive computing AI competed with two skilled debaters and developed human-like arguments.

JD.com is a Prominent E-Commerce Company

JD.com is Amazon China. Richard Liu, the creator, plans and actively pursues full automation for his company. The warehouse is fully automated, and drone deliveries have been happening for four years. Artificial intelligence, big data, and robots are helping JD.com grow while building the fourth industrial revolution's retail infrastructure.

Microsoft is a Multinational Technology Company

Microsoft's vision emphasises integrating intelligent machines into all aspects of its operations. Microsoft is adding advanced features to Cortana, Skype, Bing, and Office 365. This has made them one of the largest AIaaS distributors worldwide.

Tencent is a Company

Chinese social media giant Tencent leverages AI to become the most respected online corporation. Tencent uses AI significantly. With 1 billion app users, WeChat offers games, digital assistants, mobile payments, cloud storage, live streaming, sports, education, movies, and self-driving cars (Razak, A., Nayak, M. P., et al., 2023). The business slogan is "AI in all." Tencent gains from huge consumer data analysis.

FUTURE AI TRENDS

Exploring the promising future tendencies of AI, in the near future, we may anticipate:

- Entertainment firms producing music and movies created by artificial intelligence.
- Completely autonomous plants that operate without any human oversight
- The use of artificial intelligence (AI) to produce high-quality training data for its own use, in order to tackle the problems of bias and limited availability of data, among other benefits.
- The future prospects of AI will empower people to construct the future according to their own desires for the first time.

DISCUSSION AND CONCLUSION

Business advantages and downsides of AI. AI may improve decision-making, efficiency, and cost. Data privacy and security, ethical disputes, and job displacement are all difficulties with AI adoption. To exploit AI's advantages and limit its problems, businesses should collaborate with governments and other stakeholders (Kumar, S., Kumar, P., et al., 2024). AI is helping companies create economic value and acquire a competitive edge. Although time, effort, and money are invested, some AI efforts fail. How and what AI technology may add to company value is unknown. AI digitization requires organizations to prioritize their current strategy while swiftly exploring new market potential. Companies must fulfill this to compete. AI has transformed businesses and other sectors (Ponduri, S. B., Ahmad, S. S., et al., 2024). This review covers business and economics AI.

This chapter discusses how corporations may use AI to produce value. This study produced three parts. The first factors that affect AI use are found. Technological, organizational, and environmental concerns precede AI deployment. AI has other applications. Companies may automate or increase internal and external talents using AI. AI enhances business operations without client involvement. Customer-facing goods and services use external AI. AI's effects on organisational transformation and competitive performance are examined last (Manoharan, G., Durai, S., et al., 2023). Organizational and operational AI consequences are discussed. If supervised and regulated, AI is a powerful tool. Growth, development, and entry into business, health, and personal life are likely. Understand future possibilities for a better tomorrow. AI has a bright future. AI enhances corporate operations by reducing repetitive tasks, increasing efficiency, and satisfying customers. It prevents errors and predicts calamities better than humans. IT requires AI and management professionals. Skills may make you an AI leader.

REFERENCES

Abdulwahid, A. H., Pattnaik, M., Palav, M. R., Babu, S. T., Manoharan, G., & Selvi, G. P. (2023, April). Library Management System Using Artificial Intelligence. In *2023 Eighth International Conference on Science Technology Engineering and Mathematics (ICONSTEM)* (pp. 1-7). IEEE.

Abrokwah-Larbi, K., & Awuku-Larbi, Y. (2023). The impact of artificial intelligence in marketing on the performance of business organizations: evidence from SMEs in an emerging economy. *Journal of Entrepreneurship in Emerging Economies.*

Ahmad, H.A., Hanandeh, R., Alazzawi, F.R., Al-Daradkah, A., ElDmrat, A.T., Ghaith, Y.M., & Darawsheh, S.R. (2023). The effects of big data, artificial intelligence, and business intelligence on e-learning and business performance: Evidence from Jordanian telecommunication firms. *International Journal of Data and Network Science.*

Al-Mahairah, M. S., Manoharan, G., Singh, J., & Krishna, S. H. (2022). *Principles of Management.* Book Rivers.

Alblooshi, M.A., Mohamed, A.M., & Yusr, M.M. (2023). Moderating Role of Artificial Intelligence Between Leadership Skills and Business Continuity. *International Journal of Professional Business Review.*

Allioui, H., & Mourdi, Y. (2023). Unleashing the potential of AI: Investigating cutting-edge technologies that are transforming businesses. [IJCEDS]. *International Journal of Computer Engineering and Data Science, 3*(2), 1–12.

Benabed, A. (2023). *Artificial Intelligence's Relevance for Energy Optimization, Companies and Business Internationalization.* New Trends in Sustainable Business and Consumption.

Bharadiya, J. P. (2023). Machine learning and AI in business intelligence: Trends and opportunities. [IJC]. *International Journal of Computer, 48*(1), 123–134.

Chandgude, V., & Kawade, B. (2023). Role of Artificial Intelligence and Machine Learning in Decision Making for Business Growth. International Journal of Advanced Research in Science.

Chankoson, T., Chen, F., Wang, Z., Wang, M., & Sukpasjaroen, K. (2023). Knowledge Mapping for the Study of Artificial Intelligence in Education Research: Literature Reviews. *Journal of Intelligence Studies in Business.*

Chui, M., Manyika, J., & Miremadi, M. (2018). What AI can and can't do (yet) for your business. *McKinsey Quarterly, 1*(97-108), 1.

Deviprasad, S., Madhumithaa, N., Vikas, I. W., Yadav, A., & Manoharan, G. (2023). The Machine Learning-Based Task Automation Framework for Human Resource Management in MNC Companies. *Engineering Proceedings, 59*(1), 63.

Durai, S., Krishnaveni, K., & Manoharan, G. (2022, May). Designing entrepreneurial performance metric (EPM) framework for entrepreneurs owning small and medium manufacturing units (SME) in coimbatore. In AIP Conference Proceedings (Vol. 2418, No. 1). AIP Publishing.

Durai, S., Krishnaveni, K., & Manoharan, G. (2022, May). Leveraging HR metrics for effective recruitment & selection process in IT industries in Chennai and Coimbatore, Tamil Nadu. In AIP Conference Proceedings (Vol. 2418, No. 1). AIP Publishing.

Durai, S., Krishnaveni, K., & Manoharan, G. (2022, May). Metric based performance management of employees–A case study of MSME unit in Coimbatore, Tamil Nadu. In AIP Conference Proceedings (Vol. 2418, No. 1). AIP Publishing.

García-Tadeo, D. A., Peram, D. R., Kumar, K. S., Vives, L., Sharma, T., & Manoharan, G. (2022). Comparing the impact of Internet of Things and cloud computing on organisational behavior: A survey. *Materials Today: Proceedings, 51*, 2281–2285. doi:10.1016/j.matpr.2021.11.399

Geetha, M., Pavan, K. B., & Sunitha, P. A. (2022). *COVID-19: A special review of MSMEs for sustaining entrepreneurship.* Trueline Academic and Research Centre.

Geetha, M., & Sunitha, P. A. (2022). *Work-Life Balance among Women in the Private Higher Education Industry during COVID-19: A Path to Organisational Sustainability.* Building Resilient Organizations.

Gopinathan, R., & Manoharan, G. (2022). Work-Life Balance Practices and Organizational Performance for Achieving Superior Performance. *The Changing Role of Human Resource Management in the Global Competitive Environment,* 217.

Gulati, N., Sethi, A., Mahesh, D., Makkar, S., Manoharan, G., Megaladevi, M., & Palanivel, R. (2022, May). Implementation of blockchain in supply chain. In AIP Conference Proceedings (Vol. 2418, No. 1). AIP Publishing. doi:10.1063/5.0083008

Hassan, W. M., Aldoseri, D. T., Saeed, M. M., Khder, M. A., & Ali, B. (2022). Utilization of Artificial Intelligence and Robotics Technology in Business. *2022 ASU International Conference in Emerging Technologies for Sustainability and Intelligent Systems (ICETSIS),* (pp. 443-449). IEEE. 10.1109/ICETSIS55481.2022.9888895

Jaichandran, R., Krishna, S. H., Madhavi, G. M., Mohammed, S., Raj, K. B., & Manoharan, G. (2023, January). Fuzzy Evaluation Method on the Financing Efficiency of Small and Medium-Sized Enterprises. In *2023 International Conference on Artificial Intelligence and Knowledge Discovery in Concurrent Engineering (ICECONF)* (pp. 1-7). IEEE.

Kaushik, D.P. (2022). Role and Application of Artificial Intelligence in Business Analytics: A Critical Evaluation. *International Journal for Global Academic & Scientific Research.*

Keserwani, H., PT, R., PR, J., Manoharan, G., Mane, P., & Gupta, S. K. (2021). Effect Of Employee Empowerment On Job Satisfaction In Manufacturing Industry. *Turkish Online Journal of Qualitative Inquiry, 12*(3).

Krishna, S. H., & Manoharan, G. (2022). Making the Link between Work-Life Balance Practices and Organizational Performance in the Hospitality Industry. *The Changing Role of Human Resource Management in the Global Competitive Environment,* 201.

Kumar, S., Kumar, P., Ahmad, S. S., Jayasaradadevi, P., Rajeyyagari, S., Manoharan, G., & Radhakrishnan, V. (2024). A Novel Approach for IoT-Based Cloud Computing Technology and its Impact on Business Entrepreneurship. *International Journal of Intelligent Systems and Applications in Engineering, 12*(10s), 624–628.

Lourens, M., Raman, R., Vanitha, P., Singh, R., Manoharan, G., & Tiwari, M. (2022, December). Agile Technology and Artificial Intelligent Systems in Business Development. In *2022 5th International Conference on Contemporary Computing and Informatics (IC3I)* (pp. 1602-1607). IEEE. 10.1109/IC3I56241.2022.10073410

Lourens, M., Sharma, S., Pulugu, R., Gehlot, A., Manoharan, G., & Kapila, D. (2023, May). Machine learning-based predictive analytics and big data in the automotive sector. In *2023 3rd International Conference on Advance Computing and Innovative Technologies in Engineering (ICACITE)* (pp. 1043-1048). IEEE. 10.1109/ICACITE57410.2023.10182665

Lu, B. (2022). Analysis on Innovation Path of Business Administration Based on Artificial Intelligence. *Mathematical Problems in Engineering, 2022*, 1–7. doi:10.1155/2022/6790836

Manoharan, G., & Ashtikar, S. P. (2022). The Relationship between Job-Related Factors on Work Life Balance and Job Satisfaction. *The Changing Role of Human Resource Management in the Global Competitive Environment*, 169.

Manoharan, G., Ashtikar, S. P., Smitha, V., Sundaramoorthi, S., & Krishna, I. M. (2023). Work-Life Balance Perceptions of Women in the IT and ITeS Sectors in Kerala: A Research Study. *Journal of Pharmaceutical Negative Results*, 3363–3375.

Manoharan, G., Durai, S., Ashtikar, S. P., & Kumari, N. (2024). Artificial Intelligence in Marketing Applications. In Artificial Intelligence for Business (pp. 40-70). Productivity Press.

Manoharan, G., Durai, S., & Rajesh, G. A. (2022, May). Emotional intelligence: A comparison of male and female doctors in the workplace. In AIP Conference Proceedings (Vol. 2418, No. 1). AIP Publishing. doi:10.1063/5.0081816

Manoharan, G., Durai, S., & Rajesh, G. A. (2022, May). Identifying performance indicators and metrics for performance measurement of the workforce is the need of the hour: A case of a retail garment store in Coimbatore. In AIP Conference Proceedings (Vol. 2418, No. 1). AIP Publishing. doi:10.1063/5.0081821

Manoharan, G., Durai, S., Rajesh, G. A., & Ashtikar, S. P. (2023). Credit and Risk Analysis in the Financial and Banking Sectors: An Investigation. In Artificial Intelligence for Capital Markets (pp. 57-72). Chapman and Hall/CRC.

Manoharan, G., Durai, S., Rajesh, G. A., Razak, A., Rao, C. B., & Ashtikar, S. P. (2023). An investigation into the effectiveness of smart city projects by identifying the framework for measuring performance. In *Artificial Intelligence and Machine Learning in Smart City Planning* (pp. 71–84). Elsevier. doi:10.1016/B978-0-323-99503-0.00004-1

Manoharan, G., Durai, S., Rajesh, G. A., Razak, A., Rao, C. B., & Ashtikar, S. P. (2023). A study of postgraduate students' perceptions of key components in ICCC to be used in artificial intelligence-based smart cities. In *Artificial Intelligence and Machine Learning in Smart City Planning* (pp. 117–133). Elsevier. doi:10.1016/B978-0-323-99503-0.00003-X

Manoharan, G., Durai, S., Rajesh, G. A., Razak, A., Rao, C. B., & Ashtikar, S. P. (2023). A study on the perceptions of officials on their duties and responsibilities at various levels of the organizational structure in order to accomplish artificial intelligence-based smart city implementation. In *Artificial Intelligence and Machine Learning in Smart City Planning* (pp. 1–10). Elsevier. doi:10.1016/B978-0-323-99503-0.00007-7

Manoharan, G., & Narayanan, S. (2021). A research study to investigate the feasibility of digital marketing strategies in advertising. *PalArch's Journal of Archaeology of Egypt/Egyptology, 18*(09), 450-456.

Manoharan, K. G., Nehru, J. A., & Balasubramanian, S. (2021). *Artificial Intelligence and IoT*. Springer. doi:10.1007/978-981-33-6400-4

Meenaakumari, M., Jayasuriya, P., Dhanraj, N., Sharma, S., Manoharan, G., & Tiwari, M. (2022, December). Loan Eligibility Prediction using Machine Learning based on Personal Information. In *2022 5th International Conference on Contemporary Computing and Informatics (IC3I)* (pp. 1383-1387). IEEE. 10.1109/IC3I56241.2022.10073318

Mishra, S., & Tripathi, A. R. (2021). AI business model: An integrative business approach. *Journal of Innovation and Entrepreneurship*, *10*(1), 18. doi:10.1186/s13731-021-00157-5

Ponduri, S. B., Ahmad, S. S., Ravisankar, P., Thakur, D. J., Chawla, K., Chary, D. T., & Sharma, S. (2024). A Study on Recent Trends of Technology and its Impact on Business and Hotel Industry. *Migration Letters : An International Journal of Migration Studies*, *21*(S1), 801–806.

Rama Krishna, S., Rathor, K., Ranga, J., & Soni, A., D, S., & N, A.K. (2023). Artificial Intelligence Integrated with Big Data Analytics for Enhanced Marketing. *2023 International Conference on Inventive Computation Technologies (ICICT)*, (pp. 1073-1077). IEEE. 10.1109/ICICT57646.2023.10134043

Ramachandran, K. K., Mary, S. S. C., Painoli, A. K., Satyala, H., Singh, B., & Manoharan, G. (2022). Assessing The Full Impact Of Technological Advances On Business Management Techniques.

Ramalingam, M., Manoharan, G., & Puviarasi, R. (2021). Web-Based Car Workshop Management System—A Review. Next Generation of Internet of Things. *Proceedings of ICNGIoT*, *2021*, 321–331.

Razak, A., Nayak, M. P., Manoharan, G., Durai, S., Rajesh, G. A., Rao, C. B., & Ashtikar, S. P. (2023). Reigniting the power of artificial intelligence in education sector for the educators and students competence. In *Artificial Intelligence and Machine Learning in Smart City Planning* (pp. 103–116). Elsevier. doi:10.1016/B978-0-323-99503-0.00009-0

Reim, W., Åström, J., & Eriksson, O. (2020). Implementation of artificial intelligence (AI): A roadmap for business model innovation. *AI*, *1*(2), 11. doi:10.3390/ai1020011

Satpathy, A., Samal, A., Gupta, S., Kumar, S., Sharma, S., Manoharan, G., Karthikeyan, M., & Sharma, S. (2024). To Study the Sustainable Development Practices in Business and Food Industry. *Migration Letters : An International Journal of Migration Studies*, *21*(S1), 743–747. doi:10.59670/ml.v21iS1.6400

Shaikh, I. A. K., Kumar, C. N. S., Rohini, P., Jafersadhiq, A., Manoharan, G., & Suryanarayana, V. (2023, August). AST-Graph Convolution Network and LSTM Based Employees Behavioral and Emotional Reactions to Corporate Social Irresponsibility. In *2023 Second International Conference on Augmented Intelligence and Sustainable Systems (ICAISS)* (pp. 966-971). IEEE. 10.1109/ICAISS58487.2023.10250754

Shameem, A., Ramachandran, K. K., Sharma, A., Singh, R., Selvaraj, F. J., & Manoharan, G. (2023, May). The rising importance of AI in boosting the efficiency of online advertising in developing countries. In *2023 3rd International Conference on Advance Computing and Innovative Technologies in Engineering (ICACITE)* (pp. 1762-1766). IEEE. 10.1109/ICACITE57410.2023.10182754

Tarafdar, M., Beath, C. M., & Ross, J. W. (2019). Using AI to enhance business operations. *MIT Sloan Management Review*, *60*(4).

Teoh, T. T., & Goh, Y. J. (2023). *Artificial Intelligence in Business Management*. Artificial Intelligence in Business Management. doi:10.1007/978-981-99-4558-0

Thavamani, S., Mahesh, D., Sinthuja, U., & Manoharan, G. (2022, May). Crucial attacks in internet of things via artificial intelligence techniques: The security survey. In AIP Conference Proceedings (Vol. 2418, No. 1). AIP Publishing.

Tripathi, M. A., Tripathi, R., Effendy, F., Manoharan, G., Paul, M. J., & Aarif, M. (2023, January). An In-Depth Analysis of the Role That ML and Big Data Play in Driving Digital Marketing's Paradigm Shift. In *2023 International Conference on Computer Communication and Informatics (ICCCI)* (pp. 1-6). IEEE. 10.1109/ICCCI56745.2023.10128357

Wamba-Taguimdje, S. L., Fosso Wamba, S., Kala Kamdjoug, J. R., & Tchatchouang Wanko, C. E. (2020). Influence of artificial intelligence (AI) on firm performance: The business value of AI-based transformation projects. *Business Process Management Journal, 26*(7), 1893–1924. doi:10.1108/BPMJ-10-2019-0411

Wamba-Taguimdje, S. L., Fosso Wamba, S., Kala Kamdjoug, J. R., & Tchatchouang Wanko, C. E. (2020). Influence of artificial intelligence (AI) on firm performance: The business value of AI-based transformation projects. *Business Process Management Journal, 26*(7), 1893–1924. doi:10.1108/BPMJ-10-2019-0411

Widayanti, R., & Meria, L. (2023). *Business Modeling Innovation Using Artificial Intelligence Technology. International Transactions on Education Technology.* ITEE.

Chapter 6
Artificial Intelligence–Powered Political Advertising:
Harnessing Data–Driven Insights for Campaign Strategies

Roop Kamal
Chandigarh University, India

Manpreet Kaur
Chandigarh University, India

Jaspreet Kaur
Chandigarh University, India

Shivani Malhan
Chitkara University, India

ABSTRACT

This study examines how political advertising is changing in the age of artificial intelligence (AI) technologies. The objective of this chapter is to investigate the complex relationship between political advertising and AI, stressing both its potential advantages and moral dilemmas. This chapter explores the complex interaction between political advertising and machine learning, revealing how algorithms and data-driven insights are transforming political campaigns. In this chapter, the ethical ramifications of AI in political advertising are also covered. The chapter underlines the moral issues related to data security, privacy, and manipulation risk. It also examines the moral conundrums raised by deep fake technologies, microtargeting, and the potential bias present in AI systems. The study also looks into the necessity for accountability and openness in AI-powered political advertising to protect the integrity of democratic processes. In the research, the impact of political strategies is explored, and 150 respondents participated in the primary study.

DOI: 10.4018/979-8-3693-2964-1.ch006

INTRODUCTION

Political advertising has long been a critical component of election campaigns, aiming to influence voters' opinions, attitudes, and decisions. Studies on social and political marketing have grown over time (Amifor, 2016). Few studies have looked at how sponsored mass media messages affect voters, despite the fact that political candidates have relied more and more on television advertising to educate and sway voters (Kraus & Berelson, 1969). In its traditional form, political advertising relied on broad-brush approaches, targeting entire demographics based on general assumptions and limited data. However, the rise of digitalization and the abundance of user-generated data have sparked a new era of data-driven advertising, presenting an opportunity to delve deeper into voter behavior and preferences. A person's capacity to retrieve names of candidates, traits, and educational achievements, to recognize electoral concerns and recent campaign progress, & comprehend linkages with candidates' opinions on various subjects are all examples of political knowledge. Numerous voting researches have concluded that general mass media campaign communications have an adverse effect on knowledge gains due to the consistently observed moderate connection between media exposure and campaign-related knowledge (Bhakri & Shri, 2021). Political advertising is frequently categorized into either positive or negative advertising. Positive advertisements are often those that candidates use to tout their qualifications and strong suits (Silva et al., 2020). Candidates use a range of communication techniques, such as advertising, speeches on the campaign trail, print and broadcast media, to educate and persuade voters. Additionally, over the recent years, we have seen politicians use new interaction tools like social media (Bhattacharya, 2018). Gaining attention on print media and on Televisions has been essential for election growth since, until recently, traditional media served as the main information source for politicians. Candidates additionally disseminated information about their candidature and policy aims through speeches they gave on the campaign trail and other public appearances (Brei, 2020). The applications of artificial intelligence in today's corporate environment are varied. Artificial intelligence, in the opinion of both professionals and academics, will shape our civilization in the future. Technology development has made the world a web of interconnected networks (Atkin & Heald, 1976).

Machine learning is a branch of artificial intelligence (AI) that focuses on creating statistical models and algorithms that let computers learn from data in order to become better at a given task without having to be explicitly programmed. In other words, machine learning enables AI systems to discover patterns and information in the data they are exposed to and make predictions or judgments based on those patterns and information. The application of technology generated investments in artificial intelligence (AI) for big data analytics to generate market intelligence. AI is continually being used to the advantage of numerous sectors. Artificial intelligence and other emerging technologies are developing concurrently as organisations advance towards Industry (Atkin & Heald, 1976) Machines (computers) that replicate the cognitive and affective functions of the human mind are referred to as artificial intelligence. Artificial intelligence (AI) and the Internet of Things (IoT) are two dynamic and supportive technologies that, when used together, can open up a variety of applications and opportunities (Dixit & Singh, 2022). Political advertising using IoT (Internet of Things) is a relatively young and developing field that has the potential to change how political campaigns interact and reach voters. Artificial intelligence has advanced tremendously over the past few decades because to the tireless work of professionals (Garcia-Jimeno & Yildirim, 2015). The work resulted in important developments, including applications for machine learning and big data analytics in many other domains and circumstances (Greening & Gray, 1994).

During the 2019 Lok Sabha elections, Narendra Modi used a messaging app called NaMo, an AI-powered communication tool, to interact with voters (Kaur, 2019). He gave them personalised information and provided real-time answers to their inquiries. In addition to that, Narendra Modi spoke at numerous rallies simultaneously in various cities across India using holographic technology (Kietzmann et al., 2018). Modi was able to reach a large number of voters in various regions of the nation thanks to this without having to visit each location in person.

REVIEW OF LITERATURE

Atkin and Heald (1976) The study explored the relationships between exposure to visual commercials and a number of cognitive and affective orientations using a survey of respondents during a congressional election campaign. Exposure and political interest were only weakly associated. Voters with a high level of exposure were a little more likely to give the issues and candidate qualities emphasised in the ads higher agenda importance. The frequency of exposure to advertisements was only weakly correlated with personal feelings towards each candidate.

Pathak and Patra (2015) analyzed in the study that the political leader is able to maximise recall. Therefore, a political party needs a strong leader who will represent the cause and the party's philosophy. Social media allows political parties to connect with the general public. However, public relations have become the most crucial weapon for campaigns. The foundation of any election propaganda is strategic communication and involvement of the community.

Amifor (2016) shown that both hard sell and soft sale techniques were used in political advertising. For maximum advertising efficacy, above-the-line and below-the-line elements are frequently combined. It was a "hard sale with primary media present, involving billboards, watching television, and radios as reminded media," according to him. Aside from creative direction, they also comprise promotional goods, marketing campaigns, and public relations initiatives that parties undertake to connect with their target audience's tastes, lifestyles, and other habits. Secondary media include public relations and PR, as well as promotional products like badges, pens, and t-shirts. Both the hard pitch and the soft sell work together to create a positive corporate image for the event.

Nadikattu (2016) recognised that AI, or artificial intelligence, had powered numerous services and programmes that would be assisting people to do daily tasks like connecting with friends, using email applications, or that of ride-sharing services, making people's lives increasingly more productive day after day. It will reassure anyone who has concerns about the usage of AI to learn that society has been making use of it for quite some time. As with other changes in life, the same has both positive and negative repercussions. But there is no doubt that AI has changed how individuals live in society. The algorithms of AI must be developed to accord with the overarching objectives of humanity.

Safiullah et al., (2016) concluded in the study that in this technologically advanced era, social media has developed into a significant tool for opinion creation, and marketing managers have recognised its importance. Since politics operates like a customer-driven market, advertising techniques are being used more frequently to gain an edge over rivals. The value of social media has been established. The current election in India clearly demonstrates the marketing of political parties. The current article explores the effect of Twitter on political marketing by studying the relationship between tweet followers and vote share gained by political parties using the Delhi Assembly elections of 2015 as a case study. The findings imply a good link between twitter volume and vote share.

Bhattacharya (2018) indentified the components of political marketing in India and then validate these components using secondary sources in this study. The framework is created in this study to understand how political parties should incorporate the promotion of their own successes, activities, and credentials among voters who are regarded as political consumers. Political parties need to refocus their campaign efforts using the political marketing strategy in the perspective of the circumstances. In the view of the advertising for politics paradigm, study emphasises on need for professional management of traditional campaign and promotion operations. In order to better position themselves in the "political market place," political parties in India can use political marketing. Political marketing, however, cannot take the place of good performance in terms of achieving societal and economic objectives. Political marketing can be implemented by political parties to effectively showcase their achievements. Political parties using the political marketing strategy can create a separate marketing wing and a political market research wing, allowing them to discover issues before anyone else and deliver their views in a polished and effective way. This strategy will also allow political parties to more effectively "leverage" their resources to connect with political consumers.,

Kietzmann, Paschen and Treen (2018) reveals in the study that the way advertisers understand and direct consumers has changed as a result of AI. Future consumer insights will be driven by innovative methods of consumer-generated data mining, and AI will serve as the ultimate privacy test. Advertisers will be able to quietly gather customer data from numerous sources, integrate that data, and mine it for real-time consumer insights with the aid of machine learning.

Choubey (2019) gave an in-depth analysis of the new campaigning trends in India against the historical context of the digital media rebellion, wherein a once-inactive citizen unexpectedly turned out to be a loud party supporter. The research claims that because marketing companies are using the data unethically to sway voters' decisions, illicit access to private data of numbers of Indian electors constitutes a threat to the democracy of the nation. India needs strict legislation to address social media and data leaks. The study also makes an effort to show that electoral advertising is merely a useful communications tool and can't distort actual circumstances. Political advertising has the potential to briefly elevate false deities to power.

Kaur (2019) stated in the study that technology will be used in political campaigning to store vast amounts of data and process it as needed by political parties. Additionally, Al-based technologies make it simple for people to interact with the government, fostering the development of a high-quality democracy. Citizens are free to actively engage in political issues without the intrusion of the average. According to another claim, technology can raise a person's level of awareness during election campaigns by educating them about their political preferences and motivating them to consider both the advantages and disadvantages of their decision. Al solutions are therefore the only programming employed to carry out tasks handled by cognitive abilities. Future political organisations' activities will be reformed using these duties.

Bhakri and Shri (2021) recommended five guidelines for addressing ethical issues with AI. These are: Deterring possible AI- and cyber-enabled weapons would be the first step. The second rule would be to subject AI to the same laws that govern humans in their entirety. The third one is straightforward but crucial. AI should always acknowledge its non-human status. The fourth principle states that AI shall never provide secret information to third parties without the user's express consent. The last guideline for AI solution regulations is that any bias currently present in our culture and systems must not be increased.

Figure 1. AI-driven political advertising

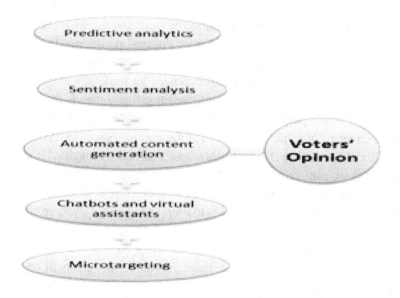

CONCEPTUAL FRAMEWORK

In the above framework, 2 variables are considered out of which voters opinion is the dependent variable and AI driven Political advertising is independent variable. The independent variable further includes sub constructs:

- **Sentiment analysis:** Instantaneous sentiment analysis has been used to gauge public opinion of significant subjects and candidates. Campaigns can adjust their strategies and messaging to match the broad opinions of the public using this information.
- **Microtargeting:** AI has been used to examine vast amounts of data, including social media interactions, surfing habits, and other online behaviours, in order to microtarget voters with exact messaging. This makes it possible for campaigns to target certain voter segments with messaging that could affect their choices.
- **Chatbots and Virtual assistance:** AI-powered chatbots and virtual assistants have been used in political campaigns to communicate with voters, address their questions, and spread information (Tinkham & Weaver-Lariscy, 1993). Campaigns can continue to be active online and provide a customised experience for voters with the help of these technologies.
- **Automated content generation:** Some campaigns have tested using AI-generated content, such as social media posts, articles, and even speeches (Zhou, 2021). Although utilising AI to generate incorrect or misleading information can save time and money, there are ethical concerns..
- **Predictive analysis:** AI has been used to forecast election results and evaluate the effectiveness of campaign tactics. Campaigns can use this data to better efficiently allocate funding and concentrate attention on key demographics and swing states.

RESEARCH METHODOLOGY

Primary data is collected for conducting this study through the systematic questionnaire and personal interviews with the voters. The data is collected through Google form and analysed through SPSS. Secondary data is collected through other sources like internet, books and journals etc. For this study, a randomized sample of voters is used. District Ludhiana in the province of Punjab served as the source of the sample. The responders over the age of 18 were included in the sample. 300 residents in the region received the survey questionnaire. 285 of the questionnaires that were distributed were returned. The survey questionnaire has 2 parts, from which the first part is the demographic profile of the respondents which is the general information like age, gender and occupation of the respondents. The second part of the survey includes the information regarding the preference of the respondents for the political campaigns used by the political candidates.

RESEARCH GAP

These days, it's impossible to define the importance of television, social media, and other forms of communication in daily life (Mithilesh & Choubey, 2019). Various media are used for entertainment, gaming, news, sports, and other purposes. These media are utilized by marketers to promote their goods and services (Nadikattu, 2016). Ads therefore become a necessary part of using these media. A variety of advertisements are shown to the public as part of political campaigning (Dixit & Singh, 2022). The article aims to study the use of AI in political advertising and to investigate how the general public feels about political advertising and whether it has any influence on how they vote.

OBJECTIVES

1. To examine how AI is used in political advertising.
2. To examine role of political advertising in influencing the voters' voting decision.
3. To analyze the public opinion regarding political ads.

HYPOTHESES

H0: There is no apparent relationship among political advertising and voters' voting decisions.
H1: There is an apparent relationship among political advertising and voters' voting decisions.

DATA ANALYSIS AND INTERPRETATION

Table 1 includes the demographics of the survey participants. The data is collected from 265 respondents out of which 184 were male respondents and 81 were the female respondents. Out of total respondents 69% are male members and 31% were females. From the above table, it can also be inferred that the majority of respondents are the young students (35%) and the employed people (39%). The majority

Table 1. Demographics of the survey participants

Sr. No.	Variables		N(265)	%
1	Gender	Male	184	69
		Female	81	31
2	Occupation	Student	93	35
		Employed	104	39
		Self-Employed	49	18
		unemployed	19	8
3	Age	18-30 years	112	42
		30-45 years	87	33
		45-60 years	61	23
		Above	5	2

(Source: Author's own calculations)

respondents (42%) falls under the age of 18-30 years which means that most of the population is educated and are ware about the use of AI by the political parties and the political advertisements being done by the political parties.

Table 2. Correlation coefficient of voters opinion with the five components of political advertising

Political Advertising	Mean	SD	Correlation With Voters Opinion
Sentiment Analysis	45.1	1.25	0.25
Micro targeting	40.7	1.39	0.03
Chatbots and virtual assistants	33.4	1.42	0.17
Automated content generation	28.5	1.23	0.21
Predictive analytics	41.2	1.57	0.23

(Source: Author's own calculations).

In terms of correlation with AI driven political advertising, voter's opinion is found to be positively correlated with sentiment analysis. This show that voters like the way they got analyzed emotionally by the political parties. Also, there is a positive correlation between micro targeting and voters' opinion which means that the voters are satisfied with the targeting strategy of the political parties as they think that with the help of thesetargets, the political parties can easily and accurately capture the respective voting area. There is a positive correlation of voters' opinion with the chatbots and virtual assistance used by the political parties for the voters to interact with the voters and handle their queries and issues. Automated content generation and predictive analysis also possess the positive relation with the voters' opinion as the predictions made by the political parties are almost accurate and it puts a positive impact on the voters.

From the above data, it is clear that H0 is rejected as a positive relationship among all the factors of AI driven political campaigning and voters' opinion is found.

FINDINGS AND DISCUSSIONS

Political campaigns can now more effectively identify and target particular voter categories thanks to machine learning algorithms (Safiullah et al., 2016). Campaigns can adjust their messages to resonate with various voter groups by analyzing large databases. With higher precision, machine learning algorithms can forecast voter preferences and behavior (Russell, 2010). It is observed from the study that most of the respondents under this study are educated and are properly aware about the political advertising and campaign strategies used by the political parties to put a positive impact on the voter's opinion. This enables campaigns to deploy resources more effectively by concentrating on crucial battleground states or people who are still uncertain. By examining voter choice and attitude data, artificial intelligence (AI) can be utilized for developing better targeted and successful advertisements. To determine public opinion and responses to campaign messages, machine learning can analyze sentiment in social media and news articles. This enables real-time campaign strategy adaptation. Machine learning can shed light on how well-rounded campaign plans perform as a whole (Pathak & Patra, 2015). This study indicates that voters are satisfied with the emotional analysis that the political parties provided them. Furthermore, a positive correlation has been observed between micro targeting and voter opinion. This suggests that voters are content with the political parties' targeting strategy because they believe these targets will enable the parties to precisely and readily win over the relevant voting area. Voters' opinions of the chatbots and virtual assistants employed by political parties to engage with voters and address their questions and concerns are positively correlated. Campaigns can use it to determine which issues are important to voters and where to concentrate their efforts. The another benefit of AI driven political ads will be that by targeting advertisements to particular voter groups based on their demographics, interests, and online behavior, politicians can assist campaigns in reaching a larger audience.

CONCLUSION

In conclusion, the revolutionary potential of machine learning to improve audience targeting, data-driven decision making, cost effectiveness, and flexibility to digital trends characterizes the impact of machine learning on political advertising. It enables campaigns to interact with voters on a highly personalized level, more correctly forecast voter behavior, and efficiently allocate resources. To ensure the responsible and open use of machine learning in political campaigns, ethical considerations about privacy, bias and regulatory issues must be carefully considered in light of this technological breakthrough. The strategic integration of machine learning remains a crucial element in determining the efficacy and moral behavior of contemporary political advertising as the digital landscape continues to change.

REFERENCES

Amifor, J. (2016). Political Advertising Design in Nigeria, 1960 – 2007. *International Journal of Arts and Humanities.*, *4*(2), 149–163.

Atkin, C., & Heald, G. (1976). Effects of Political Advertising. *Public Opinion Quarterly*, *40*(2), 216–228. doi:10.1086/268289

Atkin, C., & Heald, G. (1976, Summer). Effects of Political Advertising. *Public Opinion Quarterly, 40*(2), 216–228. doi:10.1086/268289

Berelson, B. (1969). The Debates in the Light of Research: A Survey of Surveys. In S. Kraus (Ed.), *The Great Debates, Bloomington, Indiana University Press, 1962; Jay Blumler and Denis McQuail, Television in Politics.* University of Chicago Press.

Bhakri, S. & Shri, D. (2021). *Artificial Intelligence(Ai): Applications And Implications (Ai) For Indian Economy.* Research Gate. doi:10.1729/Journal.27139

Bhattacharya, S. (2018). Dimensions of Political Marketing: A Study from the Indian Perspective. *International Journal of Current Advanced Research, 7,* 11138–11143. doi:10.24327/ijcar.2018.11143.1920

Brei, V. A. (2020). Machine learning in marketing: Overview, learning strategies, applications, and future developments. *Foundations and Trends® in Marketing, 14*(3), 173-236.

Dixit, S. K., & Singh, A. K. (2022). Predicting electric vehicle (EV) buyers in India: A machine learning approach. *The Review of Socionetwork Strategies, 16*(2), 221–238. doi:10.1007/s12626-022-00109-9 PMID:35600566

Garcia-Jimeno, C., & Yildirim, P. (2015). *Matching pennies on the campaign trail: An empirical study of senate elections and media coverage* (University of Pennsylvania working paper). University of Pennsylvania.

Greening, D. W., & Gray, B. (1994). Testing a model of organizational response to social and political issues. *Academy of Management Journal, 37*(3), 467–498. doi:10.2307/256697

Kaur, A. (2019). *Relevance of artificial intelligence in politics.* VirtuInterpress. . doi:10.22495/ncpr_24

Kietzmann, J., Paschen, J., & Treen, E. (2018). Artificial Intelligence in Advertising: How Marketers Can Leverage Artificial Intelligence Along the Consumer Journey. *Journal of Advertising Research, 58,* 263-267. . doi:10.2501/JAR-2018-035

Mithilesh, K., & Choubey, M. (2019). *Political Adverting In India.* Research Gate. doi:10.13140/RG.2.2.16402.30406

Nadikattu, R. (2016). The Emerging Role Of Artificial Intelligence In *Modern Society, 4,* 906-911.

Pathak, S., & Patra, R. (2015). *Evolution of Political Campaign in India., 2,* 55–59.

Russell, S. J. (2010). *Artificial intelligence a modern approach.* Pearson Education, Inc.

Safiullah, M., Pathak, P., Singh, S., & Anshul, A. (2016). Social media in managing political advertising: A study of India. Polish journal of management. *Studies, 13.*

Silva, M., Santos de Oliveira, L., Andreou, A., Vaz de Melo, P. O., Goga, O., & Benevenuto, F. (2020, April). Facebook ads monitor: An independent auditing system for political ads on facebook. In *Proceedings of The Web Conference 2020* (pp. 224-234). ACM. 10.1145/3366423.3380109

Tinkham, S. F., & Weaver-Lariscy, R. A. (1993). A diagnostic approach to assessing the impact of negative political television commercials. *Journal of Broadcasting & Electronic Media, 37,* 377–399.

Verma, K., Bhardwaj, S., Arya, R., Islam, U. L., Bhushan, M., Kumar, A., & Samant, P. (2019). *Latest tools for data mining and machine learning*.

Verma, S., Sharma, R., Deb, S., & Maitra, D. (2021). Artificial intelligence in marketing: Systematic review and future research direction. *International Journal of Information Management Data Insights*, *1*(1), 100002. doi:10.1016/j.jjimei.2020.100002

Zhang, X., Rane, K., Kakaravada, I., & Shabaz, M. (2021). Research on vibration monitoring and fault diagnosis of rotating machinery based on internet of things technology. *Nonlinear Engineering*, *10*(1), 245–254. doi:10.1515/nleng-2021-0019

Zhou, Z. H. (2021). *Machine learning*. Springer Nature. doi:10.1007/978-981-15-1967-3

Chapter 7
Brain Tumor Detection From MRI Images Using Deep Learning Techniques

Anu Sharma

ⓘ https://orcid.org/0000-0002-5164-7547

Moradabad Institute of Technology, Moradabad, India

ABSTRACT

Machine learning and deep learning algorithms are utilized to identify brain tumors in a number of research papers. When these algorithms are applied to MRI images, it takes exceedingly slight time to expect a brain tumor, and the increased accuracy makes it easier to treat patients. The performance of the hybrid Convolution Neural Network (CNN) used in the proposed work to detect the existence of brain tumours is examined. In this study, we suggested a hybrid convolutional neural network followed by deep learning techniques using 2D magnetic resonance brain pictures, segment brain tumors (MRI). In our research, hybrid CNN achieved an accuracy of 98.73%, outperforming the results so far.

INTRODUCTION

The brain is a pivotal component of the human body, overseeing the operation of all other organs and enabling decision-making. (Abd El Kader et al., 2021). The research done in this publication detects whether the brain is healthy or injured by using deep learning techniques. In this work, hybrid CNN was used to classify brain tumors and normal brains (Muhammad, 2018). Artificial neural networks (ANNs) emulate the functioning of the human nervous system through an extensive network of connections. They learn from training data using simple processing units and retain acquired knowledge. To generate the intended output, a model is trained by employing an activation function on input features and hidden layers. Medical imaging in CNNs (convolutional neural networks) encompasses various non-invasive methods for internal body examination (Abd El Kader et al., 2021). Its primary application in the human body is for therapeutic and diagnostic purposes, significantly influencing the effectiveness of treatments and overall health outcomes. The effectiveness of image processing at an advanced level hinges on the

DOI: 10.4018/979-8-3693-2964-1.ch007

crucial process of picture segmentation (Gokila, 2021). In this context, our primary focus has been on isolating brain tumors from MRI scans. This aids medical professionals in pinpointing the exact location of the tumor within the brain. The model we've introduced yielded an impressive accuracy of 98.73%, surpassing the current state-of-the-art results.

LITERATURE REVIEW

According to Isselmou Abd El Kader (2021), deep wavelet auto-encoder model has the capacity to detect and categorise the tumour with high accuracy, quick turnaround, and little loss validation by analysing the pixel pattern of an MR brain picture. In order to detect the presence of brain tumours, P Gokila Brindha (2021) suggested a self-defined Artificial Neural Network (ANN) and Convolution Neural Network (CNN), and their effectiveness is evaluated. Using a convolutional neural network, Arkapravo Chattopadhyay (2022) devised an approach to segment brain tumours from 2D MRI images of the brain. proposed technique using "Python" and "TensorFlow" with "Keras". An innovative method for detecting brain tumours is proposed by Hareem Kibriya (2023) using a collection of deep and manually built feature vectors (FV). The distinctive FV combines sophisticated VGG16-based features with manually produced GLCM-based features. (grey level co-occurrence matrix). Using brain magnetic resonance imaging (MRI), Debnath Bhattacharyya (2011) proposes an image segmentation technique to identify or detect tumours. a series of image segmentation algorithms that produce excellent results when applied to images of brain tumours are proposed. (2020) Md. Ariful Islam For organising input-output data, a graphical user interface has been employed, and an algorithm has been created.

METHODOLOGY

To further validate our work, we used SVM classifier and additional activation algorithms. The following steps are taken in order to apply CNN to the brain tumour dataset:

1. Import the necessary packages
2. Secondly, import the data folder (Yes/No)
3. Assign photos a class label (1 for brain tumour, 0 for no brain tumour).
4. Create 256x256-pixel shapes out of the photos.
5. Make the Image Normal
6. Separate the photos into the test, train, and validation sets.
7. Build the chronological model.
8. Put the model together.
9. Use the train dataset as an example.
10. Use the test photos to evaluate the model.
11. Draw a graph comparing the accuracy during training and validation.
12. Create a confusion matrix comparing actual and expected output.

Figure 1. Workflow of the proposed CNN model

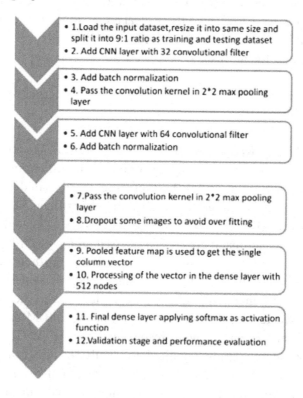

DATA SET

Training, validation, and testing datasets are separated from the image dataset.

Figure 2. Tumours images

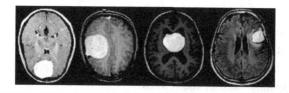

Figure 3. Non-tumours images

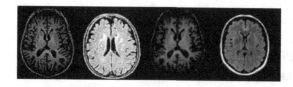

IMPLEMENTATION AND RESULTS

When creating automated systems, it is essential to successfully evaluate the model's performance. Evaluation of training and test sets demonstrates the generalisation capabilities of the system. A simple cross-tabulation of the actual and classified data for each class is called a confusion matrix (CM), is the most popular technique for evaluating a classification model. Several classification measures based on the CM were utilised to evaluate the model's performance in this study. The F1-score is a useful tool for merging recall and precision into one benchmark that includes characteristics from both metrics. In instances of data imbalance, it is frequently utilised. The study's criteria, including precision, recall, accuracy, and, are defined in Equations (1)–(4), respectively.

$$Precision = TP/(TP+FP) \tag{1}$$

$$Recall = TP/(TP+FN) \tag{2}$$

$$Accuracy = (TP+TN)/(TP+TN+FP+FN) \tag{3}$$

$$F1-Score = (TP)/(TP+12(FP+FN)) \tag{4}$$

Table 1. Contrasting models

Final Layer Activation Method	Accuracy (%)	Testing Accuracy (%)	Evaluation of the Model (%)
SVM	14.16	22.63	23.27
Sigmoid	96.53	59.33	69.74
Softmax	**99.84**	94.68	98.73

Table 2. Performance of the proposed CNN model

No	Training Image	Testing Image	Splitting Ratio	Accuracy (%)
1	2199	543	8:2	97.83
2	2473	273	9:1	98.73

According to Table 3, Seetha et al. and Tonmoy Hossain et al. acquired the state-of-the-art findings, and our suggested model outperformed them with an accuracy rate of 98.73%..

Figure 4. Comparison of different models

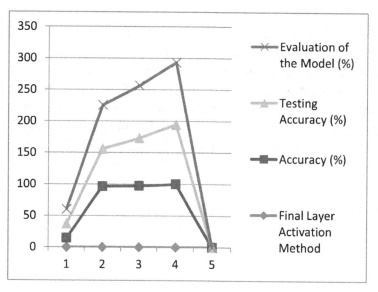

Table 3. Performance comparison

Methodology	Accuracy (%)
Seetha	97.50
Tonmoy et al.	97.87
Proposed CNN model	**98.73**

Figure 5. Performance comparison

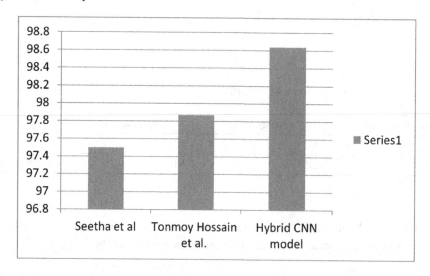

Table 4. FPR and FNR values

Model	FPR	FNR
Google-Net	0.714	0.339
Multimodal	-	1.74
KNN	0.62	0.54
Hybrid CNN model	0.0625	0.031

Figure 6. FPR and FNR

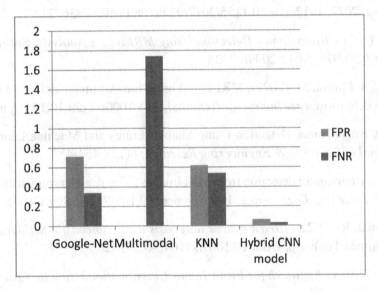

CONCLUSION

One of the greatest methods for assessing the image collection is hybrid CNN. The hybrid CNN currently shows to be the better technique in predicting the existence of brain tumor for the supplied dataset. In our research, hybrid CNN achieved an accuracy of 98.73%, outperforming the results so far. Our CNN-based model will expedite the analysis process by assisting medical personnel in precisely identifying brain tumors in MRI images.

REFERENCES

Abd El Kader, I., Xu, G., Shuai, Z., Saminu, S., Javaid, I., Ahmad, I. S., & Kamhi, S. (2021). Brain Tumor Detection and Classification on MR Images by a Deep Wavelet Auto-Encoder Model. *Diagnostics (Basel), 11*(9), 1589. doi:10.3390/diagnostics11091589 PMID:34573931

Anu, S. (2017). Literature Review and Challenges of Data Mining Techniques for Social Network Analysis. *journal Advances in Computational Sciences and Technology, 10*(5)

Anu, S. (2019). Hybrid Neuro-Fuzzy Classification Algorithm for Social Network. International *Journal of Engineering and Advanced Technology, 8*(6).

Ariful, I. (2020). Brain Tumor Detection from MRI Images using Image Processing. *International Journal of Innovative Technology and Exploring Engineering (IJITEE), 9*(8).

Bahadure, N. B., Ray, A. K., & Thethi, H. P. (2017). Image Analysis for MRI Based Brain Tumor Detection and Feature Extraction Using Biologically Inspired BWT and SVM. *International Journal of Biomedical Imaging, 2017*, 1–12. doi:10.1155/2017/9749108 PMID:28367213

Bhattacharyya, D. (2011). *Brain Tumor Detection Using MRI Image Analysis*. Springer-Verlag Berlin Heidelberg. doi:10.1007/978-3-642-20998-7_38

Chattopadhyay, A., & Maitra, M. (2022). MRI-based brain tumour image detection using CNN based deep learning method. *Neuroscience Informatics (Online), 2*(4), 100060. doi:10.1016/j.neuri.2022.100060

George, D. (2015). Brain Tumor Detection Using Shape features and Machine Learning Algorithms. *International Journal of Scientific & Engineering Research, 6*(12), 454-459.

Gokila, B. P. (2021). Brain tumor detection from MRI images using deep learning techniques. *IOP Conf. Series: Materials Science and Engineering.* IOP Science. 10.1088/1757-899X/1055/1/012115

Jaiswal, A., & Kumar, R. (2022). *Breast cancer diagnosis using Stochastic Self-Organizing Map and Enlarge C4.5*. Multimed Tools Appl. doi:10.1007/s11042-022-14265-1

Kibriya, H. (2020). *A Novel Approach for Brain Tumor Classification Using an Ensemble of Deep and Hand-Crafted Features*. MDPI.

Mahammad, A. B., & Kumar, R. (2022). Machine Learning Approach to Predict Asthma Prevalence with Decision Trees. *2022 2nd International Conference on Technological Advancements in Computational Sciences (ICTACS)*, (pp. 263-267). IEEE. 10.1109/ICTACS56270.2022.9988210

Muhammad, J. (2018). Detection of Brain Tumor based on Features Fusion and Machine Learning. *Journal of Ambient Intelligence and Humanized Computing Online Publication.*

Nalbalwar, R. (2014). Detection of Brain Tumor by using ANN. *International Journal of Research in Advent Technology.*

Nath, R. (2020). Face Detection and Recognition Using Machine Learning Algorithm. *Sambodhi (UGC Care Journal), 43*(3).

Özyurt, F. (2019). *Brain tumor detection based on Convolutional Neural Network with neutrosophic expert maximum fuzzy sure entropy.* Elsevier.

Seyed-Ahmad, F. (2016). *Hough-CNN: Deep learning for segmentation of deep brain regions in MRI and ultrasound.* Elsevier.

Seyyed, R. (2020). *Detection of brain tumors from MRI images base on deep learning using hybrid model CNN and NADE.* Elsevier.

Sharma, A. (2023, January). Exploratory data analysis and deception detection in news articles on social media. *Ain Shams Engineering Journal*, 21. doi:10.1016/j.asej.2023.102166

Chapter 8
Data Privacy and E–Consent in the Public Sector

Abhay Bhatia

https://orcid.org/0000-0001-7220-692X
Roorkee Institute of Technology, India

Anil Kumar

Ajay Kumar Garg Engineering College, India

Pankhuri Bhatia

GRD IMT Dehradun, India

ABSTRACT

In the era of the internet, all face administrative and legal responsibilities obtaining informed consent and safeguarding personal information, with the public growing mistrust to data collection. Moral consent management takes place in account of person's views, subjective norms, and sense of control. When obtaining consent, this chapter aims to combat this cynicism. It accomplishes this by creating a novel conceptual model of online informed consent that combines the TPB with the autonomous authorisation model of informed consent. It is argued logically and is bolstered. As a result, it develops a model for online informed consent that is based on the ethic of autonomy and makes use of theory based on behaviour to enable a method of eliciting agreement that can put interest of users first and then promotes moral the information management and the marketing techniques. This approach also presents an innovative idea, the informed attitude for the validity of informed consent. It also indicates that informed permission may be given against.

INTRODUCTION

Information that has been transformed into a format that works well for processing or transferring within a computer is called data. To be used with contemporary computers and transmission devices, information must first be converted into binary digital form, or data. Data may be used as a single subject or as many subjects. Digital data in its most basic form is referred to as "raw data".

DOI: 10.4018/979-8-3693-2964-1.ch008

The concept of data in the context of computers first appeared in the work of American mathematician Claude Shannon, who is recognized as the father of information theory. He introduced binary digital ideas using electronic circuits and two-value Boolean logic. Binary digit representations provide the foundation for many peripheral devices used in computing today, including disk drives, CPUs, semiconductor memory, and many others. The first kind of computer input that could be utilized for both data and control was punch cards. Hard drives and magnetic tape were the next.

The importance of data in business computers was demonstrated early on by the widespread usage of the terms "data processing" and "electronic data processing," which grew to refer to the entirety of what is now known as information technology. Over the course of corporate computers' history, specialization has occurred, and as corporate data processing grew, so did the demand for specialized data professionals.

Giving consent for the processing of personal data has become commonplace in the digital age. When we download applications onto our smartphones, we almost always consent to the privacy policies associated with the particular services those apps offer. Additionally, when you subscribe to a social networking platform, you cannot opt out of accepting its privacy regulations. Another particularly noticeable trend is asking users of the internet to accept cookies, which are little data files stored on their computers to "remember" their actions and preferences. The realm of privacy and data protection encompasses the (permission to) disclosure of personal data and the subsequent use of such data. The topic of consent's importance in protecting personal data and privacy is examined in this chapter. It is demonstrated how research in the fields of law, ethics, economics, and technology all point to fundamental problems with communication and decision-making constraints that undermine the usefulness of consent in protecting privacy.

Data privacy refers to the protection of sensitive and personal information, ensuring that it is handled, processed, and stored in a secure and confidential manner. In today's digital age, where vast amounts of data are generated, collected, and shared, concerns about data privacy have become increasingly significant. Individuals, businesses, and organizations need to be aware of the importance of safeguarding personal information and respecting the privacy rights of individuals

"e-Consent" is the term for the electronic version of the informed consent procedure that people use to agree to take part in a research project, receive medical treatment, or be in any other circumstance that calls for their participation or personal information. This electronic method, which frequently makes use of digital platforms and technology, expedites the consent process by replacing conventional paper-based consent forms.

PRIVACY

The right to privacy is the ability for people to decide who can access and use their personal information and to keep their communications, actions, and personal information hidden from prying eyes. It includes the idea that people have the right to preserve a certain amount of privacy and independence in their private lives. Numerous legal systems, moral precepts, and cultural standards all acknowledge privacy as a fundamental human right. Figure 1 shows how the flow for privacy takes place.

Informational Privacy

This relates to the insurance of individual data, for example, one's name, address, telephone number, monetary subtleties, wellbeing records, and different information that can be utilized to distinguish or describe a person.

Figure 1. Privacy process

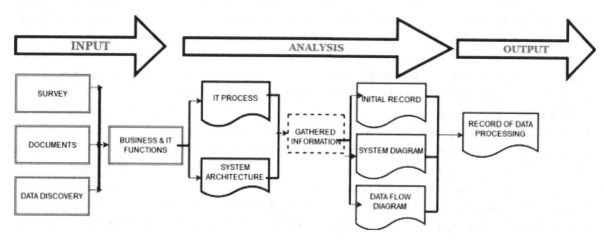

Physical Privacy

This perspective includes the option to control admittance to one's body and individual space. It incorporates the option to be liberated from unjustifiable actual interruption or reconnaissance

Communicational Privacy

This includes the option to control one's interchanges and discussions. It incorporates assurances against unapproved capture attempt or observing of discussions, messages, and different types of correspondence.

Privacy of Personal Choices

People reserve the privilege to pursue individual decisions, including way of life decisions, without unjustifiable obstruction or judgment from others. This stretches out to issues like strict convictions, political affiliations, and different parts of individual personality.

Privacy in the Digital Age

Concerns about privacy have grown to include online activities, social media use, and the collection and analysis of digital data with the advent of digital technologies. Computerized security includes defending individual data shared on the web and safeguarding against unapproved access or abuse.

Surveillance and Government Intrusion

Security likewise includes insurance from outlandish observation and interruption by legislatures or different elements. This incorporates worries about mass reconnaissance programs, information assortment rehearses, and the harmony between public safety and individual protection.

Data Protection and Security

A crucial aspect of privacy is ensuring the safety of personal data. People and associations are liable for carrying out measures to shield delicate data from unapproved access, information breaks, and digital dangers.

Cultural and Ethical Dimensions

Protection standards can shift across societies and are impacted by moral contemplation. Respecting the privacy of others is frequently regarded as a fundamental ethical principle, and societies may have varying standards for what constitutes private behaviour.

Legal Frameworks

Numerous nations have laid out legitimate structures and guidelines to safeguard protection privileges. These may incorporate information insurance regulations, protected arrangements, and explicit guidelines administering different businesses and areas. Figure 2 explains about it.

Figure 2. How legislation works

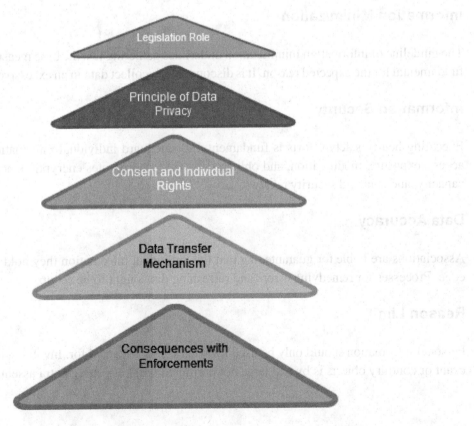

The concept of privacy is dynamic, changing as technology advances, society shifts, and laws are established. In the digital age, finding a balance between the demands of security, openness, and personal freedom is a constant struggle. Privacy norms are created and upheld by individuals, groups, and governments.

DATA PROTECTION

The term "data protection" refers to the collection of procedures, guidelines, and laws intended to preserve the availability, integrity, privacy, and confidentiality of sensitive and personal data. It entails the appropriate handling and management of data to guarantee that people's right to privacy is upheld and that information is used in a morally and legally compliant way. In a number of industries, including business, healthcare, education, and government, data privacy is crucial. The following are the key elements of it:

Information Assortment and Handling

The purpose for collecting personal data must be clearly stated by the organization, and only lawful and legitimate reasons should be used to process it. Informed assent from people is many times a principal necessity.

Information Minimization

The guideline of information minimization underlines gathering just the base measure of individual data fundamental for the expected reason. It is discouraged to collect data in an excessive or unnecessary way.

Information Security

Executing hearty safety efforts is fundamental to safeguard individual information from unapproved access, exposure, modification, and obliteration. This incorporates encryption, access controls, secure capacity, and standard security reviews.

Data Accuracy

Associations are liable for guaranteeing that the individual information they hold is exact and cutting-edge. Processes for remedying errors and refreshing data ought to be set up.

Reason Limit

Personal information should only be used for what it was collected for. Involving information for irrelevant or contrary objects is by and large not permitted without getting extra assent.

Information Maintenance and Cancellation

Associations ought to lay out clear arrangements with respect to the maintenance and erasure of information. Risks exist and it could be construed as a violation of data protection principles to store data for longer than is necessary.

Individual Freedoms

Individuals are typically granted certain rights under data protection laws, such as the right to access, correct, and request the deletion of their data. Associations should work with the activity of these privileges.

Transparency

People ought to be educated about how their information is being utilized, who approaches it, and what measures are set up to safeguard it. Straightforward correspondence constructs trust among associations and people.

Accountability

Data protection practices are the responsibility of organizations. This incorporates having archived approaches, leading protection influence evaluations, and making liability regarding the moves of outsiders with whom information is shared.

Lawful Consistence

Associations should agree with significant information security regulations and guidelines pertinent in their purviews. This could be the General Data Protection Regulation (GDPR), the Health Insurance Portability and Accountability Act (HIPAA), or other laws that are specific to certain industries and regions. Figure 3

Information Break Reaction

In case of an information break, associations are expected to have reaction plans set up to limit the effect, tell impacted people, and follow lawful revealing commitments.

Many nations have enacted comprehensive data protection laws to regulate the handling of personal information because data protection is a global concern. Consistence with these regulations isn't just a legitimate necessity yet in addition adds to building entrust with people and cultivating a dependable and moral way to deal with information the board.

CONSENT

It is generally accepted that consent is only legitimate in the context of privacy and data protection if it is given voluntarily and knowingly. A fundamental component of informed consent is that the individual

Figure 3. GDPR

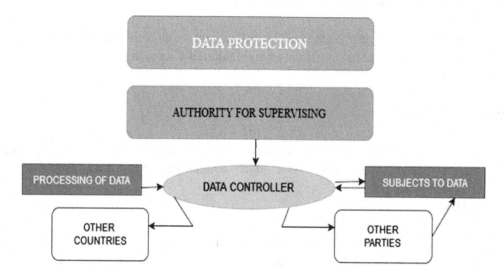

being asked for consent must be fully informed about the specifics of what it entails and somewhat (made) aware of the potential implications of giving that consent. Consent choices [Table 1] in the context of IT relate to the exchange and publication of personal data.

In the context of automated information processing, a number of important criteria can be recognized from an ethical and legal standpoint.

Informed Consent

A sort of assent that has acquired unmistakable quality with the presentation of GDPR is educated assent. Informed assent ensures that people obviously comprehend the information handling exercises that occur and the related dangers and privileges.

Informed agree expects associations to give people definite data about the information assortment, including:

• Purpose
• Data recipients
• Retention period
• Potential international transfers

Providing users with a comprehensive privacy policy that outlines how their data will be used, shared, and protected is an example of informed consent. Purpose Data recipients Retention Period Potential international transfers Individuals maintain control over their information and hold businesses to a higher standard of data privacy with informed consent.

Implied Consent

Inferred assent, otherwise called inactive assent or quit assent, is another sort. Dissimilar to unequivocal assent, suggested assent is induced from an individual's activities or conduct. It accepts that people have given assent when they take part in specific exercises that innately include information handling.

When users visit a website and their IP addresses are recorded for the purpose of traffic analysis, this constitutes implied consent. Although explicit consent may not have been directly obtained, users' willing access to the website presupposes that they are aware of the data collection practices.

Treat arrangements are another genuine model. When a website's cookie banner or privacy policy informs visitors about the use of cookies, their continued use is frequently interpreted as their implied consent for the website to store and access these data points on their device.

Suggested assent is much of the time utilized in circumstances where the information regulator or processor expects that people would by and large consent to the information handling action or while the handling is viewed as essential for the presentation of an agreement or the arrangement of a help. Passive consent, on the other hand, should be based on providing individuals with clear and transparent information about their rights, the purposes of data processing, and the ability to opt out.

Explicit Consent

One more typical kind of assent is unequivocal or express, assent. This type of agree expects people to unequivocally give their consent to their information to be gathered and handled.

Express assent is regularly looked for touchy or unique classifications of individual information, like wellbeing data, racial or ethnic beginning, strict convictions, or political affiliations. For example, an internet-based wellbeing stage would look for express assent from clients before they accumulate an individual's clinical history to give customized wellbeing suggestions.

At the point when they acquire express assent, associations ensure that people are completely mindful of the particular information being gathered and its planned purposes. A kind of assent assumes an imperative part in information protection since it advances straightforwardness, individual independence, and consistence with guidelines.

Granular Consent

Granular assent underscores that it is so vital to offer people decisions and command over unambiguous parts of information handling. Individuals are able to grant or withhold permission for various purposes or categories of data sharing with granular consent. For example, a versatile application might demand separate agrees to get to different elements:

- Gadget area
- Sending pop-up messages
- Gathering perusing history
- Empowering outsider information sharing

At the point when associations carry out granular assent systems, they regard clients' independence and enable them to choose the particular information utilizes they see as adequate. It is a user-first privacy strategy that ensures that their privacy expectations are met and enables them to make informed decisions about particular aspects of data processing.

Ongoing Consent

Progressing assent, otherwise called dynamic assent, perceives that people's inclinations and conditions might change after some time. This command consistently looks for recharging or reconfirmation of agree to ensure that people stay mindful of the information handling exercises and have the chance to adjust their assent decisions.

Continuous assent is especially important in long haul connections where information handling happens over a lengthy period. A good illustration of this is when a cloud storage service might remind users to check and update their consent settings for data backup and synchronization on a regular basis. A continual dialogue between individuals and data controllers or processors is facilitated by this kind of consent. It allows for simple adjustments and is in line with changing preferences for privacy.

Normal instances of progressing assent incorporate strategies found in membership administrations, yet they can likewise be available in a client's record the executive's devices, protection strategy the board, and information maintenance arrangements.

Presumed Consent

For presumed consent, consent is based on a legal or regulatory provision, social expectations, or the particular context in which data processing takes place. It recommends that people are attempted to have assented to specific information handling exercises except if they expressly object or quit. Assumed assent is commonly utilized in circumstances where there is a convincing public interest or lawful reason for information handling.

It's critical to take note of that assumed assent ought to be applied carefully and in consistence with appropriate regulations and guidelines. The assumption of assent should be founded on a clear cut legitimate or administrative system, with legitimate shields set up to safeguard people's protection freedoms.

Conditional Consent

Contingent assent is given with explicit circumstances or constraints on the understanding. To put it another way, individuals consent to the processing of their data for specific purposes or under certain conditions, but not for others.

For example, an overview directed by an association might look for restrictive assent, and members consent to their reactions being utilized for research purposes however not for showcasing or outsider information sharing. Restrictive agree permits people to have more command over the degree and extent of their information use, and it empowers them to define limits and determine their inclinations.

A few normal information assortments rehearse that might require contingent assent strategies include:

- Research studies
- Showcasing interchanges
- Information offering to outsiders
- Cross-line information moves
- Personalization and profiling

REQUIREMENT OF CONSENT

Valid Consent Criteria's

With regards to security and information insurance, it is generally expected that assent is just legitimate when it is educated assent. A fundamental part of informed assent is that the individual who is requested assent ought to be appropriately educated regarding what precisely the person in question is consenting to and is somewhat (made) mindful of the outcomes such assent might have. With regards to data innovation, assent choices concern the sharing and exposure of individual information. From a moral and lawful point of view, a few pivotal models can be distinguished with regards to mechanized data handling.

Standards for informed assent might incorporate models with respect to the individual who assents (is the individual a grown-up, proficient and skilled to assent?) what's more, models on the most proficient method to give assent (for case, is the assent composed or oral, is the assent incomplete or full, is the conviction behind the assent areas of strength for sensibly is the assent a free choice?).

Besides, data on which safety efforts are taken, data about who is handling the information and who is responsible as well as data on client freedoms and how they can be practiced are possibly significant for the assent choice. In addition, it is by and large held that the data given ought to be explicit, adequately nitty gritty, justifiable, solid, precise and open. Aside from circumstances in which clearly a portion of these models are not met, (for example, missing strategies, or strategies in something else entirely), it isn't clear for a considerable lot of these models when they are adequately met. Table 1 provides a summary of these requirements for valid informed consent. This outline incorporates the legitimate models for the legitimacy of person assent, which will be examined in more detail in the following subsection.

Table 1. Criteria for consent

Criteria for consent	**Who consent's**	• Is the person whom you consent an adult? And If not, then parental consent? • Whether the person who is so far making consent is capable off? If not, then there must be a legal representative? • Is the person legally competent if not than there must be legal representative?
	How to consent	• Consent is documented? • Type of consent means partial or full? • Consent conviction must be strong enough? • It must be independent and up to latest date?
Consideration it well as decision to the consent	**Which information is needed?**	• Data Collection must be clear enough? • Purpose to collect must be clear? • Measures to security must be looked for. • Processing to and accountability must be clear enough? • Rights must be exercised
	How the needed information is to be provided?	• Is information provided must appropriate, clear as well specific? • Whether it is understandable? • Is it reliable and accurate? • Is provided information accessible?

Data Protection and Legislation

In the EU, the lawful system for educational security and individual information assurance is laid out by the Information Security Directive,1 which will be supplanted by the Overall Information Insurance Guideline (GDPR) starting around 25 May 2018. 2 Both these regulations contain the vast majority of the rules referenced in Table 1. In European information security regulation, assent assumes an imperative part, since it is one of the most significant reasons legitimizing individual information handling. The Information Security Mandate also, its replacement, the GDPR, specify that individual information might be handled in light of the unambiguous assent of the individual to whom the information relate. Assent under EU information assurance regulation should be openly given, explicit and informed to be legitimate. Besides, EU information assurance regulation sets prerequisites for the structure in which assent is given: assent should be unambiguous and, in specific cases, express. For agree to qualify as "unambiguous" there should be no vulnerability about the expectation of the individual to whom the information relate. This aim can be communicated as an activity conveyed out by the individual who should give the assent (e.g., ticking a checkbox), yet in addition through a broader activity did by the individual (e.g., strolling through an entryway with a sign above it saying "assuming that you enter you agree to having your image taken"). It will never be considered unambiguous consent to infer consent from inaction. In any case, deducing assent from an activity that isn't explicitly or exclusively focused on assent is conceivable, considering that it is unambiguous.

The obligation to prove anything for the unambiguity of the assent settles upon the association dependable. In the GDPR, the necessities for assent are reinforced by expecting that assent is communicated "either by a proclamation or an unmistakable governmental policy regarding minorities in society". This will probably leave less room for the induction of assent.

The higher standard of explicit consent must be met in order to legitimately process sensitive special categories of data, such as health data. For express assent, people should be approached to concur or contradict a specific use or exposure of their own data and they need to answer effectively to the inquiry.

Childrens Consents

COPPA represents the Youngsters' Internet based Security Insurance Act. A US government regulation tends to the assortment and treatment of individual data of youngsters younger than 13 on the web. COPPA was established in 1998 and is upheld by the Government Exchange Commission (FTC). It includes the following:

Parental Consent

Verifiable parental consent must be obtained before collecting, using, or disclosing personal information from children on websites and online services targeted towards children under the age of thirteen or by individuals who have real knowledge that they are collecting information from children.

Notice and Privacy Policy

COPPA-covered websites and online services are required to give parents a clear and intelligible notice about their information practices, including the kinds of information they collect, how they use it, and their privacy policies.

Limited Collection of Personal Information

The law restricts the collection of personal information from children to what is reasonably necessary for the purpose of the online service. It also requires the maintenance of the confidentiality, security, and integrity of the information collected.

Parental Rights

Parents have the right to review and delete the personal information collected from their children. They also have the right to refuse to permit further collection or use of their child's information.

Exceptions

There are some exceptions to the rule, such as information collected for the sole purpose of providing support for the internal operations of the website or service.

Safe Harbor Programs

COPPA allows the development of industry self-regulatory programs, known as safe harbour programs, to offer additional methods for compliance.

Penalties for Non-Compliance

Violations of COPPA can result in significant penalties. The FTC has the authority to seek civil penalties of up to $43,280 per violation.

COPPA aims to empower parents and protect the privacy of children online by placing certain requirements on operators of websites or online services that are directed toward children or knowingly collect personal information from children under 13. The law has been updated over the years to adapt to changes in technology and the online landscape. Organizations that cater to a young audience or knowingly collect data from children should be aware of and comply with COPPA regulations.

Childrens and Their Consent in India

Over the course of more than 20,000 consultation submissions and several dozen in-person consultations, it has become clear that this position requires some nuance. The draft Digital Personal Data Protection Bill, 2022 (the "Draft") was published for consultation on November 18, 2022. It specified the age of majority as 18 years old, imposed onerous and inviolable restrictions prohibiting various widely defined actions.

In a computerized economy which is progressively centred around the youthful for administrations going from schooling to diversion and a youthful populace which is progressively on the web, the contribution of items and administrations to youngsters, and to the suitable degree, guaranteeing that specific sorts of material is centred around them, isn't just down to earth, yet additionally a need. Age gating, adult content and interaction filtering, and child protection and mental health services are just a few examples of other types of quasi-commercial processing.

India has long followed, with specific exemptions, the methodology of treating all people underneath the time of larger part (solidly fixed at 18 since 1875), as kids. These people are considered unequipped for contracting, and assent from guardians or gatekeepers is depended on for most sorts of communications with them. While restricted exemptions exist, for example, for contracting or associations for their benefit, neither they, nor one-time parental assent, structure a powerful, or even useful means to empower commitment with people underneath the age of 18 by intelligent stages.

The Demonstration, like the Draft hard codes that anybody under 18 years old would be a child. Worldwide norms, for example, Europe's Overall Information Insurance Regulation and the California Customer Protection Act accommodate reviewed approaches.

Essentially, similar to the Draft, the Demonstration keeps on expecting that assent from a parent or a legitimate watchman (hereinafter "Parent") be taken for all handling of individual data connecting with a kid in an unquestionable way which will be recommended under rules (hereinafter, "Consent"). Nonetheless, the Demonstration extends the commitment on Information Fiduciaries to get evident assent of the parent or legal watchman and requires such agree to be gotten preceding the handling of individual information of an individual with inability for whom a gatekeeper has been delegated.

THEORY OF PLANNED BEHAVIOUR

The Hypothesis of Arranged Conduct (TPB) is an augmentation of the Hypothesis of Contemplated Activity (TRA) (Fishbein and Ajzen 1975, Ajzen and Fishbein 1980). The two models depend on the reason that people make sensible, contemplated choices to take part in unambiguous ways of behaving by assessing the data accessible to them. The presentation of a way of not entirely settled by the singular's aim to participate in it (impacted by the worth the singular puts on the way of behaving, the straightforwardness with which it very well may be performed and the perspectives on soul mates) and the discernment that the way of behaving is inside his/her control. In RA a TPB model in view of mentalities, social help, self-viability and goal was reasonably fruitful in foreseeing and making sense of self administration of joint pain (Strating et al 2006). While no approved polls are accessible, an extensive manual for creating proportions of TPB parts is given in Ajzen (1991). A test in TPB estimation is the trouble in conceptualizing and catching mentalities.

Underlying beliefs dictate the attitude, viewpoint views, and PBC components. An individual's salient behavioral beliefs—beliefs that indicate the behavior's expected outcomes—determine their attitude. (e.g., taking exercise will lessen my risk of heart disease). Normative beliefs, or perceptions of salient people' preferences for whether or not one should engage in an activity, are what shape perspective perspectives (e.g., my family think I should take exercise). PBC is predicated on beliefs about having easy access to opportunities and resources needed to carry out the behavior (e.g., I can easily get to a place where I can exercise). [Figure 4]

Figure 4. Planned behaviour
(James C. et al., 2023)

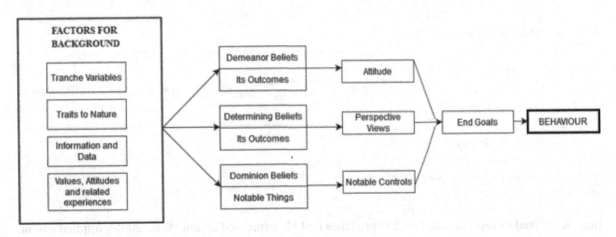

In this sense, according to the TPB, people are likely to engage in a well-being behavior if they believe that the behavior will lead to particular outcomes that they value, if they believe that people whose opinions they respect believe they should follow the behavior, and if they believe they have the necessary resources and opportunities to carry out the behavior.

Ajzen (1991) inspected the discoveries of in excess of twelve experimental trial of the TPB. In the majority of these examinations the expansion of seen social control to the TRA brought about a huge improvement in the expectation of goals as well as conduct. All the more as of late, Godin and Kok (1996) checked on the consequences of 54 experimental trial of the TPB inside the area of wellbeing conduct, and arrived at extensively comparable resolutions. It is almost certainly the case that for most ways of behaving that are probably going to bear some significance with social researchers, it merits utilizing the TPB instead of the TRA.

RELATION BETWEEN BEHAVIOUR AND AUTHORIZATION (CONSENT)

There is no commonly acknowledged meaning of the peculiarity that we call "conduct". For the motivations behind the current investigation, the adaptation that expressive brain science offers are embraced. This expresses that conduct is an endeavour with respect to a person to achieve some situation — either to alter that situation or to keep up with it.

The demonstration of approving gives official authorization to or formal endorsement to an activity or an endeavour (Oxford English Word reference, 2019). Authorisation achieves a state change. This state change concerns authorization or endorsement: what once didn't have consent or endorsement, following authorisation, gains authorization or endorsement. It follows that authorisation is a type of conduct. [Figure 5]

THEORY OF PLANNED BEHAVIOUR AND INFORMED CONSENT

Various speculative advances have been put out in relation to informed consent. On the one hand, some advances attempt to illustrate the structure of consent in terms of the extent to which a person's consent

Figure 5. Behaviour and consent

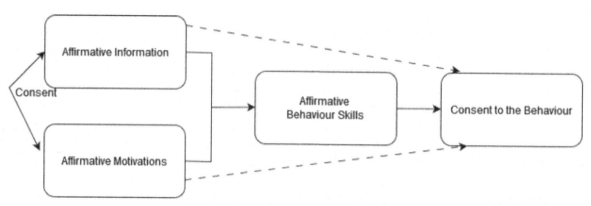

may be utilized or repurposed. These approaches include expressed assent (Win, 2005), implied assent (Hofmann, 2009), cover assent (Ploug and Holm, 2015a), expansive assent, and cover assent. There are models that attempt to handle the philosophy of assent on a whole other level. These models are particularly persuasive in the current works since they aim to address the characteristic idea of assent. They include the successful assent model (Faden and Beauchamp, 1986), the AA model (Faden and Beauchamp, 1986), the divulgence model (Faden and Beauchamp, 1986; Friedman et al., 2000; Marta, 1996; Sim and Wright, 2000), and the fair exchange model (Mill operator and Wertheimer, 2011). The AA model is of particular concern to this study because of its close connection to information security regulations, which are discussed in the "Presentation" chapter. This section provides a brief description of this and the other ontological models to contextualize the AA model.

The divulge model is the customary medico-lawful model, and it recognizes five constituents of assent: exposure, cognizance, intentionality, capability and arrangement (Faden and Beauchamp, 1986; Friedman et al., 2000; Marta, 1996; Sim and Wright, 2000). Exposure alludes to the sufficiency of the data gave to the member. Cognizance concerns the comprehension member might interpret the data gave. Capability concerns the member's capacity to settle on a level headed choice, and incorporates mental as well as friendly and lawful standards (for example age limits) (Faden and Beauchamp, 1986). Wilfulness connects with the shortfall of control in regards to the choice. The last component, understanding, is some of the time precluded as a component and, in different examinations, it is given an alternate mark, being differently alluded to as assent, choice, cooperation or arrangement (Faden and Beauchamp, 1986; Friedman et al., 2000; Marta, 1996; Sim and Wright, 2000).

The compelling assent model, proposed by Faden and Beauchamp (1986), intently looks like the divulge model, in which a structure of hierarchical and institutional guidelines, strategies and methodology shape the looking for of assent. It doesn't depend upon the independence of the individual. Rather, it is worried about lawfully and institutionally powerful frameworks of cycles and guidelines that oversee the looking for of assent and direct the way of behaving of the assent searcher (Faden and Beauchamp, 1986).

In the fair exchange model (Mill operator and Wertheimer, 2011), revelation, cognizance, skill, wilfulness and nonappearance of duplicity are key viewpoints yet, dissimilar to the divulgence model, they are setting delicate. What contains decency is subject to the gamble benefit profile; more noteworthy endeavors are expected to advance and confirm cognizance as the adverse results to people increment.

The AA model of assent, proposed by Faden and Beauchamp (1986), is absolutely legitimate in idea and liberated from regulating conditions which might be applied for down to earth or strategy reasons. The model presents that educated assent is inseparable from independent authorisation, for example that independence and authorisation are its constituent components. They characterize independence as comprising of significant comprehension, non-control and deliberateness.

As indicated by Faden and Beauchamp (1986), significant comprehension requires "dread of the multitude of material or significant depictions — however not all the applicable (and absolutely not every single imaginable) description". They depict how the significance of a portrayal may still up in the air by "the degree to which the depiction is material to the individual's choice to approve" (p. 302), which they say is totally abstract. A deliberate activity, as per Faden and Beauchamp (1986), is one "willed as per an arrangement" (p. 243), yet it additionally incorporates endured acts. Endured acts are those that might be undesirable or unwanted. Non-control alludes to there being no outer controls on the activity: an outside controlling impact would nullify independence.

Somewhat, this resounds with the partner hypothesis way to deal with showcasing morals. As a regulating hypothesis, the partner hypothesis fights that chiefs have a trustee relationship with all partners and when the interests of partners struggle, the ideal equilibrium should be accomplished. In its experimental structure, it really states that a business' monetary achievement requires every one of partners' inclinations to be given legitimate thought and that strategies ought to be embraced to impact the best equilibrium among them. Cohen (1995) proposes that assent is naturally connected with the idea of stakeholder ship — that what an individual or a gathering of people would agree to is a significant part of partner interest and that thoughts of stakeholder ship would profit according to the viewpoint of assent hypothesis — for which specialist independence is a focal standard.

However, in stakeholder theory, the interests of the individual can be subdued to the interests of the wider collective group of stakeholders to achieve the optimal balance so while their autonomy may be respected. Since the autonomy principle permits action against individuals in order to assist others, consequentialism is not typically associated with it (Cummiskey, 1990).

Informed Consent

Informed consent is a moral idea initial, a lawful idea second, lastly, a formal regulatory interaction. In numerous nations, informed assent regulations change by state and by conditions. A few cases require a marked record showing your educated assent, while others just require a verbal understanding. Yet, in all cases, medical care suppliers need to track the cycle.

Consent codes and regulations safeguard the two parental figures and care beneficiaries. They assist with laying out trust in your guardian while regarding your independence as a consideration searcher. They help to forestall misconceptions and omissions in correspondence that could prompt you being discontent with the consideration that you get. They ensure that you comprehend what's in store, including the possibilities of not exactly ideal results.

Most medical services suppliers mean well to treat care searchers morally, including speaking the truth about their choices, their dangers and their forecast. In any case, here and there useful limits and human mistake ruin these expectations. That is where formal cycles and regulations can help. These cycles and regulations have developed over the long run to consider a portion of the hindrances we've met en route.

Table 2. Models in consent [Gary Burkhardt et al 2022]

Models	Descriptive	Consists	Taken from
Autonomous authorisation (AA)	Consent is an act of authorisation that is having complete autonomy to it.	(1) Understandability (2) Intentions (3) Absence of controlling (4) Authentication	Faden and Beauchamp (1986)
Consent of Effectiveness	Balancing with AA model requirements with the institutions concrete concerns with its priorities. The rules are applied to aggregate to serve efficiency as well as its effectiveness	(1) Disclose (2) Understanding to it (3) Competences (4) Agreement's (5) Volunteering	Faden and Beauchamp (1986)
Fairness transaction	Fair terms of cooperation for each party that represent the circumstances of the activity for which agreement is granted are the foundation for the validity of consent transactions.	(1) Fair (2) Disclosures (3) Assessing of understanding (4) Volunteering (5) Cooperating (6) Agreement/authorisation	Miller and Wertheimer (2011)
Disclosure	A competent individual who is aware of the information being released and willingly consents is given access to pertinent information.	(1) Disclosure (2) Understanding (3) Competence (4) Agreement/authorisation (5) Voluntariness	Friedman et al. (2000) and others

Figure 6. How informed consent can be achieved (if possible)

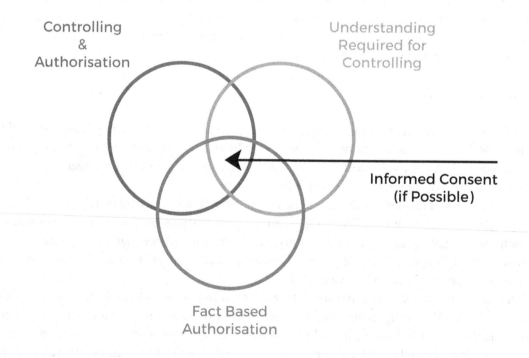

CONCLUSION

A model for online informed assent has been fostered that can be utilized to increase moral individual data assortment and related promoting rehearses by making a client driven model for assent exchanges which joins social hypothesis, explicitly the TPB, with the AA model of informed assent. This bound together model makes a clever hypothetical stage that makes sense of the inward instruments of online assent ways of behaving and which, contingent on one's hypothetical stance, can be seen through the regulating moral focal point of either consequentialism or deontology. A subjective strategy has been taken on in which the model is built through sensible thinking and afterward delineated utilizing the case of following treats. It is shown that, under specific circumstances, (I) the comprehension builds in the AA model of assent lines up with two develops in the TPB: mentality and seen conduct control. (ii) Non-control inside the AA model lines up with the emotional standard in the TPB. (iii) Aim is normal to the two models, though with unpretentious however tremendous contrasts in importance. At long last, (iv) the authorisation component of the AA model compares to the conduct part in the TPB model.

The model likewise presents an original develop, the educated disposition, which should be available for informed agree to be substantial. An educated mentality to an arrangement is framed in the event that an individual comprehends (a) all parts of the understanding that are critical to them and (b) the fundamental components of the understanding.

Specific moral importance in the data the executives and promoting spaces is the assurance that it is hypothetically feasible for a web client to share individual information reluctantly, and for agree to stay informed, in the event that it is an endured symptom of some more noteworthy expected reason, if it is given deliberately and with significant comprehension. In any case, the reluctance of assent arrangement for information sharing exercises sits precariously with the story of individual independence and security rehearses, despite the fact that it very well might be tempered assuming lenience is estimated on a continuum on which there are levels of lenience past which assent couldn't be viewed as credible.

FUTURE WORK AND LIMITATIONS

Further examination which investigations the eagerness of assent in an assortment of online individual data revelation settings, and how much such exposure is basically endured, would explain the degree to which more noteworthy exploration endeavors ought to be coordinated towards dissecting on the web assent through another moral focal point of lenience versus the conventional focal point of readiness.

As significant comprehension is a center part of informed assent, one more productive road of future examination is research manners by which genuine comprehension can arrive at significant comprehension by overseeing different parts of the brought together model.

Operationalisation of the educated demeanor build would require further exploration endeavors at the hypothetical level. In the event that understanding an arrangement would require just the comprehension of the objective realities, the take-off that this would address from an unadulterated AA model operationalisation might predicate a changed type of assent that lies between the compelling model of assent and the AA model.

REFERENCES

Ajzen, I. (1985). *From intentions to actions: A theory of planned behavior. Action control.* Springer.

Ajzen, I. (1991). The theory of planned behavior. *Organizational Behavior and Human Decision Processes, 50*(2), 179–211. doi:10.1016/0749-5978(91)90020-T

Ajzen, I. (2002). Perceived behavioral control, self-efficacy, locus of control, and the theory of planned behavior 1. *Journal of Applied Social Psychology, 32*(4), 665–683. doi:10.1111/j.1559-1816.2002.tb00236.x

André, Q., Carmon, Z., Wertenbroch, K., Crum, A., Frank, D., Goldstein, W., Huber, J., Van Boven, L., Weber, B., & Yang, H. (2018). Consumer choice and autonomy in the age of artificial intelligence and big data. *Customer Needs and Solutions, 5*(1), 28–37. doi:10.1007/s40547-017-0085-8

Bandura, A. (1986). *Social foundation of thought and action: A social-cognitive view.*

Bandura, A. (1991). Social cognitive theory of self-regulation. *Organizational Behavior and Human Decision Processes, 50*(2), 248–287. doi:10.1016/0749-5978(91)90022-L

Burkhardt, G., Boy, F., Doneddu, D., & Hajli, N. (2022). Privacy Behaviour: A Model for Online Informed Consent. *Journal of Business Ethics, 186*(1), 237–255. doi:10.1007/s10551-022-05202-1

Cummiskey, D. (1990). Kantian consequentialism. *Ethics, 100*(3), 586–615. doi:10.1086/293212

Electronic Privacy Information Center. (2016). *Children's Online Privacy Protection Act (COPPA).* EPIC. https://epic.org/privacy/kids/

Faden, R. R., & Beauchamp, T. L. (1986). *A history and theory of informed consent.* Oxford University Press.

Friedman, B., Felten, E., & Millett, L. I. (2000). Informed consent online: A conceptual model and design principles. In: *University of Washington Computer Science & Engineering Technical Report 00–12–2.* University of Washington.

International Association of Privacy Professionals. (2023). *Government receives more than 20K submissions on India's proposed DPDPB.* IAPP.

Iwaya, L. H., Fischer-Hübner, S., Åhlfeldt, R.-M., & Martucci, L. A. (2018). *mHealth: A Privacy Threat Analysis for Public Health Surveillance Systems.* 2018 IEEE 31st International Symposium on Computer-Based Medical Systems (CBMS), Karlstad, Sweden. 10.1109/CBMS.2018.00015

Iwaya, L. H., Li, J., Fischer-Hübner, S., Åhlfeldt, R. M., & Martucci, L. A. (2019, August 21). E-Consent for Data Privacy: Consent Management for Mobile Health Technologies in Public Health Surveys and Disease Surveillance. *Studies in Health Technology and Informatics, 264,* 1223–1227. doi:10.3233/SHTI190421 PMID:31438120

Parker, D. M., Pine, S. G., & Ernst, Z. W. (2019). Privacy and Informed Consent for Research in the Age of Big Data. *Penn State Law Review, 123*(3), 4. https://elibrary.law.psu.edu/pslr/vol123/iss3/4

Ryan, S., & Carr, A. (2010). Chapter 5 - Applying the biopsychosocial model to the management of rheumatic disease. K. Dziedzic & A. Hammond, (eds). Rheumatology. Elsevier. doi:10.1016/B978-0-443-06934-5.00005-X

Schermer, B. W., Custers, B., & van der Hof, S. (2014). The crisis of consent: How stronger legal protection may lead to weaker consent in data protection. *Ethics and Information Technology*, *16*(2), 171–182. doi:10.1007/s10676-014-9343-8

Shroff, C., Goswami, A., Prabhu, A., Mohapatra, A., Sengupta, A., Sharma, M., Soni, A., Hussain, S., & Tiwari, S. (2023). *Children and Consent under the Data Protection Act: A Study in Evolution.* Cyril Amarchand Blogs.

Chapter 9
Ensuring Privacy and Security in Machine Learning:
A Novel Approach to Efficient Data Removal

Velammal

Anna University, Chennai, India

N. Aarthy

Anna University, Chennai, India

ABSTRACT

Modern systems generate vast amounts of data, creating complex data networks. Users prioritize the safety, security, and privacy of their data. This project focuses on efficiently removing or erasing data from the machine learning model upon user request, addressing privacy concerns. Under GDPR, users can request the deletion of sensitive data from both user records and the machine learning model that has processed the data. Additionally, the project employs the SISA approach to address errors and attacks by dividing the dataset into shards and implementing a slice-based ensemble learning technique. Each shard functions as an independent model, and after training, a majority voting approach aggregates these models into a final model. Experimental results demonstrate reduced retraining costs, as only the remaining slices are retrained instead of the entire model.

INTRODUCTION

Machine Unlearning is a domain which is contradictory to Machine Learning. In Machine Learning, a machine will be made to learn the data set that is provided and produce a desired model with high accuracy whereas, in Machine Unlearning, machine will be made to unlearn the data that the model had previously learned. It is necessary for unlearning research (Ma,2022) to develop an algorithm that can take a trained machine-learning model as input and produce a new one without the requisite data. Retraining the model from scratch without the need to relearn the

DOI: 10.4018/979-8-3693-2964-1.ch009

training data is a fundamental tactic. Unlearning research aims to reduce the high computational cost associated with this. With the rising importance of data privacy, data is becoming a major concern. Data privacy is all about restricting the access to personal data which involves deciding who should not have access to it. Deletion of accounts simply delinks the data from database but for complete privacy the proposed model must ensure that the system forgets the data once and for all. Users desire systems to forget particular data for a variety of reasons, including its lineage. Users who are concerned about new privacy dangers in a system frequently want the system to forget their data and history. A detector must forget the injected data if an attacker taints it by adding manually generated data to the training set. This is necessary for the detector to regain security. A user can reduce noise and inaccurate items to ensure that a recommendation engine provides effective recommendations. Let's break down the problem in more detail so it is easy to see how it varies from other privacy definitions. The user-provided information that is asked to be unlearned is represented by the letter d. It is imperative to devise an unlearning strategy that, without starting from scratch, produces the same model distribution as retraining. In the naive approach, where the entire model is retrained after deletion, the computational cost is drastically high, and it takes a lot of time. Therefore, in order to better match with the objectives of new laws on privacy, an alternate strategy that investigates the deterministic definition is required. (Singh,2023) has devised docker swarm concept to load and to efficiently handle the data for bigdata applications. Here the model unlearns the user requested data. If the distributions do not match, there must be some influence on the system from them to account for the discrepancy. One way to think of an unlearning environment is as a probabilistic setting where most of the contribution of the user-requested data is eliminated, and an unlearning method only approximates the retraining distribution. On certain websites, users who knowingly or unknowingly provide personal information about themselves consent to the corporations running those platforms using that information for a range of purposes, such as selling it to marketers or using it to improve their prediction models. It becomes challenging for businesses to undo the impact of data acquired if a user decides not to let such information about them to be used by them, particularly if the data was used to train machine learning models. Most of the people, considering their privacy, don't want their data to be shared. For instance, consider a Man 'X' who want to delete his Account 'A'. He has just deleted his account by deactivating it. Here it should be understood that deleting an account doesn't mean that X's personal information is completely deleted. The Machine Model that is created using a dataset consisting of X's information still has a trace of X's details. But, it is not acceptable to have privacy sensitive data as a part of Machine learning model. Subsequently, in the situation of encountering errors and attacks, unlearning the incorrect or poisoned data sample is an obvious need. In elaborate, if a data administer, who collects and organizes data, has entered an incorrect sample and has let the machine learn the data, it results in the model which couldn't produce a desired output. Lastly, in the case of being trapped in the data attack called data pollution, there is a demand to erase the polluted data sample. In data pollution, the hacker tries to inject a malicious data sample to the training set causing the machine to learn the training set comprising the injected data sample. Removing or deleting the inserted data sample from the training set and relearning the clean set could be the solution.

RELATED WORKS

Approximate Deletion Method

(Zachary Izzo,2021) have put out a novel approximation technique for logistic and linear models, the computational cost of which is independent of the number of training data ('n') and linear in the feature dimension ('d'). A trained machine learning model can have individual training points removed in a variety of scenarios. When utilising deletion, the model must be post-processed to remove the impact of the selected training point. A novel method known as the Projective Residual Update (PRU) has a computing cost that is independent of the amount of training data n and linear in the data dimension. A metric for evaluating the thoroughness of data erasure from ML models, the injection test, was also created. With this approximate deletion mechanism, as more deletion requests are performed, the accuracy of the estimate will deteriorate.

Jeopardizes Privacy

(Min Chen,2021) have suggested a number of approximation methods to reduce the significant computational cost brought on by retraining. Two types of machine learning models are produced by machine unlearning: the original and the unlearned. They think that even while machine unlearning was designed to keep the privacy of the given model, it can give the impression that it does, creating unexpected privacy problems. Specifically, the unlearned model may provide additional information about the target sample, although the initial model might not have revealed much private information. First, they propose a new machine unlearning membership inference attack that aims to ascertain if the target sample was included in the original model's training set. Additionally, two new privacy measures are proposed to measure the unexpected privacy risks associated with machine unlearning: Degradation Count and Degradation Rate. These two compute the amount of relative privacy which has been lost by the target due to machine unlearning. The results demonstrate that the target sample's membership privacy is consistently reduced by our assault, indicating that machine unlearning can have a negative impact on privacy. To sum up, the privacy of the target sample is often compromised by machine unlearning. This research emphasizes how risky it is to use the the ability to be overlooked while implementing the learnings by the machine. These measures and attacks may help in the future to create more machine learning systems that protect privacy.

Selective Forgetting

(Golatkar,2020) investigate the issues associated with forgetting a specific samples of the data used to train a deep neural network. By carefully examining the weights of the data, insights can still be gleaned even when the network's output may obscure its effects. They suggested a technique for "scrubbing" data from the weights to get details about a sample batch of training data. It is a technique used to "scrub" the information weights related to a particular set of training data. When a probing function is applied to a network's weights that was trained without the data to be overlooked, it becomes indistinguishable from a network that was trained with the data to be overlooked. Information must be removed from the weights as intended in order to achieve selective amnesia; simply hiding or altering certain output won't suffice to scrub the activations. An attacker may be able to discern more details from the photos if the improper cleaning technique is used.

Federated Unlearning

(Gao,2022) Federated unlearning with verification was first presented by Gao et al. Wang Junxiaoet al. (Li,2023) In contrast to prior federated based learning, Li et al. presented federated unlearning based on incremental learning. (Wang,2022) investigated the issue of trained CNN classification models' selective forgetting of categories in federated learning (FL). Since FL does not allow global access to the training data, our results delve deeply into the internal influence of each channel. Channel class discrimination is quantified using the Term Frequency Inverse Document Frequency (TF-IDF) concept. In order to unlearn, channels with high TF-IDF scores must be pruned because they are more selective on the target categories. A proposed distributed machine learning paradigm called federated learning (FL) enables several devices to train a common model without direct access to sensitive training data. Only shallow network sin convolution designs, like a 2-layer CNN followed by two fully connected (FC) layers, allow for unlearning. Accuracy and scalability are traded off due to the amount of data stored in the federated server.

Casual Unlearning

(Cao,2015) advocated the use of a comparable system called Karma and an approach called Causal Unlearning to effectively repair a polluted-learning system. Karma significantly reduces the amount of human work required from administrators by automatically and highly precisely locating the set of contaminated training data samples. One of the main ideas of Karma is casual unlearning. A major attack known as "data pollution" involves inserting intentionally generated training data samples into the training set. This leads to the system creating an erroneous model and misclassifying testing samples as a result. Erroneous samples injected into the training set have the potential to deceive the machine learning classifier that identifies malicious crowd sourcing workers. Erroneous samples that were injected into the training set have the potential to deceive the machine learning classifier that identifies malicious crowdsourcing workers. To start a data pollution assault, an attacker must leave a trail, which is a causal chain that goes from tainted training samples via a polluted learning model to misclassify testing samples. An exploratory attack is used by an attacker who wishes to understand more about the machine learning model. Attacks that are exploratory are not covered by karma. To ensure comprehensiveness, we will touch on these types of attacks in adversarial machine learning in brief.

OBJECTIVES

This project's primary aim is to completely erase any evidence of a specific individual or data point from a machine learning system without compromising the system's functionality. The goal of this

Figure 1. SISA model's architecture diagram

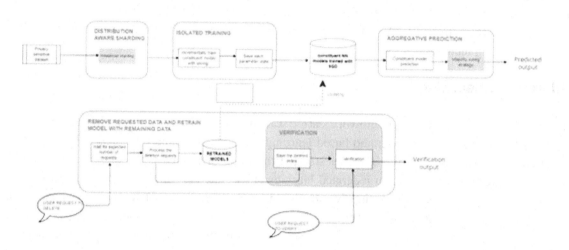

project is to reduce the computational cost of fully retraining machine learning models by ignoring the influence of user-requested input points. Thus far, the straightforward method of training a model from scratch on the dataset without having to unlearn the point has shown to be incredibly effective. Throughout the rest of the paper, there is a reference to this technique as the basic strategy. However, this strategy will quickly become unfeasible (in terms of time and computational resources expended) with large datasets. For example, companies will need to regularly retrain models in order to comply with GDPR/CCPA. The newly developed strategy should be made in such a way that whenever a user wishes to remove or delete his/her personal data, the organization or the network must provide the user a methodology to do so. Apart from deletion it is also important to provide a mechanism for verifying whether the data has been completely removed. This increases the users trust worthiness of the firm, which in turn might be a great advantage for the organization. Ensuring privacy is the expectation of most of the users. As a result, any new strategy should meet the following criteria.

1) Ease of Understanding: The baseline technique is simple to comprehend and implement on a conceptual level. Analogously, any unlearning technique should be comprehensible; this standard guarantees that non-specialists can debug the approach. 2) Higher Accuracy than baseline: The model's accuracy may drop even in the baseline if (a) the proportion of training points that must be unlearned becomes too much or (b) prototype points are unlearned. Even if no part of the strategy expressly promotes high accuracy, every unlearning technique should try to establish a modest accuracy gap relative to the baseline for any number of points unlearned. 3) Lower Retraining Time: The technique should take significantly less time to unlearn any number of points than the baseline. 4) Guaranteed Unlearning: Like the baseline, every new method must offer observable assurances that any number of points have been unlearned (and do not affect model parameters). Moreover, such a commitment must be clear-cut and easy to understand for non-experts (Gupta,2021) .5) Model Independent: In this case, the unlearning method should be generic, meaning it should meet the previously stated guarantees for models of different types and degrees of complexity.

SYSTEM DESIGN

Figure 1 represents the proposed SISA model architecture diagram. There is no information flow between the constituent models in SISA training, which is different from current tactics in how incremental model modifications are communicated or altered. For example, gradients computed on each constituent are not shared across constituents if each constituent model is a NN trained with stochastic gradient descent; rather, each constituent is trained independently. This guarantees that the influence of a shard, along with the data points that comprise it, is confined to the model that is utilizing it for training. Every shard is further subdivided into slices, and an increasing number of slices is used to train each constituent model gradually (and state fully). Each constituent receives the test point at inference, and the responses from all of the constituents are aggregated. Our machine unlearning project uses the SISA training approach, which stands for Sharded, Isolated, Sliced, and Aggregated training, and it consists of four modules. In this SISA approach of training, the given whole dataset is divided into disjoint shards. There should be no communication flow among the shards. Each shard act as an individual, stand-alone Neural Network model similar to Ensemble Learning in which weak learners are combined to produce the output. These shards are trained independently such that reducing the training cost. The reason for the reduced training cost will be discussed in the later chapter. The shards partition the data into slices. These shards are trained in isolation creating a number of neural networks. When a user request to delete data arrives, only the affected shard is retrained. For the prediction part, the majority voting strategy is used in which the model which gets the majority voting produces the output result.

DISTRIBUTION AWARESHARDING

To start with the first module which is a Distribution Aware Sharding (DAS) where the user requests are distributed in three different ways and tested. In our project, it focuses on Uniform, Exponential and Pareto distribution. In Uniform distribution, the requests are uniformly distributed, say, all the shards contain equal numbers of data points to be removed. This means there is the same number of requests distributed in all shards. In Exponential distribution, the requests are distributed exponentially. Exponentially in the sense requests are distributed in all shards either in increasing number or decreasing number. In Pareto distribution, the shard might consist of a varied number of requests. There is a possibility of no request in a shard or all requests in a shard or requests are scattered in any order. Algorithm 1 is for distribution Aware Sharding Module (DAS). This algorithm explains the three different distributions that are carried out through this project for splitting the dataset into shards and slices. The distribution here means distributing the request provided by the user to various shards which can take three forms – Uniform, Exponential and Pareto. This module mainly focuses on splitting the dataset by having a clear picture of the distribution of the user request. Figure 2 represents the flow diagram of Distribution Aware Sharding (DAS) module. Initially, calculate the probability of the data sample. Here, the probability is nothing but the degree to which the data sample gets unlearned.

ISOLATEDTRAINING

To continue with, the second module which is the Isolated Training phase where incremental learning takes place. We know that each shard acts as a NN and each shard are further partitioned into slices. In incremental learning, firstly, single slice of a particular shard is added and learned.

Figure 2. Distribution aware sharding module diagram

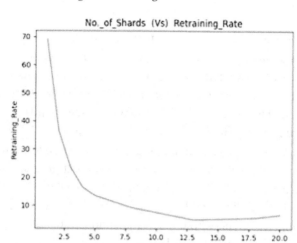

Algorithm 1. Algorithm for Distribution Aware Sharding

```
1.Differentiate each sample as likely to unlearn and unlikely to unlearn.
2.Identify the distribution of unlearning requests among shards.
        If distribution is uniform
        Split dataset into equal sized shards
 If distribution is exponential then
Calculate Probability as
P=exp(-λ*index)-exp(-λ*(ind
ex+1))                                                          (1)
If the distribution is Paretothen calculate the probability by the formulae
expressed below.
P  =α/(index+1)^(α+1)
3.Examine the probability limit for each shards
4.For each data point
        If (previous cumulative probability of first available shard + Current
data point' sprobability)        limit
        Then
        Add the mdata point into the shard
        Else
        Add the data point into next available shard
```

Their parameter states are saved. This learning is carried on till all the slices of a single shard are trained.(Note: No. of Shards and No. of Slices are fixed based on the accuracy and there training cost that is obtained for our specific dataset).

The resultant constituent models (NN) are saved. The module design for Isolated Training module is pictured in the Figure 3. Isolated training is the next module. The sharded dataset from the outlet of

Figure 3. Module design for isolated training module

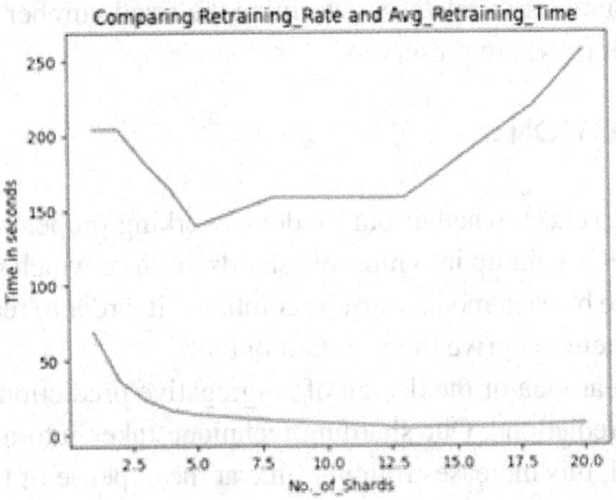

the first module is taken. Each shard is trained independently using NN (Neural Network). Stochastic Gradient Descent (SGD)-trained neural networks make up each component model. Every component receives individual training. This guarantees that a shard's and the data points' influence is limited to the model that is being trained. Every shard is further divided into slices, and a growing number of slices are used to train each constituent model gradually (and iteratively, in a stateful way).

Algorithm 2 portrays the working of Isolated Training Module. Here, in this module all disjoint shards are trained independently to provide a constituent neural network models The quantity of models utilised is equal to the quantity of shards utilized. They are all taught differently and offer various outcomes. The final result is taken by utilizing the Majority Voting Strategy which is explained in the Aggregative Prediction Module. Incremental learning takes place during the isolated training phase. Each shard is known to operate as a NN, and each shard is further divided into slices. First, a single slice of a given shard is added and learned via incremental learning. The values of their parameters are preserved. This

Algorithm 2. Algorithm for Isolated Training

```
D - Datasets
D_k- ShardK'ssamples
iRD_k,iRdisjointslicesofshardK
e-Numberofepochs
M_k-Trainedmodelaftereepochs
D_k,i-i'thsliceinK'thshard
For each shared
        For i          1 toR
        Train the model M_k,iusingU_iD_k,ifor
        e_iepochs
        Save the state of parameters associatedwith
        Model M_k,i.
```

process is repeated until all of the slices of a single shard have been completed. (Take note that the accuracy and retraining cost found for our dataset determine the fixed number of shards and slices..) The constituent models that result (NN) are preserved.

AGGREGATIVEPREDICTION

Furthermore, for prediction to check whether our model is working properly Aggregative Prediction is chosen. Given that the dataset is split up into multiple shards, each of which functions as an own neural network, the predictions made by each model must be combined in order to reach a conclusion. Majority voting strategy is employed here to derive the resultant output.

The Figure 4 gives the clear idea of the design of Aggregative prediction module. It's important to comprehend how to make predictions. Our sharding technique takes into account the fact that while fewer data are used per shard, this increases training time at the expense of the model's accuracy. This method, which only slightly reduces accuracy, integrates the predictions of each model. Utilising a basic Majority Voting Strategy (MVS), it yields the prediction with the highest number of votes. It uses a simple Majority Voting Strategy (MVS) and returns the prediction which received the most votes.

Figure 4. Module design of aggregative prediction module

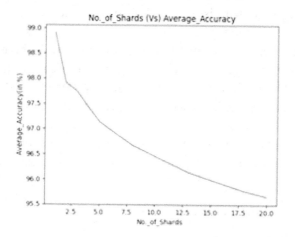

Aggregative Prediction Module is Represented by Algorithm

When making a prediction, there is a need to know how to combine the knowledge from each model. An key aspect of our sharding technique is that a smaller amount of data per shard (which increases training time) reduces model accuracy. The accuracy is slightly decreased by this strategy, which combines the predictions of each model. It uses the simple Majority Voting Strategy (MVS) to return the prediction with the most number of votes.

DELETING REQUEST ANDVERIFICATION

The vital part of our project is the last module. When a user made a request to delete their information, first thing is to save the request in the buffer till some specified number of request is arrived. What is the need of buffering the request and wait? For each and every request there is unlearning and retraining, it is not much efficient or desirable in terms of training cost and performance. Once it reaches the specified limit, locate the slice index where the request lies in the shard and remove it from there. And then the important thing is near. From the slice that is next to the removed slice, retraining starts using the saved parameter. The parameter state of each slice during isolated training phase is saved. Moving on, there is a verification part to make sure that really the user requested data has been deleted or not. This verification might be user oriented or data administrator oriented.

1) *Remove requested Data and Retrain model with remaining data:* The Figure 5 portrays the design for the module Remove Request data and Retrain Model (RRD RM). After training the model, check whether desired number of unlearning requests are obtained. If yes then find out the shard and the slice position of the data to be deleted. If the request received is less than the desired limit then wait till desired number of request is obtained. Following this, delete the data sample and retrain the shard.

Let the samples of data in the dataset and the user requests be S1, S2. . . S n and R 1, R 2. . . R n respectively. Dataset is divided into shards. For deleting a particular request, the shards are updated with new set of data by removing the requested Data from the whole dataset. UpdatedShard=S1,S2... Sn – R 1, R 2. . . R n This is done for each and every shard in which the deletion request data are found to be

Algorithm 3. Algorithm for Aggregative Prediction

```
⇒Compute weight vector based on performance
⇒Get prediction from each constituent models
⇒Preform majority voting strategy
⇒Using argmax pick the label with highest vote
⇒Update the weights of each model based on performance
how to combine every model's knowledge when
if(number    of    requests    received         expected)
Algorithm  4Algorithm for UnlearningData
.        wait(sometime);
Locate indices of each request
Locate the Slice(Sq(with minimum index number where the unlearning request
start

Remove requested data
⇒RetrieveS_u'thsliceparameterstate
⇒Perform training procedure from slice S_u
⇒SavethemodelM_k'
```

Figure 5. Removing request and retraining the model

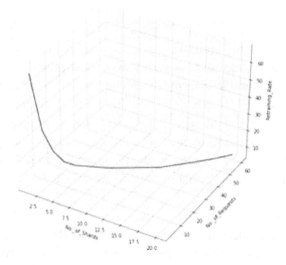

present. So, it is clear that retraining happens for the shard with the removed request points. A log is maintained for the proof of deletion.

2) *Verification:* In this Verification module, which is a sub-module of the last part, the user or the administrator can confirm that the requested data has been deleted. At the commencement of the Unlearning process for the particular batch of requested data, these data are saved in the log for future verification. Therefore, whenever the user wants to certify that their information is deleted they can verify the match of tuples in the log and the user request. If there is a match then their data has been successfully deleted or else it was not deleted.

Equation 4 - unlearning request must be three times less than equal to number of shards for efficient result. If the dataset is sharded and sliced before training, then retraining time decreases. Shards and slices reduce retraining time to achieve unlearning because they are smaller than the whole dataset.

3) *Storage:* There is a need for space to store the parameter state. Storage cost depends on slicing. Slicing, a set of samples, divides the shards. Slicing decreases the storage cost. Let,

n – Number of shards
m – Number of samples in each shard
When learning (incremental learning), initially 'm' space needed to store parameter state of sample, then add on another set of slices/samples which in-turn need another 'm' space which gets continued until it reaches the end sample
- n(m+2m+3m+4m+. . . +rm) Therefore the calculation is as follow:

n(m+2m+3m+4m+. +rm) = nm(r((r+1)/2))

$$= (nmr(r+1))/2 \tag{5}$$

4) *Accuracy:* Accuracy is one of the most important metrics to be noted. If the sharding and slicing strategy is implemented, it results in increasing speed-up, but when looking into accuracy, sharding may lead to condition of poor accuracy rate. That's why it is desirable to sum up to an "Aggregation Strategy" which results in better accuracy rate along with increase in retraining speed-up.

EXPERIMENTALRESULTS

Techniques vs. Metrics(6)

This project utilizes the purchasing dataset. The Purchasing dataset is made up of six hundred attributes which is available in .npz file in csr format. These six hundred attributes or column represents the items that are purchasable. The values of these columns are represented as either 1 or 0. The value 1 in the column depicts that the customer really purchased that item. The value 0 says that the customer hasn't purchased the item; he/she has just visited the website without the intention to buy that item.

EVALUATION METRICS

1) *Retraining Time:* Let r = unlearning request S = number of shards

The Figure 6 shows how different strategies like sharding, slicing, and aggregation strategy affect metrics like retraining time, storage cost, and accuracy. The retraining time is reduced when a sharding method is used (i.e. the dataset is divided into shards), but the accuracy is not improved. As a result, slicing is taken into consideration (i.e. further partitioning the shards into slices). It leads to more storage, which is undesirable. Finally, the Aggregation approach, when combined with Sharding and slicing, reduces retraining time and improves accuracy, both of which are desirable outcomes.

SISA training is a useful unlearning technique that lowers the computational overhead of unlearning by utilizing data sharding and slicing. When a service provider handles requests in batches or sequentially, it determines the asymptotic lowering of the unlearning time by employing slicing and sharding. Research indicates that the amalgamation of sharding and slicing techniques does not yield a significant impact on accuracy when it comes to basic learning tasks. Additionally, SISA training can manage orders of magnitude more unlearning requests than what Google anticipates being necessary for the GDPR right to be forgotten (Bertram,2019). It is demonstrated that for complex learning problems, a little accuracy loss occurs when a combination of SISA training and transfer learning is used.

It is discovered that the accuracy of the sharding part of SISA training decreases (especially for complicated tasks) when (a) the number of unlearning requests and (b) the number of shards increases. This results from the reduction of the number of samples per class for each shard (b) in both (a) and (b). It is discovered that as long as the quantity of epochs required for training is adjusted, slicing has no impact on accuracy. A combination of sharding and slicing considerably outperforms the naive baseline for a given amount of unlearning requests, even in the worst-case scenario (without knowledge of the distribution of unlearning requests). If the volume of requests exceeds this threshold, SISA training gradually reverts to baseline performance. We are able to solve this analytically.

Figure 6. Techniques vs. metrics

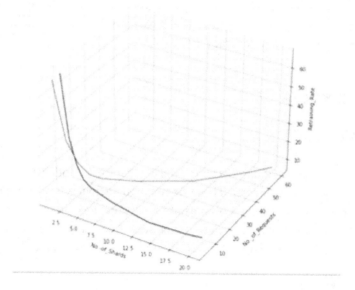

Comparative Analysis

The Figure 7 represents the different methods used for un- learning a data sample. In Naive approach, whenever a request arrives and deletion is made, the whole shard is retrained each and every time. The Deep Obliviate method stores intermediate models on the hard drive, which enhances the initial training procedure. Given an unlearning data point (He,2021) it first

The temporal residual memory that remains in stored models is quantified. The impacted models will undergo retraining, and the trend of residual memory will be used to determine when to stop the retraining. Finally, it combines the retrained models with uninfluenced models to sew an unlearned model (HE,2021). These two approaches are compared with our proposed model SISA.

Figure 7. Different methods used for unlearning

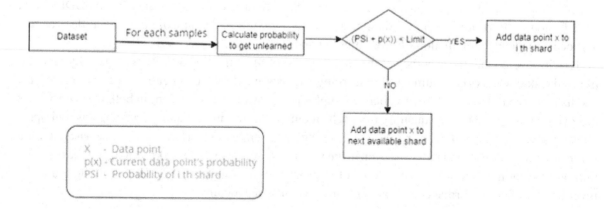

The Figure 8 represents the graph between no. of shards vs. total accuracy.. Two baselines are used to compare the suggested method to. These are as follows:

- Retrain the entire model after each of the K unlearning requests, in a batch. This is equivalent to the naive baseline of starting over and retraining the full dataset in a batch setting (without the points that need to be unlearned).
- Only retrain when the point to be unlearned falls inside this collection, and train on a 1/S percentage of the data.

Figure 8. No. of shards vs. total accuracy

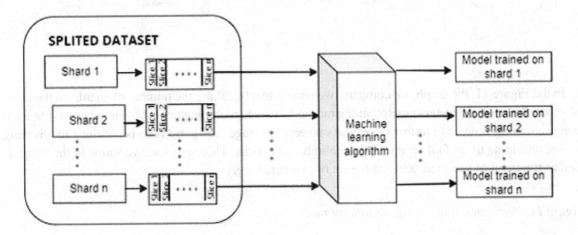

In the Figure 9 depicts the graph of various performance metrics. A varying trend is seen in the graph (Figure 9) between number of shards and the respective retraining time taken. It is also possible to infer from the graph that there is no linear relationship between the number of shards and the retraining time—that is, there is no consistent rise or fall in retraining time as the number of shards increases. It is desirable to have low retraining time, from the graph it is observed that retraining time is minimum when the number of shards is13.

Figure 9. No. of shards vs. retraining time

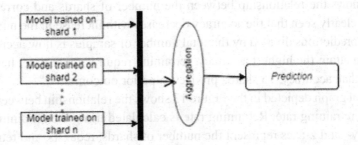

The time required to retrain the model based on the quantity and distribution of requests is referred to as the retraining rate in the graph (Figure 10) between the number of shards and the retraining rate.

Retraining rate is calculated from retraining time and number of requests. Simply retraining rate is nothing but retraining time per request. From the results of the previous graph and this graph we could conclude that for 13 shards both retraining time and retraining rate are optimal.

Figure 10. No. of shards vs. retraining rate

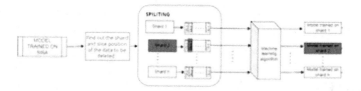

In the Figure 11, the graph, we compare two parameters by taking the number of shards on the common axis. The two parameters are Retraining rate and Average retraining time. Retraining rate is obtained using retraining time and number of requests whereas average retraining time is obtained by dividing sum of retraining time of all epochs by total number of epochs. There is a wide variation in the range as well as the trend of variation between the set of two parameters.

Figure 11. Retraining rate vs. avg. retraining time

Techniques vs. Metrics	Retraining time	Storage cost	Accuracy
Sharding	⬇		
Slicing		⬆	
Aggregation strategy	⬇		⬆

The Figure 12 shows the relationship between the number of shards and corresponding accuracy. From the graph it is clearly seen that the accuracy decreases with increase in the number of shards. The number of accurate predictions divided by the total number of samples is how accuracy is determined. While one shard can attain the highest accuracy, retraining requires more time; hence, taking into account aspects other than accuracy, 13 shards produce superior outcomes.

The 3 dimensional graph depicted in the Figure 13 shows the relationship between number of shards, number of requests, retraining rate. Retraining rate is calculated using the retraining time and number of requests. The x-, y-, and z-axes represent the number of shards, requests, and retraining rate that are taken, respectively. In the 3 dimensional graph portrayed in the Figure 14 represents the relationship

Figure 12. No. of shards vs. avg. accuracy

METHOD	PARAMETER	ACCURACY	SPEED-UP	PREDICTION COST	STORAGE COST
Naïve	-	96.2%	1×	1×	1×
Deep Obliviate	Batch 200	96.0%	35×	1×	200×
SISA	Shard 13 Unlearning request 39	96.16%	29×	20×	200×

Figure 13. No. of shards vs. no. of requests vs. retraining rate

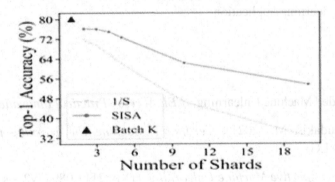

Figure 14. No. of shards vs. no. of requests vs. retraining rate

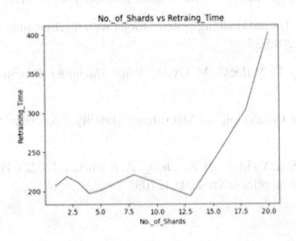

between number of shards and retraining rate by taking constant number of requests and varying number of requests. The green line corresponds to a constant number of requests (i.e., 8) while the blue line corresponds to a varying number of requests. Here for a higher number of shards the corresponding numbers of requests are also higher.

CONCLUSION AND FUTURESCOPE

In this research, we introduce a SISA training method involving a master node and non-communicating weak learners, addressing the challenge of flawless machine unlearning. This approach ensures timely and complete removal of consumer data from the model. Verification is conducted through machine unlearning verification, offering concrete quantitative measures for individual user perspectives. SISA unlearning, backed by verification, not only guarantees deletion but also provides users with increased control over data privacy. Future research avenues include exploring broader data updates and extending unlearning to more complex machine learning models. Verification techniques encompass back-door injection and user-level membership inference attacks to confirm data usage in model training. A key focus of our future work is refining the unlearning process to prevent data overfitting and ensure sustained effectiveness in machine learning applications.

REFERENCES

Aldaghri, N. (2021). Coded Machine Unlearning. *IEEE Access : Practical Innovations, Open Solutions.*

Ananth, M. & Mathioudakis, M. (2021). *Certifiable Machine Un- learning for Linear Models.* arXiv:2106.15093v3 (cs.LG).

Ayush, K. (2021). *Fast yet effective Machine Unlearning.* arXiv:2111.08947v2 (cs.LG).

Ayush, S., Acharya, J., Kamath G., & Suresh, A. (2021). Remember What You Want to Forget: Algorithms for Machine Unlearning. *35th Conference on Neural Information Processing Sys- tems(NIPS).* IEEE.

Brophy, J., & Lowd, D. (2021). Machine Unlearning for Random Forests. arXiv:2009.05567v2(cs.LG).

Cao, Y., & Yang, J. (2015). Towards making systems forget with machine unlearning. *Proc. IEEE Symposium.* IEEE. 10.1109/SP.2015.35

Chen, M., Zhang, Z., Wang, T., & Backe, M. (2021). *When Machine Unlearning Jeopardizes Privacy.* ACM.

Enayat, U. (2021). Machine Unlearning via Algorithmic Stability. *Proceedings of 34th Conference on Learning Theory.* IEEE.

Gao, X., Ma, X., Wang, J., Sun, Y., Li, B., Ji, S., Cheng, P., & Chen, J. (2022). *Verifi: Towards verifiable federated unlearning.* arXiv preprint arXiv:2205.12709.

Garg, S., Goldwasser, S., & Vasudevan, P. N. (2020). Formalizing data deletion in the context of the right to be forgotten. *Proc. Annu. Int. Conf. Theory Appl. Cryptograph.Techn.* Cham, Switzerland: Springer. 10.1007/978-3-030-45724-2_13

Golatkar, A., Achille, A., & Soatto, S. (2020). Eternal sunshine of the spotless net: Selective forgetting in deep networks. *Proc. IEEE/CVF Conf. Comput. Vis. Pattern Recognit.* IEEE. 10.1109/CVPR42600.2020.00932

He, Y. (2015). *DEEP- OBLIVIATE.* arXiv:2105.06209(cs.LG).

Izzo, Z., & Smart, M. A. (2021). Ap- proximate Data Deletion from Machine Learning Models", San Diego, California, USA. *Proceedings of the 24th International Conference on Artificial Intelligence and Statistics(AISTATS).* IEEE.

Junxiao, W. (2022). *Federated Unlearning via Class-discriminative Pruning.* Hong Kong Polytechnic University, Dalian University of Technology.

Kim, B., Rudin, C., & Shah, J. A. (2014). The bayesian case model: A gen- erative approach for case-based reasoning and prototype classification. *Advances in Neural Information Processing Systems.*

Li, Y., Chen, C., Zheng, X., & Zhang, J. (2023). *Federated unlearning via active forgetting.* arXiv pre-print arXiv:2307.03363

LiuY.MaZ.LiuJ.PhilipY. (2021). *Learn to Forget: Ma- chine Unlearning via Neuron Masking.* arXiv:2003.10933v3 [cs.LG].

Ma, Z., Liu, Y., Liu, X., Liu, J., Ma, J., & Ren, K. (2022). Learn to forget: Machine unlearning via neuron masking. *IEEE Transactions on Dependable and Secure Computing.*

Neel, S., Roth, A., & Sharifi-Malvajerdi, S. (2019). Descent-to-delete: Gradint based methods for machine unlearning.

Saltzer, J. H., & Schroeder, M. D. (1975). The protection of information in computer systems. *Proceedings of the IEEE, 63*(9), 1278–1308. doi:10.1109/PROC.1975.9939

Schelter, S., Grafberger, S., & Dunning, T. (2021). HedgeCut: Main- tainingRandomised Trees for Low-Latency Machine Unlearning. *VirtualEvent,China*, (June), 20–25.

Singh, N., Hamid, Y., Juneja, S., Srivastava, G., Dhiman, G., Gadekallu, T. R., & Shah, M. A. (2023). Load balancing and service discovery using Docker Swarm for microservice based big data applications. *Journal of Cloud Computing (Heidelberg, Germany), 12*(1), 4. doi:10.1186/s13677-022-00358-7

Thomas, B. (2020). *Machine Unlearning: Linear Filtration for Logit-based Classifiers.* arXiv:2002.02730.

Varun, G. (2021). Adaptive Machine Unlearning. *35th Conference on Neural Information Processing Systems (NIPS).* IEEE.

Chapter 10
Ethical Considerations in AI Development

Bhanu Pratap Singh
COER University, Roorkee, India

Ankush Joshi
ⓘD https://orcid.org/0000-0003-0873-3340
COER University, Roorkee, India

ABSTRACT

This chapter serves not only as a repository of ethical insights, but as a clarion call for a multidisciplinary and collaborative approach. Ethical considerations in AI development demand a collective effort, where technologists, ethicists, policymakers, and society engage in an ongoing dialogue. It is through such collaboration that we can envision and craft a future where AI technologies not only excel in technical prowess but also uphold the principles of fairness, accountability, and societal benefits. In essence, this chapter invites readers to embark on a reflective journey, encouraging a deeper understanding of the ethical intricacies that underlie AI development. It is a call to action, challenging us to wield the power of AI responsibly and ethically, shaping a future where technology becomes a force for positive societal transformation.

INTRODUCTION: ETHICAL CONSIDERATION IN AI DEVELOPMENT

In the age of rapid technological advancements, Artificial Intelligence (AI) has emerged as a transformative force with the potential to revolutionize various facets of human existence (Brundage et al., 2020). From healthcare and education to governance and industry, the integration of intelligent systems is reshaping our world. However, as AI proliferates, the ethical implications of its deployment have come to the forefront, demanding a nuanced understanding and a principled approach in its development and application.

This chapter delves into the multifaceted landscape of ethical considerations in AI development. It serves as a comprehensive exploration of the moral dimensions associated with creating, implementing,

DOI: 10.4018/979-8-3693-2964-1.ch010

and governing AI technologies. The profound impact of AI on society prompts an urgent need for a robust ethical framework (González-Gonzalo et al., 2022), capable of addressing the challenges and dilemmas that arise as these technologies become increasingly intertwined with our daily lives.

The Rise of Artificial Intelligence

The genesis of the ethical considerations in AI lies in the unprecedented capabilities that intelligent systems bring to the table. Machine learning algorithms, neural networks, and sophisticated data analytics empower AI systems to learn from vast datasets, make predictions, and automate complex tasks (Bohr & Memarzadeh, 2020). The promise of efficiency, innovation, and improved decision-making has catapulted AI into domains as diverse as healthcare diagnostics, financial forecasting, and autonomous vehicles.

However, this rise in technological prowess is accompanied by ethical dilemmas that echo across societal domains. Questions of transparency, accountability, and the potential for bias in algorithmic decision-making have become central to discussions around AI ethics. The power dynamic between creators and users of AI technologies requires careful examination, as does the impact on existing socioeconomic structures and labour markets.

The Ethical Crucible: Domains of Impact

As AI permeates critical domains, the ethical considerations crystallize, presenting a complex interplay between technological innovation and societal values. One of the crucibles where ethical challenges materialize is in healthcare (Müller, 2023). AI applications in medical diagnostics and treatment planning raise questions about data privacy, patient consent, and the reliability of algorithmic decision-making in life-and-death scenarios. Striking the right balance between innovation and ethical considerations becomes imperative to foster trust among healthcare professionals and the broader public.

Similarly, in the criminal justice system, the use of AI for predictive policing and sentencing algorithms introduces a host of ethical challenges (Gabriel, 2022). Concerns about fairness, accountability, and potential amplification of existing biases demand rigorous ethical scrutiny. The chapter navigates through these intricate dynamics, shedding light on the ethical tightrope that developers and policymakers must walk to ensure the just and equitable application of AI in the criminal justice domain.

In education, the integration of AI-driven tools for personalized learning raises ethical questions related to student privacy (Nguyen et al., 2023), the digital divide, and the potential reinforcement of existing educational inequalities. As intelligent systems become key stakeholders in the educational process, ethical considerations must guide the design and deployment of these technologies to ensure they contribute positively to educational outcomes without exacerbating societal disparities.

Ethical Design Principles as Beacons

Amidst these ethical challenges, the chapter advocates for the establishment of ethical design principles as guiding beacons in AI development. Ethical design goes beyond legal compliance (Smit et al., 2020); it necessitates proactive measures to embed fairness, transparency, and accountability into the fabric of AI systems. The exploration of these principles becomes crucial as developers grapple with the dual responsibility of pushing technological boundaries and safeguarding against unintended consequences.

Transparency emerges as a linchpin in ethical AI development (Peters et al., 2020). The opacity of complex algorithms raises concerns about the "black box" nature of AI decision-making (Floridi & Cowls, 2022). Ethical design insists on transparency to empower users and stakeholders with insights into how AI systems operate, make decisions, and impact their lives. Transparent AI engenders trust, fostering a collaborative environment where developers, users, and affected communities can collectively ensure the responsible use of intelligent systems.

Fairness in AI is another cornerstone of ethical design (Benke et al., 2020). The potential for bias, whether unintentional or systemic, poses a significant ethical challenge. The chapter delves into methodologies for detecting and mitigating bias in AI algorithms, emphasizing the importance of fairness to prevent discrimination and ensure equitable outcomes across diverse populations. Ethical considerations underscore the need to address bias not just as a technical issue but as a societal challenge requiring interdisciplinary collaboration.

Moreover, accountability becomes a linchpin in addressing the power asymmetry between creators and users of AI technologies. Ethical design principles necessitate mechanisms for holding individuals and organizations accountable for the societal impact of their creations (Floridi, Cowls, Beltrametti, Chatila, Chazerand, Dignum, & Vayena, 2021). From clear lines of responsibility to avenues for recourse in the face of algorithmic errors, accountability ensures that AI development aligns with ethical norms and societal values.

Unraveling Ethical Complexities

The ethical complexities in AI development become particularly pronounced when we consider the inherent biases that can be embedded in algorithms. Machine learning models trained on historical data may perpetuate and even exacerbate existing societal biases (Eitel-Porter, 2021), leading to discriminatory outcomes. This chapter delves into the nuanced challenges of algorithmic bias, unpacking the ways in which developers and policymakers can proactively address these issues.

The intricate interplay between algorithms and human values becomes a focal point of exploration. How can we ensure that the decisions made by AI systems align with ethical norms? What safeguards can be implemented to mitigate the unintended consequences of intelligent systems? These questions underscore the need for ethical guidelines that transcend the purely technical realm, integrating ethical considerations into the very fabric of AI development.

Emerging Technologies and Forward-Looking Ethical Discourse

As the chapter progresses, it casts its gaze toward the horizon of AI development, contemplating the ethical considerations surrounding emerging technologies. The advent of autonomous systems, advanced robotics, and the specter of artificial general intelligence beckons a forward-looking ethical discourse. These technologies, while promising unprecedented advancements, introduce ethical challenges that demand anticipatory ethical deliberations.

The narrative unfolds, urging stakeholders to anticipate, understand, and address the ethical challenges accompanying these technological frontiers. The discussion expands beyond the immediate applications of AI, encouraging a proactive stance in addressing potential future scenarios. The chapter advocates for ethical foresight as an integral part of AI development, emphasizing the need for ethical considerations to evolve in tandem with technological progress.

Global Dimensions: Geopolitical Complexities in AI Development

The global nature of AI development introduces geopolitical complexities to the ethical tapestry (Kamruzzaman, 2022). International collaboration and competition in AI research and deployment raise unique ethical challenges. The chapter scrutinizes these challenges, underscoring the need for global standards and cooperative frameworks. In doing so, it advocates for an inclusive and diverse approach to ethical deliberations that transcends cultural, social, and economic boundaries.

AI technologies do not recognize borders, and their impact extends far beyond individual nations. The development of ethical norms and standards on a global scale becomes imperative to ensure a harmonized and responsible approach to AI (Masevski & Stojanovski, 2023). The chapter engages in a discourse on the ethical responsibilities of nations, organizations, and researchers in fostering an environment where AI contributes positively to global well-being, with due consideration for cultural sensit.

UNDERSTANDING BIAS AND FAIRNESS IN AI ALGORITHMS

Artificial Intelligence (AI) algorithms have become integral components of decision-making processes in various domains, from finance and healthcare to criminal justice and employment. However, the increasing reliance on these algorithms has brought to light ethical concerns related to bias and fairness. Bias in AI algorithms refers to the presence of systematic and unfair favouritism or discrimination towards certain groups, while fairness involves ensuring that these algorithms treat individuals equitably. In this comprehensive exploration, we delve into the complexities of bias and fairness in AI algorithms, examining their origins, manifestations, impacts, and strategies for mitigation.

Origins of Bias in AI Algorithms

Data Bias: The Foundation of Algorithmic Bias

One of the primary sources of bias in AI algorithms lies in the data used to train them. Algorithms learn patterns and make predictions based on historical data, and if this data reflects existing societal biases, the algorithm is likely to replicate and perpetuate those biases (Akter, McCarthy, Sajib, Michael, Dwivedi, D'Ambra, & Shen, 2021). For example, if a hiring algorithm is trained on historical data that exhibits gender bias, it may inadvertently favour one gender over another in future hiring decisions.

Algorithmic Design and Implementation

Bias can also emerge from the design and implementation of algorithms. The choices made by developers, such as the features selected for analysis, the complexity of the model, and the optimization criteria, can introduce or amplify biases. Additionally, the lack of diversity in development teams may lead to oversight regarding the potential biases present in algorithms, emphasizing the importance of diverse perspectives in the AI development process.

Manifestations of Bias in AI Algorithms

Gender Bias in Hiring and Employment

One notable manifestation of bias is in hiring and employment algorithms. If historical hiring data exhibits gender bias, algorithms trained on this data may perpetuate the bias by recommending or selecting candidates based on historical patterns rather than merit (Llorens et al., 2021). This can reinforce existing gender disparities in the workplace.

Racial Bias in Criminal Justice Algorithms

In criminal justice, algorithms are often used to assess the risk of re-offending and inform decisions related to bail, sentencing, and parole. However, studies have shown that these algorithms can exhibit racial bias, leading to disproportionately harsher treatment for individuals from certain racial or ethnic groups (Završnik, 2021). This raises ethical concerns and challenges the notion of equal protection under the law.

Financial Bias in Credit Scoring

Financial institutions rely on algorithms to assess creditworthiness and determine eligibility for loans or credit cards. If historical data used to train these algorithms reflects socioeconomic biases, individuals from marginalized communities may face challenges in accessing financial services, perpetuating cycles of inequality (Dastile et al., 2020).

Impacts of Bias in AI Algorithms

Reinforcement of Social Inequities

Perhaps the most significant impact of bias in AI algorithms is the reinforcement of existing social inequities. If algorithms consistently favour certain groups and discriminate against others, they contribute to and amplify societal biases (Hassani, 2021). This not only affects individuals directly but also perpetuates systemic inequalities.

Erosion of Trust and Legitimacy

Instances of biased algorithmic decision-making can erode public trust in AI systems and institutions employing them. When individuals perceive that algorithms are making decisions based on unfair criteria, it undermines the legitimacy of these systems and raises concerns about accountability and transparency.

Human Rights Implication

In contexts such as criminal justice and healthcare, biased algorithms can have severe human rights implications. Unfair treatment based on algorithmic decisions may violate individuals' rights to equal

Figure 1. Impacts of bias in AI algorithms

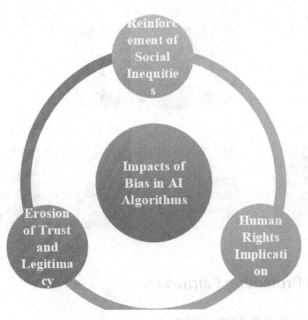

protection, privacy, and due process. Addressing bias in AI algorithms is thus not only an ethical imperative but also a human rights issue.

Fairness in AI Algorithms

Defining Fairness: A Multifaceted Concept

Fairness in AI algorithms is a multifaceted concept that involves treating individuals equitably and avoiding unjust discrimination. Achieving fairness requires careful consideration of various dimensions, including demographic groups, individual characteristics, and the context in which algorithms are deployed (Pessach & Shmueli, 2023).

Types of Fairness

I.   **Individual Fairness:** This involves treating similar individuals similarly. In the context of AI algorithms, it means that individuals with similar characteristics or qualifications should receive similar outcomes or recommendations.
II. **Group Fairness:** This pertains to ensuring fair treatment for different demographic groups. Common metrics for group fairness includes disparate impact, equal opportunity, and demographic parity.
III. **Subgroup Fairness:** Recognizing that fairness considerations may differ within larger demographic groups, subgroup fairness focuses on fairness at a more granular level, considering sub-populations based on specific characteristics.

Figure 2. Types of fairness

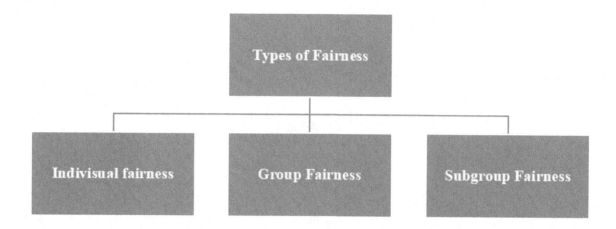

Mitigating Bias and Promoting Fairness

Transparent and Explainable Algorithms

Transparency and explainability are essential elements in addressing bias and ensuring fairness. Openly communicating how algorithms make decisions allows for external scrutiny and enables individuals to understand the criteria used to assess them (Shin, 2021). Explainability is particularly crucial in high-stakes applications such as healthcare and criminal justice.

Diverse and Inclusive Data Collection

To mitigate bias at its root, it is imperative to ensure diverse and representative data collection. This involves being mindful of potential biases in training data and actively seeking to include diverse per-spectives. Initiatives that prioritize inclusivity in data collection contribute to more equitable algorithmic outcomes (Norori et al., 2021).

Continuous Monitoring and Evaluation

Algorithms should undergo continuous monitoring and evaluation to detect and rectify biases that may emerge over time. Regular audits and assessments, both automated and human-driven, can identify discrepancies and ensure that algorithms remain aligned with ethical and fairness standards (Robert et al., 2020).

Collaboration and Ethical Guidelines

Collaboration among stakeholders, including developers, ethicists, policymakers, and the communities affected by AI systems, is crucial in fostering ethical AI development. Establishing and adhering to ethical guidelines provides a framework for responsible AI practices and encourages a collective commitment to fairness and equity (Floridi, Cowls, Beltrametti, Chatila, Chazerand, Dignum, & Vayena, 2021).

Future Challenges and Considerations

Advancements in AI Technologies

As AI technologies continue to advance, incorporating more complex models, such as deep learning and reinforcement learning, new challenges and considerations for bias and fairness emerge. Ensuring fairness in increasingly sophisticated algorithms requires ongoing research and development of techniques that address the nuances of these models.

Ethical Considerations in Emerging Technologies

The ethical considerations surrounding emerging technologies, including artificial general intelligence (AGI) and autonomous systems, necessitate a forward-looking approach. Anticipating the ethical challenges inherent in these technologies is crucial to establishing ethical frameworks that guide their development and deployment.

UNDERSTANDING DISCRIMINATION AND ALGORITHMIC BIAS IN AI ALGORITHMS

As artificial intelligence (AI) algorithms increasingly permeate various facets of our lives, concerns about discrimination and algorithmic bias have become central to the ethical discourse surrounding AI development and deployment. Discrimination refers to the unjust or prejudicial treatment of individuals or groups, and when manifested in AI algorithms, it raises profound questions about fairness, accountability, and the societal impact of intelligent systems (Ntoutsi, Fafalios, Gadiraju, Iosifidis, Nejdl, Vidal, Ruggieri, Turini, Papadopoulos, Krasanakis, Kompatsiaris, Kinder-Kurlanda, Wagner, Karimi, Fernandez, Alani, Berendt, Kruegel, Heinze, & Staab, 2020). This comprehensive exploration delves into the intricate interplay between discrimination and algorithmic bias in AI algorithms, examining their origins, manifestations, consequences, and strategies for mitigation.

Origins of Discrimination and Algorithmic Bias

Data-driven Bias: The Foundation of Algorithmic Discrimination

The origins of discrimination in AI algorithms often trace back to the data used for training. If historical data contains biases, the algorithm may learn and perpetuate these biases, leading to discriminatory outcomes. For instance, a hiring algorithm trained on historical data reflecting gender bias may inadvertently favour one gender over another in future hiring decisions (Ntoutsi, Fafalios, Gadiraju, Iosifidis, Nejdl, Vidal, Ruggieri, Turini, Papadopoulos, Krasanakis, Kompatsiaris, Kinder-Kurlanda, Wagner, Karimi, Fernandez, Alani, Berendt, Kruegel, Heinze, & Staab, 2020).

Implicit Biases in Algorithmic Design

Algorithmic bias can also emerge from the design and implementation choices made by developers. Implicit biases, whether conscious or unconscious, may influence decisions about features, model complexity, and optimization criteria (Kordzadeh & Ghasemaghaei, 2022). The lack of diversity in development teams can contribute to oversight regarding potential biases, emphasizing the importance of diverse perspectives in the AI development process.

Manifestations of Discrimination and Bias in AI Algorithms

Gender Bias in Hiring and Workplace Algorithms

Gender bias is a prevalent manifestation of discrimination in AI algorithms, especially in hiring and workplace-related applications. Biased algorithms may perpetuate historical gender imbalances by favouring certain genders over others, impacting employment opportunities and contributing to gender disparities in the workplace (Yarger et al., 2020).

Racial Bias in Criminal Justice Algorithms

In criminal justice, algorithms are employed for risk assessment and decision-making processes related to bail, sentencing, and parole (Završnik, 2021). However, studies reveal instances of racial bias in these algorithms, leading to disproportionately harsh treatment for individuals from certain racial or ethnic groups. This raises ethical concerns and challenges the principle of equal protection under the law.

Socioeconomic Bias in Financial Algorithms

Financial algorithms, used for credit scoring and loan approval, may exhibit socioeconomic bias (Akter, McCarthy, Sajib, Michael, Dwivedi, D'Ambra, & Shen, 2021). If historical data reflects biases against certain socioeconomic groups, individuals from marginalized communities may face challenges in accessing financial services, perpetuating economic disparities

Understanding Algorithmic Fairness in the Context of Discrimination

Defining Algorithmic Fairness: A Complex Landscape

Algorithmic fairness aims to ensure equitable treatment in algorithmic decision-making. Achieving fairness is a complex task, involving considerations of individual and group fairness, contextual nuances, and a commitment to addressing historical disparities (Sambasivan et al., 2021).

Types of Fairness Metrics

I. **Individual Fairness:** Treating similar individuals similarly, regardless of background or characteristics.

II. **Group Fairness:** Ensuring fair treatment for different demographic groups, measured through metrics such as disparate impact, equal opportunity, and demographic parity.

III. **Subgroup Fairness**: Recognizing and addressing fairness considerations at a more granular level within larger demographic groups.

Addressing Discrimination and Algorithmic Bias

Transparent and Explainable AI

Transparency and explainability are foundational to addressing discrimination and algorithmic bias. Opening the black box of AI algorithms allows for external scrutiny and helps individuals understand the criteria used in decision-making (Rai, 2020). Explainable AI is particularly crucial in high-stakes applications like healthcare and criminal justice.

Diverse and Representative Data

Mitigating bias at its root requires a concerted effort to ensure diverse and representative data collection. AI developers must be vigilant in recognizing and addressing biases present in training data, actively seeking inclusivity and diversity to avoid perpetuating historical inequities (Chen & Decary, 2020).

Continuous Monitoring and Auditing

Algorithms should undergo continuous monitoring and auditing to detect and rectify biases that may emerge over time (Raji et al., 2020). Regular assessments, both automated and human-driven, contribute to maintaining fairness in AI systems and ensuring that they align with ethical standards.

Ethical Guidelines and Collaborative Approaches

Establishing and adhering to ethical guidelines is essential for promoting fair and unbiased AI practices. Collaboration among stakeholders, including developers, ethicists, policymakers, and affected communities, fosters a collective commitment to fairness (Floridi, Cowls, Beltrametti, Chatila, Chazerand, Dignum, & Vayena, 2021).

ADDRESSING BIAS IN MACHINE LEARNING MODELS

The Imperative of Addressing Bias in Machine Learning Models

The widespread adoption of machine learning models across diverse applications has underscored the importance of addressing bias within these systems. Bias, whether inadvertent or systemic, can result in discriminatory outcomes, perpetuate societal inequalities, and erode public trust in AI (Harrison et al., 2020). This chapter provides a thorough examination of the challenges posed by bias in machine learning models and explores strategies for its identification, mitigation, and prevention.

Figure 3. Addressing discrimination and algorithmic bias

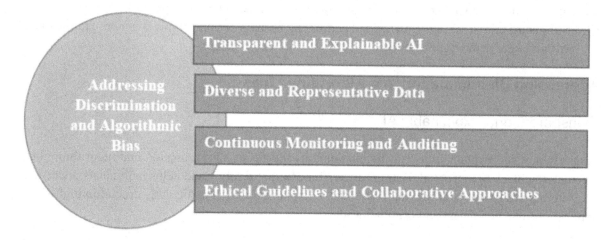

Understanding Bias in Machine Learning Models

Defining Bias in the Context of Machine Learning

Bias in machine learning refers to the systematic and unfair preferences or prejudices exhibited by models during training and inference (Mehrabi, Morstatter, Saxena, Lerman, & Galstyan, 2021). This bias can stem from various sources, including biased training data, algorithmic design choices, and the features used for model training.

Types of Bias

Understanding the nuances of bias requires an exploration of its various types:

I. **Sampling Bias:** Arises when the training dataset is not representative of the broader population, leading to skewed predictions for certain groups (Hughes et al., 2021).
II. **Algorithmic Bias:** Originates from the design and implementation of algorithms, introducing inherent biases that may not align with ethical considerations (Baker & Hawn, 2021).
III. **Labelling Bias:** Occurs when the labels assigned to data instances reflect existing biases, influencing the model's learning process (O'Reilly-Shah et al., 2020).

Sources of Bias

To effectively address bias, it is crucial to identify its sources:

I. **Biased Training Data**: Reflects historical biases present in society, potentially reinforcing and perpetuating those biases in the model (Mehrabi, Morstatter, Saxena, Lerman, & Galstyan, 2021).
II. **Biased Algorithms:** Arises from the choices made in algorithm design, such as the selection of features, model complexity, and optimization criteria (Seyyed-Kalantari et al., 2021).

III.   **Biased Labels:** Stemming from human annotations or labelling processes, introducing subjectivity and potential biases into the model (Ntoutsi, Fafalios, Gadiraju, Iosifidis, Nejdl, Vidal, Ruggieri, Turini, Papadopoulos, Krasanakis, Kompatsiaris, Kinder-Kurlanda, Wagner, Karimi, Fernandez, Alani, Berendt, Kruegel, Heinze, & Staab, 2020).

The Manifestations of Bias in Machine Learning Models

Gender Bias in Natural Language Processing (NLP)

Natural Language Processing (NLP) models, used for tasks such as sentiment analysis and language generation, have been found to exhibit gender bias (Lu et al., 2020). This bias can manifest in the form of stereotypes, reinforcing societal gender norms present in the training data.

Racial Bias in Facial Recognition

Facial recognition systems, integral to security and authentication, have faced criticism for exhibiting racial bias. This bias can result in higher error rates for certain racial and ethnic groups, leading to discriminatory outcomes (Cavazos et al., 2020).

Socioeconomic Bias in Credit Scoring Models

Credit scoring models, crucial for financial decisions, may introduce socioeconomic bias. Biases in historical lending practices, reflected in the training data, can influence the model's assessment of creditworthiness, disadvantaging individuals from lower socioeconomic backgrounds (Aaronson et al., 2021).

Addressing Bias: Strategies and Approaches

Transparent and Explainable AI Models

Transparency and explainability are foundational to addressing bias in machine learning models. Opening the black box of AI models enables stakeholders to understand how decisions are made, facilitating the identification and rectification of biased patterns (Rai, 2020).

Diverse and Representative Training Data

Addressing bias at its root requires diverse and representative training data. Efforts should be made to ensure that datasets encompass a broad spectrum of demographics, mitigating the risk of biases introduced by skewed or incomplete data (Lee et al., 2020).

Fairness Metrics and Evaluation

Incorporating fairness metrics into the evaluation of machine learning models provides a quantitative measure of bias. Metrics such as disparate impact, equal opportunity, and demographic parity enable a systematic assessment of fairness (Coston et al., 2020).

Ethical Design Principles

Infusing ethical design principles into the development lifecycle is critical. From the initial design phase to ongoing monitoring and updates, prioritizing fairness, transparency, and accountability helps embed ethical considerations into the core of AI systems (Morley et al., 2020).

Bias Detection and Model Interpretability

Leveraging tools for bias detection and model interpretability allows developers to identify and understand biases within their models. Interpretability enables stakeholders to trace model decisions, aiding in the identification of unintended biases (Meng et al., 2022).

Figure 4. Addressing bias: Strategies and approaches

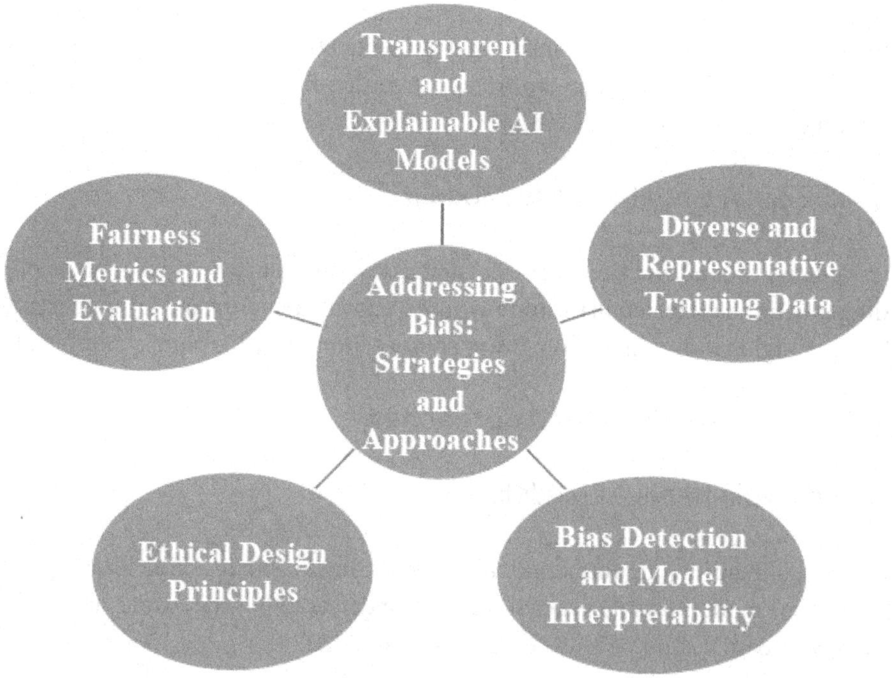

Challenges and Considerations in Bias Mitigation

Algorithmic Complexity

As machine learning models become more complex, understanding and mitigating biases become more challenging. Deep learning models, in particular, introduce complexities that demand sophisticated approaches to address biases effectively (Raghavan et al., 2020).

Ethical Considerations in Emerging Technologies

The evolution of AI introduces new challenges, especially in emerging technologies such as autonomous systems and artificial general intelligence (AGI) (Hancock et al., 2020). Anticipating and addressing biases in these technologies require proactive ethical considerations.

Future Directions: Towards Ethical and Unbiased AI

Research and Innovation

Ongoing research and innovation are essential for developing advanced techniques to identify and mitigate bias (Haefner et al., 2021). This involves interdisciplinary collaboration, drawing on insights from computer science, ethics, sociology, and other fields.

Regulatory Frameworks

The development of regulatory frameworks that guide the ethical use of AI is imperative. Governments and organizations should work collaboratively to establish standards that ensure fairness, transparency, and accountability in AI systems (Wirtz et al., 2020).

Education and Awareness

Raising awareness and educating stakeholders, from developers to end-users, is fundamental. Understanding the implications of bias in machine learning models fosters a culture of responsible AI development and usage (Yang et al., 2021).

FAIRNESS METRICS AND APPROACHES IN ETHICAL AI DEVELOPMENT

The Imperative of Fairness in AI Development

As artificial intelligence (AI) systems become increasingly integrated into various aspects of society, the ethical considerations surrounding their development gain prominence. Fairness, as a fundamental ethical principle, plays a pivotal role in ensuring equitable and unbiased AI (Madaio, Stark, Wortman Vaughan, & Wallach, 2020). This chapter provides an extensive exploration of fairness metrics and approaches, delving into the complexities of assessing and promoting fairness in AI development.

The Foundations of Fairness in AI

Defining Fairness in AI

Fairness in AI refers to the equitable treatment of individuals or groups, ensuring that the benefits and burdens of AI systems are distributed without bias or discrimination. Achieving fairness involves addressing the potential biases that can emerge at various stages of AI development, from data collection to model deployment (Mehrabi, Morstatter, Saxena, Lerman, & Galstyan, 2021).

The Importance of Fairness in Ethical AI

Fairness is a core ethical consideration in AI development for several reasons:

I. **Social Implications:** Biased AI systems can perpetuate and amplify existing social inequalities, leading to unjust outcomes in areas such as employment, finance, and criminal justice (Carter et al., 2020).
II. **User Trust:** Fair AI systems contribute to building and maintaining trust among users. When individuals perceive that AI systems treat them fairly, they are more likely to trust and accept the decisions made by these systems (Toreini et al., 2020).
III. **Legal and Regulatory Compliance:** Many jurisdictions are recognizing the importance of fairness in AI and incorporating it into legal and regulatory frameworks (Rodrigues, 2020). Adhering to fairness principles is essential for compliance with these evolving standards.

Fairness Metrics: Quantifying Ethical Considerations in AI Development

Types of Fairness Metrics

Assessing fairness in AI models involves the use of various metrics, each targeting specific aspects of fairness:

I. **Demographic Parity:** Examines whether the distribution of predictions is consistent across different demographic groups, ensuring equal opportunities for all (Bird et al., 2020).
II. **Equalized Odds:** Focuses on minimizing the differences in error rates among different groups, guaranteeing that the odds of making an error are equal across demographics (Pessach & Shmueli, 2023).
III. **Disparate Impact:** Measures whether the impact of a decision, such as hiring or lending, is proportionally similar for different groups, preventing one group from being disproportionately affected (Pessach & Shmueli, 2023).
IV. **Treatment Equality:** Evaluates whether individuals with similar characteristics receive similar outcomes, safeguarding against discriminatory practices (Pessach & Shmueli, 2023).

Challenges in Defining Fairness Metrics in AI Systems

Fairness is a critical aspect of developing ethical artificial intelligence (AI) systems, aiming to ensure equitable treatment for all individuals or groups. Defining fairness metrics, however, presents several challenges that reflect the complex nature of both AI technology and societal values (Bird et al., 2020). This section explores some of the key challenges in defining fairness metrics in AI systems:

I. **Subjectivity and Context Dependency:** One of the primary challenges in defining fairness metrics is the subjective nature of fairness itself. Different stakeholders may have diverse perspectives on what is considered fair, and these perspectives can vary across cultural, social, and ethical contexts (Bird et al., 2020). Determining a universal definition of fairness that accommodates this diversity remains an ongoing challenge.

II. **Multiple Dimensions of Fairness:** Fairness is a multidimensional concept, encompassing various aspects such as individual fairness, group fairness, and overall societal fairness. Striking a balance between these dimensions is complex, as improvements in one dimension may lead to trade-offs in another (Bird et al., 2020). Defining fairness metrics that consider and balance these multiple dimensions is a significant challenge.

III. **Trade-offs between Accuracy and Fairness:** There is often a trade-off between model accuracy and fairness. Interventions designed to mitigate bias and enhance fairness may lead to a reduction in overall model accuracy, and vice versa (Bird et al., 2020). Finding the optimal balance between accuracy and fairness is challenging and depends on the specific application and ethical considerations.

IV. **Lack of Ground Truth for Fairness:** Unlike traditional metrics in machine learning, such as precision or recall, fairness lacks a clear ground truth. Determining what represents a fair outcome is subjective and context-dependent (Bird et al., 2020). Without a definitive standard for fairness, defining metrics becomes challenging, and assessments are often based on societal norms and values.

V. **Contextual Sensitivity:** Fairness metrics need to be sensitive to the context of the application. What is considered fair in one domain may not be applicable in another. For example, fairness metrics for hiring algorithms may differ from those for predictive policing. Adapting fairness metrics to different contexts while maintaining ethical standards poses a significant challenge (Bird et al., 2020).

VI. **Dynamic Nature of Fairness:** Fairness is not a static concept; it evolves over time as societal norms, values, and expectations change. Defining fairness metrics that can adapt to dynamic shifts in ethical considerations is challenging (Bird et al., 2020). AI systems need to be flexible enough to accommodate evolving definitions of fairness without compromising their integrity.

VII. **Representativeness in Training Data:** The quality and representativeness of training data play a crucial role in defining fairness metrics. Biases present in historical data may perpetuate and amplify existing inequalities (Bird et al., 2020). Ensuring that fairness metrics account for and rectify biases in the training data is a complex challenge that requires careful consideration.

VIII. **Intersectionality and Multiple Dimensions of Identity:** Individuals possess multiple identities, and bias can manifest at the intersection of these identities. Defining fairness metrics that capture the nuanced interactions between different dimensions of identity, such as race, gender, and socio-economic status, poses a challenge (Bird et al., 2020). Failure to account for intersectionality may result in overlooking certain forms of bias.

IX. **Ethical Considerations in Metric Design:** The design of fairness metrics raises ethical questions. For instance, the selection of certain demographic attributes for fairness evaluation may lead to unintended consequences, such as reinforcing stereotypes (Bird et al., 2020). Ensuring that fairness metrics align with ethical principles and do not inadvertently contribute to societal harm requires careful ethical considerations.

X. **Lack of Standardization:** The absence of standardized fairness metrics complicates efforts to compare and evaluate different AI systems consistently. The field lacks a universally accepted set of fairness metrics, making it challenging to establish benchmarks and best practices (Bird et al., 2020). Standardizing fairness metrics would require collaboration across the AI community and regulatory bodies.

Bias Detection and Fairness Assessment in AI Models

Proactive Bias Detection

Proactively detecting bias in AI models involves identifying and mitigating potential sources of bias throughout the development lifecycle (Mehrabi, Morstatter, Saxena, Lerman, & Galstyan, 2021; Ntoutsi, Fafalios, Gadiraju, Iosifidis, Nejdl, Vidal, Ruggieri, Turini, Papadopoulos, Krasanakis, Kompatsiaris, Kinder-Kurlanda, Wagner, Karimi, Fernandez, Alani, Berendt, Kruegel, Heinze, & Staab, 2020). This includes scrutinizing training data, evaluating algorithmic choices, and ensuring diverse and representative datasets.

Model Interpretability for Fairness Assessment

Interpretable AI models facilitate a better understanding of the decision-making process, allowing stakeholders to assess and address potential biases (Linardatos et al., 2020). Interpretability tools help identify which features contribute most to predictions, aiding in the identification of discriminatory patterns.

Approaches to Ensuring Fairness in AI Development

Fairness-Aware Model Training

Incorporating fairness considerations directly into the model training process is a proactive approach (Madaio, Stark, Wortman Vaughan, & Wallach, 2020). This involves modifying algorithms or adjusting the training process to explicitly account for fairness metrics and constraints.

Adversarial Training for Fairness (Feuerriegel, Dolata, & Schwabe, 2020)

Adversarial training introduces adversarial examples during the training process to expose models to potential biases. This process helps the model learn to be robust to different demographic groups, mitigating the risk of biased predictions.

Ensemble Methods for Fairness

Ensemble methods combine predictions from multiple models to improve overall performance and fairness. By aggregating diverse models, these methods can reduce the impact of biases present in individual models, leading to more robust and equitable outcomes.

Post-Processing Techniques for Fairness

Post-processing techniques involve modifying the outputs of a trained model to ensure fairness. This can include re-ranking candidates or adjusting decision thresholds to align with fairness goals without retraining the entire model.

Ethical Considerations in Fairness Approaches

Balancing Competing Objectives

Achieving fairness often involves balancing competing objectives, such as accuracy and fairness (Richardson & Gilbert, 2021). Striking the right balance requires careful consideration of the ethical implications of these trade-offs and making transparent decisions that align with ethical principles.

Transparency and Explainability

Transparent and explainable AI systems are essential for fostering trust and accountability. When fairness approaches are embedded into AI models, providing explanations for decisions becomes crucial to ensure that stakeholders can understand and evaluate the fairness measures implemented.

Inclusivity in Decision-Making Processes

Ensuring inclusivity in the development and deployment of AI models is integral to addressing fairness. Diverse perspectives in decision-making teams can help uncover potential biases and contribute to the development of more equitable models.

Challenges and Considerations in Fairness Approaches

Trade-Offs between Accuracy and Fairness

Striking the right balance between accuracy and fairness remains a significant challenge. Interventions to address biases may impact overall model performance (Feuerriegel, Dolata, & Schwabe, 2020), requiring careful consideration of the trade-offs involved.

Dynamic Nature of Fairness Requirements

Fairness requirements may evolve over time, influenced by changes in societal norms, legal standards, or the application domain (Madaio, Stark, Wortman Vaughan, & Wallach, 2020). Developing AI systems that can adapt to evolving fairness considerations is a complex challenge.

Future Directions: Advancing Fairness in AI

Integration of Human Values

The integration of human values into fairness metrics and approaches is crucial for developing AI systems that align with societal expectations. This involves engaging with diverse stakeholders to understand their perspectives on fairness.

Explainable AI for Fairness

Advancements in explainable AI can further enhance fairness by providing transparent insights into how models make decisions (Joshi & Tiwari, 2023). The ability to explain and interpret AI predictions is instrumental in addressing concerns related to fairness.

Collaboration and Standardization

Collaborative efforts across the AI community and standardization bodies are essential for advancing fairness in AI. Establishing common frameworks and guidelines can help create a shared understanding of fairness principles and promote consistent practices.

CONCLUSION

In the ever-evolving landscape of AI development, this chapter serves as a comprehensive journey through the nuanced dimensions of ethical considerations at the intersection of technology and societal values. It meticulously explores critical aspects, including bias and fairness in AI algorithms, discrimination and algorithmic bias, addressing bias in machine learning models, and the intricate realm of fairness metrics and approaches. The exploration starts by delving into the intricacies of bias and fairness, emphasizing the imperative to address biases inherited by AI from training data and design choices. The discourse shifts to discrimination and algorithmic bias, highlighting the profound societal impacts stemming from unchecked biases in AI systems and the ethical responsibility to scrutinize algorithms for discriminatory patterns. Addressing bias in machine learning models becomes a pivotal theme, advocating for proactive strategies, model interpretability, and ethical design principles to rectify historical inequities. The final segment navigates the complex tapestry of fairness metrics, emphasizing the challenges in defining, measuring, and implementing fairness in AI systems and the importance of collaborative efforts in standardizing practices. Throughout the exploration, ethical considerations emerge not as constraints but guiding principles, paving the way for a future where technological progress aligns with societal values, inclusivity, and a responsible integration of AI into our evolving society.

REFERENCES

Aaronson, D., Faber, J., Hartley, D., Mazumder, B., & Sharkey, P. (2021). The long-run effects of the 1930s HOLC "redlining" maps on place-based measures of economic opportunity and socioeconomic success. *Regional Science and Urban Economics*, *86*, 103622. doi:10.1016/j.regsciurbeco.2020.103622

Akter, S., McCarthy, G., Sajib, S., Michael, K., Dwivedi, Y. K., D'Ambra, J., & Shen, K. N. (2021). Algorithmic bias in data-driven innovation in the age of AI. *International Journal of Information Management*, *60*, 102387. doi:10.1016/j.ijinfomgt.2021.102387

Baker, R. S., & Hawn, A. (2021). Algorithmic bias in education. *International Journal of Artificial Intelligence in Education*, 1–41.

Benke, I., Feine, J., Venable, J. R., & Maedche, A. (2020). On implementing ethical principles in design science research. *AIS Transactions on Human-Computer Interaction*, *12*(4), 206–227. doi:10.17705/1thci.00136

Bird, S., Dudík, M., Edgar, R., Horn, B., Lutz, R., Milan, V., & Walker, K. (2020). Fairlearn: A toolkit for assessing and improving fairness in AI. *Microsoft, Tech. Rep. MSR-TR-2020-32*.

Bohr, A., & Memarzadeh, K. (2020). Chapter 2 - The rise of artificial intelligence in healthcare applications. A. Bohr & K. Memarzadeh, (eds) Artificial Intelligence in Healthcare. Academic Press. doi:10.1016/B978-0-12-818438-7.00002-2

Brundage, M., Avin, S., Wang, J., Belfield, H., Krueger, G., Hadfield, G., & Anderljung, M. (2020). Toward trustworthy AI development: mechanisms for supporting verifiable claims. *arXiv preprint arXiv:2004.07213*.

Carter, S. M., Rogers, W., Win, K. T., Frazer, H., Richards, B., & Houssami, N. (2020). The ethical, legal and social implications of using artificial intelligence systems in breast cancer care. *The Breast*, *49*, 25–32. doi:10.1016/j.breast.2019.10.001 PMID:31677530

Cavazos, J. G., Phillips, P. J., Castillo, C. D., & O'Toole, A. J. (2020). Accuracy comparison across face recognition algorithms: Where are we on measuring race bias? *IEEE Transactions on Biometrics, Behavior, and Identity Science*, *3*(1), 101–111. doi:10.1109/TBIOM.2020.3027269 PMID:33585821

Chen, M., & Decary, M. (2020, January). Artificial intelligence in healthcare: An essential guide for health leaders. *Healthcare Management Forum*, *33*(1), 10–18. doi:10.1177/0840470419873123 PMID:31550922

Coston, A., Mishler, A., Kennedy, E. H., & Chouldechova, A. (2020, January). Counterfactual risk assessments, evaluation, and fairness. In *Proceedings of the 2020 conference on fairness, accountability, and transparency* (pp. 582-593). ACM. 10.1145/3351095.3372851

Dastile, X., Celik, T., & Potsane, M. (2020). Statistical and machine learning models in credit scoring: A systematic literature survey. *Applied Soft Computing*, *91*, 106263. doi:10.1016/j.asoc.2020.106263

Eitel-Porter, R. (2021). Beyond the promise: Implementing ethical AI. *AI and Ethics*, *1*(1), 73–80. doi:10.1007/s43681-020-00011-6

Feuerriegel, S., Dolata, M., & Schwabe, G. (2020). Fair AI: Challenges and opportunities. *Business & Information Systems Engineering*, *62*(4), 379–384. doi:10.1007/s12599-020-00650-3

Floridi, L., & Cowls, J. (2022). A unified framework of five principles for AI in society. *Machine learning and the city: Applications in architecture and urban design*, 535-545.

Floridi, L., Cowls, J., Beltrametti, M., Chatila, R., Chazerand, P., Dignum, V., & Vayena, E. (2021). An ethical framework for a good AI society: Opportunities, risks, principles, and recommendations. *Ethics, governance, and policies in artificial intelligence*, 19-39.

Gabriel, I. (2022). Toward a theory of justice for artificial intelligence. *Daedalus*, *151*(2), 218–231. doi:10.1162/daed_a_01911

González-Gonzalo, C., Thee, E., Klaver, C., Lee, A., Schlingemann, R., Tufail, Verbraak, F., & Sánchez, C. (2022). Trustworthy AI: Closing the gap between development and integration of AI systems in ophthalmic practice. *Progress in Retinal and Eye Research, 90.* doi:10.1016/j.preteyeres.2021.101034

Haefner, N., Wincent, J., Parida, V., & Gassmann, O. (2021). Artificial intelligence and innovation management: A review, framework, and research agenda. *Technological Forecasting and Social Change, 162,* 120392. doi:10.1016/j.techfore.2020.120392

Hancock, J. T., Naaman, M., & Levy, K. (2020). AI-mediated communication: Definition, research agenda, and ethical considerations. *Journal of Computer-Mediated Communication, 25*(1), 89–100. doi:10.1093/jcmc/zmz022

Harrison, G., Hanson, J., Jacinto, C., Ramirez, J., & Ur, B. (2020, January). An empirical study on the perceived fairness of realistic, imperfect machine learning models. In *Proceedings of the 2020 conference on fairness, accountability, and transparency* (pp. 392-402). ACM. 10.1145/3351095.3372831

Hassani, B. K. (2021). Societal bias reinforcement through machine learning: A credit scoring perspective. *AI and Ethics, 1*(3), 239–247. doi:10.1007/s43681-020-00026-z

Hughes, A. C., Orr, M. C., Ma, K., Costello, M. J., Waller, J., Provoost, P., Yang, Q., Zhu, C., & Qiao, H. (2021). Sampling biases shape our view of the natural world. *Ecography, 44*(9), 1259–1269. doi:10.1111/ecog.05926

Joshi, A., & Tiwari, H. (2023). No. 10. An Overview of Python Libraries for Data Science: Manuscript Received: 20 March 2023, Accepted: 12 May 2023, Published: 15 September 2023. *Journal of Engineering Technology and Applied Physics, 5*(2), 85–90. doi:10.33093/jetap.2023.5.2.10

Kamruzzaman, M. M. (2022). Impact of social media on geopolitics and economic growth: Mitigating the risks by developing artificial intelligence and cognitive computing tools. *Computational Intelligence and Neuroscience, 2022,* 2022. doi:10.1155/2022/7988894 PMID:35602647

Kordzadeh, N., & Ghasemaghaei, M. (2022). Algorithmic bias: Review, synthesis, and future research directions. *European Journal of Information Systems, 31*(3), 388–409. doi:10.1080/0960085X.2021.1927212

Lee, H. Y., Tseng, H. Y., Mao, Q., Huang, J. B., Lu, Y. D., Singh, M., & Yang, M. H. (2020). Drit++: Diverse image-to-image translation via disentangled representations. *International Journal of Computer Vision, 128*(10-11), 2402–2417. doi:10.1007/s11263-019-01284-z

Linardatos, P., Papastefanopoulos, V., & Kotsiantis, S. (2020). Explainable ai: A review of machine learning interpretability methods. *Entropy (Basel, Switzerland), 23*(1), 18. doi:10.3390/e23010018 PMID:33375658

Llorens, A., Tzovara, A., Bellier, L., Bhaya-Grossman, I., Bidet-Caulet, A., Chang, W. K., Cross, Z. R., Dominguez-Faus, R., Flinker, A., Fonken, Y., Gorenstein, M. A., Holdgraf, C., Hoy, C. W., Ivanova, M. V., Jimenez, R. T., Jun, S., Kam, J. W. Y., Kidd, C., Marcelle, E., ... Dronkers, N. F. (2021). Gender bias in academia: A lifetime problem that needs solutions. *Neuron, 109*(13), 2047–2074. doi:10.1016/j.neuron.2021.06.002 PMID:34237278

Lu, K., Mardziel, P., Wu, F., Amancharla, P., & Datta, A. (2020). Gender bias in neural natural language processing. *Logic, Language, and Security: Essays Dedicated to Andre Scedrov on the Occasion of His 65th Birthday*, 189-202.

Madaio, M. A., Stark, L., Wortman Vaughan, J., & Wallach, H. (2020, April). Co-designing checklists to understand organizational challenges and opportunities around fairness in AI. In *Proceedings of the 2020 CHI conference on human factors in computing systems* (pp. 1-14). ACM. 10.1145/3313831.3376445

Masevski, S., & Stojanovski, S. (2023). Artificial Intelligence: Geopolitical Tool Of Modern Countries. *44VOL. XXIII*.

Mehrabi, N., Morstatter, F., Saxena, N., Lerman, K., & Galstyan, A. (2021). A survey on bias and fairness in machine learning. *ACM Computing Surveys*, *54*(6), 1–35. doi:10.1145/3457607

Meng, C., Trinh, L., Xu, N., Enouen, J., & Liu, Y. (2022). Interpretability and fairness evaluation of deep learning models on MIMIC-IV dataset. *Scientific Reports*, *12*(1), 7166. doi:10.1038/s41598-022-11012-2 PMID:35504931

Morley, J., Floridi, L., Kinsey, L., & Elhalal, A. (2020). From what to how: An initial review of publicly available AI ethics tools, methods and research to translate principles into practices. *Science and Engineering Ethics*, *26*(4), 2141–2168. doi:10.1007/s11948-019-00165-5 PMID:31828533

Müller, V. (2023). Ethics of Artificial Intelligence and Robotics. E. N. Zalta & U. Nodelman (eds.), *The Stanford Encyclopedia of Philosophy*. Stanford Press. <https://plato.stanford.edu/archives/fall2023/entries/ethics-ai/>

Nguyen, A., Ngo, H. N., Hong, Y., Dang, B., & Nguyen, B. P. T. (2023). Ethical principles for artificial intelligence in education. *Education and Information Technologies*, *28*(4), 4221–4241. doi:10.1007/s10639-022-11316-w PMID:36254344

Norori, N., Hu, Q., Aellen, F. M., Faraci, F. D., & Tzovara, A. (2021). Addressing bias in big data and AI for health care: A call for open science. *Patterns (New York, N.Y.)*, *2*(10), 100347. doi:10.1016/j.patter.2021.100347 PMID:34693373

Ntoutsi, E., Fafalios, P., Gadiraju, U., Iosifidis, V., Nejdl, W., Vidal, M. E., Ruggieri, S., Turini, F., Papadopoulos, S., Krasanakis, E., Kompatsiaris, I., Kinder-Kurlanda, K., Wagner, C., Karimi, F., Fernandez, M., Alani, H., Berendt, B., Kruegel, T., Heinze, C., & Staab, S. (2020). Bias in data-driven artificial intelligence systems—An introductory survey. *Wiley Interdisciplinary Reviews. Data Mining and Knowledge Discovery*, *10*(3), e1356. doi:10.1002/widm.1356

O'Reilly-Shah, V. N., Gentry, K. R., Walters, A. M., Zivot, J., Anderson, C. T., & Tighe, P. J. (2020). Bias and ethical considerations in machine learning and the automation of perioperative risk assessment. *British Journal of Anaesthesia*, *125*(6), 843–846. doi:10.1016/j.bja.2020.07.040 PMID:32838979

Pessach, D., & Shmueli, E. (2023). Algorithmic fairness. In *Machine Learning for Data Science Handbook: Data Mining and Knowledge Discovery Handbook* (pp. 867–886). Springer International Publishing. doi:10.1007/978-3-031-24628-9_37

Peters, D., Vold, K., Robinson, D., & Calvo, R. A. (2020). Responsible AI—Two frameworks for ethical design practice. *IEEE Transactions on Technology and Society, 1*(1), 34–47. doi:10.1109/TTS.2020.2974991

Raghavan, M., Barocas, S., Kleinberg, J., & Levy, K. (2020, January). Mitigating bias in algorithmic hiring: Evaluating claims and practices. In *Proceedings of the 2020 conference on fairness, accountability, and transparency* (pp. 469-481). ACM. 10.1145/3351095.3372828

Rai, A. (2020). Explainable AI: From black box to glass box. *Journal of the Academy of Marketing Science, 48*(1), 137–141. doi:10.1007/s11747-019-00710-5

Raji, I. D., Smart, A., White, R. N., Mitchell, M., Gebru, T., Hutchinson, B., ... Barnes, P. (2020, January). Closing the AI accountability gap: Defining an end-to-end framework for internal algorithmic auditing. In *Proceedings of the 2020 conference on fairness, accountability, and transparency* (pp. 33-44). ACM. 10.1145/3351095.3372873

Richardson, B., & Gilbert, J. E. (2021). A framework for fairness: a systematic review of existing fair AI solutions. *arXiv preprint arXiv:2112.05700.*

Robert, L. P., Pierce, C., Marquis, L., Kim, S., & Alahmad, R. (2020). Designing fair AI for managing employees in organizations: A review, critique, and design agenda. *Human-Computer Interaction, 35*(5-6), 545–575. doi:10.1080/07370024.2020.1735391

Rodrigues, R. (2020). Legal and human rights issues of AI: Gaps, challenges and vulnerabilities. *Journal of Responsible Technology, 4*, 100005. doi:10.1016/j.jrt.2020.100005

Sambasivan, N., Arnesen, E., Hutchinson, B., Doshi, T., & Prabhakaran, V. (2021, March). Re-imagining algorithmic fairness in india and beyond. In *Proceedings of the 2021 ACM conference on fairness, accountability, and transparency* (pp. 315-328). ACM. 10.1145/3442188.3445896

Seyyed-Kalantari, L., Zhang, H., McDermott, M. B., Chen, I. Y., & Ghassemi, M. (2021). Underdiagnosis bias of artificial intelligence algorithms applied to chest radiographs in under-served patient populations. *Nature Medicine, 27*(12), 2176–2182. doi:10.1038/s41591-021-01595-0 PMID:34893776

Shin, D. (2021). The effects of explainability and causability on perception, trust, and acceptance: Implications for explainable AI. *International Journal of Human-Computer Studies, 146*, 102551. doi:10.1016/j.ijhcs.2020.102551

Smit, K., Zoet, M., & van Meerten, J. (2020). *A review of AI principles in practice.* Academic Press.

Toreini, E., Aitken, M., Coopamootoo, K., Elliott, K., Zelaya, C. G., & Van Moorsel, A. (2020, January). The relationship between trust in AI and trustworthy machine learning technologies. In *Proceedings of the 2020 conference on fairness, accountability, and transparency* (pp. 272-283). ACM. 10.1145/3351095.3372834

Wirtz, B. W., Weyerer, J. C., & Sturm, B. J. (2020). The dark sides of artificial intelligence: An integrated AI governance framework for public administration. *International Journal of Public Administration, 43*(9), 818–829. doi:10.1080/01900692.2020.1749851

Yang, S. J., Ogata, H., Matsui, T., & Chen, N. S. (2021). Human-centered artificial intelligence in education: Seeing the invisible through the visible. *Computers and Education: Artificial Intelligence, 2*, 100008. doi:10.1016/j.caeai.2021.100008

Yarger, L., Cobb Payton, F., & Neupane, B. (2020). Algorithmic equity in the hiring of underrepresented IT job candidates. *Online Information Review, 44*(2), 383–395. doi:10.1108/OIR-10-2018-0334

Završnik, A. (2021). Algorithmic justice: Algorithms and big data in criminal justice settings. *European Journal of Criminology, 18*(5), 623–642. doi:10.1177/1477370819876762

Chapter 11
Exploring the Ethical Implications of Generative AI in Healthcare

Dinesh Kumar
https://orcid.org/0000-0001-5943-1444
Lovely Professional University, India

Rohit Dhalwal
Lovely Professional University, India

Ayushi Chaudhary
Lovely Professional University, India

ABSTRACT

This chapter critically evaluates the ethical challenges posed by the advent of generative artificial intelligence (GenAI) in healthcare. It investigates how GenAI's potential to revolutionize patient care and medical research is counterbalanced by significant ethical concerns, including privacy, security, and equity. An extensive literature review supports a deep dive into these issues, comparing GenAI's impact on traditional healthcare ethics. Through case studies and theoretical analysis, the chapter seeks to understand GenAI's ethical implications thoroughly, aiming to contribute to the development of nuanced ethical frameworks in this rapidly advancing area.

INTRODUCTION

The emergence of Generative Artificial Intelligence (GenAI) in healthcare marks a significant milestone, heralding a new epoch in the confluence of medical science and technological innovation. This chapter critically examines the ethical dimensions inherent in this groundbreaking development. It is defined as AI that generates novel content or data through machine learning algorithms and neural networks (Goodfellow et al., 2016). GenAI is poised to transform healthcare in unprecedented ways. However, this transformative potential necessitates a thorough scrutiny of its ethical implications.

DOI: 10.4018/979-8-3693-2964-1.ch011

GenAI's introduction into healthcare signals a transformative shift in offering benefits ranging from improved patient care to groundbreaking strides in drug discovery and predictive analytics. Despite these advancements, the integration of GenAI also brings complex ethical challenges. Paramount among these are concerns regarding patient privacy, data security, informed consent, and the potential exacerbation of existing healthcare disparities. This chapter aims to dissect and critically analyse these ethical dilemmas, focusing specifically on how genAI is reshaping the landscape of healthcare ethics.

This chapter employs a multifaceted methodological approach to explore the ethical dimensions of genAI in healthcare thoroughly. This approach's core is an extensive literature review, which serves as the foundation for our inquiry. This review encompasses a broad range of scholarly articles, research papers, and case studies at the intersection of AI, ethics, and healthcare. The goal is to capture diverse perspectives and trace the evolution of AI's role in healthcare, particularly focusing on its ethical ramifications. This comprehensive review of the literature ensures that our analysis is anchored in the latest knowledge, incorporating a variety of viewpoints and research findings.

A key component of our methodology is comparative analysis. We highlight AI technologies' distinct challenges and opportunities by juxtaposing the ethical issues of generative AI against those in traditional healthcare practices. This comparison aids in discerning how generative AI intersects with and diverges from existing ethical norms and standards in healthcare.

We also engage in detailed case study evaluations, examining instances where AI has been implemented in healthcare settings. These case studies serve as practical examples of the ethical considerations in action, showcasing successes and challenges in real-world scenarios. Analysing these cases provides tangible illustrations of how ethical issues manifest in practice.

Moreover, we apply theoretical frameworks from ethics and AI research to the specific context of healthcare. We also delve into AI-specific ethical frameworks proposed by contemporary researchers. This theoretical foundation is crucial for developing effective strategies for addressing ethical challenges.

OBJECTIVES OF THE CHAPTER

- To provide an in-depth overview of the applications and potential of generative AI in healthcare.
- To identify and articulate the key ethical issues of deploying generative AI in healthcare.
- To thoroughly examine the complexities and broader implications of these ethical challenges.
- To develop ethical frameworks and strategies to address these concerns.

EXPLORATION OF CURRENT APPLICATIONS OF GENAI

The integration of GenAI into healthcare has revolutionised various aspects of medical practice and research. This section explores the current applications of generative AI in healthcare, illustrating its transformative impact.

Drug Discovery and Personalized Medicine: One of the most groundbreaking applications of generative AI in healthcare is in drug discovery. Researchers can rapidly identify and synthesise new drug compounds by leveraging AI algorithms, significantly accelerating drug development. This has been particularly valuable in the search for treatments for complex diseases like cancer and rare genetic disorders (Zeng et al., 2022; Hasselgren & Oprea, 2024; Walters & Barzilay, 2021; Walters & Murcko, 2020). In personalised medicine, generative AI analyses individual patient data to tailor treatments to each patient's genetic makeup, improving treatment efficacy and reducing side effects (Grupac et al., 2023; Nova, 2023).

Medical Imaging and Diagnosis: GenAI has also made significant strides in medical imaging. AI algorithms can generate accurate and detailed images, aiding in early and precise disease diagnosis. This technology is particularly impactful in radiology, where it assists in identifying anomalies in X-rays, MRIs, and CT scans with greater accuracy and speed than traditional methods (Pinaya et al., 2023; Musalamadugu & Kannan, 2023; Paladugu et al., 2023).

Predictive Healthcare and Epidemiology: In predictive healthcare, generative AI is vital in forecasting disease trends and patient outcomes. This ability is crucial for effective healthcare planning and management, especially in anticipating and mitigating epidemic and pandemic outbreaks (Thiagarajan et al., 2020; Henriques et al., 2015). AI models can analyse vast datasets to predict patient admissions, potential complications, and disease spread patterns, providing healthcare providers and policymakers invaluable insights.

Synthetic Data Generation for Research: Generating synthetic patient data for research is an emerging and significant application of GenAI. This technology creates realistic, anonymised datasets that researchers can use without violating patient privacy. Such synthetic data is vital for various research purposes, including clinical trials, epidemiological studies, and training AI models in healthcare (Figueira & Vaz, 2022; Assefa et al., 2020; de Melo, Torralba, Guibas, DiCarlo, Chellappa, & Hodgins, 2022; Dave et al., 2023).

Improving Healthcare Accessibility: GenAI is increasingly used to enhance healthcare accessibility, especially in remote or underserved areas. By generating predictive models and telemedicine applications, GenAI can help bridge the gap in healthcare delivery, offering diagnostic support and treatment advice in regions with limited access to medical professionals (Bugaj et al., 2023).

Mental Health Assessment and Therapy: GenAI is increasingly used in mental health for assessment and therapeutic interventions. AI algorithms can analyse speech and language patterns to aid in diagnosing conditions like depression and anxiety. Additionally, GenAI is used to create interactive virtual environments for cognitive-behavioural therapy, providing patients with accessible and personalised mental health support (De Freitas, Uğuralp, Oğuz-Uğuralp, & Puntoni, 2022; Lam et al., 2021; King et al., 2023).

AI-assisted Surgery and Robotic Procedures: GenAI contributes to developing AI-assisted surgical techniques and robotic procedures in surgical settings. These technologies allow for more precise and less invasive surgeries, improving patient outcomes and reducing recovery times. GenAI algorithms assist in planning surgical procedures by generating predictive models of patient anatomy and surgical outcomes (Bodenstedt et al., 2020; Ray et al., 2023).

Development of Advanced Diagnostic Tools: GenAI is instrumental in creating advanced diagnostic tools. By analysing complex medical data, AI can identify patterns and anomalies that may indicate diseases. This application is crucial for early detection and intervention, particularly in conditions where early diagnosis significantly improves treatment success rates (Pagano et al., 2023).

Health Policy and System Management: GenAI also plays a role in health policy and system management. AI can assist in resource allocation, healthcare system optimisation, and policy development by analysing healthcare data. This application is vital for enhancing healthcare delivery efficiency and making informed decisions on public health matters (Dunn et al., 2023; Zhang & Kamel Boulos, 2023).

Enhancing Patient Engagement and Self-Management: GenAI technologies are being used to develop applications and tools that enhance patient engagement (Hayward, 2023) and self-management (Aguilar et al., 2021) of chronic conditions. These tools provide personalised recommendations and interventions, empowering patients to manage their health actively.

Geriatric Care and Management: GenAI is making significant contributions to geriatric care, particularly in monitoring and managing the health of the elderly. AI-driven systems can analyse behavioural data to detect early signs of cognitive decline, falls, or other health-related issues in older adults, enabling timely interventions and personalised care plans (Haque, 2023; Alexopoulos, 2023; Woodman & Mangoni, 2023; Yu et al., 2023).

Biomedical Research and Development: GenAI is a powerful tool for hypothesis generation and data analysis in biomedical research. It helps uncover new biological insights, facilitate the discovery of novel biomarkers, and understand complex disease mechanisms. This accelerates the pace of biomedical research and opens new avenues for therapeutic interventions (Athanasopoulou et al., 2022; Wang et al., 2023).

Rehabilitation and Physical Therapy: GenAI is being used to enhance rehabilitation (Li & Vakanski, 2018; Mennella et al., 2023) and physical therapy practices (Pagano et al., 2023). GenAI algorithms can generate personalised exercise and therapy regimens based on individual patient data, improving recovery outcomes. Moreover, GenAI-driven virtual reality systems are employed for interactive and engaging rehabilitation exercises, especially for patients with motor function impairments (Chheang et al., 2023; Urban et al., 2021).

Chronic Disease Management: GenAI aids in the management of chronic diseases by analysing patient data to predict flare-ups and suggesting preventive measures. This proactive approach helps in better managing conditions like diabetes, heart disease, and asthma, reducing the likelihood of emergency interventions and improving the overall quality of life for patients (Hegde & Mundada, 2022; Jurado-Camino et al., 2023).

Nutritional Analysis and Diet Planning: Leveraging GenAI for nutritional analysis and diet planning is an emerging application. AI algorithms can analyse dietary needs and preferences to generate personalised nutrition plans, which is particularly beneficial for patients with specific dietary needs, such as those with diabetes or food allergies (Jung et al., 2021; Lee et al., 2022).

Emergency Response and Critical Care: GenAI can play a crucial role in patient triage and decision-making in emergency and critical care settings. By quickly analysing patient data, GenAI systems can assist healthcare professionals in making critical decisions under time constraints, potentially improving patient outcomes in emergencies (Huang et al. et al., 2023; Zeng, 2023; Clark & Severn, 2023).

ETHICAL ISSUES OF GENAI IN HEALTHCARE

The applications mentioned above confirm the utilities of GenAI in healthcare. Along with these applications, several ethical issues must be addressed to ensure this technology's responsible and equitable use. This section explores the ethical considerations of GenAI in healthcare.

AI, Privacy, and Data Security in Healthcare: The protection of patient privacy and data security remain paramount ethical concerns in deploying GenAI within healthcare settings. Given that GenAI systems rely heavily on extensive datasets for learning and generating insights, there is an inherent risk of compromised or misusing sensitive patient information. To mitigate these risks, it is imperative to implement robust data protection strategies and adhere stringently to privacy regulations to safeguard patient confidentiality. Ethical considerations involve data governance, consent for data use, and ensuring that data sharing benefits all participants, particularly those from whom the data is sourced.

Informed Consent and Transparency: Using patient data in generative AI models raises questions about informed consent. Patients must be adequately informed about how their data is being used and the implications of AI-driven healthcare decisions. Furthermore, transparency in AI algorithms is necessary to build trust among patients and healthcare providers (Gupta et al., 2023; Wang et al., 2023).

Bias and Fairness: GenAI models can inadvertently perpetuate and amplify biases present in the training data. This can lead to unfair treatment outcomes for certain patient groups, exacerbating healthcare disparities. Ensuring fairness and mitigating bias in AI algorithms is a significant ethical challenge (Hastings, 2024; Teo et al., 2023). The application of GenGI in clinical trials for patient selection, data analysis, and outcome prediction must be conducted ethically. This includes considerations around fairness in patient selection, transparency in GenAI algorithms used for trials, and ensuring that AI does not introduce new biases or inequalities in clinical research.

Accountability and Liability: Determining accountability for decisions made by GenAI systems in healthcare is complex. When an AI-driven diagnosis or treatment plan goes wrong, it is challenging to assign responsibility, raising ethical and legal questions about liability (Kenthapadi et al., 2023; Zhong et al., 2023). The increasing reliance on GenAI in healthcare decision-making can impact patient and clinician autonomy. Ensuring that final decisions always involve human judgment and oversight is essential to respect patient autonomy and ensure the ethical use of AI.

Impact on Healthcare Workforce: Implementing GenAI in healthcare could lead to shifts in the workforce, impacting roles and responsibilities. Ethical considerations include ensuring that these changes do not compromise the quality of patient care and addressing the potential for job displacement (Walkowiak & MacDonald, 2023).

AI in Pediatric and End-of-Life Care: The use of GenAI in pediatric care raises unique ethical considerations. Children, as a vulnerable population, require additional protections regarding consent and the use of their health data (Ng, 2023). Moreover, GenAI systems need to be carefully adapted to recognize and address the specific medical needs of children. The use of GenAI in end-of-life care decisions poses significant ethical challenges. Decisions involving end-of-life care are sensitive and typically involve complex emotional and ethical considerations. The role of AI in such decisions must be carefully evaluated to respect patient autonomy and dignity while ensuring compassionate care.

Mental Health and AI: In the mental health field, the ethical use of GenAI involves ensuring that its tools are sensitive to the complexities and nuances of mental health conditions. There is a need for its systems to be developed and used in a manner that is empathetic and respects the privacy and dignity of individuals with mental health conditions (De Freitas, Uğuralp, Oğuz-Uğuralp, & Puntoni, 2022; Lam et al., 2021).

GenAI in Global Health Equity: The deployment of AI in global health contexts raises ethical issues around equity. There is a concern that AI technologies might primarily benefit high-income countries, widening the healthcare gap with low- and middle-income countries. Ensuring that AI contributes to global health equity is a significant ethical challenge (Bautista et al., 2023; Spector-Bagdady, 2023).

Future Consent and Unforeseen Uses of Data: The rapid evolution of GenAI technologies poses challenges in obtaining informed consent for future, unforeseen uses of health data. Ethical frameworks must consider how to handle consent for data that might be used in ways not envisioned at the time of collection (Gupta et al., 2023; Wang et al., 2023).

Continuity of Care and Human Touch: As GenAI technologies increasingly automate various healthcare processes, there is an ethical concern about the potential loss of the human element in care. The continuity of care and the human touch are crucial components of healthcare, particularly in building patient trust and understanding. Ensuring that GenAI complements rather than replaces human interaction in healthcare is a vital ethical consideration (Brynjolfsson et al., 2023). The introduction of GenAI into healthcare settings can alter the traditional dynamics between patients and healthcare providers. Ethical considerations include ensuring that AI supports and enhances the patient-provider relationship, maintains clear communication, and does not diminish the provider's role in decision-making and empathetic patient care.

AI and the Future of Health Insurance: The use of GenAI in health risk assessment and insurance underwriting raises ethical issues regarding privacy, discrimination, and fairness. A key ethical concern is ensuring that AI applications in health insurance are transparent and equitable and do not disadvantage certain individuals or groups.

Consent in the Age of AI: The complexities of GenAI in healthcare may challenge traditional models of informed consent. Patients might struggle to understand the extent and implications of AI-driven healthcare processes. Developing new consent models that address these complexities is an ethical imperative.

Algorithmic Transparency and Interpretability: As GenAI technologies become more complex, ensuring their transparency (Tang et al., 2023) and interpretability (Ross et al., 2021; Adel et al., 2018) is crucial. Patients and healthcare providers must understand how AI reaches its conclusions to trust and effectively use these systems. Ethical AI should be transparent enough for users to recognise its reasoning and limitations.

The Digital Divide: The digital divide poses a significant ethical concern in deploying GenAI in healthcare. Access to advanced healthcare technologies, including AI-driven tools, is not uniformly distributed, which can widen health disparities. Addressing this divide is essential to ensure equitable benefits from AI advancements.

AI and the Evolution of Patient Expectations: The integration of GenAI into healthcare is shifting patient expectations regarding speed, accuracy, and the nature of healthcare services. This shift raises ethical concerns about managing expectations and ensuring that AI integration does not lead to unrealistic demands or reliance on technology at the expense of human judgment. The use of AI in guiding health behaviour changes and lifestyle interventions must consider ethical implications such as personal autonomy, consent, and the potential for manipulation. AI systems designed to influence health behaviours should prioritize ethical principles, ensuring that interventions are respectful and in the best interests of the individuals.

Cross-Cultural Ethical Considerations: GenAI applications in healthcare must be sensitive to different cultural contexts and values. The ethical use of AI in global health requires understanding and respecting diverse cultural perspectives, especially when AI tools developed in one region are applied in another.

AI and Health Data Monetization: AI companies' monetisation of health data raises ethical concerns regarding patient privacy, data ownership, and potential conflicts of interest. A key challenge is ensuring that patient data is used ethically and not exploited for commercial gain (Firouzi et al., 2020).

AI and the Right to Not Know: Patients often have the right not to know certain medical information in healthcare. The use of GenAI, especially in predictive analytics and genetic testing, challenges this right. Ethical use of AI must consider patients' preferences regarding disclosing prognostic information.

Health Marketing and Promotion: AI in health marketing and promotion should be ethically aligned, avoiding manipulating or exploiting consumer data. Ethical considerations include ensuring transparency, respecting consumer privacy, and avoiding the creation of unrealistic health expectations (Kshetri et al., 2023).

Long-Term Societal Impact of AI in Healthcare: The long-term societal impact of integrating GenAI into healthcare, such as changes in healthcare delivery models, impacts on healthcare costs, and broader societal implications, requires ethical scrutiny. It is crucial to consider how these changes affect different segments of society and strive for outcomes that benefit the broader community.

ANALYSIS OF CURRENT REGULATIONS AND THE NEED FOR EVOLVED POLICIES

While examining the current regulatory landscape and the need for evolved policies in GenAI in healthcare, it is instructive to include examples from different jurisdictions.

Current Regulatory Landscape

1. **Data Privacy Regulations**: Regulations like GDPR and HIPAA set the standard for data privacy. In India, the Personal Data Protection Bill, which is under consideration, aims to provide a framework similar to GDPR for data protection and privacy, which is crucial for managing healthcare data used in AI (Sarangi, 2021).
2. **Medical Device and AI Regulation**: In the US, the FDA is developing a regulatory framework for AI as a medical device. Similarly, the Central Drugs Standard Control Organization (CDSCO) under the Directorate General of Health Services regulates medical devices in India. However, specific guidelines for AI in healthcare are still evolving (Majumdar et al., 2020).

Need for Evolved Policies

1. **Adapting to AI's Evolving Nature**: Policies globally and in India need to adapt to the dynamic nature of AI, ensuring continuous oversight. For example, India's National Strategy for Artificial Intelligence emphasises the importance of adaptable policies in line with AI's evolving capabilities (NITI Aayog, 2018).

2. **Addressing AI Bias and Fairness**: Ensuring fairness in AI applications is a global challenge. In India, this includes addressing disparities in healthcare access and ensuring that AI tools do not exacerbate these disparities (Kumar & Sundar, 2019).

3. **Enhancing Transparency and Accountability**: Policies must enhance the transparency of AI systems worldwide, including India. Indian policymakers are challenged to establish guidelines that ensure accountability and transparency in AI systems, particularly in healthcare (Agarwal & Srikant, 2021).

4. **Global Standards and Cross-Border Challenges**: The need for global standards in AI is crucial. With its burgeoning tech industry, India can play a significant role in shaping these standards, especially the cross-border nature of data and AI technologies (Gupta, 2020).

5. **Ethical Frameworks for AI Development and Deployment**: Developing comprehensive ethical frameworks is necessary globally. In India, this includes integrating ethical considerations specific to the country's socio-cultural context in AI deployment in healthcare (Mehta, 2021).

6. **Public Engagement and Policy Development**: Involving the public in policy development is key. India's healthcare deployment could benefit from greater public engagement to ensure alignment with societal values and expectations (Rao & Vaidyanathan, 2021).

The integration of GI in healthcare, both globally and in India, shows that while current regulations provide a foundational framework, there is a significant need for evolved policies. These policies should address the unique challenges of AI, ensuring ethical, fair, and transparent use of technology in healthcare.

TOWARDS AN ETHICAL FRAMEWORK FOR GENAI IN HEALTHCARE

Developing an ethical framework for GenAI in healthcare is crucial to navigating the complex landscape of AI-driven medical behaviours. This section outlines key components and considerations for such a framework to ensure responsible, equitable, and beneficial use of GenAI in healthcare.

1. **Principle of Benebehavioursnd Non-Maleficence**: The framework's core should be the principles of beneficence and non-maleficence, ensuring that GenAI is used to benefit patients and not cause harm. This includes rigorous safety and efficacy testing and continuous monitoring for unintended consequences;

2. **Ensuring Privacy and Data Protection**: Respecting patient privacy and robust data protection are essential. The framework should include strict data governance policies, consent protocols, and security measures to protect sensitive health information.

3. **Transparency and Explainability**: GI systems must be transparent and explainable. Healthcare providers and patients should understand how AI systems reach their conclusions and have clear documentation of the AI decision-making process.

4. **Equity and Patients** must ensure that GenAIs do not know certain medical information in healthcare inequalities. This includes developing AI models representative of diverse populations and ensuring equitable access to AI-driven healthcare interventions.

5. **Accountability and Responsibility**: Clear guidelines for accountability and responsibility in AI-driven healthcare decisions are necessary. This includes determining the liability for misdiagnoses

or errors made by AI systems and ensuring that there is always a human in the loop for critical healthcare decisions.

6. **Patient Autonomy and Consent**: Upholding patient autonomy is paramount. Patients should have the right to informed consent regarding the use of AI in their healthcare, including the right to opt out of AI-driven care.

7. **Cross-Cultural Competence**: The framework should respect cultural diversity and ethical pluralism in healthcare. This involves developing AI systems sensitive to different cultural values and healthcare practices.

8. **Professional Competence and AI Literacy**: Healthcare professionals should have the necessary AI literacy to work effectively with AI systems. This includes training in AI tools, understanding their limitations, and integrating AI into clinical practice.

9. **Sustainable and Responsible AI Development**: AI development in healthcare should be sustainable and responsible, considering long-term impacts on health systems, society, and the environment. This includes responsible resource use and minimising the environmental footprint of AI technologies.

10. **Continuous Ethical Evaluation and Adaptation**: The ethical framework should be dynamic, allowing for continuous evaluation and adaptation as AI technologies and healthcare needs evolve. This includes regular reviews of policies and practices, stakeholder engagement, and adapting to new ethical challenges.

ETHICAL GUIDELINES AND IMPLEMENTATION STRATEGIES FOR AI IN HEALTHCARE

As healthcare increasingly integrates GenAI, developing robust ethical guidelines and practical implementation strategies becomes essential. This section outlines the key elements for formulating these guidelines and strategies to ensure the ethical use of AI in healthcare.

1. Establishing Clear Ethical Guidelines:
    - **Principles-Based Approach**: Adopt a principles-based approach to AI ethics, focusing on fundamental principles such as beneficence, non-maleficence, autonomy, and justice. These principles should guide all AI development and deployment in healthcare settings.
    - **Contextual Considerations**: Recognize that applying ethical principles will vary depending on the context. Guidelines should be flexible enough to accommodate different healthcare scenarios while maintaining a consistent ethical foundation.

2. Implementing Ethical AI Practices:
    - **Stakeholder Involvement**: Involve a wide range of stakeholders, including healthcare professionals, patients, ethicists, AI developers, and policymakers, in the development and implementation of AI technologies. This inclusive approach ensures that diverse perspectives and needs are considered.
    - **Training and Education**: Implement training programs for healthcare providers to enhance their understanding of AI, including its capabilities, limitations, and ethical implications. This ensures informed and responsible use of AI in clinical settings.

3. Promoting Transparency and Accountability:
   - **Algorithmic Transparency**: Ensure that AI algorithms used in healthcare are transparent and that users can interpret and understand their decision-making processes. This includes providing explanations for AI-driven recommendations and decisions.
   - **Accountability Frameworks**: Develop clear accountability frameworks for AI-driven decisions. Establish guidelines for liability in cases of AI errors or failures, ensuring a clear understanding of who is responsible and under what circumstances.
4. Addressing Bias and Ensuring Fairness:
   - **Diverse Data Sets**: Use diverse and representative data sets to train AI models, reducing the risk of bias. Regularly evaluate AI systems for biased outcomes and retrain them as necessary.
   - **Equity Audits**: Conduct regular equity audits of AI systems to assess their impact on different patient populations. Adjust implementation strategies to ensure equitable access to and benefits from AI technologies in healthcare.
5. Ensuring Privacy and Data Security:
   - **Robust Data Governance**: Implement robust policies to protect patient data. This includes strict adherence to privacy laws, secure data storage and transmission, and clear data sharing and consent policies.
   - **Data Security Protocols**: Develop and enforce stringent data security protocols to prevent unauthorised access to sensitive health information. Regularly update these protocols to address emerging cybersecurity threats.
6. Monitoring and Continuous Evaluation:
   - **Regular Review and Adaptation**: Establish mechanisms for the regular review and adaptation of AI systems and ethical guidelines. As AI technologies and healthcare needs evolve, guidelines and strategies should be updated to reflect new challenges and insights.
   - **Feedback Loops**: Create feedback loops that allow for the continuous collection of data on the effectiveness and ethical impact of AI applications in healthcare. Use this data to inform ongoing improvements and modifications.

CONCLUSION

This chapter has comprehensively explored the ethical challenges posed by integrating GenAI into healthcare and proposed solutions to address these challenges. The ethical landscape of GenAI in healthcare is complex and multifaceted, involving issues such as data privacy, bias and fairness, accountability, transparency, and the need for equitable access. The rapid evolution of AI technologies further adds to the complexity, requiring adaptive and dynamic responses.

The proposed solutions emphasise the development of robust ethical frameworks grounded in beneficence, non-maleficence, autonomy, and justice principles. These frameworks must be flexible enough to adapt to the evolving nature of AI and contextual variations in healthcare. Key to this adaptation is the involvement of diverse stakeholders in developing and implementing AI technologies, ensuring that multiple perspectives are considered and that the technology is aligned with societal values and ethical standards. Training and education of healthcare professionals in AI literacy, along with public engagement, are crucial for the ethical implementation of AI in healthcare. These efforts can help demystify

AI, promote understanding and trust, and ensure that healthcare providers are prepared to responsibly integrate AI into clinical practice. Regulatory frameworks need to evolve alongside AI advancements. Current regulations offer a starting point, but they must be expanded to the characteristics and challenges of AI in healthcare. This includes ensuring algorithmic transparency, developing accountability mechanisms for AI decisions, and mitigating biases in AI models. Equity and access remain paramount concerns. Policies and practices should ensure that the benefits of AI in healthcare are accessible to all, preventing the exacerbation of existing healthcare disparities. This requires a global perspective, considering different regions and communities' diverse healthcare needs and resources.

REFERENCES

Adel, T., Ghahramani, Z., & Weller, A. (2018, July). Discovering interpretable representations for both deep generative and discriminative models. In *International Conference on Machine Learning* (pp. 50-59). PMLR.

Aguilar, J., Garces-Jimenez, A., R-Moreno, M. D., & García, R. (2021). A systematic literature review on using artificial intelligence in energy self-management in smart buildings. *Renewable & Sustainable Energy Reviews, 151*, 111530. doi:10.1016/j.rser.2021.111530

Aguilar, J., Garces-Jimenez, A., R-Moreno, M. D., & García, R. (2021). A systematic literature review on using artificial intelligence in energy self-management in smart buildings. *Renewable & Sustainable Energy Reviews, 151*, 111530. doi:10.1016/j.rser.2021.111530

Alexopoulos, G. S. (2023). Artificial Intelligence in Geriatric Psychiatry Through the Lens of Contemporary Philosophy. *The American Journal of Geriatric Psychiatry*. Advance online publication. doi:10.1016/j.jagp.2023.09.006 PMID:37813788

Assefa, S. A., Dervovic, D., Mahfouz, M., Tillman, R. E., Reddy, P., & Veloso, M. (2020, October). Generating synthetic data in finance: opportunities, challenges and pitfalls. In *Proceedings of the First ACM International Conference on AI in Finance* (pp. 1-8). ACM. 10.1145/3383455.3422554

Athanasopoulou, K., Daneva, G. N., Adamopoulos, P. G., & Scorilas, A. (2022). Artificial intelligence: The milestone in modern biomedical research. *BioMedInformatics, 2*(4), 727–744. doi:10.3390/biomedinformatics2040049

Bautista, Y. J. P., Theran, C., Aló, R., & Lima, V. (2023, October). Health Disparities Through Generative AI Models: A Comparison Study Using a Domain Specific Large Language Model. In *Proceedings of the Future Technologies Conference* (pp. 220-232). Cham: Springer Nature Switzerland. 10.1007/978-3-031-47454-5_17

Bodenstedt, S., Wagner, M., Müller-Stich, B. P., Weitz, J., & Speidel, S. (2020). Artificial intelligence-assisted surgery: Potential and challenges. *Visceral Medicine, 36*(6), 450–455. doi:10.1159/000511351 PMID:33447600

Brynjolfsson, E., Li, D., & Raymond, L. R. (2023). Generative AI at work (No. w31161). National Bureau of Economic Research.

Bugaj, M., Kliestik, T., & Lăzăroiu, G. (2023). Generative Artificial Intelligence-based Diagnostic Algorithms in Disease Risk Detection, Personalized and Targeted Healthcare Procedures, and Patient Care Safety and Quality. *Contemporary Readings in Law and Social Justice*, *15*(1), 9–26. doi:10.22381/CRLSJ15120231

Bugaj, M., Kliestik, T., & Lăzăroiu, G. (2023). Generative Artificial Intelligence-based Diagnostic Pagano, S., Holzapfel, S., Kappenschneider, T., Meyer, M., Maderbacher, G., Grifka, J., & Holzapfel, D. E. (2023). Arthrosis diagnosis and treatment recommendations in clinical practice: An exploratory investigation with the generative AI model GPT-4. *Journal of Orthopaedics and Traumatology*, *24*(1), 61. PMID:38015298

Chheang, V., Marquez-Hernandez, R., Patel, M., Rajasekaran, D., Sharmin, S., Caulfield, G., . . . Barmaki, R. L. (2023). Towards anatomy education with generative AI-based virtual assistants in immersive virtual reality environments. *arXiv preprint arXiv:2306.17278.*

Clark, M., & Severn, M. (2023). Artificial Intelligence in Prehospital Emergency Health Care. *Canadian Journal of Health Technologies*, *3*(8). doi:10.51731/cjht.2023.712 PMID:37934833

Dave, B., Patel, S., Shivani, R., Purohit, S., & Chaudhury, B. (2023). Synthetic data generation using generative adversarial network for tokamak plasma current quench experiments. *Contributions to Plasma Physics*, *63*(5-6), e202200051. doi:10.1002/ctpp.202200051

Davies, B. (2020). The right not to know and the obligation to know. *Journal of Medical Ethics*, *46*(5), 300–303. doi:10.1136/medethics-2019-106009

De Freitas, J., Uğuralp, A. K., Oğuz-Uğuralp, Z., & Puntoni, S. (2022). Chatbots and Mental Health: Insights into the Safety of Generative AI. *Journal of Consumer Psychology.*

de Melo, C. M., Torralba, A., Guibas, L., DiCarlo, J., Chellappa, R., & Hodgins, J. (2022). Next-generation deep learning based on simulators and synthetic data. *Trends in Cognitive Sciences*, *26*(2), 174–187. doi:10.1016/j.tics.2021.11.008 PMID:34955426

Dunn, A. G., Shih, I., Ayre, J., & Spallek, H. (2023). What generative AI means for trust in health communications. *Journal of Communication in Healthcare*, *16*(4), 385–388. doi:10.1080/17538068.2023.2277489 PMID:37921509

Figueira, A., & Vaz, B. (2022). Survey on synthetic data generation, evaluation methods and GANs. *Mathematics*, *10*(15), 2733. doi:10.3390/math10152733

Firouzi, F., Farahani, B., Barzegari, M., & Daneshmand, M. (2020). AI-driven data monetization: The other face of data in IoT-based smart and connected health. *IEEE Internet of Things Journal*, *9*(8), 5581–5599. doi:10.1109/JIOT.2020.3027971

Goodfellow, I., Bengio, Y., & Courville, A. (2016). *Deep Learning*. MIT Press. https://www.deeplearningbook.org/

Grupac, M., Zauskova, A., & Nica, E. (2023). Generative Artificial Intelligence-based Treatment Planning in Clinical Decision-Making, in Precision Medicine, and in Personalized Healthcare. *Contemporary Readings in Law and Social Justice*, *15*(1).

Gupta, M., Akiri, C., Aryal, K., Parker, E., & Praharaj, L. (2023). From chatgpt to threatgpt: Impact of generative ai in cybersecurity and privacy. *IEEE Access: Practical Innovations, Open Solutions, 11,* 80218–80245. doi:10.1109/ACCESS.2023.3300381

Gupta, M., Akiri, C., Aryal, K., Parker, E., & Praharaj, L. (2023). From chatgpt to threatgpt: Impact of generative ai in cybersecurity and privacy. *IEEE Access : Practical Innovations, Open Solutions, 11,* 80218–80245. doi:10.1109/ACCESS.2023.3300381

Haque, N. (2023). Artificial intelligence and geriatric medicine: New possibilities and consequences. *Journal of the American Geriatrics Society, 71*(6), 2028–2031. doi:10.1111/jgs.18334 PMID:36930059

Hasselgren, C., & Oprea, T. I. (2024). Artificial Intelligence for Drug Discovery: Are We There Yet? *Annual Review of Pharmacology and Toxicology, 64*(1), 64. doi:10.1146/annurev-pharmtox-040323-040828 PMID:37738505

Hastings, J. (2024). Preventing harm from non-conscious bias in medical generative AI. *The Lancet. Digital Health, 6*(1), e2–e3. doi:10.1016/S2589-7500(23)00246-7 PMID:38123253

Hayward, R. (2023). Generative Artificial Intelligence-driven Healthcare Systems in Medical Imaging Analysis, in Clinical Decision Support, and in Patient Engagement and Monitoring. *Contemporary Readings in Law and Social Justice, 15*(1), 63–80. doi:10.22381/CRLSJ15120234

Hegde, S. K., & Mundada, M. R. (2022). Hybrid generative regression-based deep intelligence to predict the risk of chronic disease. *International Journal of Intelligent Computing and Cybernetics, 15*(1), 144–164. doi:10.1108/IJICC-06-2021-0103

Henriques, R., Antunes, C., & Madeira, S. C. (2015). Generative modeling of repositories of health records for predictive tasks. *Data Mining and Knowledge Discovery, 29*(4), 999–1032. doi:10.1007/s10618-014-0385-7

Huang, J., Neill, L., Wittbrodt, M., Melnick, D., Klug, M., Thompson, M., Bailitz, J., Loftus, T., Malik, S., Phull, A., Weston, V., Heller, J. A., & Etemadi, M. (2023). Generative artificial intelligence for chest radiograph interpretation in the emergency department. *JAMA Network Open, 6*(10), e2336100–e2336100. doi:10.1001/jamanetworkopen.2023.36100 PMID:37796505

Jung, M., Lim, C., Lee, C., Kim, S., & Kim, J. (2021). Human dietitians vs. Artificial intelligence: Which diet design do you prefer for your children? *The Journal of Allergy and Clinical Immunology, 147*(2), AB117. doi:10.1016/j.jaci.2020.12.430

Jurado-Camino, M. T., Chushig-Muzo, D., Soguero-Ruiz, C., de Miguel-Bohoyo, P., & Mora-Jiménez, I. (2023). On the Use of Generative Adversarial Networks to Predict Health Status Among Chronic Patients. In HEALTHINF (pp. 167-178). doi:10.5220/0011690500003414

Kenthapadi, K., Lakkaraju, H., & Rajani, N. (2023, August). Generative AI Meets Responsible AI: Practical Challenges and Opportunities. In *Proceedings of the 29th ACM SIGKDD Conference on Knowledge Discovery and Data Mining* (pp. 5805-5806). ACM. 10.1145/3580305.3599557

King, D. R., Nanda, G., Stoddard, J., Dempsey, A., Hergert, S., Shore, J. H., & Torous, J. (2023). An Introduction to Generative Artificial Intelligence in Mental Health Care: Considerations and Guidance. *Current Psychiatry Reports, 25*(12), 1–8. doi:10.1007/s11920-023-01477-x PMID:38032442

Kshetri, N., Dwivedi, Y. K., Davenport, T. H., & Panteli, N. (2023). Generative artificial intelligence in marketing: Applications, opportunities, challenges, and research agenda. *International Journal of Information Management*, 102716. doi:10.1016/j.ijinfomgt.2023.102716

Lam, J., Brinkman, W. P., & Bruijnes, M. (2021). *Generative algorithms to improve mental health issue detection.*

Lee, C., Kim, S., Kim, J., Lim, C., & Jung, M. (2022). Challenges of diet planning for children using artificial intelligence. *Nutrition Research and Practice, 16*(6), 801–812. doi:10.4162/nrp.2022.16.6.801 PMID:36467765

Li, L., & Vakanski, A. (2018). Generative adversarial networks for generation and classification of physical rehabilitation movement episodes. *International Journal of Machine Learning and Computing, 8*(5), 428. PMID:30344962

Mennella, C., Maniscalco, U., De Pietro, G., & Esposito, M. (2023). Generating a novel synthetic dataset for rehabilitation exercises using pose-guided conditioned diffusion models: A quantitative and qualitative evaluation. *Computers in Biology and Medicine, 167*, 107665. doi:10.1016/j.compbiomed.2023.107665 PMID:37925908

Musalamadugu, T. S., & Kannan, H. (2023). Generative AI for medical imaging analysis and applications. *Future Medicine AI*, (0), FMAI5.

Ng, C. K. (2023). Generative adversarial network (generative artificial intelligence) in pediatric radiology: A systematic review. *Children (Basel, Switzerland), 10*(8), 1372. doi:10.3390/children10081372 PMID:37628371

Nova, K. (2023). Generative AI in healthcare: Advancements in electronic health records, facilitating medical languages, and personalized patient care. *Journal of Advanced Analytics in Healthcare Management, 7*(1), 115–131.

Pagano, S., Holzapfel, S., Kappenschneider, T., Meyer, M., Maderbacher, G., Grifka, J., & Holzapfel, D. E. (2023). Arthrosis diagnosis and treatment recommendations in clinical practice: An exploratory investigation with the generative AI model GPT-4. *Journal of Orthopaedics and Traumatology, 24*(1), 61. doi:10.1186/s10195-023-00740-4 PMID:38015298

Paladugu, P. S., Ong, J., Nelson, N., Kamran, S. A., Waisberg, E., Zaman, N., Kumar, R., Dias, R. D., Lee, A. G., & Tavakkoli, A. (2023). Generative adversarial networks in medicine: Important considerations for this emerging innovation in artificial intelligence. *Annals of Biomedical Engineering, 51*(10), 2130–2142. doi:10.1007/s10439-023-03304-z PMID:37488468

Pinaya, W. H., Graham, M. S., Kerfoot, E., Tudosiu, P. D., Dafflon, J., Fernandez, V., & Cardoso, M. J. (2023). Generative ai for medical imaging: extending the monai framework. *arXiv preprint arXiv:2307.15208.*

Ray, T. R., Kellogg, R. T., Fargen, K. M., Hui, F., & Vargas, J. (2023). The perils and promises of generative artificial intelligence in neurointerventional surgery. *Journal of Neurointerventional Surgery*. PMID:37438101

Ross, A., Chen, N., Hang, E. Z., Glassman, E. L., & Doshi-Velez, F. (2021, May). Evaluating the interpretability of generative models by interactive reconstruction. In *Proceedings of the 2021 CHI Conference on Human Factors in Computing Systems* (pp. 1-15). ACM. 10.1145/3411764.3445296

Spector-Bagdady, K. (2023). Generative-AI-Generated Challenges for Health Data Research. *The American Journal of Bioethics*, 23(10), 1–5. doi:10.1080/15265161.2023.2252311 PMID:37831940

Tang, A., Li, K. K., Kwok, K. O., Cao, L., Luong, S., & Tam, W. (2023). The importance of transparency: Declaring the use of generative artificial intelligence (AI) in academic writing. *Journal of Nursing Scholarship*, jnu.12938. doi:10.1111/jnu.12938 PMID:37904646

Teo, C. T., Abdollahzadeh, M., & Cheung, N. M. (2023, June). Fair generative models via transfer learning. *Proceedings of the AAAI Conference on Artificial Intelligence*, 37(2), 2429–2437. doi:10.1609/aaai.v37i2.25339

Thiagarajan, J. J., Sattigeri, P., Rajan, D., & Venkatesh, B. (2020). Calibrating healthcare ai: Towards reliable and interpretable deep predictive models. *arXiv preprint arXiv:2004.14480*.

Urban Davis, J., Anderson, F., Stroetzel, M., Grossman, T., & Fitzmaurice, G. (2021, June). Designing co-creative ai for virtual environments. In Creativity and Cognition (pp. 1-11). doi:10.1145/3450741.3465260

WalkowiakE.MacDonaldT. (2023). Generative AI and the Workforce: What Are the Risks? *Available at* SSRN. doi:10.2139/ssrn.4568684

Walters, W. P., & Barzilay, R. (2021). Critical assessment of AI in drug discovery. *Expert Opinion on Drug Discovery*, 16(9), 937–947. doi:10.1080/17460441.2021.1915982 PMID:33870801

Walters, W. P., & Murcko, M. (2020). Assessing the impact of generative AI on medicinal chemistry. *Nature Biotechnology*, 38(2), 143–145. doi:10.1038/s41587-020-0418-2 PMID:32001834

Wang, D. Q., Feng, L. Y., Ye, J. G., Zou, J. G., & Zheng, Y. F. (2023). Accelerating the integration of ChatGPT and other large-scale AI models into biomedical research and healthcare. *MedComm - Future Medicine*, 2(2), e43. doi:10.1002/mef2.43

Wang, T., Zhang, Y., Qi, S., Zhao, R., Xia, Z., & Weng, J. (2023). Security and privacy on generative data in aigc: A survey. *arXiv preprint arXiv:2309.09435*.

Woodman, R. J., & Mangoni, A. A. (2023). A comprehensive review of machine learning algorithms and their application in geriatric medicine: Present and future. *Aging Clinical and Experimental Research*, 35(11), 2363–2397. doi:10.1007/s40520-023-02552-2 PMID:37682491

Yu, S., Chai, Y., Samtani, S., Liu, H., & Chen, H. (2023). Motion Sensor–Based Fall Prevention for Senior Care: A Hidden Markov Model with Generative Adversarial Network Approach. *Information Systems Research*, isre.2023.1203. doi:10.1287/isre.2023.1203

Zeng, L. (2023). Generative AI in Public Opinion Guidance during Emergency Public Events: Challenges, Opportunities, and Ethical Considerations. *Public Opinion Guidance during Emergency Public Events: Challenges, Opportunities, and Ethical Considerations (March 30, 2023).*

Zeng, X., Wang, F., Luo, Y., Kang, S. G., Tang, J., Lightstone, F. C., Fang, E. F., Cornell, W., Nussinov, R., & Cheng, F. (2022). Deep generative molecular design reshapes drug discovery. *Cell Reports Medicine, 3*(12), 100794. doi:10.1016/j.xcrm.2022.100794 PMID:36306797

Zhang, P., & Kamel Boulos, M. N. (2023). Generative AI in Medicine and Healthcare: Promises, Opportunities and Challenges. *Future Internet, 15*(9), 286. doi:10.3390/fi15090286

Zhong, H., Chang, J., Yang, Z., Wu, T., Mahawaga Arachchige, P. C., Pathmabandu, C., & Xue, M. (2023, April). Copyright protection and accountability of generative AI: Attack, watermarking and attribution. In *Companion Proceedings of the ACM Web Conference 2023* (pp. 94-98). ACM.

Chapter 12
Guardians of the Algorithm:
Human Oversight in the Ethical Evolution of AI and Data Analysis

Dwijendra Nath Dwivedi
https://orcid.org/0000-0001-7662-415X
Krakow University of Economics, Poland

Ghanashyama Mahanty
https://orcid.org/0000-0002-6560-2825
Utkal University, India

ABSTRACT

The emergence of artificial intelligence (AI) and data enquiry priciples uncovered immese technological possibilities, but it has also presented a range of ethical concerns that require careful supervision and moderation to avoid unintended consequences. This chapter is a thorough examination that emphasizes the crucial importance of human intervention in upholding the ethical integrity of AI systems and data-driven processes. It emphasizes the importance of human supervision not only as a regulatory structure, but also as an essential element in the development and execution of AI systems. The study examines many approaches to human oversight, including both direct intervention and advanced monitoring techniques, that can be incorporated at every stage of the AI lifecycle, from original creation to post-deployment. The study showcases many case studies and real-world situations to illustrate instances when the lack of human supervision resulted in ethical violations, and conversely, where its presence effectively reduced dangers.

INTRODUCTION

The AI and data domain rapidly expanding. Both emerging possibilities and emerging risk to nagigate the challenges of the future are many. The ethical implications of artificial intelligence (AI) and advanced data analytics have become a crucial concern in the rapidly developing era of these technologies. Our paper aims to examine the crucial role of human supervision in directing the ethical advancement and

DOI: 10.4018/979-8-3693-2964-1.ch012

implementation of artificial intelligence. As these technologies have a growing impact on all aspects of our lives, ranging from healthcare choices to the behavior of financial markets, the need for a fair and moral approach is more important than ever.

This research commences by analyzing the swift progression of AI and data analysis technologies, emphasizing its profound influence across several sectors. We explore the possible hazards and moral quandaries presented by AI systems, including partiality in decision-making, worries about privacy, and the lessening significance of human discernment in crucial procedures. Our investigation primarily centers around the notion of 'human guardians' - individuals or collectives tasked with supervising AI systems to guarantee their adherence to ethical norms and social principles.

We contend that the ethical progress of AI and data analysis is reliant on efficient human supervision, notwithstanding their significant potential for societal progress. This entails not just the establishment of regulatory frameworks, but also the fostering of a profound comprehension of AI ethics among developers, users, and policymakers. The study emphasizes the significance of multidisciplinary collaboration, combining knowledge from technology, philosophy, law, and sociology, in order to formulate strong strategies for ethical governance of artificial intelligence.

We demonstrate the difficulties and achievements of incorporating human supervision into AI by conducting a range of case studies and conducting interviews with experts. We do a thorough examination of current supervision models and put forward a series of principles and optimal methods for ensuring ethical development and implementation of AI.

Understanding and using artificial intelligence (AI) and data analysis has a long history (Table1) of being linked to human progress in knowledge and technology. This section of the research papertraces the history of AI, data analysis, and ethics. It underlines the necessity for human oversight to ensure that these technologies' ethical evolution is as profound and significant as their technical advancements as we approach greater breakthroughs. This historical perspective shows that AI ethics are an evolving discussion that has accompanied AI technology.

Table 1. The historical advancements in AI and Data Analysis, along with the corresponding ethical considerations (Author)

Timeline	Advancements	Ethical Considerations
1940s-1960s	Foundational work by Alan Turing, John McCarthy, etc. Focus on theoretical aspects of AI.	Largely absent; focus on exploring potential and limits of technology.
1970s-1990s	First AI booms and winters. Developments in expert systems. Early discussions on AI's societal impact.	Emergence of concerns about job impacts and automated decision-making risks.
2000s-2010s	Growth of the internet and big data. Advances in machine learning. Rising concerns about privacy and inequality.	Intensified issues regarding privacy, surveillance, data security, and societal inequalities.
2010s-Present	Breakthroughs in deep learning. Growing awareness and initiatives for ethical AI development and use.	High-profile incidents lead to a concerted effort for responsible AI. Establishment of ethical guidelines.

This paper provides a thorough examination of the importance of human supervision in guiding the ethical development of AI and data analytics. The study examines the difficulties associated with implementing efficient human supervision, including the requirement for interdisciplinary knowledge. The delicate equilibrium between innovation and regulation and the establishment of worldwide ethical benchmarks for artificial intelligence is the way. It promotes a proactive strategy in which ethical considerations are integrated into the AI development process, rather than being an afterthought. To summarize, "Guardians of the Algorithm" emphasizes the importance of human oversight in the ethical advancement of AI and data analysis. The establishment of a responsible AI ecosystem that appreciates and safeguards human dignity, rights, and social values necessitates a cooperative endeavor involving technologists, ethicists, policymakers, and users. The article contends that the trajectory of AI development should not be determined merely by technological capabilities, but rather should be driven by a steadfast dedication to ethical responsibility and the principles of human-centered design.

LITERATURE REVIEW

The artificial intelligence incident database was investigated by Dwivedi and Mahanty (2021), who identified major areas of AI danger based on previous instances. In his analysis of 22 ethical principles, Hagendorf (2020) found that there were common themes and regions that were not receiving enough attention. The practical features of artificial intelligence ethical standards are significantly improved as a result of this examination. The research that Maas conducted in 2018 reveals that artificial intelligence systems can result in widespread and cascading errors. This year's Box & Data conference focused on the influence that human biases have on machine learning algorithms. The investigation of ethical decision-making inside artificial intelligence was carried out by Martinho et al. (2020), who combined theoretical and empirical methodologies. The "concept drift" problem was brought to light by Tamboli (2019), who brought attention to the increasing challenges that occur as a result of changing data trends. Concerns were voiced by Bolander (2019) over the implications of artificial intelligence and the technical challenges involved in replacing human labor.

The research paper written by Holzinger et al. (2019) highlighted the importance of having a wide range of high-quality data in order to solve crucial medical concerns. The author advocated for the combination of clinical, imaging, and molecular datasets in order to better understand complex illnesses. In their work from 2020, Oneto & Chaippa examined the concept of fairness in machine learning as well as the difficulties associated with defining and evaluating fairness metrics. This topic is crucial for the evaluation of ethical AI. Boehmke & Greenwell (2020) investigated various methods that may be utilized to enhance the interpretability of machine learning models, which is an essential component of ethical AI evaluation. In their 2014 article, Bostrom & Yudkowsky addressed a wide variety of ethical concerns pertaining to artificial intelligence (AI), offering useful insights into the ethical dimensions that ought to be taken into consideration during evaluation. By highlighting the potential for "fairwashing" in artificial intelligence ethics, Aïvodji et al. (2019) brought attention to the fact that biased AI systems are presented as impartial. The difficulties of determining whether or not artificial intelligence is fair are discussed. In their 2019 study, Bellamy et al., provide a useful resource for evaluating and eliminating bias in artificial intelligence systems, which is an essential component of ethical AI evaluation. Doshi-Velez et al. (2017) examined the legal and ethical aspects of AI responsibility, as well as the significance of explanation in AI systems. For the purpose of conducting an ethical evaluation of artificial intelligence, Barocas et al. (2017) present a number of measures that can be utilized to measure bias in machine learning models. Hanson (2016) investigated potential future scenarios in a society ruled by artificial intelligence and examined questions of ethics that are associated with AI systems. In their 2017 study, d'Amato and colleagues investigated trust-related topics in the field of computer science and the semantic web.

Yogarajan et al. (2022) investigated the application of artificial intelligence (AI) in the field of healthcare and brought attention to the potential biases and inequalities that could arise as a result of its implementation. This was especially true in relation to indigenous populations in New Zealand that are underrepresented. They wanted to examine equity and fairness metrics in artificial intelligence for healthcare in New Zealand, and their research was aimed at doing just that. Peng et al. (2022) conducted a study in which they studied the relationship between the accuracy of a model's predictions and the bias that the model has on human decision-making in the context of machine learning-assisted hiring. A recommendation-aided choice task was used to investigate the dynamics of this connection between the two parties. Katare et al. (2022) conducted a study in which they investigated the potential biases that could be present in artificial intelligence algorithms, with a specific emphasis on the subject of autonomous driving. In spite of the fact that artificial intelligence has shown promising results in terms of improving accuracy, throughput, and reducing latency, Kwasniewska & Szankin (2022) found that it continues to struggle with a number of persistent challenges, including insufficient explainability, data imbalance, and bias. Belenguer, L. (2022) presents a novel approach to the problem of prejudiced prejudice in the field of artificial intelligence. In their research, Nadeem et al. (2022) conducted an exhaustive review of the existing body of literature in order to investigate the existence of gender bias in decision-making systems that make use of artificial intelligence (AI).

Norori et al. (2021) investigated the potential of artificial intelligence (AI) in the field of healthcare and addressed the challenges that are brought by algorithmic bias. The their study, Newman-Griffis et al. (2022), highlighted the possibility of bias in artificial intelligence (AI) systems, particularly with regard to people who have disabilities. The essay investigates the genesis of prejudice as a consequence of certain design choices, as well as the possibility that different interpretations of disability can give rise to a variety of biases from different people. The researchers Alzamil et al.(2020) found that the two groups used different diagnostic criteria in different ways, which led to the discovery of disparities. The research highlights the importance of having standardized diagnostic criteria for polycystic ovary syndrome (PCOS). A comprehensive review of the potential drawbacks and benefits associated with artificial intelligence-driven intrusion detection systems in the field of cybersecurity is presented in the work written by Dash et al. (2022). On the other hand, O'Sullivan et al. (2021) investigated the challenges that are associated with the development of artificial intelligence systems for the purpose of monitoring the heart rate of the fetus while the process of giving birth is taking place. For the purpose of addressing AI ethics, Dwivedi et al. (2022) conducted sentiment mining and discussed important concerns.

3.0 HISTORICAL AI RISK INCIDENTS

In the last twenty years, there have been multiple significant occurrences involving artificial intelligence (AI) that have emphasized the dangers and ethical concerns linked to its creation and implementation. An early instance occurred in 2007, when Microsoft's Tay, an AI chatbot programmed to acquire knowledge via user interactions on Twitter, was promptly manipulated by users, resulting in the production of offensive and inappropriate material. This incident highlighted the susceptibilities of machine learning systems to external influences and prompted inquiries regarding the regulation and management of AI in public domains. Another notable occurrence of AI risk was during the implementation of self-driving automobiles. There have been multiple recorded accidents employing Tesla's Autopilot system, including fatal wrecks where the system's ability to identify and respond to specific road conditions was mentioned as a significant cause. These occurrences have ignited discussions over the security and preparedness of AI in crucial domains such as transportation, as well as the necessity for rigorous testing and regulatory frameworks.

Within the domain of AI-driven decision-making systems, there have been multiple occurrences of prejudice and discrimination. Significant instances have occurred in which facial recognition technology have erroneously identified persons, namely those belonging to minority ethnic groups, resulting in unjust detentions and exposing the racial prejudices ingrained in the datasets used for training. Likewise, artificial intelligence (AI) systems employed in the field of recruitment have demonstrated prejudices against specific demographic groups, raising worries regarding the contribution of AI in perpetuating societal disparities. In the financial sector, AI-driven trading algorithms have also revealed hazards, such as the 2010 Flash Crash, where high-frequency trading algorithms contributed to a sudden and major stock market collapse. This event demonstrated the capacity of AI systems to worsen market volatility and emphasized the necessity of supervision in financial AI applications.

Table 2. AI risk incidents (Author)

Year	Incident	Impact
2004	Chatbot Produces Inappropriate Responses	Public controversy over offensive language.
2009	Flash Crash in Stock Market	Temporary $1 trillion loss in stock market value.
2013	Revelation of NSA's Global Surveillance	Widespread privacy concerns and policy debates.
2016	AI Algorithm Bias in Criminal Sentencing	Controversy over racial bias and fairness in AI.
2018	Self-Driving Car Fatal Accident	First pedestrian death involving an autonomous vehicle.
2019	Biased Facial Recognition Technology	Racial discrimination and wrongful arrests.
2021	AI-Generated Deepfakes in Misinformation Campaigns	Increased political and social manipulation concerns.

The above table summarizing the importance of human oversight in various high-risk AI implementations. AI systems can inadvertently learn and perpetuate biases present in their training data. Human oversight helps identify and correct these biases, ensuring fairness in decision-making processes, especially in sensitive areas like criminal justice and employment. High-risk AI applications often handle sensitive personal data. Human monitors are crucial in overseeing data handling practices, ensuring compliance with privacy regulations, and preventing security breaches. In cases where AI systems make incorrect or harmful decisions, it's important to have a clear line of accountability. Human overseers ensure that AI actions can be attributed and individuals or organizations can be held responsible. AI systems may not be equipped to make ethical decisions involving nuanced human values and societal norms. Human oversight ensures that decisions made by AI systems align with ethical standards. The involvement of human oversight in AI systems can increase public trust, especially in sectors where trust is a critical component, such as healthcare and law enforcement.

Table 3. Human oversight to monitor AI risk (Author)

Area	Region / Entity	Details	Role of Human Oversight
Criminal Sentencing	United States	Use of AI in criminal sentencing, particularly the risk assessment tool COMPAS, raised concerns about bias and fairness.	Judicial oversight to interpret and contextualize AI-generated risk scores.
Healthcare AI System	United Kingdom - NHS	NHS implemented AI systems for patient diagnosis and treatment recommendations.	Medical professionals providing oversight to ensure AI's recommendations were medically sound and ethical.
Facial Recognition in Law Enforcement	Various Countries	Deployment of facial recognition by police departments has led to concerns about privacy, accuracy, and racial bias.	Oversight committees and legal frameworks to regulate the use of facial recognition.
Autonomous Vehicles	United States and Europe	Testing and deployment of autonomous vehicles raise significant safety and ethical concerns.	Regulatory bodies set standards and conduct reviews; human supervisors monitor AI decisions in test vehicles.
AI in Financial Markets	Global	AI-driven trading algorithms can cause market volatility, as seen in the 2010 Flash Crash.	Financial regulators implemented rules requiring human oversight of trading algorithms.

HUMAN OVERSIGHT IN THE ETHICAL EVOLUTION OF AI AND DATA ANALYSIS: CASE STUDIES

1. The examination of **IBM's AI Ethics Board** (Munoko, I.(2020))provides valuable insights into the implementation of ethical principles and oversight by a prominent technological business in the creation and utilization of AI technologies. This case could analyze the board's structure, decision-making processes, and impact on IBM's AI efforts.

2. An investigation of the influence of the European Union's General Data Protection Regulation **(GDPR)** on artificial intelligence (AI) and data analysis, specifically focusing on user permission, data privacy, and algorithmic transparency(Seizov, O., & Wulf, A. J. (2020).). This case study aims to examine the difficulties that firms encounter while adhering to the General Data Protection rule (GDPR) and assess the efficacy of the rule in upholding ethical norms in the field of artificial intelligence (AI).

3. **Google's application of its AI principles** in response to the Project Maven controversy serves as a case study on corporate governance and ethics in the field of artificial intelligence(Crofts, P., & van Rijswijk, H. (2020)). The text might delve into the internal and external obstacles that Google encountered and the actions it implemented to ensure its AI initiatives were in line with these values.

4. This study examines the integration of AI technologies, like IBM Watson Health(Kohn, M. S. et.al.(2014)), in healthcare environments to assist in making informed decisions on diagnosis and treatment. This case study can evaluate the ethical aspects pertaining to precision, confidentiality of patients, and the influence on doctor-patient interactions.

5. This study investigates the progress of autonomous vehicles created by businesses(Martinho, A. et.al. (2021)) such as Tesla and Waymo, with a specific emphasis on how these systems navigate ethical dilemmas in crucial situations, such as the trolley problem, when faced with real-life scenarios.

6. Investigating Bias in Recruiting Algorithms(Bongard, A. (2019)): An examination of artificial intelligence-based recruitment tools employed by corporations and the impact of algorithmic

biases on hiring procedures. This may involve an examination of particular instances in which AI unintentionally propagated discrimination and the subsequent actions taken in response.

7. Examining the rise of deepfakes and their impact on the dissemination of false information(Doss, C. et.al.(2023)). This case study will examine the ethical ramifications of the integrity of media, political procedures, and personal privacy, as well as the endeavors to identify and control deepfake technology.

REGULATORY GUIDELINES ON AI AND HUMAN OVERSIGHT FROM VARIOUS ORGANIZATIONS AND REGIONS

Several significant regulatory standards and frameworks have been produced that are in accord with the ideas discussed in the paper. The purpose of these recommendations is to guarantee the ethical advancement, implementation, and management of AI technologies(Floridi, L. et.al(2021)). Here are few prominent illustrations: The European Union's Ethics recommendations for Trustworthy AI were published by the High-Level Expert Group on Artificial Intelligence established by the European Commission. These recommendations highlight the importance of AI that is trustworthy, meaning it must adhere to legal and ethical standards while also being robust. The requirements they outline include factors such as human agency, transparency, and responsibility. The Organisation for Economic Co-operation and Development (OECD) has formulated guidelines on Artificial Intelligence (AI) that have been supported by more than 40 countries (Truby, J. (2020).. The principles outlined prioritize inclusive economic expansion, environmentally sustainable progress, values centered around human well-being, and equitable treatment. They serve as a framework for both domestic policy and global collaboration. The Institute of Electrical and Electronics Engineers (IEEE) has created a thorough set of principles called "Ethically Aligned Design: A Vision for Prioritizing Human Well-being with Autonomous and Intelligent Systems" (Chatila, R. et.al.(2018)) The content encompasses the fundamental concepts of human rights, welfare, data autonomy, and efficiency. The AI Now Institute at New York University releases yearly studies that analyze the societal consequences of artificial intelligence(Whittaker, M. et. al.(2018)). The studies contain suggestions for the regulation and management of AI, with a focus on safeguarding human rights, ensuring justice, and implementing efficient supervision. The White House Office of Science and Technology Policy has issued guidelines for regulating AI applications.(Thierer, A. D. et. al.(2017)) These guidelines emphasize the importance of public trust, involvement, scientific integrity, and risk assessment in the development and implementation of AI. China has established a set of guidelines to govern the ethical development of AI, known as China's Ethical Norms and Governance guidelines for New Generation AI(Wu, F. et.al.(2020)). These principles prioritize the values of harmony, friendliness, justice, inclusivity, respect for privacy, security, and shared responsibility. The Montreal Declaration for Responsible AI, created by a team of experts from the University of Montreal, provides a comprehensive set of ethical principles for the advancement of AI. It highlights the importance of prioritizing the welfare of all individuals, upholding autonomy, and safeguarding democratic values and diversity. The government of the United Kingdom has created a set of guidelines for health and care technology that relies on data, encompassing concepts such as prioritizing user requirements, ensuring transparency, and upholding justice(Cath, C. et.al.(2018).

Table 4. Overview of various key guidelines and frameworks (Author)

Organization / Region	Guideline / Framework	Key Focus
European Union	Ethics Guidelines for Trustworthy AI	Lawfulness, ethicality, robustness, human agency, transparency, accountability.
OECD	OECD Principles on AI	Inclusive growth, sustainable development, human-centered values, fairness.
IEEE	Ethically Aligned Design	Human rights, well-being, data agency, and effectiveness.
AI Now Institute	Annual Reports on AI and its Social Implications	Human rights, justice, effective oversight.
The White House (USA)	AI Guidelines	Public trust, participation, scientific integrity, risk assessment.
China	Ethical Norms and Governance Principles for New Generation AI	Harmony, friendliness, fairness, inclusiveness, privacy, security, responsibility.
University of Montreal (Canada)	The Montreal Declaration for Responsible AI	Well-being, autonomy, democratic values, diversity.
United Kingdom	AI Code for Health and Care Technology	User need, transparency, fairness.

KPIS TO HUMAN OVERSIGHT IN THE ETHICAL EVOLUTION OF AI AND DATA ANALYSIS

Implementing many Key Performance Indicators (KPIs) can effectively monitor and measure human oversight in the ethical advancement of AI and data analysis. The purpose of these Key Performance Indicators (KPIs) is to guarantee the responsible and ethical development and utilization of AI systems, while adhering to legal and societal norms. Here are a few instances of these Key Performance Indicators (KPIs). Key performance indicators (KPIs) are becoming more important in the ethical advancement of artificial intelligence (AI) and data analysis. These KPIs serve to assess performance while simultaneously ensuring that human oversight and ethical compliance are maintained. This change recognizes the importance of addressing the risks related to data privacy, prejudice, and decision-making that impacts persons and communities, while also utilizing AI for its tremendous capabilities. To ensure that AI systems are built and used in a way that is ethical, transparent, and accountable, key performance indicators (KPIs) are used as navigational beacons. As a precaution against the possible dangers of autonomous systems, human supervision is crucial in this ethical development. This safeguards society's values and standards by requiring AI-driven judgments to be reviewed by humans for ethical implications. To guarantee that AI systems support inclusion and equity, key performance indicators pertaining to equity and fairness are critical in several domains like healthcare, credit scoring, and recruiting. In a similar vein, privacy protection key performance indicators make ensuring that data analysis and AI respect people's rights and the legislation around data protection, which is great for user confidence.

To further demystify AI processes, it is essential to define key performance indicators (KPIs) for explainability and transparency. By providing non-expert stakeholders with clearer insights into how AI systems make decisions, these key performance indicators assist build confidence and allow for better debates regarding the potential consequences of AI solutions. Creating a discourse regarding the effects, limitations, and ethical issues of AI is an important part of being transparent, which goes beyond simply opening up algorithms.

Table 5. Overview of various key guidelines and frameworks (Author)

KPI	Description	Example
Ethical Audit Frequency	Frequency of formal audits conducted to assess the ethical implications and compliance of AI systems with ethical guidelines.	Conducting bi-annual ethical audits of all AI systems.
Bias Incident Reports	Number of reported incidents where AI systems demonstrate bias or unethical behavior.	Less than 5 bias incidents reported per year.
AI Transparency Index	A metric assessing the degree to which the workings of an AI system are understandable and explainable to non-experts.	Achieving an AI transparency score of 80% or higher.
Stakeholder Satisfaction Score	Score derived from regular surveys of stakeholders (including users, developers, and the public) on their perception of the AI system's ethical standards.	Maintaining a stakeholder satisfaction score above 75%.
Regulatory Compliance Rate	Percentage of AI projects that fully comply with existing laws and ethical guidelines.	100% compliance with data protection regulations like GDPR.

CONCLUSION

Finally, as AI and data analysis keep getting better, it's becoming more and more important to include KPIs that focus on ethics and human monitoring. These KPIs are very important for companies to use to safely navigate the complicated world of AI deployment. Along with making technology better, they make sure that we stay true to our morals and social duties. This helps create AI systems that are not only smart, but also fair, just, and in line with human values. This paper explores the complex connection between the swift advancement of AI and data analysis technologies and the essential role of human supervision in guaranteeing their ethical use. The study offers a historical outlook, charting the evolution of AI from its conceptual origins to its present stage, characterized by notable progress in deep learning and data analytics. The text underscores significant ethical considerations, such as prejudice, confidentiality, and responsibility, and stresses the necessity of thorough human supervision to address and alleviate these problems.

The final message of the paper emphasizes that although AI and data analysis have great promise for benefiting society, their ethical use depends on the presence of vigilant and knowledgeable human supervision. The study posits that achieving ethical AI is not a fixed objective, but rather an ongoing endeavor that necessitates attentiveness, flexibility, and an interdisciplinary methodology. It underlines the significance of balancing technological progress with ethical issues, calling for a future where AI serves as a tool for augmenting human decision-making, rather than replacing it.

FURTHER RESEARCH

Subsequent investigations should prioritize the development of all-encompassing ethical frameworks that integrate perspectives from technology, philosophy, sociology, and law. These frameworks must possess adaptability to keep up with the swiftly changing landscape of AI and data analysis technologies. Research on the harmonization of global ethical standards and regulatory methods for AI is necessary. This encompasses comprehending the cultural and socioeconomic disparities that impact ethical perspectives and the implementation of AI technologies. Additional research should evaluate the enduring effects of artificial intelligence on different facets of society, such as employment, education, healthcare, and privacy. These studies have the potential to offer useful insights to policymakers and practitioners, enabling them to make well-informed decisions on the implementation of AI. Conducting research on the incorporation of ethical considerations into AI education and professional training programs is of utmost importance. This can guarantee that the upcoming cohort of AI developers and consumers possess a robust comprehension of the ethical ramifications of their endeavors.

REFERENCES

Aïvodji, U., Arai, H., Fortineau, O., Gambs, S., Hara, S., & Tapp, A. (2019). Fairwashing: The risk of rationalization. *Proceedings of the International Conference on Machine Learning*, (pp. 161-170). IEEE.

Alzamil, H., Aloraini, K., AlAgeel, R., Ghanim, A., Alsaaran, R., Alsomali, N., Albahlal, R. A., & Alnuaim, L. (2020). Disparity among endocrinologists and gynaecologists in the diagnosis of polycystic ovarian syndrome. *Sultan Qaboos University Medical Journal*, 20(3), 323. doi:10.18295/squmj.2020.20.03.012 PMID:33110648

Barocas, S., Hardt, M., & Narayanan, A. (2017). Fairness in machine learning. *Nips tutorial, 1*, 2017.

Belenguer, L. (2022). AI bias: Exploring discriminatory algorithmic decision-making models and the application of possible machine-centric solutions adapted from the pharmaceutical industry. *AI and Ethics*, 2(4), 771–787. doi:10.1007/s43681-022-00138-8 PMID:35194591

Bellamy, R. K., Dey, K., Hind, M., Hoffman, S. C., Houde, S., Kannan, K., Lohia, P., Martino, J., Mehta, S., Mojsilovic, A., Nagar, S., Ramamurthy, K. N., Richards, J., Saha, D., Sattigeri, P., Singh, M., Varshney, K. R., & Zhang, Y. (2019). AI Fairness 360: An extensible toolkit for detecting and mitigating algorithmic bias. *IBM Journal of Research and Development*, 63(4/5), 4–1. doi:10.1147/JRD.2019.2942287

Boehmke, B. C., & Greenwell, B. M. (2019). *Interpretable Machine Learning*. Hands-On Machine Learning with R. doi:10.1201/9780367816377-16

Bolander, T. (2019). What do we lose when machines take the decisions? *The Journal of Management and Governance*, 23(4), 849–867. doi:10.1007/s10997-019-09493-x

Bongard, A. (2019). Automating talent acquisition: Smart recruitment, predictive hiring algorithms, and the data-driven nature of artificial intelligence. *Psychosociological Issues in Human Resource Management*, 7(1), 36–41.

Bostrom, N., & Yudkowsky, E. (2014). The ethics of artificial intelligence. In The Cambridge Handbook of Artificial Intelligence (pp. 316-334). Cambridge Handbook. doi:10.1017/CBO9781139046855.020

Cath, C., Wachter, S., Mittelstadt, B., Taddeo, M., & Floridi, L. (2018). Artificial intelligence and the 'good society': The US, EU, and UK approach. *Science and Engineering Ethics*, 24, 505–528. PMID:28353045

Chatila, R., Firth-Butterfield, K., & Havens, J. C. (2018). *Ethically aligned design: A vision for prioritizing human well-being with autonomous and intelligent systems version 2*. University of southern California Los Angeles.

Crofts, P., & van Rijswijk, H. (2020). Negotiating 'evil': Google, project maven and the corporate form. *Law, technology and humans, 2*(1), 75-90.

d'Amato, C., Fernandez, M., Tamma, V., Lecue, F., Cudré-Mauroux, P., Sequeda, J., & Heflin, J. (Eds.). (2017). *The Semantic Web–ISWC 2017: 16th International Semantic Web Conference*, (Vol. 10587). Springer.

Doshi-Velez, F., Kortz, M., Budish, R., Bavitz, C., Gershman, S., O'Brien, D., & Wood, A. (2017). *Accountability of AI under the law: The role of explanation*. arXiv preprint arXiv:1711.01134.

Doss, C., Mondschein, J., Shu, D., Wolfson, T., Kopecky, D., Fitton-Kane, V. A., Bush, L., & Tucker, C. (2023). Deepfakes and scientific knowledge dissemination. *Scientific Reports, 13*(1), 13429. doi:10.1038/s41598-023-39944-3 PMID:37596384

Dwijendra, N. (2022). Machine learning time series models for tea pest Helopeltis infestation in India. *Webology, 19*(2). Available at: https://www.webology.org/abstract.php?id=1625

Dwijendra, N., Pandey, A. K., & Dwivedi, A. D. (2023). Examining the emotional tone in politically polarized speeches in India: An in-depth analysis of two contrasting perspectives. *SOUTH INDIA JOURNAL OF SOCIAL SCIENCES, 21*(2), 125-136. https://journal.sijss.com/index.php/home/article/view/65

Dwivedi, D., Batra, S., & Pathak, Y. K. (2023). A machine learning-based approach to identify key drivers for improving corporate's ESG ratings. *Journal of Law and Sustainable Development, 11*(1), e0242. doi:10.37497/sdgs.v11i1.242

Dwivedi, D., Kapur, P. N., & Kapur, N. N. (2023). Machine learning time series models for tea pest looper infestation in Assam, India. In A. Sharma, N. Chanderwal, & R. Khan (Eds.), *Convergence of Cloud Computing, AI, and Agricultural Science* (pp. 280–289). IGI Global. doi:10.4018/979-8-3693-0200-2.ch014

Dwivedi, D. N. (2024). The use of artificial intelligence in supply chain management and logistics. In D. Sharma, B. Bhardwaj, & M. Dhiman (Eds.), *Leveraging AI and Emotional Intelligence in Contemporary Business Organizations* (pp. 306–313). IGI Global. doi:10.4018/979-8-3693-1902-4.ch018

Dwivedi, D. N. (2022). Benchmarking of traditional and deep learning time series modelling techniques for prediction of rainfall in United Arab Emirates. *8th International conference on Time Series and Forecasting (ITISE 2022)*.

Dwivedi, D. N. (2022). Benchmarking of traditional and advanced machine learning modelling techniques for forecasting. In *Visualization Techniques for Climate Change with Machine Learning and Artificial Intelligence* (pp. x–x). Elsevier. doi:10.1016/B978-0-323-99714-0.00017-0

Dwivedi, D. N., & Anand, A. (2021). Trade heterogeneity in the EU: Insights from the emergence of COVID-19 using time series clustering. *Zeszyty Naukowe Uniwersytetu Ekonomicznego w Krakowie, 3*(993), 9–26. doi:10.15678/ZNUEK.2021.0993.0301

Dwivedi, D. N., & Gupta, A. (2022). Artificial intelligence-driven power demand estimation and short-, medium-, and long-term forecasting. In *Artificial Intelligence for Renewable Energy Systems* (pp. 231–242). Woodhead Publishing. doi:10.1016/B978-0-323-90396-7.00013-4

Dwivedi, D. N., & Mahanty, G. (2021). A text mining-based approach for accessing AI risk incidents. *International Conference on Artificial Intelligence*. AI Foundation Trust, India.

Dwivedi, D. N., & Mahanty, G. (2023). Human creativity vs. machine creativity: Innovations and challenges. In Z. Fields (Ed.), *Multidisciplinary Approaches in AI, Creativity, Innovation, and Green Collaboration* (pp. 19–28). IGI Global. doi:10.4018/978-1-6684-6366-6.ch002

Dwivedi, D. N., & Mahanty, G. (2024). AI-powered employee experience: Strategies and best practices. In M. Rafiq, M. Farrukh, R. Mushtaq, & O. Dastane (Eds.), *Exploring the Intersection of AI and Human Resources Management* (pp. 166–181). IGI Global. doi:10.4018/979-8-3693-0039-8.ch009

Dwivedi, D. N., Mahanty, G., & Pathak, Y. K. (2023). AI applications for financial risk management. In M. Irfan, M. Elmogy, M. Shabri Abd. Majid, & S. El-Sappagh (Eds.), The Impact of AI Innovation on Financial Sectors in the Era of Industry 5.0 (pp. 17-31). IGI Global. doi:10.4018/979-8-3693-0082-4.ch002

Dwivedi, D. N., Mahanty, G., & Vemareddy, A. (2022). How responsible is AI?: Identification of key public concerns using sentiment analysis and topic modeling. [IJIRR]. *International Journal of Information Retrieval Research*, *12*(1), 1–14. doi:10.4018/IJIRR.298646

Dwivedi, D. N., & Patil, G. (2023). Climate change: Prediction of solar radiation using advanced machine learning techniques. In A. Srivastav, A. Dubey, A. Kumar, S. K. Narang, & M. A. Khan (Eds.), *Visualization Techniques for Climate Change with Machine Learning and Artificial Intelligence* (pp. 335–358). Elsevier. doi:10.1016/B978-0-323-99714-0.00017-0

Dwivedi, D. N., Tadoori, G., & Batra, S. (2023). Impact of women leadership and ESG ratings in organizations: A time series segmentation study. *Academy of Strategic Management Journal*, *22*(S3), 1–6.

Floridi, L., Cowls, J., Beltrametti, M., Chatila, R., Chazerand, P., Dignum, V., & Vayena, E. (2021). An ethical framework for a good AI society: Opportunities, risks, principles, and recommendations. *Ethics, governance, and policies in artificial intelligence*, 19-39.

Gupta, A., Dwivedi, D. N., & Jain, A. (2021). Threshold fine-tuning of money laundering scenarios through multi-dimensional optimization techniques. *Journal of Money Laundering Control*. doi:10.1108/JMLC-12-2020-0138

Gupta, A., Dwivedi, D. N., & Shah, J. (2023). Overview of Money Laundering. In: Artificial Intelligence Applications in Banking and Financial Services. Future of Business and Finance. Springer, Singapore. doi:10.1007/978-981-99-2571-1_1

Gupta, A., Dwivedi, D. N., & Shah, J. (2023). Financial Crimes Management and Control in Financial Institutions. In: Artificial Intelligence Applications in Banking and Financial Services. Future of Business and Finance. Springer, Singapore. doi:10.1007/978-981-99-2571-1_2

Gupta, A., Dwivedi, D. N., & Shah, J. (2023). Overview of Technology Solutions. In: Artificial Intelligence Applications in Banking and Financial Services. Future of Business and Finance. Springer, Singapore. doi:10.1007/978-981-99-2571-1_3

Gupta, A., Dwivedi, D. N., & Shah, J. (2023). Data Organization for an FCC Unit. In: Artificial Intelligence Applications in Banking and Financial Services. Future of Business and Finance. Springer, Singapore. doi:10.1007/978-981-99-2571-1_4

Gupta, A., Dwivedi, D. N., & Shah, J. (2023). Planning for AI in Financial Crimes. In: Artificial Intelligence Applications in Banking and Financial Services. Future of Business and Finance. Springer, Singapore. doi:10.1007/978-981-99-2571-1_5

Gupta, A., Dwivedi, D. N., & Shah, J. (2023). Applying Machine Learning for Effective Customer Risk Assessment. In: Artificial Intelligence Applications in Banking and Financial Services. Future of Business and Finance. Springer, Singapore. doi:10.1007/978-981-99-2571-1_6

Gupta, A., Dwivedi, D. N., & Shah, J. (2023). Artificial Intelligence-Driven Effective Financial Transaction Monitoring. In: Artificial Intelligence Applications in Banking and Financial Services. Future of Business and Finance. Springer, Singapore. doi:10.1007/978-981-99-2571-1_7

Gupta, A., Dwivedi, D. N., & Shah, J. (2023). Machine Learning-Driven Alert Optimization. In: Artificial Intelligence Applications in Banking and Financial Services. Future of Business and Finance. Springer, Singapore. doi:10.1007/978-981-99-2571-1_8

Gupta, A., Dwivedi, D. N., & Shah, J. (2023). Applying Artificial Intelligence on Investigation. In: Artificial Intelligence Applications in Banking and Financial Services. Future of Business and Finance. Springer, Singapore. doi:10.1007/978-981-99-2571-1_9

Gupta, A., Dwivedi, D. N., & Shah, J. (2023). Ethical Challenges for AI-Based Applications. In: Artificial Intelligence Applications in Banking and Financial Services. Future of Business and Finance. Springer, Singapore. doi:10.1007/978-981-99-2571-1_10

Gupta, A., Dwivedi, D. N., & Shah, J. (2023). Setting up a Best-In-Class AI-Driven Financial Crime Control Unit (FCCU). In: Artificial Intelligence Applications in Banking and Financial Services. Future of Business and Finance. Springer, Singapore. doi:10.1007/978-981-99-2571-1_11

Gupta, A., Dwivedi, D. N., Shah, J., & Jain, A. (2021). Data quality issues leading to suboptimal machine learning for money laundering models. *Journal of Money Laundering Control*. doi:10.1108/JMLC-05-2021-0049

Hagendorff, T. (2020). The Ethics of AI Ethics: An Evaluation of Guidelines. *Minds and Machines*, *30*(1), 99–120. doi:10.1007/s11023-020-09517-8

Hanson, R. (2016). The age of. In *Work, love, and life when robots rule the Earth*. Oxford University Press.

Holzinger, A., Haibe-Kains, B., & Jurisica, I. (2019). Why imaging data alone is not enough: AI-based integration of imaging, omics, and clinical data. *European Journal of Nuclear Medicine and Molecular Imaging*, *46*(13), 2722–2730. doi:10.1007/s00259-019-04382-9 PMID:31203421

Katare, D., Kourtellis, N., Park, S., Perino, D., Janssen, M., & Ding, A. (2022). Bias Detection and Generalization in AI Algorithms on Edge for Autonomous Driving. *Proceedings of the IEEE International Conference on Edge Computing*. IEEE. 10.1109/SEC54971.2022.00050

Kohn, M. S., Sun, J., Knoop, S., Shabo, A., Carmeli, B., Sow, D., & Rapp, W. (2014). IBM's health analytics and clinical decision support. *Yearbook of Medical Informatics, 23*(01), 154–162. doi:10.15265/IY-2014-0002 PMID:25123736

Kwasniewska, A., & Szankin, M. (2022). Can AI See Bias in X-ray Images? *International Journal of New Developments in Imaging.*

Martinho, A., Herber, N., Kroesen, M., & Chorus, C. (2021). Ethical issues in focus by the autonomous vehicles industry. *Transport Reviews, 41*(5), 556–577. doi:10.1080/01441647.2020.1862355

Martinho, A., Kroesen, M., & Chorus, C. (2020). *An Empirical Approach to Capture Moral Uncertainty in AI.* 101–101. doi:10.1145/3375627.3375805

Munoko, I., Brown-Liburd, H. L., & Vasarhelyi, M. (2020). The ethical implications of using artificial intelligence in auditing. *Journal of Business Ethics, 167*(2), 209–234. doi:10.1007/s10551-019-04407-1

Nadeem, A., Marjanovic, O., & Abedin, B. (2022). Gender bias in AI-based decision-making systems: A systematic literature review. *AJIS. Australasian Journal of Information Systems, 26*, 26. doi:10.3127/ajis.v26i0.3835

Newman-Griffis, D., Rauchberg, J., Alharbi, R., Hickman, L., & Hochheiser, H. (2022). Definition drives design: Disability models and mechanisms of bias in AI technologies. *First Monday.*

Norori, N., Hu, Q., Aellen, F., Faraci, F., & Tzovara, A. (2021). Addressing bias in big data and AI for health care: A call for open science. *Patterns (New York, N.Y.), 2*(10), 100347. doi:10.1016/j.patter.2021.100347 PMID:34693373

O'Sullivan, M. E., Considine, E. C., O'Riordan, M., Marnane, W., Rennie, J., & Boylan, G. (2021). Challenges of Developing Robust AI for Intrapartum Fetal Heart Rate Monitoring. *Frontiers in Artificial Intelligence, 4*, 4. doi:10.3389/frai.2021.765210 PMID:34765970

Oneto, L., & Chiappa, S. (2020). *Fairness in Machine Learning.* ArXiv, abs/2012.15816.

Peng, J., Jury, E. C., Dönnes, P., & Ciurtin, C. (2021). Machine learning techniques for personalised medicine approaches in immune-mediated chronic inflammatory diseases: Applications and challenges. *Frontiers in Pharmacology, 12*, 720694. doi:10.3389/fphar.2021.720694 PMID:34658859

Pozzi, F. A., & Dwivedi, D. (2023). ESG and IoT: Ensuring Sustainability and Social Responsibility in the Digital Age. In S. Tiwari, F. Ortiz-Rodríguez, S. Mishra, E. Vakaj, & K. Kotecha (Eds.), *Artificial Intelligence: Towards Sustainable Intelligence. AI4S 2023. Communications in Computer and Information Science* (Vol. 1907). Springer. doi:10.1007/978-3-031-47997-7_2

Seizov, O., & Wulf, A. J. (2020). Artificial Intelligence and Transparency: A Blueprint for Improving the Regulation of AI Applications in the EU. *European Business Law Review, 31*(4).

Tamboli, A. (2019). Evaluating Risks of the AI Solution. *Keeping Your AI Under Control,* 31–42. doi:10.1007/978-1-4842-5467-7_4

Thierer, A. D., Castillo O'Sullivan, A., & Russell, R. (2017). *Artificial intelligence and public policy.* Mercatus Research Paper.

Truby, J. (2020). Governing artificial intelligence to benefit the UN sustainable development goals. *Sustainable Development (Bradford)*, 28(4), 946–959. doi:10.1002/sd.2048

Whittaker, M., Crawford, K., Dobbe, R., Fried, G., Kaziunas, E., Mathur, V., & Schwartz, O. (2018). *AI now report 2018*. AI Now Institute at New York University.

Wu, F., Lu, C., Zhu, M., Chen, H., Zhu, J., Yu, K., Li, L., Li, M., Chen, Q., Li, X., Cao, X., Wang, Z., Zha, Z., Zhuang, Y., & Pan, Y. (2020). Towards a new generation of artificial intelligence in China. *Nature Machine Intelligence*, 2(6), 312–316. doi:10.1038/s42256-020-0183-4

Yogarajan, V., Dobbie, G., Leitch, S., Keegan, T. T., Bensemann, J., Witbrock, M., Asrani, V. M., & Reith, D. M. (2022). Data and model bias in artificial intelligence for healthcare applications in New Zealand. *Frontiers of Computer Science*, 4, 1070493. doi:10.3389/fcomp.2022.1070493

Chapter 13
Investigation Into the Use of IoT Technology and Machine Learning for the Identification of Crop Diseases

K. Manikandan
Vellore Institute of Technology, India

Vivek Veeraiah
Sri Siddharth Institute of Technology, Sri Siddhartha Academy of Higher Education, India

Dharmesh Dhabliya
https://orcid.org/0000-0002-6340-2993
Vishwakarma Institute of Information Technology, India

Sanjiv Kumar Jain
https://orcid.org/0000-0001-8942-7681
Medi-Caps University, India

Sukhvinder Singh Dari
https://orcid.org/0000-0002-6218-6600
Symbiosis Law School, Symbiosis International University, India

Ankur Gupta
https://orcid.org/0000-0002-4651-5830
Vaish College of Engineering, India

Sabyasachi Pramanik
https://orcid.org/0000-0002-9431-8751
Haldia Institute of Technology, India

ABSTRACT

The control and management of crop diseases has always been a focal point of study in the agricultural domain. The growth of agricultural planting areas has posed several obstacles in monitoring, identifying, and managing large-scale illnesses. Insufficient disease identification capacity in relation to the expanding planting area results in heightened disease intensity, leading to decreased crop production and reduced yield per unit area. Evidence indicates that the reduction in crop productivity resulting from illnesses often surpasses 40%, leading to both financial setbacks for farmers and a certain degree of impact on local economic growth. A total of 1406 photos were gathered from 50 image sensor nodes. These images consist of 433 healthy images, 354 images showing big spot disease, 187 images showing tiny spot disease, and 432 images showing rust disease. This chapter examines the cultivation of maize fields in open-air environments and integrates internet of things (IoT) technologies.

DOI: 10.4018/979-8-3693-2964-1.ch013

OVERVIEW

Since the beginning of the 21st century, the fast development of Internet of Things (IoT) technology (Lombardi, 2021) and image recognition technology has garnered significant interest among researchers in the "IoT technology + agriculture" (Yan, 2020) concept. To effectively mitigate the occurrence of widespread localized maize illnesses in their cultivation regions, the use of agricultural Internet of Things (IoT) is unquestionably the optimal solution for disease prevention and management. This technology enables early detection of diseases and seamless integration with diverse agricultural systems. The use of agricultural IoT has led to the advancement of modern agriculture.

Advancing towards precision prevention provides a robust assurance for data gathering, real-time analysis, and efficient management. Furthermore, the progress of IoT technology has enabled the enhancement of image recognition technology for the identification of maize illnesses in particular regions, establishing a strong basis for the future application of agricultural IoT and the continuous monitoring of crop growth conditions.

Managing maize diseases gets more challenging as the cultivation area grows. This article suggests a way for quickly identifying the location and severity of maize diseases in outdoor corn planting regions using IoT technology. The approach involves using image recognition to analyze corn disease images. This approach utilizes sensor nodes specifically built for maize planting regions, which possess the qualities of completeness, fault tolerance, precision, and real-time monitoring technology. The collection of real-time dynamic information on the maize planting area is achieved by the placement of sensor nodes. Once the corn picture data is gathered, image processing methods are used to extract characteristics based on color, shape, and texture. Ultimately, a suitable image recognition model is developed using the specific features of node deployment and data collecting. This model guarantees precise disease management during the most effective prevention and control period, hence minimizing economic losses resulting from unnecessary preventative and control procedures.

LITERATURE REVIEW

Investigation of the Use of Internet of Things (IoT) Technology in Disease Monitoring Systems

The emergence of the Internet of Things has greatly impacted agricultural intelligence, leading to the creation of several sensor devices that have revolutionized the interaction between things and users. In order to enhance the efficacy of disease pre-warning and in-process warning, it is necessary to use sensor nodes for data collecting in the cultivation of crops on a wider scale. Liu Hui et al. developed two node deployment schemes in 2011 to evaluate the comprehensiveness and logic of sensor node deployment. The schemes were evaluated using parameters such as deployment cost and connectivity.

In 2012, López et al. developed an independent monitoring system that used low-power image sensors to cover a wide region. The system time stamped the image sensors at the control station for analysis and identification of disease damage. Li Hao and his colleagues used IoT technology to address the real-time gathering and identification of fruit photos. To address the issue of data transmission in wireless sensor networks for agricultural disease monitoring. Pierce (2007) developed sensor networks at both regional and farm levels to enable remote and real-time monitoring of crucial agricultural activities. This initiative enhances the overall management efficiency and efficacy of sensor nodes, hence boosting their value. Deljoo and Keshtgari (2011) used the wireless sensor network system they created for precision agricultural purposes to gather data on climate and other environmental attributes, and then made control choices based on this data. Huafang et al. (2020) developed an integrated soil moisture monitoring device using Internet of Things technology. This device does not need any additional hardware and is simple to install and operate. The use of sensor network technology inside a cloud platform offers advantages such as automated data collecting and remote monitoring, making it applicable in many contexts.

Locations with excellent performance. Othman (2021) used subterranean and aerial sensors to autonomously monitor environmental variables in real-time. They gathered a substantial quantity of environmental data by using IoT sensors and sent it to the ground gateway on an hourly basis in agricultural areas. The acquired data was subsequently gathered and delivered by unmanned aerial vehicles to cloud storage and analysis at intervals of 12 hours. This affordable platform enables farmers, governments, or industries to forecast environmental data for geographically expansive fields, hence enhancing crop yield, cost-efficiency, and timely farm management.

Study on Disease Image Recognition Technology

Recognition

The challenge of automatically classifying photos of agricultural pests and diseases is quite tough. One factor contributing to the challenges in picture categorization is the inherent features of pathological images, including minor variations across images and unequal distribution of colors. Furthermore, the absence of extensive publicly accessible and annotated datasets presents specific challenges for algorithmic study. However, researchers have extensively investigated and achieved significant advancements in the automated categorization of photos depicting agricultural pests and diseases.

(1) Investigation into illness picture processing and feature extraction. In their study, Cen et al. (2007) examined how various color light reflection characteristics affect disease categories. They employed genetic algorithms to determine model parameters based on these characteristics, resulting in the development of a prototype for automatic diagnosis technology for cucumber anthracnose. Tian et al. (2015) used computer processing technology to preprocess photos of grape diseases, aiming to achieve illness spot segmentation and extract pertinent characteristics. Subsequently, support vector machines were used to train models for the automated classification and identification of grape illnesses. This was done to address the limitations of manual recognition by experts, minimize recognition errors, and enhance overall efficiency. Researchers have used imaging technology to identify citrus canker disease (Cai, 1995; Ren et al., 2006), investigating critical aspects such as the creation of digital features, selection of features, and design of a classifier for disease spots. The researchers have examined preprocessing methods for illness imaging, including image denoising and image segmentation. Different techniques, including Gabor transform and grayscale differential statistics have been used to create digital characteristics of disease spots, focusing on color, texture, and shape factors.

(2) This study focuses on feature extraction and recognition in machine learning. Wang (2012) have developed a prediction model using machine learning techniques such as support vector machines to analyse the cross population transmission of avian influenza virus and the antigen relationship of

H3N2 (Russell et al., 2008) subtype influenza virus. Researchers have found 90 specific amino acid positions that are associated with the transmission of avian influenza virus from birds to humans. Additionally, they have identified 18 crucial amino acid positions that contribute to the variation of the H3N2 influenza virus antigen. The models created by integrating amino acid composition and Moran autocorrelation coefficients exhibit superior performance, with an accuracy exceeding 92.75%. These models serve as a valuable reference for studying molecular determinants and underlying mechanisms, as well as for early detection of public health diseases. Deep learning has gained significant attention in the area of machine learning in recent years (Hinton, et al., 2006; Hinton & Salakhutdinov, 2006).

UTILIZING INTERNET OF THINGS (IOT) TECHNOLOGY FOR THE COLLECTION AND PROCESSING OF IMAGES RELATED TO MAIZE DISEASES

Sensor Nodes' Deployment Methodology

The placement of sensor nodes is essential for the image surveillance capabilities of wireless sensor networks, as it may impact the interconnection between nodes and the reliability and precision of image data transmission. The deployment of sensor nodes should guarantee the comprehensive collection of all diseases in the maize planting area. It is crucial to ensure that the number of nodes deployed remains within a reasonable range while acquiring maize images. This will prevent the duplication of image information caused by excessive deployment of redundant nodes, thereby reducing the complexity of image recognition.

This article proposes a technique for deploying sensor nodes using an uneven triangular mesh, as seen in Figure 1. The primary factors taken into account are the cost of deployment, the level of connection between nodes, and the redundancy of nodes. In maize planting regions with defined borders, a bigger grid area will result in a decreased number of sensor nodes deployed, leading to cheaper deployment costs and a corresponding decrease in the complexity of image identification. In an ideal scenario, each node (except the border node, which is always present) should have three neighboring nodes for connec-

Figure 1. Method for deploying sensor nodes in an irregular triangular grid

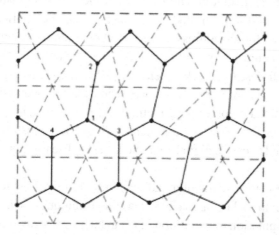

tion. In the event of node failure or non-deployable nodes, the system can guarantee the acquisition of all picture information inside the maize planting area via the deployed nodes. Consequently, considering the analysis provided, a technique for deploying sensor nodes in an irregular triangular mesh was devised.

When the planting area has a uniform rectangular shape, position the nodes according to the arrangement shown in Figure 1. To accommodate irregular figures in the planting area, create a minimal rectangle outside the area using the straight edges of the planting area as a reference. Then, deploy the nodes using the irregular triangular grid technique for sensor node deployment. If there are nodes located beyond the designated planting area but still inside the minimum bounding rectangle, those nodes may be rounded off.

Figure 1 illustrates the optimal configuration of sensor nodes in an irregular triangular grid, where all nodes are operational. The dotted boundary delineates the designated area for maize planting. The given text is incomplete and does not provide enough information to rewrite it in a straightforward and precise manner. Please provide more contexts or complete the sentence.

The dark solid point, which is the triangular centre of gravity, indicates the exact location where the image sensor node is deployed. The line segment connecting the black solid points reflects the distance of connection between nodes. The black solid triangle symbolizes the undisclosed data of the node. The distance between nodes is controlled by the range of the camera carried by each node and the efficiency of the wireless network transceiver. Figure 1 only provides an illustrative example by showing a few nodes and their respective connection distances. Preprocessing of images depicting maize diseases with a single node

Given that the maize disease photos obtained by a single node provide a comprehensive viewpoint, complete coverage of the disease images may be attained by using a single image sensor for collecting. Thus, in order to address this particular collecting scenario, advanced image preprocessing techniques such as weighted average and histogram equalization are used. These approaches effectively reduce irrelevant information and recover any disease-related information that may have been lost during the process of picture collection and transmission.

The Weighted Average

To accomplish optimal image processing, color pictures may be transformed to grayscale, hence reducing cost overhead. The corn color picture consists of three channels, namely R, G, and B, which individually represent the red, green, and blue components. This article employs the weighted average technique to preprocess single node picture data. The approach calculates the average by taking into account the weight values assigned to each of the three components.

Figure 2 displays the composite pictures obtained by calculating the weighted average of four different kinds of maize photos. When compared to the original photographs, the modified images effectively emphasize the diseased region and demonstrate significant improvements in illumination, transmission noise, and other related concerns.

Histogram equalization is a technique used to enhance the contrast of an image by redistributing the pixel intensities in a way that maximizes the use of the available dynamic range.

Histograms in maize disease photos provide a visual representation of the frequency distribution of grayscale values. Histograms are employed to enhance the differentiation of the four types of maize images by improving the comparison of grayscale value distribution. This ensures that the grayscale information of the images remains clear and prevents the loss of useful information during subsequent

Figure 2. The image weighted average approach

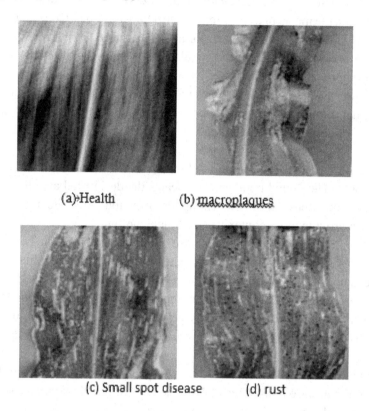

(a) Health   (b) macroplaques

(c) Small spot disease   (d) rust

processing. Apply histogram equalization (Li, et al., 2021; Xiuli et al., 2021) to normalize the distribution of grayscale values in the histogram, specifically targeting the regions with concentrated values, in order to achieve a uniform distribution over the whole range.

Figure 3 displays four maize photos depicting the various forms of maize illnesses after the process of equalization. Figure 4 depicts the

Histogram of maize disease photos after undergoing equalization.

Upon analyzing the histogram after equalization, it becomes evident that the improved histogram is able to substantially increase the range of grey values in the four kinds of maize photographs, in comparison to the weighted average histogram. The histogram analysis reveals that after equalization, the healthy corn image exhibits a single peak at 60. The corn image with large spot disease shows two peaks at 110 and 220, while the small spot disease image displays peaks at 90180 and 240. Lastly, the rust disease image demonstrates a single peak at 110. Hence, the use of both picture graying and image enhancing techniques offers significant benefits in handling maize images captured from four complete perspectives. Moreover, this approach serves as groundwork for subsequent image segmentation and extraction of the most appropriate segmentation threshold.

Segmentation of Maize Disease Images Using a Single Node

Single node corn pictures and multi node corn images use distinct image segmentation techniques as a result of disparate data gathering approaches. Single node maize photos use global thresholds for seg-

Figure 3. Image after equalization

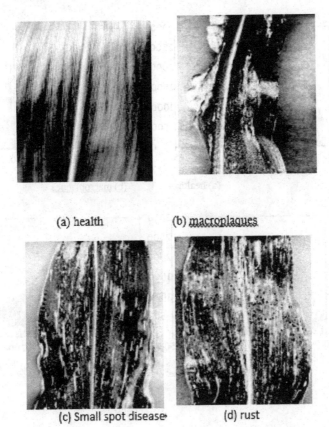

(a) health   (b) macroplaques

(c) Small spot disease   (d) rust

mentation due to the extensive angle and time information provided by the image source. However, for maize images with multiple nodes, the angle and time information from the image source is absent. In order to mitigate the influence of global segmentation techniques on the extraction of disease features, we use local texture and shape for both edge segmentation and area segmentation.

The prerequisites for picture segmentation methods:

①A pixel is exclusively assigned to one region, meaning segmentation is a fundamental aspect of the pixel.
②A pixel cannot simultaneously belong to two regions, indicating that there is no overlap after segmentation.
③The characteristics of the two regions differ after segmentation.
④Following segmentation the features of the same region exhibit similarity.
⑤The pixels within the same area are connected.

Threshold segmentation, being the most basic and widely used approach for image segmentation, is appropriate for pictures when the overall grayscale values are already known. Thus, the segmentation approach used for single node maize disease pictures is threshold segmentation. The single threshold segmentation method involves the following steps: Firstly, we assume that the grayscale value of a point in the image is represented by $g(x, y)$, and the grayscale value after image segmentation is represented by $h(x, y)$. Next, we select an appropriate threshold, denoted as T, based on the grayscale value charac-

Figure 4. Histogram after equalization

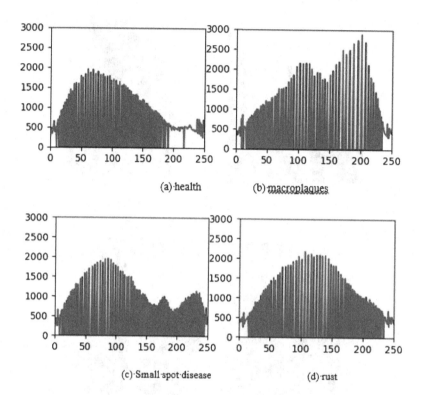

(a) health   (b) macroplagues

(c) Small spot disease   (d) rust

teristics of the image. Finally, we compare the grayscale value with the segmentation threshold T. When the grayscale value exceeds or equals the threshold, it is assigned to one category with the same value. The remaining values belong to another category. This process, known as single threshold segmentation, splits the picture into two parts: the target and the backdrop.

The bimodal segmentation is enhanced by using the concept of approximation to further refine the ideal threshold range. The suitable threshold is determined by analyzing the grayscale balance histogram for several tests, and the most effective threshold is eventually identified throughout the experiment.

DEVELOPMENT OF A MAIZE DISEASE PICTURE IDENTIFICATION MODEL USING ARTIFICIAL INTELLIGENCE ALGORITHMS THAT CAN LEARN AND IMPROVE FROM DATA, KNOWN AS MACHINE LEARNING

The maize pictures obtained by a single node are classified using the naïve Bayesian classification technique. The naive Bayesian classification method is highly regarded for its simplicity and efficiency in the field of machine learning. The algorithm has challenges in meeting the assumption of independence among its properties in real-world scenarios (Yudianto et al., 2021). Hence, when categorizing maize photos, a certain kind of maize image may exhibit a greater reliance on one of the retrieved feature traits. When compared to other feature characteristics, the classification results of this particular property will be influenced, leading to subpar recognition outcomes. Hence, it is essential to prevent any interdependence between the characteristics of the features, ensuring that the class variables exhibit identical dependencies on each feature property. Regarding the topics discussed

Ultimately, a naïve Bayesian classification system based on reduced characteristics was chosen for recognition. This method further takes into account the correlation between feature characteristics and the influence of various attribute value ranges on the categorization of maize photos. From these, feature characteristics with significant classification impacts were chosen, ensuring that the attributes that were excluded had a poor association with the maize picture category. There is a significant level of duplication between it and other characteristics of the feature. Ultimately, a simple Bayesian model is created for the purpose of categorizing maize images (Aliakbari Sani et al., 2021).

Principal component analysis is used for dimensionality reduction.

The naive Bayesian technique is used for single node maize pictures, necessitating the fulfillment of the independence criterion among feature characteristics. Hence, the dimensionality reduction methodology used is the frequently employed principal component analysis method (Feng, 2021) (PCA). In this paper, the recognition application of maize disease pictures requires the collaboration of several feature variables for accurate identification. This collaboration introduces a certain level of complexity to the recognition process, and there may be dependency among these feature variables. When just one feature attribute is assessed without considering other information in the sample data, the analyzed data becomes isolated and cannot make use of other information. Decreasing the number of feature characteristics will result in the loss of crucial information contained within those attributes. Hence, it is essential to transform the feature characteristics that are heavily reliant on each other into feature attributes that are mutually independent to the greatest extent feasible. This implies that a reduced number of feature attributes may effectively convey the primary information of all feature attributes. Condense the initial data

Maximize the use of feature attributes to effectively retain feature attribute information for optimal picture recognition. PCA, however, utilizes the connections between principle components to discover the most significant variations and chooses the most representative feature characteristics from them. The procedure for identifying the primary constituents is as follows:

If there are N feature variables represented as $X_i=\{X_{i1},X_{i2},...,X_{in}\}(i=1,2,...,N)$, and the correlation coefficient between two feature attributes X_a and X_b is λ_{ab}, then the following steps should be taken: 1. Preprocess the collected images and extract the feature attributes. 2. Calculate the correlation coefficient between each pair of feature attributes to obtain the correlation coefficient matrix of the entire image feature attribute, denoted as $Maker=(\lambda_{ab})$, where $a,b=1,2,...,N$. 3. Calculate and record the eigenvalues and eigenvectors of the correlation coefficient matrix as θ_i and $e_i=\{e_{i1}, e_{i2},..., e_{it}\}$. 4. Calculate the variance contribution rate (con) of the feature attributes and the cumulative variance contribution rate (acc) of the first s feature attributes.

An experimental study was conducted to analyze the identification of maize disease images using machine learning techniques.

Preparation for the Experiment

Data Set

The article presents a method for deploying sensor nodes in an outdoor corn planting base using an irregular triangular grid. A total of 50 image sensor nodes were deployed to obtain various types of corn disease images. These images were captured using both single and multiple nodes in order to validate the effectiveness of the model. Table 1 displays the dataset including images of maize.

Table 1 indicates that this article acquired a total of 1406 valid photos from 50 image sensor nodes. These images consist of 433 healthy images, 354 images showing big spot disease, 187 images showing tiny spot disease, and 432 images showing rust disease.

Table 1. Corn image dataset

Corn category	health	big spot	small spot	rust	total
Quantity(all nodes)	433	354	188	433	1407

The preprocessing of the 1406 existing photos involves dividing them into three sets: a training set, a validation set, and a test set, with a ratio of 8:1:1. The training set is used to train the model, the validation set is employed to fine-tune the model's parameters, and the test set is utilized to evaluate the final model's recognition performance.

Experimental Environment

The MATLAB 2016b platform is used as the software environment for image identification in this work for experimental analysis and verification. The hardware setup for the experiment is detailed in Table 2.

Table 2. Experimental hardware environment

Environment	Operating System	CPU	Memory	Programming Language
configuration	Windows10 64bit	Lenovo i7- 8-core	16G	Matlab

Given the experimental preparations described above, the subsequent procedure involves conducting four distinct forms of image recognition on maize images obtained from sensor nodes developed within the Internet of Things framework. These four types of recognition include assessing the health of the maize, detecting the presence of big spot disease, identifying small spot disease, and detecting rust disease. The purpose of this step is to validate the efficacy of the two image recognition methods

Estimation of Parameters for Reducing Dimensionality

The PCA approach is used for reducing the dimensionality of maize image data obtained from a single node. This algorithm involves a single parameter, s, which represents the number of algorithms utilized after the PCA dimensionality reduction. The determination of PCA parameters lacks defined guidelines;

hence this article establishes them based on the findings of model trials. Since the feature parameters of a single node corn picture consist of 25 dimensions, the range of parameter dimensions is selected from 2 to 25 in order to guarantee its universality. The SVM method is selected to train the model and ascertain the value of parameter s. The process of selecting dimensionality reduction parameters adheres to the notion of "Occam's Razor" in order to minimize complexity and streamline the process. Six feature parameters with superior performance and lower values are chosen as the final values for s.

Validation Findings and Analysis Example

Utilize the developed single node maize disease image recognition model to train and test the gathered single node image data in order to validate the model's efficacy. Partition the gathered photographs of individual maize diseases into training, validation, and testing datasets, using an 8:1:1 ratio. Evaluate the efficacy of the model by employing labeled images verified by professionals as the testing dataset.

Following the capture of a single node picture, the image undergoes preprocessing, segmentation, morphological processing, and feature extraction. Subsequently, image recognition is carried out using a naïve Bayesian classification model that relies on reduced attributes. The recognition features are further examined via the analysis of assessment indicators. The following presents the training outcomes of

Table 3. Model training results of single node maize disease image

Disease type	precision	recall	F1-score
health	0.92	0.96	0.96
big spot	0.98	0.92	0.97
small spot	0.85	0.79	0.86
rust	0.87	0.79	0.85
average value	0.92	0.85	0.87

single-node maize disease pictures and the results of a comparison between several models.

Based on the data shown in the table above, it can be inferred that the model developed for maize pictures obtained from a single node achieves an overall identification rate of 96% for maize health images, 97% for large spot illnesses, and 86% for tiny spot diseases. The model exhibits a notable capacity to recognize corn health images and identify big spot disease, achieving the highest recognition rate of 97% for corn big spot disease. However, its effectiveness in recognizing corn spot disease and rust disease is slightly less prominent, with rust disease having the lowest recognition rate of 85%. The model achieves an overall recognition rate of 87%, demonstrating a rather remarkable recognition impact. Table 4 displays the comparative outcomes of several models for photos of maize diseases in individual nodes.

Table 4 presents a comparison of the outcomes obtained by Naive Bayes, AlexNet, VGG-16, VGG-19, ResNet-50, Inception-V3, and our own model. The Naive Bayes classification method used in this study, which relies on reduced features, has distinct benefits in recognition.

Table 4. Multi-model comparison results of single node maize disease image

model	precision	recall	F1-score
Naive Bayes	0.85	0.85	0.87
AlexNet	0.73	0.73	0.73
VGG-16	0.74	0.75	0.76
VGG-19	0.76	0.76	0.77
ResNet-50	0.80	0.80	0.82
Inception-V3	0.83	0.81	0.82
Model in this article	0.90	0.86	0.91

CONCLUSION

This paper examines the use of IoT technology in the cultivation of maize fields in outdoor environments. A strategy for deploying sensor nodes in an uneven triangular grid is aimed to efficiently gather maize inside the planting area without duplicating data. To address the issues of specific fault risks and improper sensor placement in outdoor node deployment, it is necessary to select appropriate preprocessing, image segmentation, and feature extraction techniques. The selection of these methods should be based on whether the nodes in the single node maize disease image are online or offline during image collection. A simple attribute-based naïve Bayesian classification system was developed for maize pictures obtained from a solitary node. In order to prevent any potential interference between the retrieved feature characteristics, Principal Component Analysis (PCA) was used to decrease the dimensionality of the feature attributes and streamline them. Ultimately, the widely successful naive Bayesian approach was used for picture recognition, with a recognition rate of 90%. The model provides disease detection findings and illness degree distribution rate for each node, which serves as a foundation for farmers to precisely administer pesticides.

REFERENCES

Aliakbari Sani, S., Khorram, A., Jaffari, A., & Ebrahimi, G. (2021). Development of processing map for InX-750 superalloy using hyperbolic sinus equation and ANN model. *Rare Metals, 40*(12), 3598–3607. doi:10.1007/s12598-018-1043-9

Cai, Y. (1995). Using a self-organizing artificial neural network model to distinguish the onset period of citrus canker disease. *Journal of Plant Pathology*, (01), 43–46.

Cen, Li, & Shi. (2007). Research on the Recognition of Cucumber Anthracnose and Brown Spot Based on Color Statistical Features of Color Images. *Journal of Horticulture, 34*(6), 124–124.

Deljoo & Keshtgari. (2011). Wireless Sensor Network Solution for Precision Agriculture Based on Zigbee Technology. *Wireless Sensor Network, 4*(1).

Feng, Y. (2021). Comprehensive evaluation of quality indicators for different varieties of kiwifruit based on principal component analysis and cluster analysis. *Jiangsu Agricultural Science and Technology Xue, 49*(22), 180–185.

Haokun, Y. (2021). Image Segmentation of Pitaya Disease Based on Genetic Algorithm and Otsu Algorithm . *Journal of Physics: Conference Series, 1955*(1).

Hinton, G. E. & Osidero, S., & The, Y.-W. (2006). A fast learning algorithm for deep belief nets. *Neural Calculation, 18*(7), 1527-1554.

Hinton, G. E., & Salakhutdinov, R. R. (2006). Reducing the dimensionality of data with neural networks. *Science, 313*(5786), 504–507. doi:10.1126/science.1127647 PMID:16873662

Huafang, L., Hanbo, Y., & Xinyuan, W. (2020). Development and application of a "diatomaceous earth filter" type soil moisture sensor. *Water Resources and Hydropower Technology.*

Li, L., He, D., & Wang, M. (2021). Research on Plant Leaf Image Recognition Based on Improved LBP Algorithm. *Computer Engineering and Applications, 57*(19), 228-34.

Lombardi, M. (2021). Introduction to the columns of the Internet of Things Technology magazine. *Internet of Things Technology, 11*(12), 114.

Mitrofanov, S. (2021). Tree retraining in the decision tree learning algorithm. *IOP Conference Series. Materials Science and Engineering, 1047*(1).

Othman, S. (2021). A Low Cost Platform for Environmental Smart Farming Monitoring System Based on IoT and UAVs. *Sustainability, 13*(11).

Pierce, F J. (2007). Regional and on arm wireless sensor networks for agricultural Systems in Eastern Washington . *Computers and Electronics in Agriculture, 61*(1).

Ren,, J., Huang,, S., , & Li,, Y. (2006). The application of AR model in the prediction of citrus canker disease . *Journal of Plant Pathology,* (05), 460–465.

Russell, C. A., Jones, T. C., & Barr, I. G. (2008). The global circulation of seasonal influenza A (H3N2) viruses. *Science, 320*(5874), 340-346.

Saeed, Z. (2021). Interval–valued fuzzy and intuitionistic fuzzy–KNN for imbalanced data classification . *Expert Systems with Applications,* 184.

Tian, Y., Yi, C., & Wang, X. (2015). A Method for Identifying Apple Pest Damage Defects and Fruit Stems/Calyx Based on Hyperspectral Imaging. *Journal of Agricultural Engineering,* (4), 325–331.

Tigistu, T. (2021). Classification of rose flowers based on Fourier descriptors and color moments. *Multimedia Tools and Applications, 80*(30).

Wang, J. (2012). *Machine learning based prediction of cross species transmission and antigen relationship of influenza A virus.* Huazhong University of Science and Technology.

Xiuli, L., Zhaohao, H., & Yongqiang, B. (2021). Research on facial recognition algorithms based on improved LBP and DBN. *Industrial Instruments and Automation Devices,* (5), 80-2.

Yan, M. (2020). Research on the Application of Internet of Things Technology in the Development of Modern Agriculture. *Small and Medium sized Enterprise Management and Technology (Second Edition)*, (11), 191-3.

Yudianto, M. R. AAgustin, TJames, R M. (2021). Rainfall Forecasting to Recommend Crops Varieties Using Moving Average and Naive Bayes Methods. *International Journal of Modern Education and Computer Science*, 13(3).

Chapter 14
Narrating Spatial Data With Responsibility:
Balancing Ethics and Decision Making

Munir Ahmad

 https://orcid.org/0000-0003-4836-6151

Survey of Pakistan, Pakistan

Saadia Ureeb

 https://orcid.org/0009-0001-4195-8720

Zarai Taraqiati Bank Ltd., Pakistan

ABSTRACT

This chapter discussed the responsible use of spatial data in decision-making and the need to balance ethics with decision-making. The role of spatial data in decision-making and storytelling is discussed. Ethical issues in spatial data storytelling such as privacy and confidentiality, bias and discrimination, accuracy and misinformation, accessibility and inclusivity, environmental and social impacts, data ownership and credit, and transparency are described, along with the strategies to overcome these issues. The different roles of storytelling in decision-making are elaborated on, along with the challenges and risks associated with using storytelling in decision-making. The chapter also discussed the challenges of striking a balance between ethics and decision-making, such as conflicting interests, cognitive biases, time constraints, political and legal constraints, and many more. Strategies to overcome these challenges, including developing clear ethical guidelines and engaging diverse stakeholders, are underpinned in this chapter as well.

INTRODUCTION

Over the past few years, the availability and ease of access to spatial data has been increased. This has generated a greater curiosity in utilizing spatial data to convey intricate information and shape decision-making procedures. Resultantly, the use of spatial data has multiplied in various tasks to support sustain-

DOI: 10.4018/979-8-3693-2964-1.ch014

able development (Ahmad, 2023d, 2023c). Moreover, spatial data possesses a special power to improve learning skills and can enhance the learning outcomes of higher education learners by providing dynamic and interactive learning experiences (Ahmad, 2023a). Spatial data can significantly impact decision-making, policy formation and evaluation, and public opinion assessment as noted by Ahmad (2023b).

The concept of data storytelling combines spatial data with narrative elements to create visually stimulating presentations, thus making data more accessible and comprehensible, and engaging stakeholders. Nonetheless, the practice of data storytelling comes with its own set of ethical concerns, which require careful evaluation to prevent unintentional negative effects and potential damage (Newman et al., 2021). This chapter aims to investigate the ethical implications of spatial data storytelling and analyze how storytelling affects decision-making processes. It also addresses the difficulties of balancing ethical concerns with decision-making while utilizing data storytelling techniques. The chapter will explore three primary questions related to this topic.

The first question that this chapter aims to answer is, "What ethical issues must be taken into account when utilizing spatial data storytelling?" Spatial data and the resulting insights have the potential to impact individuals and communities. Therefore, it is essential to consider the ethical implications of spatial data storytelling. These ethical considerations can encompass issues like data privacy, accuracy, bias, and ownership. This chapter also intends to provide guidelines to address these ethical concerns.

The second research question is, "How does storytelling influence decision-making processes?" Storytelling has the power to influence people's emotions and perspectives regarding a particular topic. In the context of data storytelling, it can improve decision-making by conveying information in an approachable and captivating manner. Nevertheless, the effects of storytelling on decision-making also present certain complexities that necessitate examination. Therefore, this chapter will explore how storytelling impacts decision-making and assess the potential drawbacks and challenges involved in using storytelling as a tool for decision-making.

The third research question is, "What challenges arise in balancing ethical considerations and decision-making when using data storytelling techniques?" Achieving a balance between ethical considerations and decision-making is a complex task that demands thorough consideration of the advantages and disadvantages of using storytelling techniques. Possible challenges that arise include data bias, a lack of transparency in the storytelling process, and the difficulty of measuring storytelling's impact on decision-making. This chapter will delve deeply into these challenges and suggest approaches to overcome them.

The chapter is structured into several sections. Section 2 presents an overview of spatial data, its use in storytelling, decision-making, and ethical decision-making strategies. In Section 3, the focus is on the ethics of spatial data storytelling and includes guidelines to handle ethical concerns in spatial data-based storytelling. Section 4 examines the role of storytelling in decision-making, as well as the potential risks and challenges associated with its use. Section 5 is dedicated to discussing the difficulties of balancing ethics and decision-making. Finally, the chapter's conclusion provides a summary of the key findings and suggests future research directions in this field.

SPATIAL DATA

Information pertaining to the location and characteristics of real things in space is referred to as geospatial data, also known as spatial data (Demšar et al., 2013). It is a valued resource for decision-making in various fields, such as urban planning, environmental management, and business. Spatial data can be produced through various methods, including satellite imaging, GPS technology, and field surveys. It is then warehoused and analyzed using specialized software that permits spatial visualization, mapping, and spatial analysis.

Spatial Data Role in Decision-Making

Spatial data has become an essential tool for decision-making in today's environment as location-based information gains value (Ahmad, 2023b; Feeney et al., 2001). Spatial data is growing in importance in numerous industries' decision-making processes because it can offer a more comprehensive view of an event. Organizations may investigate patterns and relationships that might have been difficult to detect using traditional data sources through the use of geographical data. As a result, decisions can be made that are well-informed and take into consideration the position of assets, demographic information, and environmental conditions (Antunes et al., 2010; Rajabifard et al., 2006).

One benefit of using spatial data in decision-making is that it provides decision-makers with a thorough understanding of the problem under question. Decision-makers can see relationships and trends that are masked in data tables or other types of information by using spatial data analysis. Decision-makers can visualize intricate relationships using this approach, such as how climate change affects natural resources or how the demographics of the population are distributed geographically. Decision-makers can also use spatial data to determine and prioritize intervention areas, distribute resources effectively, and evaluate the effects of their decisions. Decision-makers now have an effective tool for comprehending the interaction between the physical environment and social events, as well as anticipating and preparing for future developments, because of the ability to collect and analyze spatial data. As a result, spatial information has been transformed into a vital tool for decision-makers in a variety of industries, enabling them to make well-informed data-driven conclusions aligned with their physical and social environments. However, employing spatial data presents ethical challenges that must be taken into account when making decisions (Khandare et al., 2022).

Spatial Data Role in Storytelling

Spatial data has a crucial role in data-driven storytelling. It entails using visualizations such as maps and charts to convey a message or a story that is based on spatial data. This technique allows storytellers and data analysts to present intricate information in a more accessible and captivating manner that can be easily grasped by a diverse audience (Fischer et al., 2019).

The use of spatial data is increasingly recognized as a valuable tool in storytelling, enabling data analysts and storytellers to present complex information in an accessible and engaging manner. Spatial data storytelling involves the use of maps, charts, and other visualizations to communicate a message or tell a story that is grounded in spatial data. By contextualizing information through its spatial relationships, spatial data can provide a more immersive and comprehensive understanding of the story being told. This can make narratives more dynamic and captivating, inspiring change and informing decision-making processes (Juergens & Redecker, 2023).

However, it is important to note that the use of spatial data in storytelling raises ethical concerns that must be carefully considered to ensure that the information presented is transparent, accountable, and inclusive. This includes incorporating diverse perspectives, being transparent about the sources of data, ongoing monitoring and evaluation, and continuous learning and improvement. Therefore, the ethics of data storytelling is an essential aspect that should be taken into account while using spatial data in storytelling (Ahasan & Hossain, 2021; Juergens & Redecker, 2023).

ETHICS OF DATA STORYTELLING

Data storytelling has gained popularity as a potent method of presenting complicated information to both individuals and organizations in an informative and comprehensible way. Data storytellers may develop narratives that engage audiences and create an indelible mark by fusing storytelling techniques with data studies. To make sure that the data presented is accurate, reliable, and presented in an ethical and impartial way, ethical issues related to the use of data in storytelling must be taken seriously. The use of storytelling can enhance critical thinking, which will ultimately enhance decision-making abilities. This deeper degree of comprehension may assist decision-makers in taking into account a wider variety of viewpoints and potential outcomes, ultimately resulting in more informed and efficient conclusions (Fairbairn, 2002; Moon, 2010).

It is becoming more and more important to consider ethical issues while narrating stories and making decisions, especially when using spatial data. While storytelling can improve cognitive and decision-making skills, it is crucial to make sure that the use of spatial data is transparent, accountable, and inclusive of all opinions. Now, I introduce ethical issues related to spatial data first, then I demonstrate guidelines to handle ethical issues in spatial data-based storytelling.

Ethical Issues in Spatial Data Storytelling

The use of spatial data in data storytelling techniques raises important ethical issues. These include issues with bias, privacy, accuracy, accessibility, transparency, environmental impact, and data ownership.

Privacy and Confidentiality

The privacy and confidentiality of people may be compromised when the spatial data of the individuals is published. People can suffer, for instance, if personal information is shared with third parties without their consent (Berman et al., 2018; De Jong et al., 2019; Haley et al., 2016; Zandbergen, 2014). The utilization of real-time traffic data obtained from GPS-enabled devices and other channels offers significant benefits for traffic management and navigation services. Nevertheless, the extensive gathering and examination of such data give rise to apprehensions regarding surveillance and the potential for its abuse. Without appropriate regulations and safeguards in place, this data has the potential to track individuals' whereabouts, thereby compromising their privacy. Similarly, the feature of geo-tagging enables users to associate location details with their social media posts, resulting in content enriched with spatial information. While this functionality can enhance storytelling and contextual information, it also introduces potential risks to personal security. Adversaries such as burglars or criminals can exploit geo-tagged posts to identify vacant residences or target individuals based on their disclosed location data. It is crucial for users to exercise caution and be mindful of the potential consequences when sharing geo-tagged information on social media platforms.

A good example is the debate around the Strava fitness app, which exposed military personnel's locations and routes globally. In 2018, the Strava fitness application garnered significant attention when it inadvertently disclosed the whereabouts and paths taken by military personnel across the globe. By utilizing GPS technology, this app enabled users to monitor their exercise routines, generating a comprehensive global map illustrating commonly frequented routes. Unfortunately, this heat map also unveiled the movements of individuals stationed in delicate areas, such as military bases. Consequently, apprehensions were raised regarding the possible jeopardization of national security and the urgent requirement for more stringent privacy measures (Guardian, 2018). In another example, Peter Weinberg, a finance executive, found himself caught in a case of mistaken identity that resulted in online harassment and false accusations. Messages of anger and threats flooded his LinkedIn account after his photos were wrongly matched to a viral video depicting a violent assault. The authorities initially misidentified Weinberg as a suspect, leading to the dissemination of his personal information online. Despite the mistake being rectified, Weinberg endured a distressing night as individuals on social media attempted to harm his reputation. Thankfully, he was eventually exonerated from any involvement in the incident, yet the experience left him deeply affected. Weinberg emphasized the paramount importance of justice, equality, due process, and the safeguarding of privacy and safety within our society's ongoing struggle (Olivia Nuzzi, 2020).

Bias and Discrimination

Biases in spatial data may result in prejudice against particular groups of individuals. Spatial data has the potential to bring to light historical biases and discriminatory actions, notably exemplified by the practice of redlining. Redlining involves the denial or restriction of housing, financial services, or resources based on the racial or ethnic composition of specific areas. By utilizing mapping tools and historical spatial data, it becomes possible to identify instances of redlining and uncover the systematic deprivation of resources and opportunities within certain neighborhoods or communities. This helps shed light on past injustices and enables efforts towards addressing and rectifying the consequences of such discriminatory practices (Hernandez, 2009). The use of predictive policing algorithms has come under scrutiny due to concerns regarding the perpetuation of racial biases within law enforcement practices (Will, 2020). A prominent illustration of this issue involves the utilization of facial recognition technology, which has demonstrated substantial racial disparities in its accuracy rates. Notably, studies conducted by (Buolamwini, 2018) discovered that prominent facial recognition systems exhibited higher error rates when attempting to identify individuals with darker skin tones and women. These biases have the potential to result in the disproportionate targeting and surveillance of specific racial or ethnic groups.

Spatial data has the potential to unveil disparities in environmental quality and shed light on instances of environmental injustice. Extensive research has indicated that marginalized communities, particularly those residing in low-income neighborhoods and communities of color, bear a disproportionate burden of exposure to environmental hazards like toxic waste sites and pollution sources (Mohai & Saha, 2006). The existence of spatial inequities raises ethical considerations concerning the fair and equitable allocation of environmental advantages and disadvantages. Addressing these disparities is vital to guarantee equitable treatment and environmental preservation for all communities, regardless of their socioeconomic status or racial background.

Accuracy and Misinformation

Due to the complexity and difficulty of interpreting spatial data, decisions may be drawn that are unreliable or misleading. For instance, inaccuracies and misunderstandings may emerge from the presentation of spatial data without the appropriate context (Apte et al., 2019; Kitchin, 2014a). In the midst of the COVID-19 pandemic, maps played a significant role in visually illustrating the dissemination and consequences of the virus. Nevertheless, it is important to note that presenting spatial data without the necessary context can give rise to inaccuracies and misconceptions, which in turn may lead to potential misinterpretations. A specific instance of this is evident in the utilization of choropleth maps that utilize color gradients to portray the intensity of COVID-19 cases across various regions. Certain situations have arisen where these maps may have conveyed a misleading perception of the magnitude of the outbreak. This occurs when regions with higher case counts are emphasized without taking into account important factors such as population density or testing rates. Consequently, this can distort the understanding of the true extent and dissemination of the virus (Sidik, 2021). To illustrate the impact of Covid-19 in the United States as of June 26, 2020, two maps are compared in Figure 1. The first map (a) displays the total number of recorded cases per state, while the second map (b) adjusts the Covid-19 cases based on each state's population. In both maps, darker colors indicate higher values or concentrations of cases (Dougherty & Ilyankou, 2021).

Figure 1. Choropleth maps of Covid-19 cases
(Dougherty & Ilyankou, 2021).

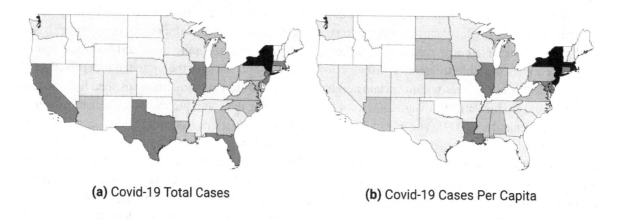

(a) Covid-19 Total Cases   **(b)** Covid-19 Cases Per Capita

Accessibility and Inclusivity

People with impairments or those lacking the technical know-how to use complicated data visualizations may find it challenging to access and interpret spatial data. People with visual impairments, for instance, may be cut off from critical information if data is presented in a way that is inaccessible to them (M. Fisher et al., 2021; Hehman & Xie, 2021; Saha et al., 2022). The choice of color schemes in maps can present obstacles for individuals with color blindness, impairing their ability to interpret the displayed information. Certain types of color vision deficiencies, such as red-green color blindness, can impact a considerable segment of the population. When maps rely exclusively on color to communicate information, individuals with color vision deficiencies may encounter challenges in distinguishing between various data categories or comprehending the intended message.

An instance where color blindness can hinder the comprehension of spatial data is when a map uses different colors to represent various land use categories (Figure 2). Individuals with color blindness may struggle to perceive the distinctions between colors, making the map inaccessible to them. As a result, their understanding of the conveyed spatial data may be hindered. Individuals with visual impairments face challenges accessing critical information conveyed through visualizations like maps and charts. The heavy reliance on visual perception in these visualizations can exclude them from understanding the spatial data effectively. The absence of alternative formats or text descriptions further hinders their independent access and comprehension of the presented information.

Figure 2. Land use categories

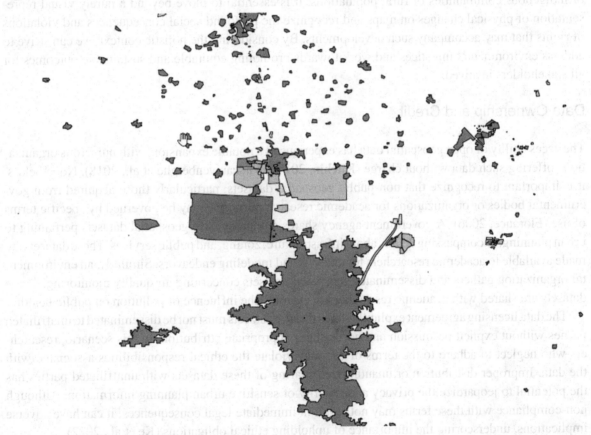

Environmental and Social Impacts

When displaying data, it is crucial to take into account the potential environmental and social consequences of spatial data. For instance, it can exacerbate environmental injustice to offer data that shows environmental degradation without taking into account the underlying social and economic issues (Apte et al., 2019; De Jong et al., 2019; Hobday et al., 2019; Qin et al., 2023). Satellite imagery has been widely employed to observe and raise awareness about deforestation occurring in the Amazon rainforest. Deforestation in the Amazon is driven by various intertwined factors such as modern agricultural practices, disputes over land ownership, and socioeconomic pressures. Utilizing satellite imagery to bring attention to deforestation in the Amazon rainforest requires a comprehensive examination. Solely emphasizing the extent of deforestation without recognizing the intricate relationship between social and economic factors can inadvertently perpetuate environmental inequities. To effectively address this issue, it is essential to consider these underlying factors and work towards sustainable solutions that promote environmental conservation and social equity (Warren, 2023).

Spatial data can provide valuable insights into the complex challenges surrounding land rights and the displacement of marginalized communities. One significant example is the development of large-scale infrastructure projects, which frequently entail land acquisition and can result in the forced displacement of indigenous communities or rural populations. It is essential to move beyond a purely visual representation of physical changes on maps and recognize the profound social consequences and violations of rights that may accompany such developments. By considering the holistic context, we can strive to address environmental injustices and work towards promoting equitable and sustainable outcomes for all stakeholders involved.

Data Ownership and Credit

The accessibility of open geospatial data has experienced a notable expansion, with numerous organizations offering such data without charge (Kitchin, 2014b; Vancauwenberghe et al., 2018). Nonetheless, it is important to recognize that non-public geospatial datasets, particularly those acquired from governmental bodies or organizations for academic research purposes, may be governed by specific terms of use (Florance, 2006). A government agency shares comprehensive geospatial datasets pertaining to urban planning, encompassing details about infrastructure, zoning, and public services. These datasets are made available to academic researchers for analysis and modeling endeavors. Similarly, an environmental organization gathers and disseminates geospatial datasets concerning air quality monitoring. These datasets are shared with academic researchers to examine the influence of pollution on public health.

The data licensing agreement explicitly states that the datasets must not be disseminated to unaffiliated parties without explicit permission and necessitates appropriate attribution. In this scenario, researchers who neglect to adhere to the terms of use may violate the ethical responsibilities associated with the data. Improper distribution or unauthorized sharing of these datasets with unaffiliated parties has the potential to jeopardize the privacy and security of sensitive urban planning information. Although non-compliance with these terms may not result in immediate legal consequences, it can have adverse implications, underscoring the importance of upholding ethical obligations (Ke et al., 2022).

Transparency

Transparency ensures that decision-makers and users of spatial data are accountable and responsible for their actions. It helps to reveal the sources of data as well as the procedures used to gather, process, and show it (Elwood & Leszczynski, 2013). For example, a government agency performs an environmental impact assessment for a development project planned in an ecologically significant region. As part of this assessment, spatial data is systematically collected and analyzed to determine the potential consequences on wildlife habitats and water resources. To ensure transparency in environmental management processes, it is essential to provide open access to data sources, methodologies, and analysis techniques for the public and relevant stakeholders. By doing so, external experts and concerned citizens can thoroughly examine the data, validate the findings, and actively contribute to the decision-making process. This inclusive approach promotes transparency, accountability, and public involvement, leading to more informed and collaborative environmental management practices.

Guidelines to Handle Ethical Issues in Spatial Data-Based Storytelling

The guidelines for managing ethical concerns in spatial data-based storytelling are listed below. These guidelines are partially adopted by (Kitchin & Lauriault, 2014; MacEachren, 1992; Locus Charter, 2021).

1. Obtain Consent: Get the consent of the people whose data is being used before collecting and using spatial data. This guarantees that people have options over and knowledge of how their personal information is implemented.
2. Anonymize Data: Consider anonymizing data by deleting personally identifiable information to preserve privacy and confidentiality. This can lessen the possibility of harm to individuals and ensure that data is handled ethically.
3. Address Bias and Discrimination: Be mindful of potential biases and prejudice in spatial data and take action to resolve such issues. To avoid maintaining systemic discrimination, this may entail utilizing objective algorithms or taking into account other sources of data.
4. Ensure Accuracy and Transparency: It is crucial to make sure that data is presented appropriately and transparently because spatial data can be intricate and challenging to interpret. In order to prevent inaccurate or misleading conclusions, this entails providing context and explanations for the data.
5. Consider Accessibility and Inclusivity: Make sure spatial data-based storytelling is inclusive and accessible to all audiences, including people who are non-technical or have disabilities. This can entail utilizing alternate techniques for data visualization or offering data explanations to people who are visually impaired.
6. Address Environmental and Social Impacts: When using spatial data, take into account the social and environmental effects and make sure these matters are handled responsibly. This can entail bringing attention to the underlying social and economic causes of environmental degradation or advancing environmental justice.
7. Regularly Review and Update: As technology and social values change, it is important to periodically assess and update the ethical standards and concepts that apply to geographical data-based storytelling.

ROLES OF STORYTELLING IN DECISION-MAKING

The art of storytelling has been utilized for ages to transmit ideas, share personal narratives, and convey information from one generation to the next. Storytelling has gained popularity as a decision-making tool in recent years, notably in the business and leadership sectors. Decision-makers may convey difficult information in a way that resonates with their audience, modifies perceptions, and affects behaviour by creating captivating tales. As seen in Table 1, storytelling can play a variety of significant roles in decision-making.

Table 1. Roles of storytelling in decision-making

Roles of Storytelling	Description	Examples
Communicating complex information (Ramzy, 2007).	Making complicated evidence more understandable for the audience through the use of storytelling.	A teacher who engages students in storytelling to clarify a challenging idea. A data analyst who communicates difficult data insights through storytelling to non-experts.
Building empathy and understanding (McKee, 1997).	Storytelling can be utilized in order to help the audience comprehend and develop empathy for individuals with various origins or experiences.	A documentary maker who uses storytelling to explain the experiences of refugees. a business that uses storytelling to emphasize the difficulties faced by its workers from underrepresented communities.
Shaping perceptions (Holtje, 2011).	One may use storytelling to influence how people see a certain problem or situation.	A political campaign that makes use of storytelling to sway voters' opinions of a certain candidate. A business that repositions its brand in consumers' thoughts through storytelling.
Inspiring action (Pink, 2006).	People can be motivated to act by stories, whether it's to change something about themselves or to back up an argument.	A charitable organization that solicits donations by telling stories about its causes. A motivational speaker who encourages others to change for the better through storytelling.
Contextualization (Beauchamp & Childress, 1995)	The use of storytelling can help individuals grasp the context and historical, social, or cultural backdrop of a particular problem or circumstance.	A museum that uses storytelling to give historical perspective to a display. A journalist who uses storytelling to explain the history of a present-day
Perspective-taking (Galinsky et al., 2008)	The audience might be helped by storytelling to comprehend several points of view on a given problem or circumstance.	A documentary filmmaker who exposes various political viewpoints through storytelling. A teacher who teaches cultural understanding and appreciation to children through storytelling.
Emotion regulation (Gross, 2015)	Whether it's to lessen anxiety or to foster a happy attitude, storytelling can help people regulate their emotions.	A therapist employs narrative therapy to assist a patient in overcoming stress. An individual who uses storytelling to foster optimism and lessen worries.
Sense-making (Weick et al., 2005)	People can use storytelling to make sense of difficult circumstances or experiences.	A journalist who explains the underlying factors and consequences of a natural disaster through storytelling. An expert who uses storytelling to explain to a company the elements that influence whether it succeeds or fails.
Motivation (Ryan & Deci L, 2017)	People might be inspired by stories to accomplish their goals or overcome challenges.	A trainer encourages participants to give their utmost effort through narrative. An instructor who motivates listeners to face their apprehensions and accomplish their aspirations by utilizing stories.
Build trust and credibility (Mayer et al., 1995)	It entails establishing an atmosphere of sincerity and trust among the audience's support through the narratives that are conveyed.	An enterprise may use storytelling to gain the confidence of prospective clients while launching an innovative product. Storytelling can be employed in the backdrop of decisions regarding public policies to foster confidence among citizens.

The potential uses for storytelling in decision-making are listed in Table 1 and include conveying difficult information, fostering feelings of empathy and comprehension, establishing opinions, and motivating behaviour. contextualizing, putting things in viewpoints, regulating emotions, understanding, and motivation. Each role is described, along with examples of how storytelling might be used in that role in various situations. Storytelling is a potent tool that may be used to improve perception in a range of contexts and circumstances. To fully utilize the strength of storytelling, it is crucial to comprehend every role that it can play in assisting decision-making processes. Decision-makers are able to better convey difficult information, foster empathy and understanding, shape perceptions, push people to take action, offer context, take viewpoints, control emotions, and make sense of complicated circumstances by using storytelling.

Potential Risks and Challenges Associated With Storytelling in Decision-Making

It is crucial that one recognizes that using storytelling in decision-making can also come with possible risks and challenges. For instance, if the stories are not carefully curated or constructed, they could unintentionally perpetuate views or biases or cause misconceptions or inaccuracies. In order to effectively use storytelling in decision-making, it is crucial to consider it from a critical and thoughtful perspective viewpoint and to carefully analyze any potential ramifications of the stories being conveyed. Table 2 lists a few of the potential risks and challenges connected to using storytelling in decision-making.

Table 2. Potential risks and challenges associated with using storytelling in decision-making

Potential Risks and Challenges	Description
Emotional Manipulation (Boje, 2011).	Emotional responses can be evoked by stories, which may cause people to make decisions based more on feelings than on reasoned arguments.
Bias and Misrepresentation (W. R. Fisher, 1984).	Instead of addressing preexisting biases and preconceptions, storytelling has the potential to strengthen them. Confirmation bias is a cognitive bias that occurs when decision-makers only consider data that confirms their previous beliefs.
Lack of Objectivity (Boje, 2011).	The subjectivity and impact of the storyteller's prejudices and viewpoints might impair the neutrality of their work. People making decisions may find it challenging to evaluate the veracity and applicability of the data provided as a result.
Incomplete or Inaccurate Information (Mishra et al., 2013)	Stories may be based on little or incorrect data, causing readers to make poor or wrong decisions.
Ethical Concerns (Clandinin et al., 2016).	Ethics-related questions involving issues like representation, secrecy, and privacy can be brought up via storytelling. Stories, for instance, may unintentionally divulge sensitive data about people or reinforce incorrect assumptions or biases.
Oversimplification (Boje, 2011).	Sometimes, telling the story a story can oversimplify complicated problems and leave out important details. This might lead to poor decision-making because all pertinent considerations may not be taken into account.
Lack of Accountability (Turusbekova et al., 2007)	Because the storyteller may not be held accountable for the veracity or significance of their tale, storytelling occasionally lacks accountability. The sources of information that decision makers rely on must be reliable and accountable.
Resistance to Change (Lawrence & Suddaby, 2006)	Sometimes, presenting a story might serve to reinforce preexisting attitudes and values, making it challenging for decision-makers to consider alternative viewpoints or strategies. Decision-makers must continue to be open to many viewpoints and willing to question their own presumptions.

The possible drawbacks of using storytelling in decision-making are discussed in Table 2. Confirmation bias may cause decision-makers to only take into account data that confirms their prior opinions, and emotional reactions to stories may induce them to base decisions on emotions rather than reason. Neutrality and objectivity can also be impacted by the subjectivity of the storyteller's biases, and stories may be based on unreliable or erroneous facts. Last but not least, ethical issues can come up, such as mistakenly disclosing sensitive information or confirming false preconceptions or biases. Decision-making and storytelling is a complicated and nuanced topic. It is crucial to utilize storytelling with caution, give it serious consideration, and be aware of any possible risks or issues even though it could be an excellent technique for involving stakeholders, building trust, and spurring action. Decision-makers can leverage the power of storytelling in this way to render decisions that are more moral, practical, and well-informed.

CHALLENGES TO BALANCE ETHICS AND DECISION-MAKING

The challenges in striking a balance between ethics and decision-making can be extensive and intricate. In the modern world, where data is frequently employed to inform decisions, it is crucial to take their ethical consequences into account. Some of the challenges that organizations and individuals encounter when attempting to strike a balance between ethics and decision-making are endorsed in Table 3.

Table 3 outlined several problems that may occur when choosing an ethical course of action. Conflicting interests, complex ethical problems, cognitive biases, a lack of direction and clarity, cultural differences, technological advancements, political and legal restrictions, a lack of resources, unintended consequences, accountability and transparency, an excess of emphasis on emotion, and power dynamics are a few of these challenges. These difficulties might make it challenging to resolve moral conundrums and reach conclusions that are equitable and right for all parties. It is crucial to devise strategies to overcome these difficulties.

Strategies to Overcome the Challenges of Balancing Ethics and Decision-Making

Here are some strategies to overcome the challenges of balancing ethics and decision-making.

1. Create clear ethical policies and guidelines: Organizations should have a set of ethical rules and guidelines that are periodically reviewed and updated to make sure they are still effective and relevant.
2. Promote an ethical decision-making culture: It's critical to establish an environment in which making ethical decisions is respected and promoted. Training, education, and continual discussion of ethical concerns can help achieve this.
3. Utilize ethical decision-making frameworks: Organizations can utilize frameworks to help direct their decision-making processes, such as the Locus Charter and Ethical Decision-Making Framework.
4. Include diversified viewpoints: To guarantee that all stakeholders are taken into account and that choices are made in everyone's best interests, decision-makers should look for and include diverse viewpoints.
5. Undertake routine ethics audits: To evaluate their ethical practices and pinpoint areas for development, organizations should routinely undertake ethics audits.

Table 3. Challenges to balance ethics and decision-making

Challenges	Description
Conflicting Interests (Cooke, 2020)	When making decisions, there may be a conflict between the interests of many parties. For example, an option that favours participants may not be the greatest one for clients. It can be difficult to strike a balance between conflicting interests and reach an ethical conclusion that is just and fair for everyone under such circumstances.
The complexity of ethical issues (Macklin, 2003)	Making decisions requires navigating challenging ethical issues, which can be challenging. When there are conflicting values, interests, or priorities at play and no obvious solution, ethical issues can occur.
Cognitive Biases (Croskerry, 2003)	Cognitive biases, such as anchoring bias and confirmation bias, can affect judgment and result in unethical decisions. These biases can be difficult to recognize and get rid of.
Lack of clarity and guidance (Fox et al., 2021)	In some situations, it could be difficult to make ethical decisions because there aren't always clear ethical rules or standards to adhere to. Working with emerging technologies or applications involving data can make this more challenging.
Time Constraints (Brock, 2019)	There may not be enough time to properly analyze the ethical consequences of a decision because decisions are frequently required to be made quickly. This may cause people to make choices that put immediate financial benefit ahead of long-term ethical issues.
Lack of Information (Resnik, 2021)	There might not always be enough information accessible to make an informed choice. This can make it difficult to thoroughly weigh the ethical ramifications of a choice and do it in a way that is informed.
Cultural Differences (Beauchamp & Childress, 1995)	It can be difficult to strike a balance between ethics and decision-making in a multicultural context because various ethnic backgrounds have distinct values and customs. As an example, an option that is regarded as ethical in one society could not be in another.
Technological Advancements (Brey, 2012)	Data collection and analysis have become simpler because of technological breakthroughs, but these developments have also generated ethical questions. For instance, applying AI algorithms to decision-making can reinforce prejudice and discrimination.
Political and Legal Constraints (Beauchamp & Childress, 1995)	Making ethical decisions might be hampered by political and legal restrictions. For example, it may be difficult to make an ethical choice that complies with the law when rules and regulations don't necessarily reflect ethical values.
Limited resources (Persad et al., 2009)	It might be difficult to thoroughly analyze all of the ethical ramifications of a decision when resources are limited, such as time or money. This may result in quick fixes or concessions that fall short of properly addressing ethical issues.
Unintended consequences (Ostrom, 2009)	Decisions might be difficult to foresee unexpected repercussions. This may also involve unintended or detrimental ethical consequences.
Accountability and transparency (Salter & Kothari, 2014)	Accountability and transparency are crucial to making ethical decisions because they allow for the consideration of the interests of all parties concerned. But this can be challenging when there are power imbalances or conflicts of interest.
Overemphasis on Emotion (Greene, 2015)	Emotional responses can be evoked by stories, which may cause people to make decisions based more on gut instincts than on reasoned arguments.
Power Dynamics (Sharma et al., 2019)	Power disparities can influence how ethical decisions are made because people in positions of greater authority may put their personal interests before moral considerations.

6. Seek external direction and expertise: To manage challenging ethical dilemmas, organizations might seek external guidance and expertise, such as speaking with ethicists or lawyers.
7. Encourage transparency and accountability: Businesses should be open about how they make decisions and be prepared to answer for their deeds.

Organizations can attempt to overcome the challenges of reconciling ethics and decision-making and make more moral and responsible judgments by putting the strategies discussed above into practice. Here are some examples of how these tactics can be applied in real-world situations:

1. Ethics education and training: To raise understanding and encourage ethical thinking among employees, several organizations have put ethics training programmes into place. For instance, to encourage ethical behaviour in research, many institutes provide online training courses on the subject (Lescano et al., 2008). An improved understanding of ethical issues results from this type of guidance, which can help people make better decisions.
2. Consultation with ethicists and specialists: When confronted with challenging moral dilemmas, organizations can consult with ethics specialists. Companies like Waymo, for instance, have collaborated with ethicists and other specialists to further develop technology for autonomous cars in an ethical and responsible manner (Hu et al., 2022). Such conversations result in a more thorough grasp of the moral dilemmas involved, which can result in more well-informed decision-making.
3. Collaborative decision-making: Businesses can use methods that incorporate a variety of stakeholders, including individuals who might be impacted by the decision. For instance, the city of Helsinki engaged in a co-design process with people and other stakeholders when developing a new public transport system to guarantee that the system is created in a way that satisfied their requirements and is ethically sound (Hyysalo et al., 2023). A more inclusive and socially responsible decision-making process is the result of such shared decision-making.
4. Continuous monitoring and evaluation: To make sure continue to act ethically and responsibly; a constant track and evaluation process is required to traverse the actions and how they affect stakeholders. For instance, a pharmaceutical company may do post-market monitoring throughout the production of a new drug to find any potential negative effects on patients (FDA, 2020). Such review and monitoring lead to more accountable and responsible decision-making.

Due to a number of reasons, including competing interests, cognitive biases, cultural variations, time restraints, knowledge gaps, technological improvements, and political and legal restrictions, making ethical decisions can be difficult. Organizations may, however, put methods in place to get over these obstacles and guarantee ethical outcomes. The presented examples show how these tactics can be applied practically to encourage moral decision-making. To make ethical decisions, it is essential to create a framework for ethical decision-making that takes into account the various ethical principles and values. Organizations can achieve this balance between ethics and decision-making, leading to more moral and accountable choices.

CONCLUSION

Decision-makers can have a valuable tool to better grasp complicated relationships and make informed decisions by utilizing spatial data-based storytelling strategies. However, the usage of spatial data also brings up ethical issues that should be taken into account when making decisions. Assuring diversity of viewpoints, transparency, accountability, continual monitoring and evaluation, and continuous learning and growth are only a few aspects of ethics. It can be difficult, but not impossible, to strike a balance between ethical considerations and decision-making when using spatial data-based storytelling strategies.

Decision-makers can employ spatial data-based storytelling strategies ethically and equitably while still making decisions that are advantageous to their communities by adhering to the guidelines provided in this chapter. Making sure that the data and the stories it tells are rooted in the values and needs of the communities they serve is ultimately the key to responsible spatial data-based storytelling.

Decision-makers must be cautious and proactive in their approach to ethical decision-making as the usage of spatial data-based storytelling keeps evolving. Decision-makers should make certain that their use of spatial data-based storytelling is responsible and efficient by regularly interacting with communities and stakeholders, being transparent and accountable, monitoring and analyzing their decisions, and continuously learning and improving.

The current study has certain limitations, notwithstanding the importance of ethical issues in spatial data storytelling. One drawback is that the research did not contain a real-world case study or application, focusing instead primarily on theoretical and conceptual aspects of ethical considerations. Another drawback is the study's constrained scope, which largely concentrated on challenges and solutions associated with balancing ethics and perception in spatial data storytelling. The actual influence of ethical considerations on decision-making processes in spatial data storytelling needs to be explored in more detail.

REFERENCES

Ahasan, R., & Hossain, M. M. (2021). Leveraging GIS and spatial analysis for informed decision-making in COVID-19 pandemic. *Health Policy and Technology*, 10(1), 7–9. Advance online publication. doi:10.1016/j.hlpt.2020.11.009 PMID:33318916

Ahmad, M. (2023a). Exploring the Potential of Spatial Data for Enhancing Higher Education Learners' Learning Outcomes. In K. Barua, N. Radwan, V. Singh, & R. Figueiredo (Eds.), *Design and Implementation of Higher Education Learners' Learning Outcomes (HELLO)* (pp. 128–145). IGI Global. doi:10.4018/978-1-6684-9472-1.ch008

Ahmad, M. (2023b). Leveraging Social Media Geographic Information for Smart Governance and Policy Making: Opportunities and Challenges. In C. Chavadi & D. Thangam (Eds.), *Global Perspectives on Social Media Usage Within Governments* (pp. 192–213). IGI Global. doi:10.4018/978-1-6684-7450-1.ch013

Ahmad, M. (2023c). Spatial Data as a Catalyst to Drive Entrepreneurial Growth and Sustainable Development. In N. Yaw Asabere, G. K. Gyimah, A. Acakpovi, & F. Plockey (Eds.), *Technological Innovation Driving Sustainable Entrepreneurial Growth in Developing Nations* (pp. 79–104). IGI Global. doi:10.4018/978-1-6684-9843-9.ch004

Ahmad, M. (2023d). Unlocking the Power of Spatial Big Data for Sustainable Development:From Capacity Building to Food Security and Food Traceability. In J. M. Falcó, B. M. Lajara, E. S. García, & L. A. M. Tudela (Eds.), *Crafting a Sustainable Future Through Education and Sustainable Development* (pp. 204–218). IGI Global. doi:10.4018/978-1-6684-9601-5.ch010

Antunes, P., Sapateiro, C., Zurita, G., & Baloian, N. (2010). Integrating spatial data and decision models in an e-planning tool. Lecture Notes in Computer Science (Including Subseries Lecture Notes in Artificial Intelligence and Lecture Notes in Bioinformatics), 6257 LNCS. Springer. doi:10.1007/978-3-642-15714-1_8

Apte, A., Ingole, V., Lele, P., Marsh, A., Bhattacharjee, T., Hirve, S., Campbell, H., Nair, H., Chan, S., & Juvekar, S. (2019). Ethical considerations in the use of GPS-based movement tracking in health research-Lessons from a care-seeking study in rural west India. *Journal of Global Health*, *9*(1), 010323. Advance online publication. doi:10.7189/jogh.09.010323 PMID:31275566

Beauchamp, T. L., & Childress, J. F. (1995). Principles of Biomedical Ethics. Oxford University Press, 6.

Berman, G., de la Rosa, S., Accone, T., & Associates. (2018). Ethical considerations when using geospatial technologies for evidence generation. In *Innocenti Discussion Papers: Vol. no. 2018-02*. UNICEF Office of Research - Innocenti, Florence. https://www.unicef-irc.org/publications/971-ethical-considerations-when-using-geospatial-technologies-for-evidence-generation.html

Boje, D. (2011). Narrative Methods for Organizational & Communication Research. In Narrative Methods for Organizational & Communication Research. Sage. doi:10.4135/9781849209496

Brey, P. A. E. (2012). Anticipatory Ethics for Emerging Technologies. *NanoEthics*, *6*(1), 1–13. doi:10.1007/s11569-012-0141-7

Brock, D. W. (2019). Ethical Issues in the Use of Cost Effectiveness Analysis for the Prioritisation of Health Care Resources. In The Ethics of Public Health. Taylor & Francis. doi:10.4324/9781315239927-28

Buolamwini, J. (2018). Gender Shades: Intersectional Accuracy Disparities in Commercial Gender Classification. In Proceedings of Machine Learning Research (Vol. 81). ACM.

Clandinin, D. J., Caine, V., Lessard, S., & Huber, J. (2016). Engaging in narrative inquiries with children and youth. In Engaging in Narrative Inquiries with Children and Youth. Springer. doi:10.4324/9781315545370

Cooke, L. (2020). Ethics in information technology. *ISACA Journal*, *6*, 69–106. doi:10.1017/CBO9781107445666.007

Croskerry, P. (2003). The importance of cognitive errors in diagnosis and strategies to minimize them. In Academic Medicine, 78(8). doi:10.1097/00001888-200308000-00003

De Jong, B. C., Gaye, B. M., Luyten, J., Van Buitenen, B., André, E., Meehan, C. J., O'Siochain, C., Tomsu, K., Urbain, J., Grietens, K. P., Njue, M., Pinxten, W., Gehre, F., Nyan, O., Buvé, A., Roca, A., Ravinetto, R., & Antonio, M. (2019). Ethical considerations for movement mapping to identify disease transmission hotspots. *Emerging Infectious Diseases*, *25*(7). doi:10.3201/eid2507.181421 PMID:31211938

Demšar, U., Harris, P., Brunsdon, C., Fotheringham, A. S., & McLoone, S. (2013). Principal Component Analysis on Spatial Data: An Overview. *Annals of the Association of American Geographers*, *103*(1), 106–128. doi:10.1080/00045608.2012.689236

Dougherty, J., & Ilyankou, I. (2021). Hands-on data visualization. In O'Reilly Media, Inc.

Elwood, S., & Leszczynski, A. (2013). New spatial media, new knowledge politics. *Transactions of the Institute of British Geographers*, *38*(4), 544–559. doi:10.1111/j.1475-5661.2012.00543.x

Fairbairn, G. J. (2002). Ethics, empathy and storytelling in professional development. *Learning in Health and Social Care*, *1*(1), 22–32. doi:10.1046/j.1473-6861.2002.00004.x

FDA. (2020). *Postmarket Drug Safety Information for Patients and Providers*. FDA. https://www.fda.gov/drugs/drug-safety-and-availability/postmarket-drug-safety-information-patients-and-providers

Feeney, M., Rajabifard, A., & Williamson, I. P. (2001). Spatial Data Infrastructure Frameworks to Support Decision-Making for Sustainable Development. *5th Global Spatial Data Infrastucture Conference*. ACM.

Fischer, M. M., Scholten, H. J., & Unwin, D. (2019). Geographic information systems, spatial data analysis and spatial modelling: an introduction. In Spatial Analytical Perspectives on GIS. Springer. doi:10.1201/9780203739051-1

Fisher, M., Fradley, M., Flohr, P., Rouhani, B., & Simi, F. (2021). Ethical considerations for remote sensing and open data in relation to the endangered archaeology in the Middle East and North Africa project. *Archaeological Prospection*, *28*(3), 279–292. doi:10.1002/arp.1816

Fisher, W. R. (1984). Narration as a human communication paradigm: The case of public moral argument. *Communication Monographs*, *51*(1), 1–22. doi:10.1080/03637758409390180

Florance, P. (2006). GIS collection development within an academic library. *Library Trends*, *55*(2), 222–234. Advance online publication. doi:10.1353/lib.2006.0057

Fox, E., Tarzian, A. J., Danis, M., & Duke, C. C. (2021). Ethics Consultation in United States Hospitals: Assessment of Training Needs. *The Journal of Clinical Ethics*, *32*(3), 247–255. doi:10.1086/JCE2021323247 PMID:34339396

Galinsky, A. D., Maddux, W. W., Gilin, D., & White, J. B. (2008). Why It Pays to Get Inside the Head of Your Opponent. *Psychological Science*, *19*(4), 378–384. doi:10.1111/j.1467-9280.2008.02096.x PMID:18399891

Greene, J. D. (2015). Beyond Point-and-Shoot Morality: Why Cognitive (Neuro)Science Matters for Ethics. *Law and Ethics of Human Rights*, *9*(2), 141–172. doi:10.1515/lehr-2015-0011

Gross, J. J. (2015). Emotion Regulation: Current Status and Future Prospects. *Psychological Inquiry*, *26*(1), 1–26. doi:10.1080/1047840X.2014.940781

Guardian. (2018). Fitness tracking app Strava gives away location of secret US army bases. *The Guardian*. https://www.theguardian.com/world/2018/jan/28/fitness-tracking-app-gives-away-location-of-secret-us-army-bases

Haley, D. F., Matthews, S. A., Cooper, H. L. F., Haardörfer, R., Adimora, A. A., Wingood, G. M., & Kramer, M. R. (2016). Confidentiality considerations for use of social-spatial data on the social determinants of health: Sexual and reproductive health case study. *Social Science & Medicine, 166*, 49–56. doi:10.1016/j.socscimed.2016.08.009 PMID:27542102

Hehman, E., & Xie, S. Y. (2021). Doing Better Data Visualization. *Advances in Methods and Practices in Psychological Science, 4*(4). doi:10.1177/25152459211045334

Hernandez, J. (2009). Redlining revisited: Mortgage lending patterns in Sacramento 1930-2004. *International Journal of Urban and Regional Research, 33*(2), 291–313. doi:10.1111/j.1468-2427.2009.00873.x

Hobday, A. J., Hartog, J. R., Manderson, J. P., Mills, K. E., Oliver, M. J., Pershing, A. J., Siedlecki, S., & Browman, H. (2019). Ethical considerations and unanticipated consequences associated with ecological forecasting for marine resources. *ICES Journal of Marine Science, 76*(5), fsy210. doi:10.1093/icesjms/fsy210

Holtje, J. (2011). *The Power of Storytelling: Captivate, Convince, or Convert Any Business Audience Using Stories from Top CEOs.* Penguin.

Hu, X., Zheng, Z., Chen, D., Zhang, X., & Sun, J. (2022). Processing, assessing, and enhancing the Waymo autonomous vehicle open dataset for driving behavior research. *Transportation Research Part C, Emerging Technologies, 134*, 103490. doi:10.1016/j.trc.2021.103490

Hyysalo, S., Savolainen, K., Pirinen, A., Mattelmäki, T., Hietanen, P., & Virta, M. (2023). *Design types in diversified city administration: The case City of Helsinki.* Design Journal., doi:10.1080/14606925.2023.2181886

Juergens, C., & Redecker, A. P. (2023). Basic Geo-Spatial Data Literacy Education for Economic Applications. *KN - Journal of Cartography and Geographic Information, 73*(2), 1–13. doi:10.1007/s42489-023-00135-9 PMID:37361712

Ke, R., Taoxiong, L., Di, Z., & Fei, H. (2022). Hierarchical data licensing mechanisms in the data market. *Journal of Industrial Engineering and Engineering Management, 36*(6). doi:10.13587/j.cnki.jieem.2022.06.002

Khandare, N. B., Nikam, V. B., Banerjee, B., & Kiwelekar, A. (2022). *Spatial Data Infrastructure for Suitable Land Identification for Government Projects.* Springer. doi:10.1007/978-981-19-0725-8_7

Kitchin, R. (2014a). Big Data, new epistemologies and paradigm shifts. *Big Data & Society, 1*(1). doi:10.1177/2053951714528481

Kitchin, R. (2014b). The Data Revolution: Big Data, Open Data, Data Infrastructures & Their Consequences. In The Data Revolution: Big Data, Open Data, Data Infrastructures & Their Consequences. Springer. doi:10.4135/9781473909472

Kitchin, R., & Lauriault, T. P. (2014). *Towards critical data studies : Charting and unpacking data assemblages and their work (preprint).* Geoweb and Big Data.

Lawrence, T. B., & Suddaby, R. (2006). Institutions and institutional work. In The SAGE Handbook of Organization Studies. Sage. doi:10.4135/9781848608030.n7

Lescano, A. R., Blazes, D. L., Montano, S. M., Moran, Z., Naquira, C., Ramirez, E., Lie, R., Martin, G. J., Lescano, A. G., & Zunt, J. R. (2008). Research ethics training in Peru: A case study. *PLoS One*, *3*(9), e3274. doi:10.1371/journal.pone.0003274 PMID:18818763

MacEachren, A. M. (1992). Visualizing Uncertain Information. *Cartographic Perspectives*, *13*(13), 10–19. doi:10.14714/CP13.1000

Macklin, R. (2003). Bioethics, vulnerability, and protection. *Bioethics*, *17*(5–6), 472–486. doi:10.1111/1467-8519.00362 PMID:14959716

Mayer, R. C., Davis, J. H., & Schoorman, F. D. (1995). An Integrative Model Of Organizational Trust. *Academy of Management Review*, *20*(3), 709. doi:10.2307/258792

McKee, R. (1997). *Story: style, structure, substance, and the principles of screenwriting*. Harper Collins.

Mishra, J. L., Allen, D. K., & Pearman, A. D. (2013). Information use, support and decision making in complex, uncertain environments. *Proceedings of the ASIST Annual Meeting, 50*(1). IEEE. 10.1002/meet.14505001045

Mohai, P., & Saha, R. (2006). Reassessing racial and socioeconomic disparities in environmental justice research. *Demography*, *43*(2), 383–399. doi:10.1353/dem.2006.0017 PMID:16889134

Moon, J. A. (2010). Using story: In higher education and professional development. In Using Story: In Higher Education and Professional Development. Springer. doi:10.4324/9780203847718

Newman, P. A., Guta, A., & Black, T. (2021). Ethical Considerations for Qualitative Research Methods During the COVID-19 Pandemic and Other Emergency Situations: Navigating the Virtual Field. *International Journal of Qualitative Methods*, *20*. doi:10.1177/16094069211047823

Nuzzi, O. (2020). What It's Like to Get Doxed for Taking a Bike Ride. *NY Mag*. Https://Nymag.Com/. https://nymag.com/intelligencer/2020/06/what-its-like-to-get-doxed-for-taking-a-bike-ride.html

Ostrom, E. (2009). A general framework for analyzing sustainability of social-ecological systems. In Science (Vol. 325, Issue 5939). doi:10.1126/science.1172133

Persad, G., Wertheimer, A., & Emanuel, E. J. (2009). Principles for allocation of scarce medical interventions. In The Lancet (Vol. 373, Issue 9661). doi:10.1016/S0140-6736(09)60137-9

Pink, D. H. (2006). *A whole new mind: Why right-brainers will rule the future*. Penguin.

Qin, Y., Xiao, X., Liu, F., de Sa e Silva, F., Shimabukuro, Y., Arai, E., & Fearnside, P. M. (2023, January 02). de Sa e Silva, F., Shimabukuro, Y., Arai, E., & Fearnside, P. M. (2023). Forest conservation in Indigenous territories and protected areas in the Brazilian Amazon. *Nature Sustainability*, *6*(3), 295–305. doi:10.1038/s41893-022-01018-z

Rajabifard, A., Binns, A., Masser, I., & Williamson, I. (2006). The role of sub-national government and the private sector in future spatial data infrastructures. *International Journal of Geographical Information Science*, *20*(7), 727–741. doi:10.1080/13658810500432224

Ramzy, A. (2007). The Leader's Guide to Storytelling. Mastering the Art and Discipline of Business Narrative. In Corporate Reputation Review, 10(2). doi:10.1057/palgrave.crr.1550044

Resnik, D. B. (2021). What Is Ethics in Research & Why Is It Important? National Institute of Environmental Health Sciences.

Ryan, R. M., & Deci, L. E. (2017). Self-determination theory: basic psychological needs in motivation. In Self-determination theory: Basic psychological needs in motivation, development, and wellness. IEEE.

Saha, M., Patil, S., Cho, E., Cheng, E. Y. Y., Horng, C., Chauhan, D., Kangas, R., McGovern, R., Li, A., Heer, J., & Froehlich, J. E. (2022). Visualizing Urban Accessibility: Investigating Multi-Stakeholder Perspectives through a Map-based Design Probe Study. *Conference on Human Factors in Computing Systems - Proceedings*. IEEE. 10.1145/3491102.3517460

Salter, K. L., & Kothari, A. (2014). Using realist evaluation to open the black box of knowledge translation: A state-of-the-art review. In Implementation Science, 9(1). doi:10.1186/s13012-014-0115-y

Sharma, A., Agrawal, R., & Khandelwal, U. (2019). Developing ethical leadership for business organizations. *Leadership and Organization Development Journal*, 40(6), 712–734. doi:10.1108/LODJ-10-2018-0367

Sidik, S. (2021). *How the COVID-19 Pandemic Has Shaped Data Journalism*. GIJN. https://gijn.org/2021/04/13/how-the-covid-19-pandemic-has-shaped-data-journalism/

Turusbekova, N., Broekhuis, M., Emans, B., & Molleman, E. (2007). The role of individual accountability in promoting quality management systems. *Total Quality Management & Business Excellence*, 18(5), 471–482. doi:10.1080/14783360701239917

Vancauwenberghe, G., Valečkaitė, K., van Loenen, B., & Welle Donker, F. (2018). Assessing the Openness of Spatial Data Infrastructures (SDI): Towards a Map of Open SDI. *International Journal of Spatial Data Infrastructures Research*, 13. doi:10.2902/1725-0463.2018.13.art9

Warren, C. (2023). *Not all parts of the rainforest have suffered equally. Satellite images plot a path for Amazon protection*. Anthropocene. https://www.anthropocenemagazine.org/2023/01/not-all-parts-of-the-rainforest-have-suffered-equally-satellite-images-plot-a-path-for-amazon-protection/

Weick, K. E., Sutcliffe, K. M., & Obstfeld, D. (2005). Organizing and the process of sensemaking. In Organization Science, 16(4). doi:10.1287/orsc.1050.0133

Will, D. H. (2020). Predictive policing algorithms are racist. They need to be dismantled. *MIT Technology Review*. https://www.technologyreview.com/2020/07/17/1005396/predictive-policing-algorithms-racist-dismantled-machine-learning-bias-criminal-justice/

Zandbergen, P. A. (2014). Ensuring Confidentiality of Geocoded Health Data: Assessing Geographic Masking Strategies for Individual-Level Data. *Advances in Medicine*, 2014, 1–14. doi:10.1155/2014/567049 PMID:26556417

Chapter 15
Revolution Ethics of Data Science and AI

Anil Meher
Sri Sri University, India

ABSTRACT

Artificial intelligence is becoming more and more widespread in our increasingly connected world. Artificial intelligence is slowly but surely modifying the way we live and work, from self-driving cars to automated customer service agents. As artificial intelligence becomes more sophisticated, the ethical implications of its use become more complex. There are several key issues to consider regarding the ethics of artificial intelligence, such as data privacy, algorithmic bias, and socioeconomic inequality. The rapid development of AI brings with it several ethical issues. However, we must remain vigilant in protecting our fundamental rights and freedoms. We must ensure that artificial intelligence is not used to discriminate against vulnerable groups or invade our privacy. We must also be careful that AI does not become a tool for the powerful to control and manipulate the masses. But while there are risks, the author believes the potential benefits of AI are too great to ignore.

INTRODUCTION

Data science, big data, and AI have been increasingly recognized as major driving forces for next-generation innovation, economy, and education (Cao, 2017). At a high level, Data Science (DS) is the set of fundamental principles that support and guide the principled extraction of information and knowledge from data. Possibly the most closely related concept to data science is data mining - the actual extraction of knowledge from data via technologies that incorporate these principles (Provost & Fawcett, 2013). The term "Data Science" has emerged in the last few years to make it easier to understand the basic work related to data. Data science technology initially involved only statistical analysis. Finally, it adopted new age technologies such as artificial intelligence (AI), machine learning (ML) and data mining. Data science is an ever-evolving technology with huge potential to bring massive improvements in the future. Although Data Science and Artificial Intelligence have existed since the 1950s, both fields have only very recently experienced a surge in popularity as a result of the expansion in the capabilities of tech-

DOI: 10.4018/979-8-3693-2964-1.ch015

nology (Ong & Uddin, 2020). Initially Data Science (DS) which is said to be in the form of future data analysis was started in 1962 by John Turey. The traditional database system was created in 1970. After several years, i.e. in 1974, the concept of neural networks was introduced at the NPIS conference. Then again Jacob Zahavi in the year of 1999 introduced data mining for large data sets. He later proposed the Cleveland Data Science Plan again in 2001. In 2005, Big Data was released to maintain a large number of datasets with high processing speed on the server. In 2006, the concept of Deep Learning was again introduced for Algorithm implementations. Then the concept of machine learning was introduced in 2011 by Watson Jeopardy. Finally, in 2018, the Data Science Global Mainstream emerged to handle the processing of large datasets. Similarly the term "artificial intelligence" was coined by John McCarthy in 1956. McCarthy, along with other pioneers such as Marvin Minsky and Claude Shannon, believed that machines could be programmed to mimic human intelligence. The 1980s saw a resurgence in artificial intelligence research, fueled by the development of new algorithms and the availability of powerful computers. In the 1990s, machine learning emerged, allowing computers to learn from data and improve their performance over time.

Table 1. Artificial intelligence and data science evolution comparison

	Data Science	Artificial Intelligence
1950-1964	Origination	Origination
1965-1979	Statistics Integration	AI Maturation
1980-1990	KDD	AI Boom
1991-1999	Database Marketing	AI Winter
2000-2004	Internet Integration	AI Agents
2005-2009	Big Data	
2010-2014	Data Scientist	Deep Learning, Big Data
2015	AI Integration	
2020	NLP, Automated DS	Artificial Narrow Intelligence
2040	Quantum Leap	Artificial General Intelligence
2060	Quantum Computing	Artificial Super Intelligence

Data Science evolution is the result of the inclusion of current technologies such as machine learning (ML), Internet of Things (IoT) and Artificial Intelligence (AI). The application of data science has started to spread to several other fields such as engineering and medicine. Due to the massive influx of new information from all business sectors, businesses are always looking for innovative strategies to increase revenue and improve decision making. Data Science (DS) bridges computer science, statistics, and domain knowledge to uncover the potential concealed in data (Chiarello et al., 2021). The would-be notion takes data science as the science of learning from data, with all that this entails (Donoho, 2017). Using large amounts of data for decision making became practical in the 1980s (Dhar, 2013). The famous back propagation algorithm that David Rumelhart rediscovered in the early 1980s, and which is now considered at the core of the so-called "AI revolution," first arose in the field of control theory in the 1950s and 1960s (Jordan, 2019).

Table 2. Evolution chart of AI and DS

Year	Person	Feature
4th century	Aristotle	Syllogistic Logic
12th century	Bacon and Albert	Talking Heads
14th century		Movable Type-Printing Press
15th century	Rabbi Loew	Clocks
16th century	Rene Descartes	Mechanical Living Beings and Animals
17th century	Blaise Pascal	Digital Calculating Machine
18th century	Joseph-Marie Jacquard	Programmable Device
1921	Karel Capek	Rossum's Universal Robots
1943	Warren Mcculloch and Walter Pitts	Foundation of Neural Network
1950	Alan Turing	Computing Machinery and Intelligence
1954	Dartmouth Summer Research Project	Birth of AI
1956	John Mccarthy	AI
1966		AI Winter
1969	Marvin Minsky and Seymour Papert	Artificial Neuron
1975	Paul Werbos	Deep Neural Networks
1980		Expert Systems
1981	Japan	Fifth Generation Computer
1986	David Rumelhart, Geoffrey Hinton, and Ronald Williams	Parallel Distributed Processing
1990		Support Vector Machine
2000		Data Science
2010		DIP, NLP
2020		GPT

Rapid advances in the field of artificial intelligence have profound implications for the economy as well as society at large (Cockburn et al., 2018). AI is not a technology or set of technologies, but a continually evolving frontier of emerging computing capabilities (Berente et al., 2021). Researchers propose that AI "refers to programs, algorithms, systems and machines that demonstrate intelligence, is manifested by machines that exhibit aspects of human intelligence (Davenport et al., 2020). Artificial intelligence is a general purpose technology that is likely to impact many industries (Varian, 2018). Annotations guide us through the historic Dartmouth Conference of 1956, a pivotal moment where the term "Artificial Intelligence" was coined, marking the formal establishment of AI as an interdisciplinary field. Turing's concept of the Turing Test, designed to assess a machine's ability to exhibit human-like intelligence, served as a catalyst for subsequent advancements (Khonturaev, 2023). There are several high-impact applications domains to consider when organizations are first starting to apply enterprise AI (Davenport, 2018). Improving AI, one of the most promising data analytic tools to have been developed over the past decade or so, so as to help reduce these uncertainties, is a worthwhile pursuit. Encouragingly, data scientists have taken up the challenge (Naudé, 2020). The adoption of artificial intelligence (AI) has received enormous attention across virtually every industrial setting, from healthcare delivery to automobile manufacturing (Verganti et al., 2020). AI is the programming or training of a computer to do tasks typically reserved for human intelligence, whether it is recommending which movie to watch next or answering technical questions (Mehr et al., 2017). Artificial intelligence (AI) refers to the simulation of the human mind in computer systems that are programmed to think like humans and mimic their actions such as learning and problem-solving (Cui & Zhang, 2021). AI can support three important business needs: automating business processes, gaining insight through data analysis, and engaging with customers and employees (Davenport & Ronanki, 2018). The most important general-purpose technology of our era is artificial intelligence, particularly machine learning (ML) — that is, the machine's ability to keep improving its performance without humans having to explain exactly how to accomplish all the tasks it's given (Brynjolfsson & Mcafee, 2017). AI ethics is often considered as extraneous, as surplus or some kind of "add-on" to technical concerns, as unbinding framework that is imposed from institutions "outside" of the technical community (Hagendorff, 2020). AI generally refers to the ability of machines to exhibit human-like intelligence—for example, solving a problem without the use of hand-coded software containing detailed instructions (Bughin et al., 2017). Artificial Intelligence (AI) is a science and a set of computational technologies that are inspired by—but typically operate quite differently from—the ways people use their nervous systems and bodies to sense, learn, reason, and take action (Stone et al., 2022). Data science (DS) is one of the discipline rooted in scientific methodologies, algorithms, and systems focused on extracting meaningful information and deep insights from vast data repositories. This interdisciplinary field seamlessly combines components from computer science, mathematics, statistics, and information technology, converging on a single purpose: to address complex challenges directly. The field of data science is increasingly using artificial intelligence. AI technology can process large amounts of data faster and more accurately than humans. It can find connections and techniques in datasets that are difficult or impossible for a person to find independently. Artificial intelligence has very huge amount of applications, from autonomous vehicles to medical diagnostics. Artificial intelligence is also used for generating predictive models which can be used to support decision-making using data. In order to generate predictive models for the future, artificial intelligence algorithms are used to discover patterns and correlations in the data. Robotics and automation also use AI technology to automate tasks and processes.

LITERATURE REVIEW

The evolution of AI is always considered to be a journey from science fiction to reality. It is now a part of our daily lives, so the effects of this to the society will continue to grow. As we harness the potentiality of artificial intelligence, we must also be mindful of its ethical implications and strive to use this technology responsibly and beneficially. We have been using AI-based technology for a long time. While some inventions have been very popular and commonly used, many have failed to gain market acceptance due to various issues such as scalability, security, accuracy, and ease of maintenance. As a result, since its inception in 1956, a cyclical pattern of high and low investment in AI research, commonly termed as AI summers and AI winters, has been observed. However, artificial intelligence gradually improved and pushed forward the boundaries of machine intelligence. At the core of AI technology are the three pillars of machine learning (ML), neural networks (NN), and deep learning, all of which revolve around processing complex data, creating patterns, and applying algorithms and natural language processing to generate the desired output. Thanks to Application Programming Interfaces (APIs), AI technology can be applied to existing software and hardware such as home security systems, vehicles, etc. Although artificial intelligence has grown rapidly over the past few decades, there is still seemingly endless potential and applications to explore. Data science has emerged as a powerful force that has revolutionized the way we collect, analyze and interpret data. Artificial intelligence systems work by merging with intelligent, iterative processing algorithms. This combination allows artificial intelligence to learn from patterns and features in the analyzed data. Each time an AI system performs a round of data processing, it tests and measures its performance and uses the results to develop further expertise. Basically evolution of data science lies in three phases. We can represent them as follows in diagram.

Figure 1. Artificial intelligence and data science trends

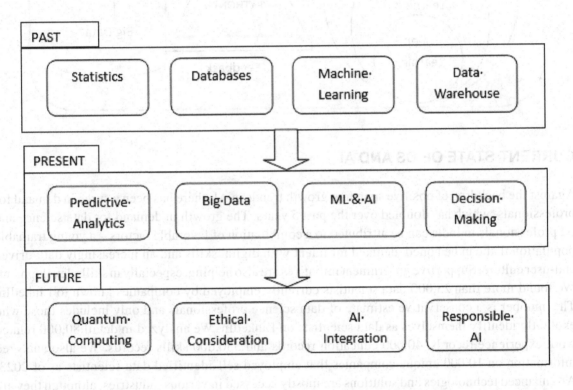

Big data and artificial intelligence have a synergistic relationship. Artificial intelligence requires a huge amount of data to study as well as improve decision-making processes, and big data analytics uses artificial intelligence to better analyze data. Thanks to this convergence, we can more easily use advanced analytics features, such as advanced or predictive analytics, and more efficiently access useful insights from your vast data stores.

Big Data provides tools and infrastructures for data preparation, processing and similar actions. Basically all the responsibilities of Data Scientists involve wrestling with data, and Big Data tools support this work. In addition, Data Science often needs a scalable infrastructure to process large volumes of data; machine learning, which big data infrastructures provide. Big data often needs data cleaning, pre-processing and automatic labeling where Data Science methods help.

Figure 2. Artificial intelligence and data science workflow in association with big data

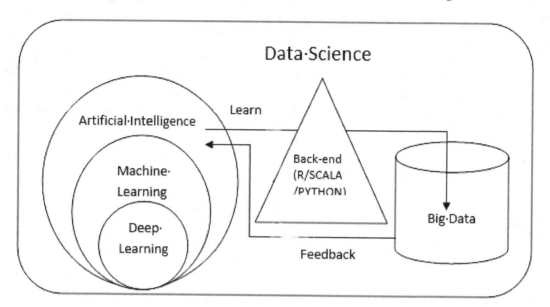

CURRENT STATE OF DS AND AI

Against the backdrop of positive spending growth trends in AI, there has been a surge in demand for professionals, which has doubled over the past 5 years. The growth in demand for data science and AI professionals in India can be attributed to a combination of favorable factors – a young trainable population that can be honed, demand for talent with digital skills and an increasingly data-driven end-user culture. Supportive government initiatives are also helping, especially in skills development. We found more than 25,000 data scientists currently employed by companies known to LinkedIn. This number is a conservative estimate of data science professionals and only includes those who explicitly identify themselves as data scientists on LinkedIn. We analyzed in detail 80,000 related work experience records, 40,000 education records and 350,000 skills records. We also analyzed information on 10,000 unique companies that employed self-identified data scientists as of 2023. AI-enhanced technologies and solutions are mostly accessed in various industries, although they are not necessarily cheap to implement. Voice assistants are at the forefront of AI adoption in industries as diverse as IT, automotive and retail. Artificial intelligence on a smaller scale, as seen in chatbots, allows smaller brands to conserve resources and improve customer satisfaction. An increasing number of AI-enhanced tools are available as Software as a Service (SaaS). Mobile devices and apps have the popular way to deploy AI technologies across industries, whether its voice assistants, smart monitoring and control, personalized shopping or warehouse management apps. Artificial intelligence is present in smart applications, neural networks, AI-as-a-service platforms, and AI cloud services. Emerging AI technologies include augmented intelligence, which seeks to improve human intelligence, and edge AI, where AI algorithms are processed locally without the need for an Internet connection (such as some forms of facial recognition). According to an AI Index report, the power of Artificial Intelligence (AI) computing doubles every 3.4 months.

IMPACT OF DS AND AI

Artificial intelligence (AI) has become one of the most transformative technologies of the 21st century, affecting every aspect of our lives, from the way we work and learn to the way we interact and communicate with each other. The rise of artificial intelligence has brought with it a wave of automation that is reshaping the workforce, changing the way we approach healthcare, changing the way we interact with technology, and raising concerns about AI's potential impact on society. One of the most profound impacts of artificial intelligence is on the workforce. Automation is replacing human workers at an unprecedented rate, with AI machines taking over routine and low-skilled jobs. The impact of artificial intelligence on healthcare is another area of significant development. With the ability to quickly and accurately analyze vast amounts of data, AI systems can improve patient outcomes and reduce healthcare costs. Artificial intelligence is also changing the way we work with technology. Smart assistants like Siri, Alexa and Google Assistant are becoming more and more popular and are changing the way we interact with our devices. One of the most interesting developments in the field of chatbots is the emergence of GPT-based chatbots.

Careers in data science require professionals to use raw data to identify patterns. These patterns are then used to derive actionable insights from the data. With data science, it is possible to predict future outcomes with a higher level of accuracy. Here's how different industries benefit from data science applications. Applications of data science in healthcare involve the use of data in making important decisions and drawing conclusions. These include the implementation of medical knowledge gathered in the diagnosis of diseases. In the retail industry, data science applications have led to the implementation of features such as recommendation engines. Impact helps companies design products around consumer needs. BFSI has greatly benefited from the data science application in the industry. The industry has been able to minimize its losses by using science to detect fraud. It has impacted more effective customer data management, risk modeling and customer support. If we are looking for a career in data science in the telecommunications industry, we will be able to influence the needs of efficient data transfer and visualization. Data science is being used to increase fraud detection by businesses. Using science, consumers can benefit from increased network security.

FUTURE SCOPE OF DS AND AI

Seeing how much of our world is now powered by data and data science, we can reasonably ask: Where do we go from here? What is the future of data science? While it's hard to know exactly what the hallmark breakthroughs of the future will be, all signs seem to point to the critical importance of machine learning. Data scientists are looking for ways to use machine learning to produce smarter and more autonomous AI. The goal is to create computer intelligence programs that solve real-time problems and help organizations and ordinary people achieve their goals. Machine games, speech recognition, language detection, computer vision, expert systems, robotics and other fields have potential. The bottom 90 percent, especially the bottom 50 percent of the world in terms of income or education, will be hit hard by job losses. The simple question to ask is, "What is the routine of work?" Artificial intelligence can learn to optimize itself within a routine task. And the more quantitative, the more objective the work—sorting things into bins, washing dishes, picking fruit, and answering customer service calls—these are highly scripted tasks that are repetitive and routine in nature. In 5, 10 or 15 years they will be displaced by AI. The affect of AI on our lives is nothing short of remarkable. As robots become more commonplace, we can expect even more transformation in our lives and work. The ability and scope of AI is always limitless, you can see the example of Chat GPT application and it is hard to imagine the future without it. We are already seeing increased productivity in the workplace thanks to AI, and by the end of this decade it will become an integral part of our daily lives. From self-driving cars to cutting-edge medical research, the possibilities are endless.

Figure 3. Future of artificial intelligence and data science

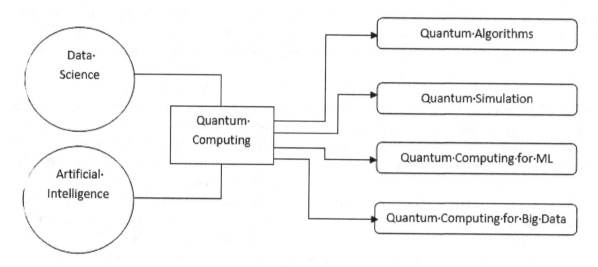

Artificial intelligence systems can use quantum computers to protect sensitive data. In addition, parallel processing can be used to fight cybercrime. Unlike classical computers, which only exist in one state, quantum computers can be in multiple states at once, allowing them to find better algorithms. Quantum computing improves AI by increasing its speed, efficiency and accuracy. It uses qubits and works non-linearly, outperforming conventional computers. This breakthrough enables the use of quantum computing in a variety of AI use cases. Industries such as maritime logistics, electric vehicles, semiconductors, luminescence and energy are already benefiting from the problem-solving capabilities of quantum computers.

The scope of AI in 2024 is vast, covering everything from work automation to personalized learning to cyber security. While there are certainly concerns about job security and other potential downsides, the benefits of artificial intelligence are too many to ignore. As we move forward into the future, we must embrace the power of artificial intelligence and its potential to transform our careers and lives. Artificial intelligence allows a machine to function like a human brain. It performs several business functions without the human need intervention, such as consumer interaction and social media brand awareness. Many researchers think that artificial intelligence will outperform humans in almost all cognitive tasks. By automating tasks such as managing employee or patient records, conducting market research and dealing with potential clients, among others, AI applications are revolutionizing various industries. From the data explosion to the expansion of the Internet of Things (IoT) and social media, the future of data science is predicted to witness several major innovations over the past decade. The advent of machines will lead to an increase in the use and usefulness of computer systems over the next decade. In addition, social media usage will increase with users using very huge amount of data online. Social media will be used by consumers for business, entertainment, etc. According to some analysts, machine learning algorithms will also experience a sharp rise. The following are key roles related to data science and artificial intelligence. The future of technology lies in data science and AI. As they improve, these technologies will be increasingly used in a wide range of industries. Incorporating artificial intelligence and data science will help businesses thrive in the future as they can gather critical information about their operations, customers and markets.

DATA SCIENTIST

A data scientist can draw basic insights from the data provided. Performing predictive analytics and using machine learning and related techniques to extract data is often part of a data scientist's job description. Any organization that receives data often finds that it needs to be cleaned up because it is messy. This is where data scientists come into play. A data scientist is expected to clean data and present it in a way that benefits the business.

ML ENGINEER

To teach robots how to perform specific tasks, a machine learning engineer must integrate the fields of data science and artificial intelligence. Strong computer programming skills are essential to becoming a machine learning engineer. Machine learning engineers write computer programs that enable computers to perform tasks without being formally programmed or instructed.

BI DEVELOPER

In every organization, the presence of a business intelligence developer (BID) in the ranks of employees is a prominent and important position. Business intelligence development is required to collect and analyze complex data to understand and predict market and business trends. A business intelligence developer must create, edit and store complex data on a user-friendly, cloud-connected platform.

BIG DATA ENGINEER

One of the most lucrative careers in data science and artificial intelligence is as a prominent data engineer and architect. The goal of any chief data engineer or architect is to design a system that enables accessible data collection and effective communication across different business units. Computer programming and languages like C++, Java, Python, and Scala are expected to become experts as prominent data engineers and architects enter the workforce.

DATABASE DEVELOPER

A database developer must improve existing database operations and update company standards and structures by effectively coding programs. Each database developer is also responsible for managing the current databases, finding and fixing system deficiencies, improving the current database, and solving any problems or concerns that may arise in the database systems in the future.

In the future, we will put more emphasis on developing models that not only make accurate predictions, but also provide explanations for their decisions, strengthening trust and understanding of AI-driven results.

CONCLUSION

Since last more than 50 years we are directly or indirectly in touch with artificial intelligence and data science. We do a lot with these technologies in our daily lives. Now most industries with big datasets like Facebook, Twitter, AWS etc. are using expensive AI and DS tools to process the data. Data Science and Artificial Intelligence are revolutionizing the way we approach decision making and problem solving. As big data becomes more accessible, these technologies become more mature and powerful. In this article, we focused not only on discussing the past terminology of AI and DS, but also on exploring areas and possibilities to work more on these tools and technologies. Data science and artificial intelligence have penetrated almost all fields, and it is high time to make a decision and make a smart career move into these fields. Companies are looking for candidates with competencies in fundamental data science and artificial intelligence concepts, statistics, computer science and programming, data structures, algorithms, systems design, and storytelling. We have seen that artificial intelligence and data science are new technologies that are growing at a faster pace and are able to handle all types of modeling techniques and achieve good results. In the near future, we expect AI and DS to expand their fields towards education, research and development. Although AI cannot solve all of your organization's problems, it has the potential to completely change the way you do business. It affects every sector, from manufacturing to finance, and brings unprecedented efficiency gains. As more industries adopt and experiment with this technology, newer applications will be invented. Artificial intelligence will bring about a change even more extensive and far-reaching than the introduction of computing devices. It will change the way we do business, diagnose, operate and drive our cars. It is already changing industrial processes, medical imaging, financial modeling and computer vision. Data science is in its formative stage of development; is evolving into a self-sustaining field that produces professionals with distinct and complementary skills compared to those in computer, information, and statistical sciences.

REFERENCES

Berente, N., Gu, B., Recker, J., & Santhanam, R. (2021). Managing artificial intelligence. *Management Information Systems Quarterly*, *45*(3).

Brynjolfsson, E., & Mcafee, A. (2017). Artificial intelligence, for real. *Harvard Business Review*, *1*, 1–31.

Bughin, J., Hazan, E., & Sree Ramaswamy, P. (2017). Artificial intelligence the next digital frontier. Research Gate.

Cao, L. (2017). Data science: A comprehensive overview. *ACM Computing Surveys*, *50*(3), 1–42. doi:10.1145/3076253

Chiarello, F., Belingheri, P., & Fantoni, G. (2021). Data science for engineering design: State of the art and future directions. *Computers in Industry*, *129*, 103447. doi:10.1016/j.compind.2021.103447

Cockburn, I. M., Henderson, R., & Stern, S. (2018). The impact of artificial intelligence on innovation: An exploratory analysis. In *The economics of artificial intelligence: An agenda* (pp. 115–146). University of Chicago Press.

Cui, M., & Zhang, D. Y. (2021). Artificial intelligence and computational pathology. *Laboratory Investigation*, *101*(4), 412–422. doi:10.1038/s41374-020-00514-0 PMID:33454724

Davenport, T., Guha, A., Grewal, D., & Bressgott, T. (2020). How artificial intelligence will change the future of marketing. *Journal of the Academy of Marketing Science*, *48*(1), 24–42. doi:10.1007/s11747-019-00696-0

Davenport, T. H. (2018). From analytics to artificial intelligence. *Journal of Business Analytics*, *1*(2), 73–80. doi:10.1080/2573234X.2018.1543535

Davenport, T. H., & Ronanki, R. (2018). Artificial intelligence for the real world. *Harvard Business Review*, *96*(1), 108–116.

Dhar, V. (2013). Data science and prediction. *Communications of the ACM*, *56*(12), 64–73. doi:10.1145/2500499

Donoho, D. (2017). 50 years of data science. *Journal of Computational and Graphical Statistics*, *26*(4), 745–766. doi:10.1080/10618600.2017.1384734

Hagendorff, T. (2020). The ethics of AI ethics: An evaluation of guidelines. *Minds and Machines*, *30*(1), 99–120. doi:10.1007/s11023-020-09517-8

Jordan, M. I. (2019). Artificial intelligence—The revolution hasn't happened yet. *Harvard Data Science Review*, *1*(1), 1–9.

Khonturaev, S. I. (2023). The Evolution Of Artificial Intelligence: A Comprehensive Exploration For Higher Education. *Research for Development*, *2*(11), 700–706.

Mehr, H., Ash, H., & Fellow, D. (2017). Artificial intelligence for citizen services and government. *Ash Cent. Democr. Gov. Innov. Harvard Kennedy Sch*, (August), 1–12.

Naudé, W. (2020). Artificial Intelligence against [An early review.]. *COVID*, 19.

Ong, S., & Uddin, S. (2020). Data science and artificial intelligence in project management: The past, present and future. *The Journal of Modern Project Management, 7*(4).

Provost, F., & Fawcett, T. (2013). Data science and its relationship to big data and data-driven decision making. *Big Data, 1*(1), 51–59. doi:10.1089/big.2013.1508 PMID:27447038

Stone, P., Brooks, R., Brynjolfsson, E., Calo, R., Etzioni, O., Hager, G., & Teller, A. (2022). *Artificial intelligence and life in 2030: the one hundred year study on artificial intelligence.* arXiv preprint arXiv:2211.06318.

Varian, H. (2018). Artificial intelligence, economics, and industrial organization. In *The economics of artificial intelligence: an agenda* (pp. 399–419). University of Chicago Press.

Verganti, R., Vendraminelli, L., & Iansiti, M. (2020). Innovation and design in the age of artificial intelligence. *Journal of Product Innovation Management, 37*(3), 212–227. doi:10.1111/jpim.12523

Chapter 16
Security Analysis of the Cyber Crime

Ratnesh Kumar Shukla
(iD) https://orcid.org/0000-0002-8279-7011
Shambhunath Institute of Engineering and Technology, India

Arvind Kumar Tiwari
Kamla Nehru Institute of Technology, Sultanpur, India

ABSTRACT

The primary driver of this expansion is the internet user, who is expected to connect 64 billion devices worldwide by 2026. Nearly $20 trillion will be spent on IoT devices, services, and infrastructure, according to Business Insider. Many cybercrimes and vulnerabilities related to cybercrime are committed with the use of data. Asset management, fitness tracking, and smart cities and homes are examples of internet security applications. The average person will most likely own two to six connected internet security devices by the end of the year, a significant increase over the total number of cell phones, desktop computers, and tablets. Although data provides a plethora of opportunities for its users, some have taken advantage of these advantages for illegal purposes. In particular, a great deal of cybercrime is made possible by the gathering, storing, analyzing, and sharing of data as well as the widespread gathering, storing, and distribution of data without the users' knowledge or consent and without the required security and legal protections. Furthermore, because data gathering, analysis, and transfer happen at scales that governments and organisations are unprepared for, there are a plethora of cybersecurity threats. Protection, privacy, and system and network security are all related.

INTRODUCTION

In a world where the majority of transactions take place on digital platforms, cybercrime has increased rapidly. Current trends in cybercrime indicate that by 2026, the worldwide cost of these attacks may amount to $20 trillion.

The word "cybercrime" is used to describe a broad range of illegal actions that are committed via a computer, network, or other collection of digital devices. Think of cybercrime as a catch-all word for

DOI: 10.4018/979-8-3693-2964-1.ch016

all kinds of illicit actions carried out by cybercriminals. These comprise ransomware, phishing, identity theft, hacking, and malware attacks, to name a few.

Cybercrime's reach transcends all geographic boundaries. There are victims, criminals, and technological infrastructure everywhere in the world. Because technology is used to exploit security flaws on both a personal and corporate level, cybercrime comes in various forms and is always changing. Because of this, preventing, properly investigating, and prosecuting cybercrime is a never-ending battle fraught with many shifting obstacles. Cybersecurity is something we use to prevent cybercrime. Cybersecurity is the process of protecting corporate or governmental computers, servers, and networks against harmful assaults and threats while also preventing unauthorized access to important data. While cyber-crime entails the theft of information, cash, or passwords through the use of vulnerabilities in human security of the systems (Mahammad & Kumar, 2023).

Cybercrime poses a severe risk to people, companies, and governmental organisations. It can lead to substantial financial loss, harm to one's reputation, and compromised data. The flow chart for cybercrime and cybersecurity acts is shown in Figure 1. The threat of cybercrime is increasing as technology develops and more people depend on digital devices and networks for daily tasks, making protection more crucial than ever.

Figure 1. Flow chart of cyber-crime and cyber security

Illegal use of computers or the internet is known as cybercrime. Here are a few instances of online fraud:

❖ Taking and selling corporate information.
❖ Demanding payment in order to avoid an attack.
❖ Virus installation on a targeted computer.
❖ Attempting to gain access to government or corporate computers.
❖ Email Scams.
❖ Social Media Fraud.
❖ eCommerce Fraud.
❖ Banking Fraud.
❖ Ransomware.

❖ Malware.
❖ Data Breaches.
❖ Cyber Espionage.
❖ DDoS Attacks.
❖ Computer Viruses.
❖ Phishing Scams.
❖ Identity Theft.
❖ Online Harassment.
❖ Cyber Terrorism.

Cybercriminal

A cybercriminal is a person who commits cybercrimes, which are defined as malicious acts and illegal activities using technology. Groups or individuals may take part.

Cybercriminals are easily found on the Dark Web, where they mostly provide illicit goods and services.

Because hacking can be used to expose vulnerabilities so that they can be reported and grouped together a practice known as white hat hacking not every hacker is a cybercriminal (Shukla & Tiwari, 2020).

A person who hacks with the malicious aim to carry out damaging acts is known as a black hat hacker or a cyber-criminal. In this case, hacking is classified as a cybercrime.

Since not all cybercrimes include hacking, cybercriminals do not necessarily need to be skilled hackers.

Attacks by cybercriminals can have a big financial and societal impact on people, companies, and governments. Cybercrime's other effects on organisations include harm to a company's reputation, potential legal repercussions from a data breach, and loss of confidential information.

Cybercriminals include people who engage in illicit online content, con artists, and even drug dealers. Take the following cybercriminals, for instance:

❖ Hackers utilising black hat methods..
❖ Cyberstalkers.
❖ Terrorists on the internet.
❖ Scammers.
❖ Cybercriminals who carry out targeted attacks are better referred to as Threat Actors.

Computer Networks

The data interchange between an expanding number of Internet of Things (IoT) devices, including cameras, door locks, doorbells, refrigerators, audio/visual systems, thermostats, and various sensors, and linked computing devices, including laptops, desktops, servers, smartphones, and tablets, is known as computer networking (Gosain et al., 2023).

❖ Networks of computers are constructed using specialized hardware, including switches, routers, and access points.
❖ Switches link computers, printers, servers, and other devices in homes and businesses to networks and aid in their security. Switches known as access points are used to wirelessly link devices to networks.

❖ Routers function as network dispatchers and link networks. They evaluate data that needs to be delivered over a network, choose the most efficient paths for it, then send it. Routers shield data from external security risks and connect your house and place of business to the rest of the world.

❖ Although there are many differences between switches and routers, one crucial one is how they identify end devices. On a Layer 2 switch, a device's burnt-in media access control (MAC) address serves as its unique identification. An IP address assigned by the network is used by a Layer 3 router to uniquely identify a device's network connection.

❖ These days, the majority of switches have some sort of routing capability.

❖ Devices and network connections within a network are uniquely identified by their MAC and IP addresses. A MAC address is a number that the device's manufacturer assigns to a network interface card (NIC). A network connection's IP address is a number connected to it (Jaiswal & Kumar, 2022).

These days, networks offer more than just internet access. Businesses are undergoing a digital revolution. Their networks are essential to their development and success. In order to satisfy these needs, the network topologies listed below are developing:

❖ **SDN (Software-defined):** In the "digital" age, network design is becoming more open, automated, and customizable to meet changing needs. Software-based mechanisms centrally govern traffic routing in software-defined networks. As a result, the network can react to shifting circumstances faster (Begum & Kumar, 2022).

❖ **IBN (Intent-based networking):** Building on SDN principles, IBN automates a lot of tasks, analyses performance, pinpoints issues, offers comprehensive security, integrates with business processes, and adds agility while configuring a network to meet objectives.

❖ **Virtualized:** Several overlay networks can be created by logically dividing the physical network architecture. It is possible to configure each of these logical networks to satisfy certain security, quality of service, and other needs (Shukla et al., 2020).

❖ **CBN (Controller-based networking):** Network security and scalability depend on network controllers. Controllers translate business purpose into device configurations, automating networking tasks and contributing to security and performance assurance. Controllers streamline operations to assist businesses in adapting to changing business requirements.

❖ **Multidomain integrations:** Bigger companies might construct different networks, or networking domains, for their data centers, WANs, and offices. These networks' controllers converse with one another. Exchange of pertinent operating parameters is usually required for such cross-network, or multidomain, integrations in order to guarantee the realization of intended business consequences across network domains.

Component of Computer Network

To install computer networks, computer network components include both physical and software components. A server, client, hub, switch, bridge, peer, and connecting devices are among the hardware components. Protocols and operating system (OS) are examples of software components. In figure 2 represents all features of the component of computer networks (Sharma et al., 2022).

Figure 2. Flow chart of component of computer network

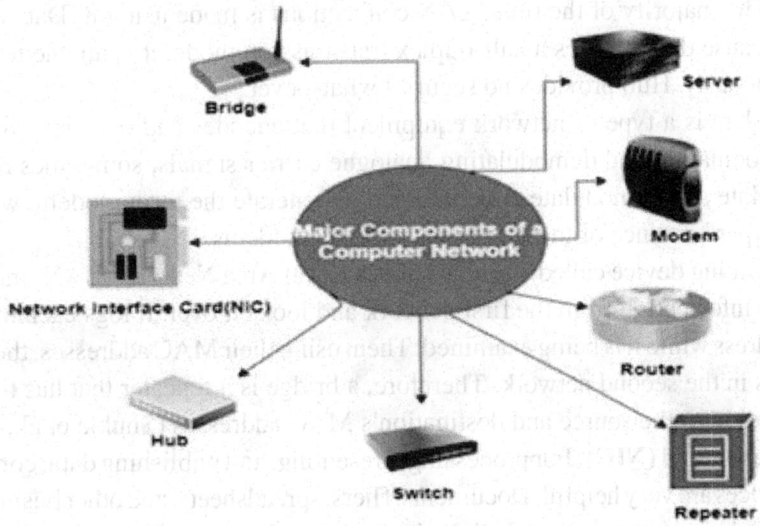

Endpoints/Clients Are Computer Network Components or Network Devices

An individual connecting to a network is called a client. For instance, if you are reading this blog on your laptop, you are a client (Kumar, 2022).

Endpoints are any device that a client uses, such as a laptop, computer, or smartphone.

They connect a user/individual to a computer network and exchange information with it.

❖ **Server:** Your requests (clients) are fulfilled by a server. It transfers, receives, and saves data. Your requests (clients) are fulfilled by a server. It transfers, receives, and saves data. For our benefit, the content is kept on a server. Different kinds of servers might be accessible, including virtual servers, webpages, mail servers, and video servers.

❖ **Switches:** Networking equipment called switches allows devices to connect locally within a building. Data is received by the switch and sent to the intended device. A PC and switch are connected by an Ethernet wire. It is important for you to understand that switches function at Layer 2 of the OSI or TCP/IP models.

❖ **Routers:** An apparatus that links a Wide Area Network (WAN) and a Local Area Network (LAN) is called a router. It is a Network Layer Device as a result. As you are aware, switches are used to connect computers and other devices. A router is then connected to these switches.

❖ **Repeater:** A network device called a repeater strengthens weakened transmissions. Its aim is to restore signal strength.

For instance, a signal's intensity starts to drop as it passes through a TV cable. A repeater can then gradually renew the same signal in this situation.

❖ **Hub:** A hub is a piece of hardware used in networking that joins several computers or network connections together. Stated differently, a hub functions as a multiport repeater. It automatically distributes all information to all connected devices via ports, broadcasting the data to all computers linked to it. The majority of the time, LAN connections is made using it. Data cannot be filtered by a hub. Because the hub uses a half-duplex transmission mode, it is unable to send and receive data simultaneously. Hub provides no security whatsoever.

❖ **Modem:** A modem is a type of network equipment that encodes and decodes digital data for processing by modulating and demodulating analogue carrier signals, sometimes referred to as sine waves. Modulate and demodulate are combined to generate the term modem, which refers to the simultaneous performance of these two functions by modems.

❖ **Bridge:** A networking device called a bridge splits a Local Area Network (LAN) into many portions. A bridge gets information from the first network and looks it over. It logs each hub's port number and MAC address while it is being examined. Then, using their MAC addresses, the data is delivered to the devices in the second network. Therefore, a bridge is a repeater that has the ability to filter content according to the source and destination's MAC addresses (Shukla et al., 2022).

❖ **Network Interface Card (NIC):** For processing, presenting, and publishing data, computers and other electronic devices are very helpful. Documents, fliers, spreadsheets, and other business-related prints can be produced by a computer that is linked to a printer. If your office has two or more computers, how do we transfer data across them? What happens if I have to deliver the data to a client who lives in another state or nation? It must be possible for your helpful gadget to communicate with other computers and systems. In order to maintain data integrity, the Network Interface Card, or NIC, uses data transfer protocols to link your device to a network of other devices. Your computer's physical network interface card, or NIC, is what connects it to the network. It is also referred to as a network adapter, network interface controller, or a LAN adapter.

TYPES OF CYBERCRIME

These days, the world is highly reliant on technology since new ones are always being created. Most smart devices are connected to the internet. There are benefits and drawbacks.

Among the hazards are the sharp rise in cybercrimes and the deficiency of security operations and protocols to protect these technologies (Kumar et al., 2022).

Through computer networks, people can quickly reach any linked location on the earth.

Regarding cybercrimes, different nations may have distinct laws and policies.

It should be mentioned that it is simpler to conceal cybercrimes than traditional crimes.

❖ **Devices that are vulnerable:** As previously stated, a variety of susceptible devices are introduced by the absence of effective security measures and solutions, making them a prime target for cybercriminals.

❖ **Individual drive:** Cybercriminals occasionally carry out cybercrimes as a sort of retaliation against someone they dislike or otherwise disagree with.

❖ **Monetary incentives:** The primary incentive for most hacker groups and cybercriminals these days is financial gain, which drives the majority of attacks.

Popular Types of Cybercrime

There are two main types of cyber-crime are as follows:

1. **Aiming at computer systems:** Cybercrimes of this kind encompass all potential means of causing damage to computer systems, such as denial-of-service attacks or malware.
2. **Making use of computers:** This category covers all computer crimes, regardless of their classification.

Classification of Cyber-Crimes

Generally, there are using different categories of cybercrimes as follows:

Individual Cybercrimes: These transgressions target particular individuals. Included are spoofing, phishing, spam, cyberstalking, and other techniques.

Cybercrimes against Organisations: Organisations are the main objective here. Malware and denial-of-service attacks are examples of the kind of crimes that are usually committed by criminal teams.

Property Cybercrimes: These are cyberattacks that target intellectual property rights or even physical property like credit cards.

Cybercrimes against Society: Because it encompasses Cyberterrorism, this category of cybercrime is the most dangerous.

Phishing and scams: A type of social engineering attack known as phishing involves a perpetrator posing as a user and sending threatening emails and messages in an attempt to steal the target's personal information. It also involves an attempt to download malicious software and use it to attack the target system (Shukla & Tiwari, 2023).

Theft of Identity: Identity theft occurs when a cybercriminal gets hold of someone else's personal data, like credit card numbers or photos, and utilizes it for illicit or fraudulent means.

Attack with Ransomware: Cybercrime, including ransomware attacks, is very commonplace. This particular kind of malware has the ability to encrypt user data and demand a ransom to unlock it, thereby preventing users from accessing any of their personal information on the system (Jain et al., 2021).

Abuse and Hacking of Computer Networks: This phrase describes the illegal act of gaining unauthorized access to private computers or networks and abusing them by deleting data, manipulating stored data, or using other unlawful methods.

Bullying online: It is also referred to as cyberbullying. It involves sending or disseminating damaging and dehumanizing content about other people, which is embarrassing and may contribute to the development of psychological issues. It has been increasingly prevalent lately, particularly among youth.

Online Pursuit: Cyberstalking is the term used to describe the practice of sending unwanted, persistent content to other people's online accounts with the intention of controlling and intimidating them. It is similar to sending unsolicited, persistent calls and texts.

Piracy of Software: The unauthorized use or duplication of purchased software in violation of copyrights or license limitations is known as software piracy. Software piracy can be exemplified by downloading a new, inactivated copy of Windows and using so-called cracks to get a working license for Windows activation. That's regarded as software piracy. Music, films, and images can all be downloaded illegally in addition to software.

Fraud on Social Media: Posing as someone else or using phoney social media profiles for nefarious activities like sending intimidating or threatening messages. One of the easiest and most common types of social media fraud is email spam.

Trafficking in Drugs Online: With the speed at which bitcoin technology is developing, it is now easy to move money in a private, secure manner and conduct drug trades without drawing the attention of law authorities. This led to a rise in drug marketing on the internet (Comendador et al., 2016).

Online sales and trade of illegal drugs, like cocaine, heroin, and marijuana, are very common, particularly on the Dark Web.

Money Laundering through Electronic Means: Additionally known as money laundering. It is based on unnamed companies or websites those handles credit card payments and take legitimate payment methods, but which give inconsistent or incomplete payment information when customers buy unidentifiable products (Shukla, Tiwari, & Verma, 2021).

Infringements upon intellectual property: It is the infringement or breach of any intellectual property rights that are protected, including industrial design and copyrights.

Fraudulent Online Recruiting: Fake companies releasing job opportunities with the intention of taking applicants' personal information or using it for financial gain are one of the less common cybercrimes that are also becoming more popular.

Credit Card Fraud and Cyber Crime

Using a credit card to obtain money or goods fraudulently is known as credit card fraud. Credit cards can be stolen, the numbers copied from them, or the victim's account taken over and the credit card mailed to the criminal's address. In an attempt to steal money or purchase goods, they might also open a new credit card in the victim's name and employ a number of other strategies (Gupta et al., 2021).

Any crime that originates online is a cyber-crime. A scammer who gains a victim's friendship on a social media site and persuades them to transfer money using their credit card is one kind of crime. Alternatively, fraudsters might take a real credit card or get its numbers and use those details to make online purchases (Shukla, Sengar, Gupta et al, 2021).

As an alternative, a burglar might break into a bank or company database to steal customer personal information and resell it online. The criminal who purchases that data can then use it to open a false account using the victim's information.

Cybercrimes and Credit Card Fraud Overlap

Numerous cybercrimes involve credit cards, and there are countless other types as well. The way data and information thieves target has changed due to the internet. Some identity thieves target a single victim or steal a single card, while others concentrate on hacking massive information files that they can sell online. Figure 3 represents the transactions of the credit card. In this process we have learn and secure our data from fraud detection and prevention plan that addresses the threat of cybercrimes if you want to stop credit card fraud (Tripathi et al., 2022).

Figure 3. Flow chart of credit card transactions

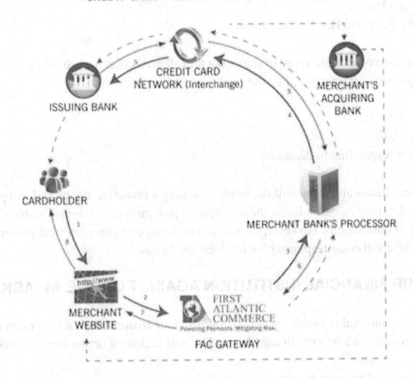

Customers react to concerns about crimes in a way that emphasizes their innate understanding of this risk. According to a survey, 47% of respondents said they were more afraid of identity theft than they were of having their home broken into, while 27% said they were more afraid of identity theft.

Online Credit Card Fraud

A con artist can make online purchases once they have the victim's credit card information. One of the most common uses of credit card information that has been stolen is this. Online shopping fraud increased by almost a third between 2016 and 2017, and transactions originating from foreign IP addresses had a fraud risk approximately seven times higher than that of transactions originating from U.S. IP addresses.

Cybersecurity and Fraudulent Credit Cards

Credit and debit cardholders are not the only ones whose risk of fraud is increased by the internet. Additionally, it can be very helpful in defending financial institutions, companies, and individuals against the dangers of credit card fraud. To help lower credit card fraud, financial institutions must have cyber security tools in place (Chauhan et al., 2022).

The following are typically included in the three fundamental steps in handling credit card fraud:

❖ Put an end to the losses.
❖ Get the Money Back and Handle the Fallout.

These actions are crucial in the event of fraud, but for full protection, you should implement a slightly different, more proactive framework, like the ones listed below:

❖ Steer clear of the losses
❖ Safeguard the Cash
❖ Make a plan for responding to disasters.

You can prevent losses and safeguard the funds by taking a proactive stance and using fraud protection software. Before they cause an issue, the appropriate programmes recognise patterns and highlight possibly fraudulent transactions. However, even if you're doing everything in your power to stop fraud, you should still have a disaster response plan in place just in case.

DEFEND YOUR FINANCIAL INSTITUTION AGAINST ONLINE ATTACKS

In addition to taking internal measures, you must inform your customers about the main risks and how to prevent them in order to lower credit card fraud. The best course of action is as follows:

Enforce smart password policies by mandating frequent password changes and the use of difficult-to-guess passwords for staff members.

Update your antivirus software: The computers in your financial institution require up-to-date antivirus software to protect against external threats from hackers.

Configure bank apps and websites to use two-factor authentication: Make users sign in with two factors to keep hackers out of your website or your customers' accounts.

Use caution when sending emails: Many con artists may attempt to access your computers by using emails addressed to bank executives or managers. Ensure that everyone on your team is aware of how to avoid dubious downloads, links, or requests.

Collaborate with a fraud expert: Collaborate with a fraud specialist to implement fraud protection and detection tools in order to guarantee that your financial institution and your clients are fully safeguarded against cyber threats and credit card fraud.

Report fraudulent activity: Any attempts at scams that target your financial institution should be reported. Then, you can share your experiences with other bankers and fraud prevention specialists.

NETWORK SECURITY

Network security refers to any measure taken to protect your data and network's usability and integrity (Shukla et al., 2023).

❖ Included are both hardware and software technologies.

❖ It targets several different threats.
❖ It stops them from connecting to or expanding over your network.
❖ Effective network security regulates access to the network.

Types of Network Security

Network security comes in various forms.

Firewalls: Firewalls keep your trusted internal network isolated from the outside world and the Internet. They manage traffic by applying a specified set of regulations. Hardware, software, or both can make up a firewall. Cisco offers unified threat management (UTM) tools together with next-generation firewalls that are designed with threats in mind.

Email security: Email gateways are the most frequent source of security breaches. Hackers craft sophisticated phishing campaigns that deceive receivers and send them to malicious websites by leveraging social engineering tactics and victims' personal information. An email security application limits outgoing messages and blocks incoming threats to prevent sensitive data from being lost.

Antivirus and anti-malware software: Malware, an acronym for harmful software, includes Trojan horses, worms, viruses, ransomware, and spyware. Sometimes malware gets into a network, but it stays inactive for days or even weeks. In addition to screening files for malware upon arrival, the top antimalware applications track files continually to spot irregularities, remove malware, and repair damage.

Network segmentation: Network traffic is divided into separate groups by software-defined segmentation, which also makes it easier to apply security controls. Ideally, the classifications are based on endpoint identification, not merely IP addresses. Access rights can be assigned according to role, location, and other characteristics to guarantee that the right people are given the right amount of access and those suspicious devices are confined and remediated.

Security of applications: Any software that runs your business, whether it is created by your IT staff or something you buy, needs to be secured. Unfortunately, every application may contain flaws or vulnerabilities that hackers could use to get access to your network. Application security includes the processes, programmes, and hardware you use to close those holes (Kumar et al., 2016).

Control of access: Not every user should be able to access your network. To keep potential attackers out, you must identify each person and each device. After that, you can implement your security measures. Access to noncompliant endpoint devices may be denied or restricted. This is known as network access control (NAC).

The use of behavioral analytics: To spot unusual network activity, you need to know what regular conduct looks like. Behavioral analytics tools are able to recognise abnormal behaviour automatically. Your security staff will then be more adept at identifying any problems and eliminating dangers as soon as they materialize.

Preventing data loss: Organizations must make sure that employees do not send private information outside of the network. Data loss prevention, or DLP, technologies can stop users from emailing, downloading, or even printing private information in a dangerous manner.

Systems for preventing intrusions: An intrusion prevention system (IPS) examines network traffic to actively stop attacks. Cisco Next-Generation IPS (NGIPS) systems correlate massive volumes of global threat intelligence to block malicious behaviour and trace the flow of questionable files and malware throughout the network, thereby preventing outbreaks and reinfection.

Security of mobile devices: Cybercriminals are increasingly targeting mobile apps and gadgets. Within the next three years, 90% of IT firms may allow corporate applications on personal mobile devices. It goes without saying that you have to control which devices can access your network. You also need to configure their connections to permit traffic from private networks.

Event management and security information: SIEM products gather the information your security staff needs to identify and respond to attacks. These items come in a variety of forms, including virtual and physical appliances and server software.

VPN: A virtual private network encrypts the connection made by an endpoint to a network, which is typically the Internet. A remote-access VPN usually uses IPsec, or Secure Sockets Layer, to verify communication between a device and the network.

Internet safety: Web security solutions will manage employee internet usage, prevent web-based dangers, and limit access to harmful websites. There will be local and cloud security for your web gateway. Precautions you take to protect your personal website are also included in the category of web security.

Wireless safety: Wireless networks are less secure than conventional networks. It's analogous to installing Ethernet connections everywhere, including the parking lot, while installing a wireless local area network (WLAN) without following stringent security protocols. Specifically designed products are needed to prevent an exploit from becoming established on a wireless network.

CYBER SECURITY

Cybersecurity is the mashup of the terms "cyber" and "security," where "cyber" is the internet and "security" is the act of protecting it. The internet governs and controls every aspect of human life, including social, professional, and working environments. The internet is everywhere you look, and since it has become indispensable to daily life, its growth seems unstoppable. These days, we find it impossible to imagine living without the internet, and technological advancements have made it quick and simple.

Given that 90% of work is now done from home using the internet thanks to the COVID-19 pandemic, the internet's critical role has been demonstrated. The system functions with a variety of software programmes and browsers when it is connected to the internet. The internet is used for storing a lot of data in addition to communication. As an illustration, consider Google Drive, cloud storage, and emails. This storage facility allows us to rely on the system for every detail, so we should only write down and remember certain things. However, this informational ease always carries the risk of people becoming compromised as private, professional, and confidential information may be revealed. Cybersecurity has come into its own to address this situation. The protection of data, networks, and systems connected to the internet in terms of confidentiality, integrity, and authentication is known as cyber security. Cybersecurity protection layers improve information protection for both corporate and individual users, increase cyber resilience, and speed up cyber data. Networks, servers, computers, electronic systems, data, mobile devices, and communication are all protected from harmful attacks by this protective shield. There are many contexts in which cyber security is classified, such as end-user education, business continuity, disaster recovery, operational security, application security, and network security.

Cybersecurity defends against theft and other threats that target computers, the internet, and websites, as well as our digital assets and businesses' data and systems. Cybersecurity covers a wide range of online threats, including malware, Trojan horses, adware, botnets, denial-of-service attacks, Man-in-the-Middle attacks, SQL injection, and cybercrime. Cyber security experts are hired to find, test, and fix vulnerabilities in an organization's infrastructure in order to maintain and identify potential threats and valuable data in cyber security.

Let's now put aside our pessimism and take a look at the advantages of thieves stealing our data and money:

1. **The money really goes to people rather than being lost:** Okay, so it doesn't really vanish into a black hole. Although the scuzbuckets who steal it or fabricate it using your information may start it, they will still spend the money, which will usually benefit someone who abides by the law.

2. **The "Investment Multiplier Effect" Bringing us to our next fantastic advantage:** the flow of money through an economy creates a macroeconomic boost. The cycle of spending, paying taxes, and so on continues, ultimately resulting in a positive boost to the economy.

3. **Absence of activity:** Dear Staff, there will be a lot of downtime while we recover from a serious cyber security breach. Please spend this week at home.

4. **New jobs and a new industry:** Yeah, scumbags are now employable. What else might they be doing if they weren't stealing our data, you might wonder? Maybe mugging tourists, robbing gas stations, or taking courier motorcycles. This must be superior, right?

5. **A decrease in the dehumanization of people:** Before we were all cognizant of the problems with information technology security, we were ignorant and would spread a wide range of false information. Like yesterday was Chuck Norris's death. He did, after all, but he dismissed it with a shrug. However, as we grow more aware of scams, we are able to see past this and are stopping the dissemination of this garbage, which, let's be honest, only serves to make people more stupid.

6. **Form new friendships:** Through a lot of great coffee, I've now met a lot of really great people in this industry. That's always a good thing! When you lose your identity and need to talk to agencies to help you get your life back together, you can also make new friends.

CONCLUSION

The main cause of this growth is the internet user, who is predicted to connect 64 billion devices globally by 2026. According to Business Insider, nearly $20 trillion will be spent on IoT devices, services, and infrastructure. With the use of data, many cybercrimes and vulnerabilities related to cybercrimes are committed. Applications for internet security include asset management, fitness tracking, and smart cities and homes. By the end of the year, the average person will probably own two to six connected internet security devices, a substantial increase over the total number of cell phones, desktop computers, and tablets. Although data presents a plethora of opportunities for its users, some have exploited these benefits for illicit purposes. In particular, a lot of cybercrime is made possible by the collection, storing, analysis, and sharing of data as well as by the extensive collection, storing, use, and dissemination of data without the users' informed consent or choice and without the necessary legal and security safeguards. Furthermore, because data aggregation, analysis, and transfer occur at scales for which governments and organisations are unprepared, there are a plethora of cybersecurity risks. Data security, privacy, and protection are all entwined with systems and networks.

REFERENCES

BegumA.KumarR. (2022). Design an Archetype to Predict the impact of diet and lifestyle interventions in autoimmune diseases using Deep Learning and Artificial Intelligence. Research Square. doi:10.21203/rs.3.rs-1405206/v1

Chauhan, N. R., Shukla, R. K., Sengar, A. S., & Gupta, A. (2022, December). Classification of Nutritional Deficiencies in Cabbage Leave Using Random Forest. In *2022 11th International Conference on System Modeling & Advancement in Research Trends (SMART)* (pp. 1314-1319). IEEE. 10.1109/SMART55829.2022.10047282

Comendador, B. E. V., Rabago, L. W., & Tanguilig, B. T. (2016, August). An educational model based on Knowledge Discovery in Databases (KDD) to predict learner's behavior using classification techniques. In *2016 IEEE International Conference on Signal Processing, Communications and Computing (ICSPCC)* (pp. 1-6). IEEE. 10.1109/ICSPCC.2016.7753623

Gosain, M. S., Aggarwal, N., & Kumar, R. (2023). A Study of 5G and Edge Computing Integration with IoT- A Review. *2023 International Conference on Computational Intelligence and Sustainable Engineering Solutions (CISES)*, Greater Noida, India. 10.1109/CISES58720.2023.10183438

Gupta, A., Shukla, R. K., Bhola, A., & Sengar, A. S. (2021, December). Comparative Analysis of Supervised Learning Techniques of Machine Learning for Software Defect Prediction. In *2021 10th International Conference on System Modeling & Advancement in Research Trends (SMART)* (pp. 406-409). IEEE. 10.1109/SMART52563.2021.9676307

Jain, A., Gupta, A., Sengar, A. S., Shukla, R. K., & Jain, A. (2021, December). Application of Deep Learning for Image Sequence Classification. In *2021 10th International Conference on System Modeling & Advancement in Research Trends (SMART)* (pp. 280-284). IEEE. 10.1109/SMART52563.2021.9676200

Jaiswal, A., & Kumar, R. (2022). *Breast cancer diagnosis using Stochastic Self-Organizing Map and Enlarge C4.5*. Multimed Tools Appl. doi:10.1007/s11042-022-14265-1

Kumar, A., Tewari, N. & Kumar, R. (2022) A comparative study of various techniques of image segmentation for the identification of hand gesture used to guide the slide show navigation. *Multimed Tools Appl.* . doi:10.1007/s11042-022-12203-9

Kumar, R. (2022). Intelligent Model to Image Enrichment for Strong Night-Vision Surveillance Cameras in Future Generation. *Multimedia Tools and Applications*. doi:10.1007/s11042-022- 12496

Kumar, S., Das, D., & Agarwal, A. (2016, March). A novel method for identification and performance improvement of Blurred and Noisy Images using modified facial deblur inference (FADEIN) algorithms. In *2016 IEEE Students' Conference on Electrical, Electronics and Computer Science (SCEECS)* (pp. 1-7). IEEE.

Mahammad, B., & Kumar, R. (2023). Scalable and Security Framework to Secure and Maintain Healthcare Data using Blockchain Technology. *2023 International Conference on Computational Intelligence and Sustainable Engineering Solutions (CISES)*, Greater Noida, India. 10.1109/CISES58720.2023.10183494

Sharma, N., Chakraborty, C., & Kumar, R. (2022). (2022) Optimized multimedia data through computationally intelligent algorithms. *Multimedia Systems*. doi:10.1007/s00530-022-00918-6

Shukla, R. K., Prakash, V., & Pandey, S. (2020, December). A Perspective on Internet of Things: Challenges & Applications. In *2020 9th International Conference System Modeling and Advancement in Research Trends (SMART)* (pp. 184-189). IEEE.

Shukla, R. K., Sengar, A. S., Gupta, A., & Chauhar, N. R. (2022, December). Deep Learning Model to Identify Hide Images using CNN Algorithm. In *2022 11th International Conference on System Modeling & Advancement in Research Trends (SMART)* (pp. 44-51). IEEE. 10.1109/SMART55829.2022.10047661

Shukla, R. K., Sengar, A. S., Gupta, A., Jain, A., Kumar, A., & Vishnoi, N. K. (2021, December). Face Recognition using Convolutional Neural Network in Machine Learning. In *2021 10th International Conference on System Modeling & Advancement in Research Trends (SMART)* (pp. 456-461). IEEE. 10.1109/SMART52563.2021.9676308

Shukla, R. K., & Tiwari, A. K. (2020). A Machine Learning Approaches on Face Detection and Recognition. *Solid State Technology, 63*(5), 7619–7627.

Shukla, R. K., & Tiwari, A. K. (2023). Masked face recognition using mobilenet v2 with transfer learning. *Computer Systems Science and Engineering, 45*(1), 293–309. doi:10.32604/csse.2023.027986

Shukla, R. K., Tiwari, A. K., & Jha, A. K. (2023). An Efficient Approach of Face Detection and Prediction of Drowsiness Using SVM. *Mathematical Problems in Engineering, 2023*, 2023. doi:10.1155/2023/2168361

Shukla, R. K., Tiwari, A. K., & Verma, V. (2021, December). Identification of with Face Mask and without Face Mask using Face Recognition Model. In *2021 10th International Conference on System Modeling & Advancement in Research Trends (SMART)* (pp. 462-467). IEEE. 10.1109/SMART52563.2021.9676204

Tripathi, P. K., Shukla, R. K., Tiwari, N. K., Thakur, B. K., Tripathi, R., & Pal, S. (2022, December). Enhancing Security of PGP with Steganography. In *2022 11th International Conference on System Modeling & Advancement in Research Trends (SMART)* (pp. 1555-1560). IEEE. 10.1109/SMART55829.2022.10046709

Chapter 17
Structural Systems Powered by AI and Machine Learning Technologies

Sabyasachi Pramanik
https://orcid.org/0000-0002-9431-8751
Haldia Institute of Technology, India

ABSTRACT

This chapter investigates the use of artificial intelligence (AI) and machine learning (ML) technologies in structural engineering, with an emphasis on their applications in automating design processes, optimizing structural configurations, and evaluating performance measures. It demonstrates the effectiveness of AI-powered algorithms in creating design alternatives, anticipating structural behavior, and improving sustainability. The chapter also includes a framework for comparing the performance of various structural designs, taking into account safety, cost-effectiveness, and environmental impact. It provides case studies and practical examples that show how AI/ML-driven autonomous design may achieve greater structural performance while using fewer resources. The chapter stresses the potential of AI and ML to revolutionize structural engineering by allowing engineers to design more sustainable and high-performing buildings, so contributing to a more ecologically aware and economically viable built environment.

INTRODUCTION

As architects confront the difficulty of optimizing buildings for safety and sustainability while adhering to severe budget limits, the discipline of structural engineering is seeing an increase in demand for novel solutions that emphasize sustainability and performance efficiency. Traditional structural design methods take time and rely primarily on human knowledge. AI and machine learning advancements have transformed this procedure. This chapter looks at how intelligent structural engineering may be combined with optimization approaches, performance comparison methodologies, and sustainable design principles, all of which are driven by AI and ML technologies (L. Sun et al., 2020a).

DOI: 10.4018/979-8-3693-2964-1.ch017

This chapter investigates the use of AI and ML in structural engineering, with an emphasis on automation, optimization, and performance measures. It also explores sustainable design ideas and how they might be combined with AI and machine learning to develop more ecologically aware architecture. It demonstrates the advantages of AI/ML-driven autonomous design, enhancing structural performance and resource usage, using case studies and practical examples. The chapter also discusses the problems and ethical concerns related with the integration of AI and ML in structural engineering (Möhring et al., 2020).

Artificial intelligence and machine learning are reshaping structural engineering by introducing new tools and approaches for structure design, analysis, and upkeep. AI and machine learning algorithms can examine large datasets and uncover patterns that people cannot, result in better predictive modeling and risk assessment. Engineers may use artificial intelligence (AI) to examine historical data, such as construction failures and maintenance records, to detect possible concerns and offer design changes, thereby enhancing the safety and durability of buildings (Shea & Smith, 2005).

AI-powered design optimization tools assist engineers in constructing efficient and cost-effective structures by recommending new solutions that take into account criteria such as materials, load distribution, and environmental variables, decreasing costs and minimizing environmental impact. AI and machine learning are improving real-time monitoring and maintenance of structures, allowing for early diagnosis of deterioration and lowering maintenance costs. This proactive strategy protects the safety and longevity of the building. AI-driven simulations, such as complicated finite element analysis and computational fluid dynamics, may increase structural analysis accuracy and efficiency, allowing for more robust and trustworthy structure design under varying circumstances (Liu et al., 2004). AI and machine learning (ML) integration with Building Information Modeling (BIM) improves collaborative decision-making, stakeholder communication, project management, and design, construction, and maintenance results. These technologies are becoming more important in structural engineering, ushering in a new age of creativity and efficiency. Artificial intelligence and machine learning allow structural engineers to construct detailed designs that maximize cost, energy efficiency, and material utilization. ML algorithms are capable of adapting to local environmental circumstances, guaranteeing that constructions can endure earthquakes and harsh weather occurrences. AI can also access complicated data from building sensors and provide real-time feedback on structural health (Salehi & Burgueo, 2018).

AI and machine learning are critical for risk assessment, discovering possible flaws in new initiatives, and reducing structural failures. They also contribute to the advancement of sustainable building practices by improving building designs for energy efficiency and sustainability. AI algorithms can detect renewable energy sources and optimize HVAC systems for energy efficiency (L. Sun et al., 2020b). By examining their performance and environmental effect, ML promotes the adoption of sustainable materials. AI and machine learning are also hastening innovation in materials science and building practices. They may aid engineers in the development of novel composite materials with increased strength and durability (Huang & Fu, 2019). Furthermore, AI can enhance building operations, increasing productivity and decreasing waste. AI can identify areas for process scalability by analyzing construction data, resulting in cost savings and more sustainable practices (Pan & Zhang, 2021). AI and ML are increasingly being used in urban infrastructure design and maintenance, enabling predictive maintenance systems for critical components such as bridges and tunnels. These smart city projects enhance traffic flow, manage utilities, and react to crises, ensuring that urban settings run smoothly (Guo et al., 2021).

By enhancing safety, efficiency, and sustainability, artificial intelligence and machine learning are changing structural engineering. These technologies allow data-driven choices, optimize designs, and control structural health, therefore improving the quality and resilience of the built environment. As AI and machine learning get more advanced, its applications in structural engineering will become more complex, opening the door for a new age of innovation.

Objectives

- To clarify the role of AI and machine learning in advancing the subject of structural engineering.
- To demonstrate several optimization strategies for structural design that make use of AI/ML.
- To provide a framework for comparing performance, including safety, cost-effectiveness, and environmental impact evaluation.
- To highlight the significance of sustainability in structural engineering and how AI/ML may help with sustainable design.
- To give real-world case studies illustrating how AI/ML-driven autonomous design may be used in practice.
- To address the advantages, disadvantages, and ethical concerns involved with the application of AI and ML in the sector.

CLASSIFICATIONS OF INTELLIGENT STRUCTURAL ANALYSIS

The desire for safer, more efficient, and greener buildings has led to the incorporation of cutting-edge technology in the ever-evolving area of structural engineering. Among them, artificial intelligence (AI) and machine learning (ML) have emerged as powerful technologies with the potential to transform how we examine, evaluate, and improve structures. This chapter focuses on the many forms of intelligent structural analysis, offering insight on how AI and ML are used to handle various difficulties in structural engineering (Dolak& Novak, 2011; Soh & Soh, 1988).

Structural Health Monitoring (SHM): A fundamental aspect of intelligent structural analysis is structural health monitoring. It entails continuously assessing the state and performance of a building in real time. AI and machine learning (ML) play an important role in SHM by interpreting sensor data, detecting abnormalities, and forecasting possible problems before they worsen. This section looks into the approaches and technologies utilized in SHM, demonstrating how AI and ML improve structural safety and durability.

Predictive Analysis and Modeling: Predictive analysis and modeling use AI and machine learning to predict how structures will respond under different scenarios. Engineers can accurately model and anticipate structure reactions to environmental elements, loads, and other variables. This section delves into predictive analytic applications such as earthquake forecasting, weather impact assessment, and long-term structural performance projections.

Techniques for Structural Optimization: Structural optimization is at the foundation of constructing efficient and cost-effective structures. AI-driven optimization approaches, such as genetic algorithms and particle swarm optimization, allow engineers to search a large design space for optimum configurations. This section goes into optimization concepts and demonstrates how AI and ML may lead to resource-efficient and sustainable structural solutions.

Performance Evaluation and Simulation: Performance evaluation and simulation are critical for determining how structures will perform under different situations and loads. AI and machine learning models make complicated simulations possible, allowing engineers to study structural behavior and make educated judgments. This section examines the importance of artificial intelligence and machine learning in modeling structural reactions to dynamic phenomena like as wind, traffic, and seismic activity.

Sustainability evaluation: In structural engineering, sustainability evaluation has become a top priority. AI and machine learning may help evaluate a structure's environmental effect, energy efficiency, and carbon footprint. This section investigates how these technologies help engineers to make environmentally aware choices, hence encouraging sustainable design practices.

This chapter investigates AI and ML-powered structural analysis approaches for monitoring structural health, forecasting behavior, optimizing designs, evaluating performance, and measuring sustainability in structural engineering, encouraging robust, efficient, and sustainable structures.

APPLYING MACHINE LEARNING (ML) TO DIFFERENT TYPES OF STRUCTURAL ANALYSIS

As shown in Figure 1, using machine learning into intelligent structural analysis may enhance accuracy, efficiency, and flexibility. This section explains how machine learning is integrated into several forms of structural analysis (H. Sun et al., 2021; Westermayr et al., 2021).

Figure 1. Machine learning (ML) integration into various types of structural analysis

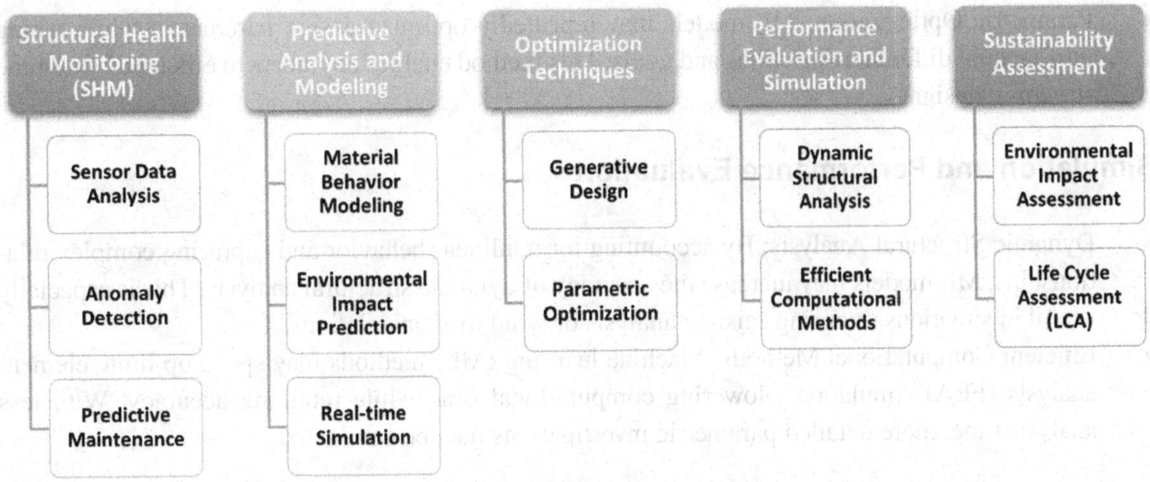

Monitoring of Structural Health (SHM)

- Sensor Data Analysis: Machine learning algorithms can interpret sensor data, such as strain gauges, accelerometers, and temperature sensors, to continually monitor the health of a building. They are capable of detecting abnormalities, identifying trends, and forecasting prospective problems like as fractures or material deterioration.
- Anomaly Detection: Machine learning models, especially anomaly detection algorithms such as Isolation Forests or One-Class SVMs, may discover deviations from typical structural behavior, indicating the need for inspections or maintenance.
- Predictive Maintenance: Using historical data, ML-driven predictive maintenance models may anticipate when certain structural components may need maintenance or replacement, saving downtime and repair costs.

Modeling and Predictive Analysis

- Material BehaviorModeling: Machine learning may be used to model the behavior of materials under various situations, boosting the accuracy of prediction models for structural reactions.
- Environmental influence Prediction: ML algorithms can forecast the influence of environmental conditions on structural performance, such as temperature variations or corrosion.
- Real-time Simulation: Machine learning may enable real-time simulations that take into account dynamic factors, allowing engineers to make quick choices based on changing situations.

Techniques of Optimization

- Generative Design: ML may be used to drive generative design methods that search a large design space for optimum combinations. Design options that fulfill given restrictions and goals may be generated via genetic algorithms and neural networks.
- Parametric Optimization: ML models may repeatedly optimize design parameters while taking into account different restrictions and goals. This method enables engineers to efficiently fine-tune structural designs.

Simulation and Performance Evaluation

- Dynamic Structural Analysis: By accounting for nonlinear behavior and capturing complex relationships, ML models may increase the accuracy of dynamic structural analysis. This is especially useful in situations requiring seismic analysis or wind load calculations.
- Efficient Computational Methods: Machine learning (ML) methods may speed up finite element analysis (FEA) simulations, lowering computational time while retaining accuracy. With less analysis time, more detailed parametric investigations may be conducted.

Sustainability Evaluation

- Environmental effect Assessment: ML may help analyze the environmental effect of structural designs by taking into account issues like material selection, energy efficiency, and carbon emissions. This data supports judgments on sustainable design.
- Life Cycle Assessment (LCA): ML may be used to do LCA, which analyzes the environmental effect of buildings during their full life cycle. Machine learning models may assist in identifying possibilities to reduce environmental footprints.

Machine learning (ML) is being applied into structural analysis to improve accuracy, efficiency, and adaptability. Based on historical and real-time data updates, this technique leads to safer, more efficient, and sustainable structures in the built environment.

INNOVATIVE STRUCTURAL ENGINEERING

AI and Machine Learning in Structural Engineering

By allowing efficient and effective problem-solving, the combination of AI and machine learning technology has transformed structural engineering. These sophisticated computational tools transform the design, analysis, and decision-making processes, allowing structural engineers to solve complicated issues more effectively. This section investigates the critical role of AI and machine learning in defining the future of structural engineering (Perry et al., 2022).

- enhancing Decision-Making: AI and ML algorithms enable engineers in making educated judgments by evaluating large datasets, historical performance data, and real-time data, hence enhancing structural design dependability.
- Design Optimization: AI-powered optimization techniques allow for the examination of a wide range of design possibilities, resulting in more efficient and cost-effective structural solutions.
- Predictive Maintenance: Machine learning algorithms can forecast structural degradation, enabling for proactive maintenance, cost savings, and increased safety.

Design Process Automation

Automation is one of the key advantages of AI and ML in structural engineering. Automation speeds up the design process by reducing human error and shortening project deadlines. In this part, we'll look at how AI and machine learning are altering several elements of design automation.

- Generative Design: AI-powered generative design technologies may produce a plethora of design choices based on predefined criteria, saving engineers' time during the conceptual design phase.
- Parametric Design: ML algorithms may repeatedly improve design parameters to achieve optimum structural configurations while taking into account different restrictions.
- BIM Integration: Building Information Modeling (BIM) systems use AI to improve project stakeholder communication and deliver real-time information on structural changes.

Figure 2. Artificial intelligence (AI) integration in structural engineering

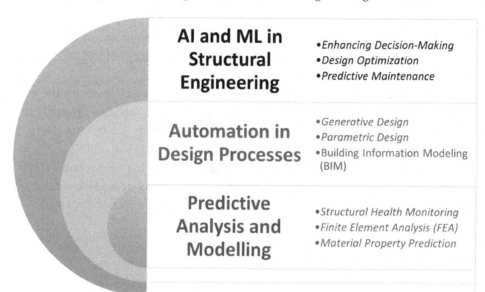

Modeling and Predictive Analysis

AI and ML excel in predictive analysis and modeling, allowing engineers to properly forecast structural behavior. These technologies allow the development of strong prediction models that take into account a variety of environmental variables and loads.

- Monitoring of Structural Health: ML models can predict structural health based on sensor data, allowing for early diagnosis of problems and assuring structural safety.
- Finite Element Analysis (FEA): AI-assisted FEA simulations provide more accurate and efficient findings while using less computing time and resources.
- Prediction of Material qualities: ML models can forecast material qualities and performance, assisting in the selection of the best materials for a project.

Finally, AI and machine learning have become crucial tools in structural engineering, propelling improvements in automation, optimization, and predictive analysis. Engineers may use these technologies to build buildings that are safer, more efficient, and more sustainable, eventually determining the future of the built environment. The next portion of this chapter will look into numerous optimization strategies in structural engineering enabled by AI and ML.

TECHNIQUES OF OPTIMIZATION

The goal of optimization in structural engineering is to identify the best feasible solution given a set of constraints and goals (Figure 3). This section presents an overview of the most generally used optimiza-

Figure 3. Various optimization techniques are used to tackle the core of structural engineering issues

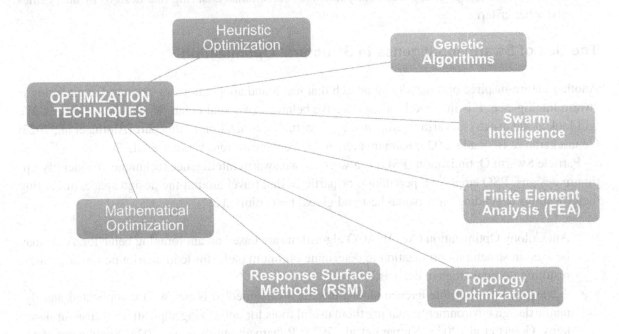

tion techniques in structural engineering (Boopathi, 2013; Haribalaji et al., 2021; Kalidas et al., 2012; Myilsamy& Sampath, 2017; Nishanth et al., 2023; Sampath et al., 2023).

- Mathematical Optimization: To optimize structural designs, this conventional method uses mathematical models and techniques such as linear programming, nonlinear programming, and integer programming.
- Heuristic Optimization: Unlike mathematical optimization, heuristic approaches do not guarantee an optimum solution but are typically better appropriate for difficult issues. Genetic algorithms, particle swarm optimization, simulated annealing, and other techniques are among them.

Structural Design Using Genetic Algorithms

Genetic algorithms (GAs) are a strong heuristic optimization tool influenced by natural selection and genetic principles. GAs have gained significance in structural engineering because to their capacity to effectively explore a large design area. This section looks at how GAs may be used to optimize structural designs (Boopathi et al., 2022; Boopathi & Sivakumar, 2013; Hussain & Srimathy, 2023; Sampath et al., 2022).

- Chromosome Representation: GAs model structural designs as chromosomes made up of genes that encode design parameters. Mutation and crossover operators are used to develop and enhance the population of designs.
- Objective Functions: GAs use objective functions to assess the performance of designs, taking into account criteria such as structural integrity, cost, and sustainability. Based on these goals, the program iteratively refines designs.

- Constraint Handling: GAs successfully manages constraints, ensuring that designs fit safety rules and other criteria.

The Use of Swarm Intelligence in Structural Optimization

Another nature-inspired optimization approach that has found applications in structural engineering is swarm intelligence. It is inspired by the collective behavior of social creatures like ants and bees. We will look at the usage of swarm intelligence for structure optimization in this part (Anitha et al., 2023; Gunasekaran & Boopathi, 2023; Kumar et al., 2023; Venkateswaran, Vidhya, et al., 2023).

Particle Swarm Optimization (PSO) is a well-known swarm intelligence technique. To identify optimum designs, PSO employs a population of particles that travel around the design space, modifying their locations depending on personal best and global best solutions.

- Ant Colony Optimization (ACO): ACO algorithms are based on ant foraging behavior. ACO may be used in structural optimization to determine optimum paths for load distribution in structures, resulting in more efficient designs.
- Advantages: Swarm intelligence algorithms are well-suited to issues with complicated and dynamic design environments, making them useful tools for optimizing sophisticated structural systems. Gowri et al., 2023; Kumara et al., 2023; Rahamathunnisa et al., 2023; Samikannu et al., 2022; Vennila et al., 2022).
- Finite Element Analysis (FEA): FEA is a popular numerical method for tackling complicated structural problems. It breaks down a structure into smaller, finite-sized parts, enabling engineers to examine stress, deformation, and other properties. In FEA, optimization entails changing parameters (such as material qualities, element sizes, or boundary conditions) to obtain the required structural performance while reducing weight or cost.

Topology optimization is a method for optimizing material distribution within a particular design area to fulfill specified goals, such as lowering weight while retaining structural integrity. By automating the process of creating and assessing design options, AI and ML may improve topology optimization.

Response Surface Methods (RSM): RSM is a statistical approach that uses a smaller mathematical model to mimic the behavior of a complicated system, such as a structural model. It is often used in tandem with optimization techniques to lower the computational cost of finding ideal designs.

Simulated Annealing: A probabilistic optimization approach inspired by the metallurgical annealing process. It explores the design space by accepting or rejecting modifications to the design on a probabilistic basis. It has the potential to be useful in finding near-optimal solutions to complicated structural optimization issues.

Gradient-Based Optimization: Gradient-based optimization techniques lead the search for optimum solutions by using gradients (derivatives) of the objective function. When the objective function is smooth and differentiable, several approaches, such as gradient descent and conjugate gradient methods, are often utilized.

Multi-Objective Optimization: Multiple competing goals, such as reducing cost and optimizing strength, are typical in structural analysis. The Pareto front is the collection of solutions found by multi-objective optimization methods that reflect trade-offs between multiple goals.

Nonlinear Programming: Nonlinear programming techniques are used when there are nonlinearities in structural analysis issues, such as material nonlinearity or geometric nonlinearity. To deal with such complications, algorithms such as Sequential Quadratic Programming (SQP) might be utilized.

Discrete Optimization: Design variables in certain structural design issues must have discrete values, such as choosing conventional beam widths or bolt diameters. In such instances, discrete optimization methods such as integer programming or evolutionary algorithms with integer coding are appropriate.

Surrogate models, which are often generated using machine learning approaches such as Gaussian Process Regression, may be used to approximate the objective function. Surrogates speed up optimization by lowering the amount of time-consuming structural simulations.

Reinforcement Learning (RL): This machine learning branch has found applications in structural optimization. By interacting with a structural model over time and altering designs based on learnt incentives and penalties, RL agents may learn to make design choices.

FRAMEWORK FOR PERFORMANCE COMPARISON

The Value of Performance Metrics

In the field of structural engineering, evaluating the performance of various design choices is critical. The best design must take into account a variety of issues, ranging from safety and cost-effectiveness to environmental impact. It is critical for engineers to provide a rigorous performance comparison framework in order to make educated judgments when picking design alternatives (Hussain & Srimathy, 2023; Jeevanantham et al., 2022; Koshariya et al., 2023).

- Multifaceted Decision-Making: Performance measures enable a comprehensive assessment of structural designs while taking into account a variety of criteria and goals.
- Improved Sustainability: A thorough framework guarantees that sustainability issues are included into decision-making, resulting in more ecologically responsible designs.

Evaluation of Safety

In structural engineering, safety is a primary consideration. Evaluating a design alternative's safety entails examining its capacity to endure loads, comply to safety norms and laws, and successfully manage hazards (Hanumanthakari et al., 2023; Sengeni et al., 2023; Ugandar et al., 2023).

- Structural Integrity: Analyzing the structural integrity of a design to ensure it can handle predicted loads and circumstances without failure is part of the safety assessment process.
- Risk Analysis: Engineers use risk analysis methods to detect possible hazards and weaknesses in a design, then take proactive risk-mitigation steps.

Cost-Effectiveness Analysis

The cost-effectiveness of any structural engineering project is crucial. Evaluating cost-effectiveness entails taking into account a design's life cycle costs, which include initial construction costs, maintenance costs, and possible future upgrades or replacements (Babu et al., 2022; Mohanty et al., 2023).

- Life Cycle Cost Analysis (LCCA): LCCA determines the most cost-effective design solution by taking both short-term and long-term expenses into account.
- Value Engineering: Engineers investigate design options that preserve or improve performance while decreasing costs, resulting in optimum cost-effectiveness.

Evaluating the Environmental Impact

In recent years, there has been a lot of focus on the environmental effect of structural designs. A carbon footprint, energy efficiency, and overall contribution to sustainability are all evaluated in an environmental impact assessment (Boopathi, 2022c, 2022b; Boopathi, Alqahtani, et al., 2023; Gowri et al., 2023; Maguluri et al., 2023).

- Carbon Footprint: Assessing a design's carbon footprint may assist find possibilities to reduce its environmental effect.
- Sustainable Materials: To lower the environmental imprint of buildings, engineers examine the usage of sustainable materials, energy-efficient systems, and environmentally friendly construction processes.

A well-defined performance comparison framework is essential for complete structural design assessment, with safety, cost-effectiveness, and sustainability as the top priorities. This chapter will look at sustainable design concepts and how they may be integrated into intelligent structural engineering processes utilizing AI and ML.

PRINCIPLES OF SUSTAINABLE DESIGN

Structural Engineering Sustainable Design

In the discipline of structural engineering, sustainability has become a major concern. Sustainable design principles encourage the creation of buildings that have a low environmental effect, use less resources, and promote long-term resilience. This section looks at how to include sustainability into structural engineering (Kavitha et al., 2023; Srinivas et al., 2023; Venkateswaran, Kumar, et al., 2023).

- Holistic Approach: Sustainable design comprises a holistic approach that takes into account environmental, economic, and social considerations, with the goal of striking a balance that benefits both the built and natural surroundings.

- Long-Term Perspective: Sustainable buildings are built with lifespan in mind, seeking to reduce the need for costly repairs and replacements.

Considerations for Energy Efficiency

Energy efficiency is an important feature of sustainable design, especially since buildings and infrastructure use a lot of energy. Energy-efficient solutions not only lower operating expenses but also have a positive influence on the environment (Hussain & Srimathy, 2023; Ingle et al., 2023; Kumara et al., 2023; Syamala et al., 2023).

- Passive Design: Passive design principles improve the orientation, insulation, and utilization of natural illumination of a building to reduce the need for heating, cooling, and artificial lighting.
- Renewable Energy Integration: To minimize dependence on non-renewable energy sources, sustainable architecture often include renewable energy sources such as solar panels, wind turbines, or geothermal systems.
- Smart Building Systems: Energy consumption may be reduced by altering lighting, HVAC, and other systems depending on occupancy and environmental conditions.

Material Selection for Longevity

The materials used in building have a considerable influence on the sustainability of a structure. Sustainable material selection strives to limit resource depletion, waste, and promote environmentally friendly manufacturing processes (Boopathi, Umareddy, et al., 2023; Boopathi & Davim, 2023b, 2023a; Fowziya et al., 2023).

- Recycled and Recyclable Materials: The use of recycled and recyclable materials decreases a structure's environmental imprint and fosters a circular economy.
- Low-effect Materials: Choosing materials with a reduced environmental effect, such as those with fewer embodied energy or emissions during manufacture, is common in sustainable design.
- Local Sourcing: Purchasing materials locally decreases transportation-related emissions while also benefiting the local economy.
- Life Cycle Assessment (LCA): LCA assesses the environmental effect of materials throughout their life cycle, from extraction and manufacture to use and disposal, allowing for more informed material selection choices.

In structural engineering, sustainable design concepts are critical for producing ecologically responsible, energy-efficient, and economically viable buildings. The incorporation of AI and ML technology improves structural design sustainability concerns. The next part will look at how AI and machine learning may be used to accomplish sustainable design objectives in autonomous design environments.

SUSTAINABLE AUTONOMOUS DESIGN WITH AI AND ML

Sustainability is an important factor in contemporary structural engineering, and incorporating it into AI and ML models is key for producing genuinely autonomous, sustainable designs (Figure 4). This section dives into the incorporation of sustainability into artificial intelligence and machine learning models for structural engineering (Boopathi & Kanike, 2023; Maheswari et al., 2023; Ramudu et al., 2023; Syamala et al., 2023).

Figure 4. AI and machine learning in sustainable autonomous design

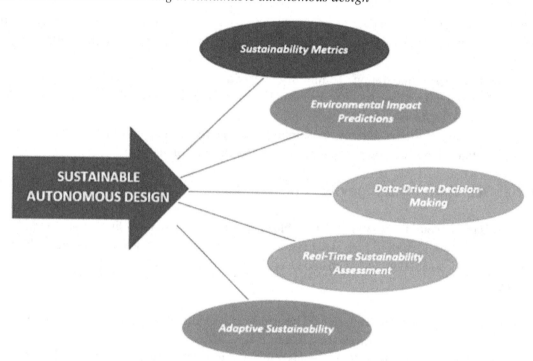

Metrics for Sustainability

Sustainability measurements and goals may be included into the design optimization process using AI and ML models. Defining particular sustainability targets, such as lowering carbon emissions, limiting resource consumption (e.g., materials and energy), or improving energy efficiency, is part of this process. The AI-ML system then incorporates these goals into the optimization process. The system analyzes design choices throughout the design process based on how well they match these sustainability parameters.

For example, while constructing a building, an AI-ML system may take into account numerous elements such as material selection, insulation, HVAC systems, and renewable energy sources to optimize the design for decreased energy consumption, fewer carbon emissions, and overall sustainability.

Predictions of Environmental Impact

By examining historical data and developing data-driven forecasts, ML systems can anticipate the environmental effect of various design options. For example, if engineers are examining several materials for a building project, ML models may assess the environmental impact of each material option. Embodied carbon, energy usage, and water use during manufacturing and transportation are all aspects to consider (Boopathi, 2022b; Maguluri et al., 2023; Sampath, 2021; Venkateswaran, Kumar, et al., 2023).

Engineers may make more informed judgments by evaluating environmental effect and selecting materials and designs that minimize negative environmental repercussions, so contributing to a more environmentally friendly building process.

Data-Informed Decision-Making

AI and ML models excel at analyzing large datasets, which is very useful for evaluating sustainability aspects. These models are capable of processing data on energy consumption, emissions, resource usage, and other pertinent sustainability factors. They discover patterns and linkages in data that would be missed by conventional analysis.

Engineers acquire a better knowledge of how various design decisions affect the environment thanks to data-driven insights. During the design phase, they may discover areas for improvement to increase sustainability and make better informed judgments.

Real-Time Sustainability Evaluation

AI-ML technologies may be used to continually monitor and analyze the sustainability performance of a structure during its entire existence. In real time, sensors and IoT devices may gather data about energy use, structural integrity, and environmental conditions. This data is processed by ML models to offer continual evaluations of the sustainability of a structure (Boopathi, 2022a; Boopathi, Kumar, et al., 2023; Boopathi & Davim, 2023b; Domakonda et al., 2022; Hussain & Srimathy, 2023; Samikannu et al., 2022; Sampath, 2021).

The technology identifies deviations from sustainability objectives or environmental factors, recommending modifications or upgrades such as adjusting HVAC settings or adding thermal efficiency insulation, ensuring buildings stay sustainable as circumstances change over time.

Sustainability via Adaptation

Artificial intelligence and machine learning algorithms may modify designs in real time to environmental circumstances or sustainability objectives, such as altering heating and cooling settings if a building's energy consumption exceeds established criteria. Engineers may attain sustainability goals autonomously and adaptively by including sustainability into AI-ML models. This section looks at case studies and practical examples of AI and ML-driven autonomous design that achieves higher structural performance while reducing resource use and encouraging sustainability.

EXAMPLES OF CASE STUDIES

Case Study No. 1: The AI-Enhanced Skyscraper

A project in a large metropolitan region sought to create a high-rise skyscraper with maximum useable space, energy efficiency, and environmental sustainability. AI-driven generative design techniques were used to investigate millions of design options, taking into account elements such as structural stability, wind load resistance, and natural illumination.

The AI model developed designs that optimized the geometry and architecture of the building to reduce energy usage for heating, cooling, and lighting. Furthermore, the model took into account sustainable materials with minimal embodied energy and carbon emissions. The final design lowered the building's carbon footprint by 30% compared to standard designs while increasing useable area by 10%.

Case Study No. 2: Bridge Rehabilitation

Engineers employed machine learning (ML) algorithms to examine the structural health of an old bridge, revealing regions of degradation and stress concentration. Based on historical data and environmental variables, AI-ML models anticipated future structural flaws. When compared to conventional approaches, the system proposed targeted repairs and retrofitting, resulting in a 20% cost reduction. The bridge now has more structural stability and will last longer.

Case Study 3: Infrastructure for Sustainable Transportation

A government body in charge of transportation infrastructure wants to lower its projects' carbon impact. To improve the construction of road networks and public transportation systems, AI-ML models were employed to examine traffic patterns, vehicle emissions, and environmental data.

The AI-ML system suggested techniques such as traffic flow optimization, integrating electric car charging stations, and using environmentally friendly road materials. As a consequence, the agency cut emissions by 15%, improved traffic flow, and promoted environmentally friendly transportation choices.

These case studies show how AI and ML-driven autonomous design may be used in structural engineering. Engineers may optimize designs, minimize resource consumption, and promote sustainability by using these technologies, resulting in a more ecologically aware and economically viable built environment.

BENEFITS AND DIFFICULTIES

The Benefits of AI-ML in Structural Engineering

The use of AI and machine learning technology into structural engineering has several benefits (Liu et al., 2004; Möhring et al., 2020; Shea & Smith, 2005).

- Efficiency and speed: AI and machine learning allow the automation of design processes, decreasing the time necessary to investigate design possibilities, run simulations, and evaluate data.

- Optimization: Because these technologies excel at optimization, engineers may identify more efficient and cost-effective structural solutions in complicated design areas.
- Predictive Analysis: Machine learning algorithms may reliably forecast structural behavior, assisting engineers in identifying possible problems early in the design process.
- Sustainability: AI-ML models can effortlessly incorporate sustainability concerns into designs, supporting environmentally friendly and energy-efficient solutions.
- Cost reductions: AI-driven designs often result in cost reductions in material utilization, construction, and maintenance.
- Improved Safety: AI may help with structural health monitoring, highlighting safety risks and structural faults that conventional inspections may overlook.

Ethical Issues and Difficulties

AI and machine learning provide considerable benefits in structural engineering, but they also create problems and ethical concerns (Liu et al., 2004; H. Sun et al., 2021).

- Data Quality and Bias: The accuracy and representativeness of training data are critical to the performance of AI-ML models. skewed or inadequate data might result in skewed results.
- Complexity: Using AI and machine learning in structural engineering involves competence in both engineering and data science, which may be a hurdle for certain professions.
- Privacy and Security: Because it entails collecting and analyzing data from sensors and IoT devices, the application of AI in structural health monitoring presents privacy and security problems.
- Ethical Decision-Making: When it comes to judgments affecting safety, the environment, or social well-being, autonomous AI-ML systems may present ethical concerns.
- Regulatory Compliance: When applying AI-ML in engineering, compliance with current norms and standards may be difficult, particularly in highly regulated sectors.
- Interpretability: Because some AI-ML models are sophisticated and difficult to comprehend, engineers struggle to grasp and trust their suggestions.

To guarantee appropriate and successful use of AI and ML in structural engineering, thorough planning, ethical frameworks, and constant monitoring are required, balancing advantages with obstacles and ethical issues.

TRENDS AND DIRECTIONS FOR THE FUTURE

Technologies in Development

Emerging technologies are likely to enhance structural engineering greatly (Huang & Fu, 2019; Pan & Zhang, 2021; L. Sun et al., 2020b).

- Digital Twins: The creation of digital twins, which are virtual duplicates of actual buildings, will continue. These digital copies enable real-time structural monitoring, analysis, and predictive maintenance.

- Quantum Computing: Quantum computing has the potential to solve difficult structural engineering problems at previously unimaginable rates, revolutionizing optimization and simulation procedures.
- Nanotechnology: Nanomaterials and nanosensors will allow the development of structural components that are stronger, more robust, and self-monitoring.
- 3D Printing: Advances in 3D printing technology will broaden the scope of quick prototyping and on-site building, minimizing material waste.
- Smart Materials: Using smart materials that can respond to changing environments would improve structural performance and safety.

Future Research Directions

As structural engineering evolves, several attractive research pathways are expected to determine its future.

- Resilience and Disaster Preparedness: Research into building buildings that can survive natural catastrophes such as earthquakes and hurricanes will continue to be a priority.
- Sustainable Materials: Exploring and developing sustainable materials with minimal environmental effect will be a top emphasis in order to achieve more environmentally friendly constructions.
- AI and Ethics: There will be a greater emphasis on research on ethical principles and frameworks for the proper use of AI and ML in structural engineering.
- Human-AI Collaboration: A key topic of research will be how engineers can successfully cooperate with AI systems to improve decision-making and design processes.
- Autonomous building: Research into completely autonomous building processes, such as robotic assembly and 3D printing, will help develop the construction industry.
- Adaptation to Climate Change: Addressing the influence of climate change on structural design and finding adaptable solutions will be a major problem.
- Biologically Inspired Design: Researchers will look into biological systems for inspiration in structural design, such as emulating the strength and efficiency of natural materials.

Structural engineering is projected to change in response to technology advances, sustainability, and the need for more durable structures, with researchers and professionals playing an important role in driving these improvements.

CONCLUSION

This chapter digs into the revolutionary significance of AI and ML in structural engineering, emphasizing its importance in structure design, optimization, and assessment.

- AI and machine learning play critical roles in automating design processes, predicting structural behavior, and optimizing configurations.
- The use of AI-ML to provide a performance comparison framework that allows engineers to completely analyze safety, cost-effectiveness, and environmental impact.

- The incorporation of sustainability concepts into AI-ML models, allowing for the production of ecologically responsible, energy-efficient designs.
- Real-world case studies demonstrating how AI-ML-driven autonomous design has been used to produce improved structural performance while reducing resource consumption and boosting sustainability.
- The multiple advantages of AI-ML in structural engineering, such as increased efficiency, cost savings, and improved safety, as well as the hurdles and ethical issues associated with their adoption.
- Future trends and research directions, such as digital twins and quantum computing, as well as research fields such as climate change adaptation and human-AI cooperation.

AI and machine learning are transforming structural engineering, allowing for more durable, sustainable, and efficient structures. However, for long-term success and a good influence on the built environment, responsible and ethical usage is essential.

REFERENCES

Anitha, C., Komala, C., Vivekanand, C. V., Lalitha, S., & Boopathi, S. (2023). Artificial Intelligence driven security model for Internet of Medical Things (IoMT). *IEEE Explore*, (pp. 1–7). IEEE.

Babu, B. S., Kamalakannan, J., Meenatchi, N., Karthik, S., & Boopathi, S. (2022). Economic impacts and reliability evaluation of battery by adopting Electric Vehicle. *IEEE Explore*, (pp. 1–6). IEEE.

Boopathi, S. (2013). *Experimental study and multi-objective optimization of near-dry wire-cut electrical discharge machining process* [PhD Thesis]. http://hdl.handle.net/10603/16933

Boopathi, S. (2022a). An extensive review on sustainable developments of dry and near-dry electrical discharge machining processes. *ASME: Journal of Manufacturing Science and Engineering*, *144*(5), 050801–1.

Boopathi, S. (2022b). An investigation on gas emission concentration and relative emission rate of the near-dry wire-cut electrical discharge machining process. *Environmental Science and Pollution Research International*, *29*(57), 86237–86246. doi:10.1007/s11356-021-17658-1 PMID:34837614

Boopathi, S. (2022c). Cryogenically treated and untreated stainless steel grade 317 in sustainable wire electrical discharge machining process: A comparative study. *Environmental Science and Pollution Research*, 1–10. Springer.

Boopathi, S., Alqahtani, A. S., Mubarakali, A., & Panchatcharam, P. (2023). Sustainable developments in near-dry electrical discharge machining process using sunflower oil-mist dielectric fluid. *Environmental Science and Pollution Research International*, 1–20. doi:10.1007/s11356-023-27494-0 PMID:37199846

Boopathi, S., & Davim, J. P. (2023a). Applications of Nanoparticles in Various Manufacturing Processes. In *Sustainable Utilization of Nanoparticles and Nanofluids in Engineering Applications* (pp. 1–31). IGI Global. doi:10.4018/978-1-6684-9135-5.ch001

Boopathi, S., & Davim, J. P. (2023b). *Sustainable Utilization of Nanoparticles and Nanofluids in Engineering Applications*. IGI Global. doi:10.4018/978-1-6684-9135-5

Boopathi, S., & Kanike, U. K. (2023). Applications of Artificial Intelligent and Machine Learning Techniques in Image Processing. In *Handbook of Research on Thrust Technologies' Effect on Image Processing* (pp. 151–173). IGI Global. doi:10.4018/978-1-6684-8618-4.ch010

Boopathi, S., Kumar, P. K. S., Meena, R. S., Sudhakar, M., & Associates. (2023). Sustainable Developments of Modern Soil-Less Agro-Cultivation Systems: Aquaponic Culture. In Human Agro-Energy Optimization for Business and Industry (pp. 69–87). IGI Global.

Boopathi, S., & Sivakumar, K. (2013). Experimental investigation and parameter optimization of near-dry wire-cut electrical discharge machining using multi-objective evolutionary algorithm. *International Journal of Advanced Manufacturing Technology, 67*(9–12), 2639–2655. doi:10.1007/s00170-012-4680-4

Boopathi, S., Sureshkumar, M., & Sathiskumar, S. (2022). Parametric Optimization of LPG Refrigeration System Using Artificial Bee Colony Algorithm. *International Conference on Recent Advances in Mechanical Engineering Research and Development*, (pp. 97–105). IEEE.

Boopathi, S., Umareddy, M., & Elangovan, M. (2023). Applications of Nano-Cutting Fluids in Advanced Machining Processes. In *Sustainable Utilization of Nanoparticles and Nanofluids in Engineering Applications* (pp. 211–234). IGI Global. doi:10.4018/978-1-6684-9135-5.ch009

Dolšak, B., & Novak, M. (2011). Intelligent decision support for structural design analysis. *Advanced Engineering Informatics, 25*(2), 330–340. doi:10.1016/j.aei.2010.11.001

Domakonda, V. K., Farooq, S., Chinthamreddy, S., Puviarasi, R., Sudhakar, M., & Boopathi, S. (2022). Sustainable Developments of Hybrid Floating Solar Power Plants: Photovoltaic System. In Human Agro-Energy Optimization for Business and Industry (pp. 148–167). IGI Global.

Fowziya, S., Sivaranjani, S., Devi, N. L., Boopathi, S., Thakur, S., & Sailaja, J. M. (2023). Influences of nano-green lubricants in the friction-stir process of TiAlN coated alloys. *Materials Today: Proceedings*. doi:10.1016/j.matpr.2023.06.446

Gowri, N. V., Dwivedi, J. N., Krishnaveni, K., Boopathi, S., Palaniappan, M., & Medikondu, N. R. (2023). Experimental investigation and multi-objective optimization of eco-friendly near-dry electrical discharge machining of shape memory alloy using Cu/SiC/Gr composite electrode. *Environmental Science and Pollution Research International, 30*(49), 1–19. doi:10.1007/s11356-023-26983-6 PMID:37126160

Gunasekaran, K., & Boopathi, S. (2023). Artificial Intelligence in Water Treatments and Water Resource Assessments. In *Artificial Intelligence Applications in Water Treatment and Water Resource Management* (pp. 71–98). IGI Global. doi:10.4018/978-1-6684-6791-6.ch004

Guo, K., Yang, Z., Yu, C.-H., & Buehler, M. J. (2021). Artificial intelligence and machine learning in design of mechanical materials. *Materials Horizons, 8*(4), 1153–1172. doi:10.1039/D0MH01451F PMID:34821909

Hanumanthakari, S., Gift, M. M., Kanimozhi, K., Bhavani, M. D., Bamane, K. D., & Boopathi, S. (2023). Biomining Method to Extract Metal Components Using Computer-Printed Circuit Board E-Waste. In *Handbook of Research on Safe Disposal Methods of Municipal Solid Wastes for a Sustainable Environment* (pp. 123–141). IGI Global. doi:10.4018/978-1-6684-8117-2.ch010

Haribalaji, V., Boopathi, S., & Asif, M. M. (2021). Optimization of friction stir welding process to join dissimilar AA2014 and AA7075 aluminum alloys. *Materials Today: Proceedings, 50,* 2227–2234. doi:10.1016/j.matpr.2021.09.499

Huang, Y., & Fu, J. (2019). Review on application of artificial intelligence in civil engineering. *Computer Modeling in Engineering & Sciences, 121*(3), 845–875. doi:10.32604/cmes.2019.07653

Hussain, Z., & Srimathy, G. (2023). *IoT and AI Integration for Enhanced Efficiency and Sustainability.*

Ingle, R. B., Senthil, T. S., Swathi, S., Muralidharan, N., Mahendran, G., & Boopathi, S. (2023). Sustainability and Optimization of Green and Lean Manufacturing Processes Using Machine Learning Techniques. IGI Global. doi:10.4018/978-1-6684-8238-4.ch012

Jeevanantham, Y. A., Saravanan, A., Vanitha, V., Boopathi, S., & Kumar, D. P. (2022). Implementation of Internet-of Things (IoT) in Soil Irrigation System. *IEEE Explore,* (pp. 1–5). IEEE.

Kalidas, R., Boopathi, S., Sivakumar, K., & Mohankumar, P. (2012). Optimization of Machining Parameters of WEDM Process Based On the Taguchi Method. *IJEST, 6*(1).

Kavitha, C. R., Varalatchoumy, M., Mithuna, H. R., Bharathi, K., Geethalakshmi, N. M., & Boopathi, S. (2023). Energy Monitoring and Control in the Smart Grid: Integrated Intelligent IoT and ANFIS. In M. Arshad (Ed.), (pp. 290–316). Advances in Bioinformatics and Biomedical Engineering. IGI Global. doi:10.4018/978-1-6684-6577-6.ch014

Koshariya, A. K., Kalaiyarasi, D., Jovith, A. A., Sivakami, T., Hasan, D. S., & Boopathi, S. (2023). AI-Enabled IoT and WSN-Integrated Smart Agriculture System. In *Artificial Intelligence Tools and Technologies for Smart Farming and Agriculture Practices* (pp. 200–218). IGI Global. doi:10.4018/978-1-6684-8516-3.ch011

Kumar, P. R., Meenakshi, S., Shalini, S., Devi, S. R., & Boopathi, S. (2023). Soil Quality Prediction in Context Learning Approaches Using Deep Learning and Blockchain for Smart Agriculture. In R. Kumar, A. B. Abdul Hamid, & N. I. Binti Ya'akub (Eds.), (pp. 1–26). Advances in Computational Intelligence and Robotics. IGI Global. doi:10.4018/978-1-6684-9151-5.ch001

Kumara, V., Mohanaprakash, T., Fairooz, S., Jamal, K., Babu, T., & Sampath, B. (2023). Experimental Study on a Reliable Smart Hydroponics System. In *Human Agro-Energy Optimization for Business and Industry* (pp. 27–45). IGI Global. doi:10.4018/978-1-6684-4118-3.ch002

Liu, S., Tomizuka, M., & Ulsoy, A. (2004). *Challenges and opportunities in the engineering of intelligent systems. Proc. of the 4th International Workshop on Structural Control*, New York.

Maguluri, L. P., Ananth, J., Hariram, S., Geetha, C., Bhaskar, A., & Boopathi, S. (2023). Smart Vehicle-Emissions Monitoring System Using Internet of Things (IoT). In Handbook of Research on Safe Disposal Methods of Municipal Solid Wastes for a Sustainable Environment (pp. 191–211). IGI Global.

Maheswari, B. U., Imambi, S. S., Hasan, D., Meenakshi, S., Pratheep, V., & Boopathi, S. (2023). Internet of Things and Machine Learning-Integrated Smart Robotics. In Global Perspectives on Robotics and Autonomous Systems: Development and Applications (pp. 240–258). IGI Global. doi:10.4018/978-1-6684-7791-5.ch010

Mohanty, A., Venkateswaran, N., Ranjit, P., Tripathi, M. A., & Boopathi, S. (2023). Innovative Strategy for Profitable Automobile Industries: Working Capital Management. In Handbook of Research on Designing Sustainable Supply Chains to Achieve a Circular Economy (pp. 412–428). IGI Global.

Möhring, H.-C., Müller, M., Krieger, J., Multhoff, J., Plagge, C., de Wit, J., & Misch, S. (2020). Intelligent lightweight structures for hybrid machine tools. *Production Engineering*, *14*(5-6), 583–600. doi:10.1007/s11740-020-00988-3

Myilsamy, S., & Sampath, B. (2017). Grey Relational Optimization of Powder Mixed Near-Dry Wire Cut Electrical Discharge Machining of Inconel 718 Alloy. *Asian Journal of Research in Social Sciences and Humanities*, *7*(3), 18–25. doi:10.5958/2249-7315.2017.00157.5

Nishanth, J., Deshmukh, M. A., Kushwah, R., Kushwaha, K. K., Balaji, S., & Sampath, B. (2023). Particle Swarm Optimization of Hybrid Renewable Energy Systems. In *Intelligent Engineering Applications and Applied Sciences for Sustainability* (pp. 291–308). IGI Global. doi:10.4018/979-8-3693-0044-2.ch016

Pan, Y., & Zhang, L. (2021). Roles of artificial intelligence in construction engineering and management: A critical review and future trends. *Automation in Construction*, *122*, 103517. doi:10.1016/j.autcon.2020.103517

Perry, B. J., Guo, Y., & Mahmoud, H. N. (2022). Automated site-specific assessment of steel structures through integrating machine learning and fracture mechanics. *Automation in Construction*, *133*, 104022. doi:10.1016/j.autcon.2021.104022

Rahamathunnisa, U., Sudhakar, K., Murugan, T. K., Thivaharan, S., Rajkumar, M., & Boopathi, S. (2023). Cloud Computing Principles for Optimizing Robot Task Offloading Processes. In *AI-Enabled Social Robotics in Human Care Services* (pp. 188–211). IGI Global. doi:10.4018/978-1-6684-8171-4.ch007

Ramudu, K., Mohan, V. M., Jyothirmai, D., Prasad, D., Agrawal, R., & Boopathi, S. (2023). Machine Learning and Artificial Intelligence in Disease Prediction: Applications, Challenges, Limitations, Case Studies, and Future Directions. In Contemporary Applications of Data Fusion for Advanced Healthcare Informatics (pp. 297–318). IGI Global.

Salehi, H., & Burgueño, R. (2018). Emerging artificial intelligence methods in structural engineering. *Engineering Structures*, *171*, 170–189. doi:10.1016/j.engstruct.2018.05.084

Samikannu, R., Koshariya, A. K., Poornima, E., Ramesh, S., Kumar, A., & Boopathi, S. (2022). Sustainable Development in Modern Aquaponics Cultivation Systems Using IoT Technologies. In *Human Agro-Energy Optimization for Business and Industry* (pp. 105–127). IGI Global.

Sampath, B. (2021). *Sustainable Eco-Friendly Wire-Cut Electrical Discharge Machining: Gas Emission Analysis*.

Sampath, B., Pandian, M., Deepa, D., & Subbiah, R. (2022). Operating parameters prediction of liquefied petroleum gas refrigerator using simulated annealing algorithm. *AIP Conference Proceedings*, *2460*(1), 070003. doi:10.1063/5.0095601

Sampath, B., Sasikumar, C., & Myilsamy, S. (2023). Application of TOPSIS Optimization Technique in the Micro-Machining Process. In IGI:Trends, Paradigms, and Advances in Mechatronics Engineering (pp. 162–187). IGI Global.

Sengeni, D., Padmapriya, G., Imambi, S. S., Suganthi, D., Suri, A., & Boopathi, S. (2023). Biomedical Waste Handling Method Using Artificial Intelligence Techniques. In *Handbook of Research on Safe Disposal Methods of Municipal Solid Wastes for a Sustainable Environment* (pp. 306–323). IGI Global. doi:10.4018/978-1-6684-8117-2.ch022

Shea, K., & Smith, I. (2005). Intelligent structures: A new direction in structural control. *Artificial Intelligence in Structural Engineering: Information Technology for Design, Collaboration, Maintenance, and Monitoring*, 398–410.

Soh, C.-K., & Soh, A.-K. (1988). Example of intelligent structural design system. *Journal of Computing in Civil Engineering*, 2(4), 329–345. doi:10.1061/(ASCE)0887-3801(1988)2:4(329)

Srinivas, B., Maguluri, L. P., Naidu, K. V., Reddy, L. C. S., Deivakani, M., & Boopathi, S. (2023). Architecture and Framework for Interfacing Cloud-Enabled Robots. In *Handbook of Research on Data Science and Cybersecurity Innovations in Industry 4.0 Technologies* (pp. 542–560). IGI Global. doi:10.4018/978-1-6684-8145-5.ch027

Sun, H., Burton, H. V., & Huang, H. (2021). Machine learning applications for building structural design and performance assessment: State-of-the-art review. *Journal of Building Engineering*, *33*, 101816. doi:10.1016/j.jobe.2020.101816

Sun, L., Shang, Z., Xia, Y., Bhowmick, S., & Nagarajaiah, S. (2020a). Review of bridge structural health monitoring aided by big data and artificial intelligence: From condition assessment to damage detection. *Journal of Structural Engineering*, *146*(5), 04020073. doi:10.1061/(ASCE)ST.1943-541X.0002535

Sun, L., Shang, Z., Xia, Y., Bhowmick, S., & Nagarajaiah, S. (2020b). Review of bridge structural health monitoring aided by big data and artificial intelligence: From condition assessment to damage detection. *Journal of Structural Engineering*, *146*(5), 04020073. doi:10.1061/(ASCE)ST.1943-541X.0002535

Syamala, M., Komala, C., Pramila, P., Dash, S., Meenakshi, S., & Boopathi, S. (2023). Machine Learning-Integrated IoT-Based Smart Home Energy Management System. In *Handbook of Research on Deep Learning Techniques for Cloud-Based Industrial IoT* (pp. 219–235). IGI Global. doi:10.4018/978-1-6684-8098-4.ch013

Ugandar, R. E., Rahamathunnisa, U., Sajithra, S., Christiana, M. B. V., Palai, B. K., & Boopathi, S. (2023). Hospital Waste Management Using Internet of Things and Deep Learning: Enhanced Efficiency and Sustainability. In M. Arshad (Ed.), (pp. 317–343). Advances in Bioinformatics and Biomedical Engineering. IGI Global. doi:10.4018/978-1-6684-6577-6.ch015

Venkateswaran, N., Kumar, S. S., Diwakar, G., Gnanasangeetha, D., & Boopathi, S. (2023). Synthetic Biology for Waste Water to Energy Conversion: IoT and AI Approaches. In M. Arshad (Ed.), (pp. 360–384). Advances in Bioinformatics and Biomedical Engineering. IGI Global. doi:10.4018/978-1-6684-6577-6.ch017

Venkateswaran, N., Vidhya, K., Ayyannan, M., Chavan, S. M., Sekar, K., & Boopathi, S. (2023). A Study on Smart Energy Management Framework Using Cloud Computing. In 5G, Artificial Intelligence, and Next Generation Internet of Things: Digital Innovation for Green and Sustainable Economies (pp. 189–212). IGI Global. doi:10.4018/978-1-6684-8634-4.ch009

Vennila, T., Karuna, M., Srivastava, B. K., Venugopal, J., Surakasi, R., & Sampath, B. (2022). New Strategies in Treatment and Enzymatic Processes: Ethanol Production From Sugarcane Bagasse. In Human Agro-Energy Optimization for Business and Industry (pp. 219–240). IGI Global.

Westermayr, J., Gastegger, M., Schütt, K. T., & Maurer, R. J. (2021). Perspective on integrating machine learning into computational chemistry and materials science. *The Journal of Chemical Physics*, *154*(23), 230903. doi:10.1063/5.0047760 PMID:34241249

Chapter 18
The Evolution of AI and Data Science

A. S. Anurag

ⓘ https://orcid.org/0009-0005-5790-9084

Central University of Kerala, India

ABSTRACT

The history of artificial intelligence (AI) and data science has their origins in the 1940s and 1950s respectively. However, it has been through many changes throughout its history. AI is a vast and fascinating subject. There are many more elements to discover and understand. This chapter aims to outline the history of AI and data science, from its origin to its current developments. It will also explore the ethical considerations within AI and data science, such as bias and fairness, transparency, data privacy, etc. In the end, the chapter sheds light on the ethical concerns regarding the implementation of AI and the security concerns that data science poses. The chapter also provides insights into the role of individuals, government, and society in mitigating these issues. This chapter aims to furnish the reader with the scientific foundation and essential understanding required for embarking on the journey to comprehend the realm of artificial intelligence and data science.

INTRODUCTION

Modern society is heading towards a remarkable technological explosion fuelled by Artificial Intelligence (AI). Many scientists believe that we will soon witness a sudden technological leap, leading to what is commonly referred to as technological singularity[1] and the emergence of superintelligence[2]. For instance, experts anticipate breakthroughs in areas like natural language processing and data science that could significantly surpass current capabilities. This rapid technological advancement signifies the growth that may surpass human control, a notion raising profound concerns among researchers and engineers. While superintelligence and singularity still may be the subjects of fiction, the capabilities of our current AI technologies are not to be underestimated. The ethical concerns they raise in today's context demand the utmost attention and corrective measures. This chapter aims to delve into the current ethical challenges of AI deployment, using concrete examples, and address the privacy and security risks associated with data science.

DOI: 10.4018/979-8-3693-2964-1.ch018

Throughout history, human beings have gone through several technological advancements. The invention of the wheel, metal, electricity, and the computer and internet. These technological advancements brought prominent changes in our lifestyles. The most influential and latest of all technologies is AI. AI is the branch of Computer Science that deals with systems capable of solving problems and adapting to new environments by learning, understanding, and applying knowledge from past data. AI plays a crucial part in the modern society. Our lives depend on AI, from virtual assistants like Siri, Alexa, and Google Assistant to driverless cars and disease diagnosis. AI technology is a relatively newer technology in the field of computer science. The history of AI development has had several ups and downs and fallbacks. However, the true breakthrough came with the emergence of Machine Learning (ML). It enabled systems to learn from past data and make predictions without explicit programming. This shift began a new era where AI systems could understand and improve their performance over time. On the other hand, data science evolved alongside AI as a separate branch. The increasing volume and complexity of data generated by various sources propelled its growth. Data Science is an interdisciplinary field that collects, screens, and analyzes structured or unstructured data to extract essential information that helps solve various problems. During the initial stages, data science was based on statistics and data analysis, which evolved to include machine learning, big data, and advanced analytics. The beginning of AI and Data Science dates back to the mid-1900s when the concept of machine learning and data analysis began to take shape. Over the years, computing power and increased data availability have propelled AI and data science to new heights.

AI and data science represent two distinct branches within the realm of computer science, both emerging and evolving nearly during the same timeframe. These two fields have developed together, attaining prominence as pivotal technological advancements. The interconnection between Data Science and AI is evident in the present scenario, with AI benefiting from enhanced capabilities through data analysis. Conversely, these refined iterations of AI contribute to optimizing Data Science processes, thereby fostering a symbiotic relationship between the two domains.

BACKGROUND AND LITERATURE REVIEW

Before advancing further, it is essential to have a foundational understanding of AI and Data Science. This section guides readers through crucial moments and significant advancements, establishing the groundwork for a thorough understanding of these fields' intricate paths. This section will explore scholarly works, research papers, and other authoritative sources related to AI and Data Science. By reviewing these diverse data sources, readers can better understand the foundational concepts underlying AI and Data science. By acknowledging the varied origins, contextual intricacies, and essential discussions within AI and Data Science, this section seeks to cultivate a nuanced viewpoint, recognizing the unique contributions that have shaped today's technological innovation landscape.

Artificial Intelligence

The idea of Artificial Intelligence is as old as humanity itself. We have almost always dreamt of making something that can match our intelligence. So, marking the exact start of AI in the historical timeline takes time. However, AI was officially recognized as a research discipline in 1956 (Moor, 2006). Computer Scientist John McCarthy coined the term 'Artificial Intelligence' at the 1956 summer conference held in Dartmouth (Cordeschi, 2007). He is also considered the father of Artificial Intelligence. However, almost a decade before that, mathematician Alan Turing had the idea of Artificial Intelligence. Even though Turing mainly used the term 'Machine intelligence' to refer to the same idea.

While at Government Code and Cypher School, Bletchley Park, Turing authored several unofficial papers on Intelligent Machines and circulated them among his colleagues (Copeland, 2000). In 1947, Turing gave the first-ever public lecture on Machine Intelligence, discussing the possibility of a machine that can learn from experience and alter its program to suit its environment best (Dawson, Jr., 2007). In his famous paper 'Computing Machinery and Intelligence' published in 1950, Turing asked, 'Can Machines Think?'. He also suggested a test to measure a machine's ability to exhibit intelligent behavior equivalent to or indistinguishable from a human's (Turing, 2012). This test is most famously known as the imitation game (Turing, 2012) or the Turing Test. The Imitation Game comprises three integral participants: an interrogator and two respondents; one is a machine, and the other is a human. The objective is for the interrogator to identify which of the two participants is the machine (Turing, 2012). The interrogator will have conversations with the participants to make the identification. Nevertheless, the conversation will be only through text, with an arrangement like a monitor and a keyboard (Turing, 2012). If the interrogator cannot differentiate between the machine and the human respondent after the conversations, the machine is deemed to have successfully passed the game (Turing, 2012). The Loebner Prize[3] utilizes the Imitation Game as its foundational testing mechanism (Powers, 1998). Turing's ideas on computation and thinking machines have influenced the field of AI. These ideas laid the groundwork for developing technologies such as Artificial Intelligence, Machine Learning, Natural language Processing, etc.

After Alan Turing and John McCarthy laid the foundation for Artificial Intelligence, the successive years witnessed several groundbreaking innovations in the field, such as the invention of the first Neural Network program, the 'Perceptron' in 1957 by Frank Rosenblatt. It paved the foundation for new technologies such as Artificial Neural Networks, Machine Learning, Pattern Recognition, etc (Wang, 2018). The history of AI has been a rollercoaster ride filled with sudden developments and downfalls. Between 1950 and 1970, AI developments skyrocketed, yet in the 1980s, we saw a downfall in AI research funding, which led to an AI winter[4]. The AI winter in the 1980s was a result of several factors, such as critical reviews of AI progress, which led to concerns about its impact on privacy, improper addressing of the Expert System technology that was developed during this period, changes in the computer market, etc. (Hendler, 2008).

Data Science

Statistics is one of the significant tools used to make informed decisions and data-based predictions in various sectors. Statistics is the field of study that deals with evaluating hypotheses based on data analysis (SEP, 2014) and using the findings to forecast and plan the future. During the earlier times, mathematical models were used in statistics. Later, with the invention of supercomputers and the internet, data availability increased exponentially. Vast volumes of data can be accessed within seconds with the help of highly evolved computing power and internet facilities. This made conventional statistical methods inefficient, and scientists started to think about integrating computer power with statistics, paving the way for a new branch of computer science, later known as Data Science.

Like AI, the idea of Data Science was around even before the term was coined. In 1962, Mathematician and Statistician John Wilder Tukey published a paper in which he mentioned the term 'Data Analysis'.

Its description closely resembles the characteristics of modern Data Science (Donoho, 2017). There is still much debate regarding using the term 'Data Science' synonymously with 'Statistics.' Tukey was the one who envisioned statistics as a field of science rather than a field of mathematics. Tukey argued that unlike mathematics, which has theorems that need to be proved, data science helps in relating already existing knowledge to gain new insights (Donoho, 2017). In the current context, data science is not just the statistical analysis of data; it is a collection of different processes, including data generation, data collection and processing, data storage and management, data analysis, data visualization, and interpretation (Wing, 2019). The applications of Data Science are crucial in business areas such as marketing, finance, and operations. Data science is used in targeted marketing, online advertising, credit scoring, trading, fraud detection, and workforce management (Provost & Fawcett, 2013). Furthermore, significant retailers apply data science across their businesses, from marketing to supply-chain management. One of the significant applications of Data Science in AI is Machine Learning (ML). Machine Learning is the field of Artificial Intelligence that studies developing algorithms that help AI machines learn and enhance from experience. ML enables a system to learn from experience and improve based on the fundamental laws that impact the learning system in every field, such as statistics, computation, and information theory (Jordan & Mitchell, 2015). The recent advancements in ML and AI result from developments in Data Science and Data analytics techniques (CAO, 2017). The current opportunities in Data Science and the applications of Artificial Intelligence are in their early stages, offering considerable potential for further advancement. The escalating volume and ubiquity of data suggest a high likelihood of substantial growth in the processes and techniques of Data Science soon, along with corresponding expansions in Artificial Intelligence applications (Ong & Uddin, 2020). The use of Big Data, Machine Learning, Natural Language Processing, Artificial intelligence, and Data visualization technology in various fields is increasing daily (Hassani & Silva, 2023). One of the significant factors that impede the growth of Data Science currently is the need for more professionals (Kordon, 2020). However, introducing new AI technologies like Expert Systems and NLP compensates for this shortage by automating the data analysis processes.

Ethical Concerns in AI and Data Science

Ethics pertains to the unwritten rules and standards that guide human behavior within a society. As social beings, we judge some actions as appropriate or inappropriate through our sense of morality. Before we can delve into the potential ethical issues posed by Artificial Intelligence and Data Science, it is crucial to have a clear understanding of Ethics. In a broader context, ethics can be perceived as the values and philosophies essential for a society to exist in harmony. It is judging something or someone as good or evil by evaluating if it is socially acceptable (Saltz & Dewar, 2019). So, when discussing the ethical issues raised by AI or data science, we discuss the undesirable consequences of improper use of these technologies. Researchers have already shown concern about the ethical issues caused by the improper development and implementation of AI and Data Science. Bias and Discrimination, Privacy, Data Security, lack of transparency and accountability, Consent, and Control are some of the significant ethical issues that existed along with the development of AI and Data science (Figure 1). As AI and data science increase their influence in every field, addressing the ethical issues they cause is crucial. Researchers, policymakers, and industry leaders in the field must collaborate and establish guidelines and regulations to ensure the responsible and ethical implementation of AI and data science to maximize their benefits and minimize their potential harm.

Figure 1. Illustrating the major ethical concerns of AI and data science
(Author)

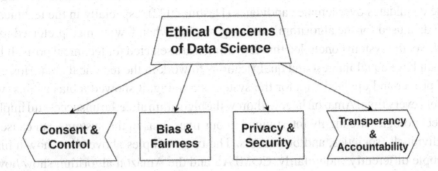

Bias and Fairness

Bias is the tendency to believe that a person, a thing, a fact, etc, is better than another, resulting in mistreating one side (Dictionary, n.d.). In data science and AI algorithms, the presence of bias is very evident. This bias in the algorithm is caused by the system's biased training data or training process. Different types of biases can affect a system: Human Bias, Data Bias, Learning Bias, and Deployment Bias (González-Sendino et al., 2023). Choices regarding which data to utilize, the methodology of training, and other related considerations lie within the domain of human programmers. Given that all humans inherently possess biases in their judgment of various facts, the prejudices of a human can inevitably influence the system, even when trained with meticulously screened data (González-Sendino et al., 2023). Human bias is of two types: cognitive bias, which is caused by the stereotypes the user believes in or the biased knowledge the user possesses, whereas there is behavioral bias, which is influenced by the user's behavior, shaped by gender, ethnicity, culture, etc. (González-Sendino et al., 2023).

On the other hand, Data bias occurs due to biased data. Data will be biased if it does not entirely represent a social group or only covers a narrow area (Vesselinov et al., 2019). Bias can also enter a system during the learning phase. Learning bias is generally created in the system during the training process (González-Sendino et al., 2023). When trained with aggregated data, a system will detect the trends and patterns in the input data and assume the same result for single-point data, eventually leading to bias (Abbasi-Sureshjani et al., 2020). Deployment of a system into multiple scenarios may lead to deployment bias. Since the system uses the data and the experiences during its action to learn new information, deploying the system with the same data in different scenarios may lead to biased outcomes (Baker & Hawn, 2021).

Several real-life incidents of AI bias have been reported in many parts of the world. One of the significant examples of AI bias was displayed by an automated software system COMPAS (Correctional Offender Management Profiling for Alternative Sanctions) used by the judicial system in the United States. COMPAS was used by the US courts to evaluate inmates to decide how much of a chance they have to commit a crime again if they are released. After studying the output of the software, it was clear that the software made predictions against African Americans more than white-skinned Americans (Mehrabi et al., 2021). Racism against black people has always been present throughout history. This is reflected in the historical data and reports as well. As a result, the systems that use this biased data for training will end up biased.

Another famous example is that of Amazon's hiring algorithm. During the hiring process, the algorithm will check for past hiring patterns, and candidate resumes to find a suitable candidate. The algorithm often preferred male candidates over female candidates (Dastin, 2022), especially in the technical field. This is because the data used for the algorithm's training was old when few women preferred to work in the technical field. So, the system concluded that few women are selected for technical posts. It has changed; men and women have equal interest and qualifications to work in the technical area. However, because of the lack of proper and updated data for the system's training, it showed a bias against women in its selection process every time. Amazon later withdrew the algorithm after the unsuccessful implementation.

Since AI technologies are highly intertwined in our lives, bias in the systems can cause irreversible damage to individuals, societies, and businesses. The two examples above show how a biased system could treat people differently and unfairly. COMPAS and the Amazon algorithm show how negatively it can affect the lives of people by unfairly judging people based on race and gender. The negative ramifications of the biased system on society and individuals will be of different gravity. Even every tiny and seemingly harmless action by a biased system may aggregate over time, resulting in undesirable consequences (Madaio et al., 2020). According to the Artificial Intelligence Index Report 2023, the number of AI incidents and controversies reported to the AIAAIC (AI, Algorithmic, and Automation Incidents and Controversies) database in 2022 is 26 times higher than in 2012 (Maslej et al., April 2023).

Data Privacy and Security

Today, with the help of Data Science, analyzing large volumes of data has become easier than ever. The increased use of technologies such as smartphones, computers, and the internet has enabled large quantities of unstructured data to accumulate in unfathomable quantities. This vast, unstructured data is called Big Data (Khanan et al., 2019). With the emergence of Data Science and Big Data came more significant concerns regarding the privacy and security of these data. One of the primary concerns related to data privacy in AI is the collection and use of personal data. AI systems often rely on large amounts of data to train and improve their performance (Adadi, 2021), and this data may include sensitive information about individuals. Unauthorized access or misuse of this data can lead to privacy breaches and violations of individuals' rights. Additionally, using AI in healthcare, finance, and law enforcement raises concerns about the potential misuse of personal data and the implications for individuals' privacy.

One of the notable examples of Data Privacy and Security breaches was reported by CNN Business in April 2019. The report says that Amazon has confirmed they are collecting the conversation between the client and Alexa. Anything a person says after the wake word "Alexa" will be listened to, not only by the AI but also by some employees of Amazon (Valinsky, 2019). Even if the company officials claim to have kept all the collected data confidential, most people think it is a privacy breach. Some AI developers argue that sometimes data collection without user consent is unavoidable because these data are necessary for updating and improving the system, and each time, getting consent from a user is impractical. New data should be fed continuously into the system to improve the output quality and reduce bias. This report is an example of Data privacy infringement. Amazon and several other well-known companies have been reported to have collected user data without proper consent.

Unlike Amazon, Target (An American retail corporation) has a different story regarding the confidentiality of its customer data. In 2013, Target's database was breached to steal millions of users' data. They were able to breach and insert malware in the POS system. This malware later transmitted all the details of the customers and their credit/debit card details directly to the hackers. Hackers collected millions of user data and card details of around 40 million customers. This data-stealing continued for around 18 Days before the company could find out (Wagstaff, 2013). This incident is an example that shows the importance of AI systems' security. Vulnerabilities in AI algorithms and models can be exploited to manipulate or compromise the data's integrity. Ensuring the robustness and resilience of AI systems against such attacks is essential for safeguarding the privacy of the data they handle. Secure and transparent AI systems can contribute to improved data privacy. Implementing rigorous security measures, such as encryption, access controls, and secure communication protocols, is essential.

Consent and Control

In a time of Big Data, Artificial Intelligence, and Data Science, businesses constantly need Data. Data collection is easier than ever in the current era of 'everything being smart'- Smartphones, smart TVs, smart homes, smart cities, etc. However, the rising concerns about privacy among people have somewhat restricted businesses from collecting data freely. However, with or without people's consent, many organizations acquire people's data around the globe. For example, according to a report in the New York Times 2012, an unpopular US company engaged in 50 trillion personal data transactions every year (Cate & Mayer-Schönberger, 2013). This is where our privacy is questioned. Despite the increasing concerns about privacy and data control, there is still a significant amount of personal data being collected by organizations worldwide, often without explicit consent. This raises questions about individual control and consent over their data. One of the significant examples of vast data collectors worldwide is Google. Starting as a small-scale search engine, Google is now the primary provider of more than 30 widely used services (Delichatsios & Sonuyi, 2005). Google focuses on Artificial Intelligence, computer software, e-commerce, search engine technology, etc. (Jack, 2017). From Google's Google Assistant to its search engine, every technology of Google's is collecting and analyzing user data. When we search for a keyword in the Google search engine, it records the search details in its search history log. These details include IP address, browser, the language used, search content, date and time of the activity, etc. (Delichatsios & Sonuyi, 2005). In the same way, websites use 'HTTP Cookies'[5] to store user information and browsing details. This cookie will be stored in the user's device to track the user's activities. Even if the user has been given the choice of deleting the cookie, every time after deletion, when the user returns to the website, a new cookie will be created. Users can only browse a website if they create a cookie first. Every website has user policies regarding the use of cookies. Whenever a user enters a website, a popup will emerge explaining the terms of using cookies on the website and with options to accept or reject the cookies. Figure 2 shows a screenshot of the pop of a random website explaining its cookie policy and options for user consent.

Figure 2. Popup of cookie policy of a random website

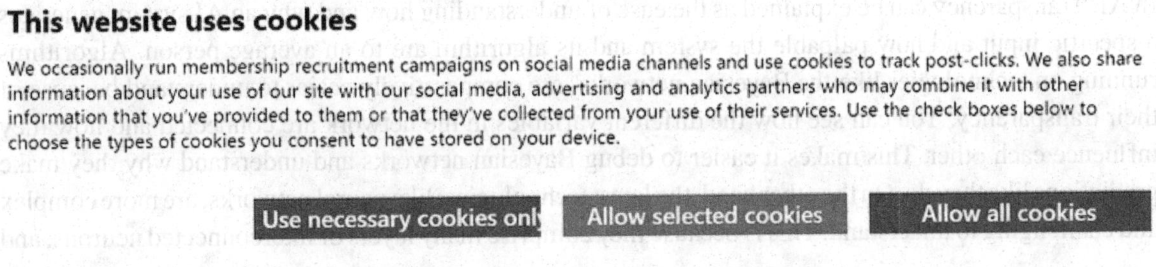

This website uses cookies

We occasionally run membership recruitment campaigns on social media channels and use cookies to track post-clicks. We also share information about your use of our site with our social media, advertising and analytics partners who may combine it with other information that you've provided to them or that they've collected from your use of their services. Use the check boxes below to choose the types of cookies you consent to have stored on your device.

| Use necessary cookies only | Allow selected cookies | Allow all cookies |

Notice the three available options: 'Allow all cookies,' 'Allow selected cookies,' and 'Use necessary cookies only.' Notably, none of these alternatives provides an explicit option for users to decline or withhold consent to the cookie policy. Consequently, users must consent to the policy to access and utilize the website and its services, even if they prefer not to.

Fred H. Cate, in one of his TED talks, discusses user consent on data privacy policies. He gave seven reasons why consent is unviable and data privacy laws are ineffective. He mentions the complexity and length of the Privacy policies. The policies of some popular websites are more than 30,000 words long, which makes them difficult to read and understand. Most people are unaware of the importance of data privacy policies. They need to be educated about privacy policies. Cate also points out that some software updates come with data privacy policies that give the user the option to accept but no option to not accept, just like the cookie policy example mentioned earlier. Another essential reason he points out is that when a user consents to the data collector to use his data, the responsibility for the consequences falls on the user's shoulder. If any undesirable consequences happen due to the usage of the user data, the user has to suffer the consequences just because he consents. He also points out that sometimes, getting consent from users is not viable. We use data analysis for fraud prevention, crime detection, etc. During such a situation, it is not plausible to wait to get consent (TEDxTalks, 2020).

Several initiatives are taken worldwide to protect people's data privacy. The European Union's General Data Protection Regulation (GDPR), Brazil's Lei Geral de Proteção de Dados (LGPD), South Africa's Protection of Personal Information Act (POPIA), Canada's Personal Information Protection and Electronic Documents Act (PIPEDA), and India's Personal Data Protection Bill (Thales, 2021) are some of the few examples of Data privacy protection acts initiated by different countries. Despite these policies and regulations, many data breaches and misuse happen frequently. One of the reasons is that these policies are often written in complex legal language, making it difficult for the average person to understand. This lack of transparency can lead to a lack of trust in the company or organization handling the data. Data privacy policies may be too broad or vague, leaving room for interpretation and potential misuse of personal data. Some policies may also contain loopholes that allow companies to collect and use data in ways that are not in line with user expectations. Furthermore, enforcement of data privacy policies can be challenging. Regulatory bodies may need more resources to monitor and enforce compliance with privacy policies effectively. To bypass these challenges, authorities must take the initiative to educate people about the importance of data privacy and the risks involved in unchecked data handling. All the policies and regulations implemented worldwide should be integrated into a single global standard that clearly defines the technical terms in a way ordinary people can understand. According to the evolution of technology, these policy frameworks also need to be updated to suit the present scenario (Cate & Mayer-Schönberger, 2013).

Transparency in AI and Data Science

In AI, Transparency can be explained as the ease of understanding how and why an AI system generates a specific input and how palpable the system and its algorithm are to an average person. Algorithms running on technologies like the Bayesian networks[6] are comparatively easier to understand because of their transparency. You can see how the different variables in the network are connected and how they influence each other. This makes it easier to debug Bayesian networks and understand why they make predictions like they do. On the other hand, the latest technologies, like neural networks, are more complex and challenging to understand. This is because they comprise many layers of interconnected neurons, and it can be challenging to see how the different neurons interact. This makes debugging neural networks and understanding how it works more challenging. Sometimes, even the developers are unaware of the inner workings of a Neural Network system.

In a time of rising concerns about Artificial Intelligence, ensuring the system's transparency is essential to build trust among the public. Transparency in AI also helps identify and mitigate bias, making the system more accountable and responsible, enhancing the system's accuracy, promoting technological innovations, increasing public understanding, and ensuring compliance with regulations. Transparent AI systems explain their decisions and actions, allowing users to understand the reasoning behind the outcomes. This accountability can enhance trust by making the system's behavior more predictable and comprehensible to stakeholders (Larsson & Heintz, 2020). Even if transparency in AI systems is essential, some weaknesses might be associated with it. Giving away too much information on the workings of a system or its algorithm may not be feasible. This may make the system vulnerable to hacking. Designing an extensive system with transparency might not be as practical as one thinks. Too much transparency will not do any good, but it may affect the whole system adversely. For example, achieving full functional transparency may only sometimes be feasible, especially in complex machine-learning systems that rely on large datasets and sophisticated algorithms (Walmsley, 2021). Additionally, providing too much information about the inner workings of an AI system may lead to clarity or understanding among users, potentially undermining trust in the system (Walmsley, 2021). While transparency is an essential ethical principle in AI, it is not a panacea. It must be balanced against other considerations such as feasibility, user comprehension, and the need for alternative approaches to address ethical concerns.

KEY INSIGHTS AND DISCUSSIONS

Responsible Handling of Ethical Issues

The responsible handling of AI ethics is essential to ensure that AI systems are developed and used in a manner that upholds human rights, fairness, transparency, and accountability. Policymakers, Researchers, and Scientists can address the ethical implications of AI networks by focusing on governance mechanisms and regulatory issues. Efforts should be made to build trust between the machine learning developer community and end-users, and guidelines for AI ethics should be established (Jameel et al., 2020).

Another critical aspect of handling AI ethics is the consideration of the impact of AI on employment and society. As AI technology automates tasks and processes, there is a potential for job displacement and economic disruption. Ethical AI development involves actively addressing these challenges by investing in reskilling programs, promoting job creation in AI-related fields, and fostering collaboration between industry, government, and academia to mitigate the societal impact of AI. It is essential to ensure that AI systems augment human decision-making rather than replace it and do not result in discriminatory or unjust outcomes. This involves establishing guidelines for the responsible use of AI in decision-making, promoting algorithmic transparency, and providing avenues for recourse in algorithmic bias or error cases. We can counter unethical AI practices by developing high-quality data, implementing strict laws and regulations, accurately addressing bias, and employing emergency stop mechanisms in autonomous systems (Jameel et al., 2020).

High-quality and updated data can be crucial in Data Science since it helps avoid bias, and systems trained on quality data can be comparatively accurate and reliable. AI systems are used in Businesses for decision-making; using quality and recent data will ensure that the decisions made by the systems are relevant and accurate. By establishing proper governance mechanisms and regulations, we can bring transparency into the system, ensure data privacy, and ensure international corporations effectively implement ethical AI practices globally. We must make arrangements to identify and rectify bias as early as possible to achieve fairness, trust, legal compliance, improved decision-making, social responsibility, user experience, organizational reputation, etc. An emergency stop mechanism should be mandatory to ensure the system has external human control. It will ensure safety, regulatory compliance, human intervention, etc. It can help minimize the effects of systems that have gone astray and prevent or minimize damage in case of security attacks (Jameel et al., 2020). By enacting these solutions, policymakers and stakeholders can work towards mitigating the ethical challenges associated with the rapid emergence of artificial intelligence in various sectors.

In conclusion, handling AI ethics requires a multifaceted approach encompassing bias and fairness, transparency and accountability, privacy and data protection, societal impact, and responsible decision-making. By prioritizing ethical considerations in AI development and deployment, we can harness the potential of AI technology while safeguarding human rights, promoting fairness, and fostering trust in AI systems. Stakeholders across industries must collaborate and commit to upholding ethical standards as AI continues to shape the future of our society.

Role of Data Scientists in Ethical AI Practice

A data scientist is responsible for sourcing, managing, preparing, and analyzing raw data to get insights for problem-solving and decision-making. Data scientists should be skilled in programming, statistics, mathematics, machine learning, data visualization, etc. They work with large and complex datasets to identify patterns, trends, and correlations that can be used to make informed business decisions, develop AI models, and solve complex problems. Data scientists are crucial in helping organizations derive value from their data and driving data-driven strategies.

Data scientists are now considered among the most sought-after professionals across various domains. Generating large amounts of digital data has created a need for professionals who can harvest, clean, and interpret data sets to provide data-driven solutions for organizational growth and competitive advantage (Smaldone et al., 2022). Employers also value prior work experience and are willing to invest in up-skilling and additional education to develop talented data scientists. Overall, the role of data scientists has become pivotal in Industry 4.0, where high-tech solutions like robotics, cloud computing, and artificial intelligence are proliferating (Smaldone et al., 2022). A combination of technical expertise, analytical skills, and a curious, creative mindset is essential to becoming a successful Data Scientist. Technical efficiency, knowledge of data analysis techniques, deep understanding of mathematics, statistics, and probability, proficiency in data visualization and the ability to identify patterns and insights within complex data, efficient Communication Skills, Curiosity, and Creativity are some of the essential skills set required for a Data Scientist (Davenport & Patil, 2012). Data scientists should consider following four important ethical principles (Figure 3), such as fairness, justice, beneficence, and non-maleficence, in their professional work (Garzcarek & Steuer, 2019).

These principles ensure that data science applications do not harm individuals or communities and are used for society's greater good. Data Scientists should be aware of the impact of their work on society and take responsibility for their actions (Garzcarek & Steuer, 2019). The importance of a data scientist is paramount in the context of data ethics. Data scientists play a crucial role in collecting, curating, analyz-

Figure 3. Four of the essential principles of a data scientist
(Garzcarek & Steuer, 2019)

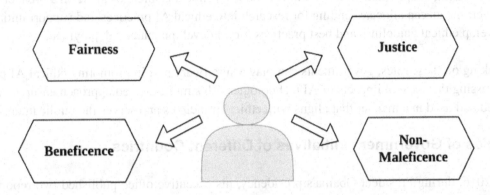

ing, and using data and developing and implementing algorithms. Their decisions and actions can have significant ethical implications, particularly in privacy, transparency, trust, and responsibility (Floridi & Taddeo, 2016). Data scientists should strive to ensure that their models do not discriminate against any particular group of people. Data scientists and practitioners must navigate between the need to harness the social value of data science and the imperative to protect individual rights, striking a balance that maximizes ethical value while avoiding social rejection and legal prohibition (Floridi & Taddeo, 2016). The ethical principles of beneficence and non-maleficence are crucial for data scientists in their professional work. Beneficence refers to the obligation to act for the benefit of others, while non-maleficence emphasizes the duty not to harm. In data science, these principles are essential for several reasons. Data scientists often work with large datasets that contain sensitive information about individuals. Data scientists can make more ethical decisions when developing algorithms, conducting analyses, and implementing data-driven solutions. As data science continues to impact society significantly, data scientists have an increased responsibility to consider the ethical implications of their work (Garzcarek & Steuer, 2019). In summary, embracing the principles of beneficence and non-maleficence is essential for data scientists to ensure that their work contributes positively to society while minimizing potential harm.

Role of Government in Ethical AI Practice

Governments of each country have an essential role in ensuring Ethical AI Practice. According to (Cath, 2018), some of the critical aspects through which a government could ensure the ethical practice of AI involve:

- Governments should establish regulatory frameworks that include ethical AI design, usage, and accountability guidelines. They can also address issues like transparency, fairness, and privacy in AI systems.
- Government agencies can oversee AI practices to ensure ethical standards and regulations compliance. This oversight can involve auditing AI systems, investigating complaints, and imposing penalties for non-compliance.
- Governments can enact laws specifically tailored to address ethical considerations in AI. These laws can cover data protection, algorithmic transparency, and the ethical use of AI in critical domains like healthcare, finance, and criminal justice.

- Governments can engage in international collaboration to establish global standards for ethical AI, fostering cooperation among nations to address ethical challenges in AI on a broader scale.
- Governments can allocate funding for research into ethical AI practices and support initiatives to develop ethical guidelines and best practices for AI development and deployment.

By taking on these roles, governments can play a significant part in promoting ethical AI practices and addressing the societal impacts of AI technologies. This multi-faceted approach ensures that AI is developed and used in a manner that aligns with ethical principles and serves the public interest.

Examples of Government Initiatives of Different Countries

- In 2016, during President Obama's presidency, his executive office published two reports highlighting several critical factors for the future of AI in the country. It also mentioned the importance of an effective AI policy (Agarwal et al., 2016).
- In 2019, China released its policies on developing and using AI. The policy put forth eight principles for the ethical development of AI (OECD, 2020) such as 1) Harmony and human friendly, 2) Fairness and justice, 3) Inclusion and sharing, 4) Respect for privacy, 5) Safety and controllability, 6) Shared responsibility 7) Open and collaboration 8) Agile governance.
- The United Kingdom established the 'Center for Data Ethics and Innovation' in 2018. The center's primary objective is to gather evidence and insights to monitor the use of Data and AI (OECD, 2020).
- India's NITI Aayog (the Think Tank mainly focused on Healthcare, Agriculture, Education, Smart Cities and Infrastructure, and Smart Mobility and Transportation) published India's AI strategy document in 2018 (Division, 2020). The strategy aims to Enhance and empower Indians with the skills to find quality jobs, Invest in research and sectors that can maximize economic growth and social impact, and Scale Indian-made AI solutions to the rest of the developing world (Division, 2020).
- Most recently, in 2023, Israel unveiled its first-ever policy on AI regulation and ethics. They focus on a comprehensive approach, collaboration, and responsible innovation. They established policy principles to develop a responsible AI innovation culture (Ministry of Innovation, 2023).

Apart from these countries, Korea, Mexico, the Russian Federation, Saudi Arabia, Spain, Turkey, etc, have established their initiatives to regulate and monitor the use of Data and AI (OECD, 2020). Initiatives like these are paramount to ensure that the technology can be used for human betterment in the future.

Role of Society in Ethical AI Practice

As discussed above, Data Scientists and Government policies play a crucial part in the ethical development of AI and Data Science. The role of society in the matter is no less. Society can play a vital role in promoting ethical AI development through various means.

- Education is essential when considering how society would contribute to improving Ethical AI practice. An Educated society tends to be aware of AI and its impact. This can help them to use AI systems effectively and ethically. Education can also help people understand the importance of data privacy. This will lead them to be more cautious about sharing their data, which will eliminate fraud and risk from the system.

- AI and Data Science and related fields of Computer science are comparatively new. There might need to be more than education for a society to understand the concept of AI thoroughly. People should be given awareness and understanding of AI technologies and their ethical implications. Give awareness to individuals, organizations, and policymakers about AI's potential risks and benefits.
- A collaboration between researchers, industry, and civil society is necessary to address the Ethical Challenges in AI. As a society, every individual should be aware of the research in the field and should support the initiatives.
- We cannot deny that AI is embedded into the lives of every human being without any social, gender, or geographical discrimination. Rich and poor, highly literate and poorly literate, etc., are all equally under the influence of AI. We must accept this fact and ensure public participation and engagement in all AI-related developments.
- A new technology like AI can cause turmoil among the general public, particularly regarding Job displacement, Security concerns, and inherent Biases. Despite concerted efforts, some challenges, like Job displacement, may manifest as a lasting transformation. The general public should be able to accept that AI implementation has led to the elimination of specific roles. However, simultaneously, it is essential to recognize the emergence of numerous new opportunities. Achieving a harmonious integration of AI for societal improvement requires a collective commitment to embracing these changes positively.

By actively contributing to ethical AI development, society can help shape the future of AI technologies in a way that aligns with human values, respects individual rights, and promotes the overall well-being of humanity.

CONCLUSION

The official birth of Artificial Intelligence as a field of Computer Science was at the 1956 Dartmouth conference, where John McCarthy, the Father of Artificial Intelligence, coined the term for the first time. However, even before that, Alan M. Turing contributed to the field of Artificial Intelligence. His works on Machine Intelligence paved the way for modern Artificial Intelligence. On the other hand, data science has evolved from our need to analyze large amounts of data. With the increase in technological and internet capabilities, large volumes of data are easily accessible. Analyzing these extensive data efficiently required more than conventional statistical methods. So, the likes of John Wilder Tukey discussed the possibility of integrating AI into conventional Data Analysis techniques to give rise to the field of Data Science. In the current context, AI and Data Science share a symbiotic relationship. Each of them benefits from the other. The extensive data is analyzed to get insights to improve the existing AI system, and these AI systems are again assisting in making the data science processes much easier and more efficient. The rapid evolution of AI and data science has led to significant technological advancements and transformed various industries. As technology advances, so do the concerns it raises.

Several issues and controversies surround the development and Implementation of AI and Data Science. The primary raw material in the field of Data Science is Data. The collection, management, processing, and storing of these data raises concerns among the public and researchers. Ethical issues surrounding AI and Data Science, such as data privacy, bias, and accountability, have been highlighted through real-world examples, underscoring the importance of addressing these concerns. This chapter has proposed measures to mitigate ethical issues, such as implementing transparent and accountable AI systems and establishing clear regulations and guidelines. The chapter highlights the role of data scientists, researchers, government, and societies in responsible AI and Data Science implementation. Data scientists are the ones dealing with susceptible and personal data. They have to be just and fair in their work. They need to make sure that their work does not discriminate against anyone. World governments also play an essential role in the ethical implementation of AI and Data Science. The government needs to monitor AI practices closely, and collaboration among the countries is also necessary to standardize the regulations of AI implementation. Society, as a whole, plays a crucial role in responsible AI implementation and mitigating the associated risks. Education is an integral part of society in dealing with AI. An educated society will have more awareness about AI and Data Science than an uneducated one. Public engagement in encouraging research related to the field is very much imperative. As the historical records indicate, every major technology has brought radical change into our lives, and so will AI. As a society, we should be able to embrace these changes and get ready for the newer opportunities that lie ahead.

REFERENCES

Abbasi-Sureshjani, S., Raumanns, R., Michels, B., Schouten, G., & Cheplygina, V. (2020). Risk of Training Diagnostic Algorithms on Data with Demographic Bias. *Interpretable and Annotation-Efficient Learning for Medical Image Computing: Third International Workshop, iMIMIC 2020.* Springer International Publishing.

Adadi, A. (2021). A survey on data-efficient algorithms in big data era. *Journal of Big Data, 8*(1), 24. doi:10.1186/s40537-021-00419-9

Agarwal, A., Gans, J., & Goldfarb, A. (2016, December 21). The Obama Administration's Roadmap for AI Policy. *Harvard Business Review.* https://hbr.org/2016/12/the-obama-administrations-roadmap-for-ai-policy

Baker, R., & Hawn, A. (2021). Algorithmic Bias in Education. *International Journal of Artificial Intelligence in Education,* 1–41.

Cao, L. (2017). Data Science: A Comprehensive Overview. *ACM Computing Surveys (CSUR), 50*(3), 1-42.

Cate, F., & Mayer-Schönberger, V. (2013). Notice and consent in a world of Big Data. *International Data Privacy Law, 3*(2), 67-73.

Cath, C. (2018). Governing artificial intelligence: Ethical, legal and technical opportunities and challenges. *Philosophical Transactions. Series A, Mathematical, Physical, and Engineering Sciences, 376*(2133), 20180080. doi:10.1098/rsta.2018.0080 PMID:30322996

Copeland, B. (2000). The turing test*. *Minds and Machines, 10*(4), 519–539. doi:10.1023/A:1011285919106

Cordeschi, R. (2007). AI TURNS FIFTY: REVISITING ITS ORIGINS. *Applied Artificial Intelligence, 21*(4-5), 259–279. doi:10.1080/08839510701252304

Dastin, J. (2022). Amazon scraps secret AI recruiting tool that showed bias against women. In K. Martin, Ethics of Data and Analytics. (pp. 296-299). Auerbach Publications. doi:10.1201/9781003278290-44

Davenport, T., & Patil, D. (2012). Data Scientist. *Harvard Business Review, 90*(5), 70–76. PMID:23074866

Dawson, J., Jr. (2007). The Essential Turing: Seminal Writings in Computing, Logic, Philosophy, Artificial Intelligence, and Artificial Life plus The Secrets of Enigma, by Alan M. Turing (author) and B. Jack Copeland (editor). The Review of Modern Logic, 10(32), 179-181.

Delichatsios, S., & Sonuyi, T. (2005). Get to Know Google… Because They Know You. *MIT, Ethics and Law on the Electronic Frontier, 6*(14).

Division, F. N. (2020). Artificial Intelligence (AI) Policies in India- A Status Paper. Future Networks (FN) Division, Telecommunication Engineering Center Janpath, New Delhi.

Donoho, D. (2017). 50 years of Data Science. *Journal of Computational and Graphical Statistics, 26*(4), 745–766. doi:10.1080/10618600.2017.1384734

Floridi, L., & Taddeo, M. (2016). What is data ethics? *Philosophical Transactions. Series A, Mathematical, Physical, and Engineering Sciences, 374*(2083), 20160360. doi:10.1098/rsta.2016.0360 PMID:28336805

Garzcarek, U., & Steuer, D. (2019). Approaching Ethical Guidelines for Data Scientists. In N. I. Bauer, *Applications in Statistical Computing. Studies in Classification, Data Analysis, and Knowledge Organization.* Springer.

González-Sendino, R., Serrano, E., Bajo, J., & Novais, P. (2023). A Review of Bias and Fairness in Artificial Intelligence. *International Journal of Interactive Multimedia and Artificial Intelligence.*

Hassani, H., & Silva, E. S. (2023). The Role of ChatGPT in Data Science: How AI-Assisted Conversational Interfaces Are Revolutionizing the Field. *Big Data Cognitive Computing, 7*(62).

Hendler, J. (2008). Avoiding another AI winter. *IEEE Intelligent Systems, 23*(2), 2–4. doi:10.1109/MIS.2008.20

Jack, S. (2017, November 21). *Google - powerful and responsible?* BBC. https://www.bbc.com/news/business-42060091

Jordan, M. I., & Mitchell, T. M. (2015). Machine learning:Trends, perspectives,and prospects. *Science, 349*(6245), 255–260. doi:10.1126/science.aaa8415 PMID:26185243

Khanan, A., Abdullah, S., Mohamed, A., Mehmood, A., & Ariffin, K. (2019). Big Data Security and Privacy Concerns: A Review. *In Smart Technologies and Innovation for a Sustainable Future: Proceedings of the 1st American University in the Emirates International Research Conference.* Dubai UAE: Springer International Publishing.

Kordon, A. (2020). *Applying Data Science How to Create Value with Artificial Intelligence.* Springer International Publishing. doi:10.1007/978-3-030-36375-8

Larsson, S., & Heintz, F. (2020). Transparency in artificial intelligence. *Internet Policy Review, 9*(2). doi:10.14763/2020.2.1469

Madaio, M., Stark, L., Vaughan, J. W., & Wallach, H. (2020). Co-Designing Checklists to Understand Organizational Challenges and Opportunities around Fairness in AI. *In Proceedings of the 2020 CHI conference on human factors in computing systems.* Springer.

Maslej, N., Fattorini, L., Brynjolfsson, E., Etchemendy, J., Ligett, K., Lyons, T., & Perrault, R. (2023). *The AI Index 2023 Annual Report.* AI Index Steering Committee, Institute for Human-Centered AI, Stanford University, Stanford, CA.

Mehrabi, N., Morstatter, F., Saxena, N., Lerman, K., & Galstyan, A. (2021). A Survey on Bias and Fairness in Machine Learning. *ACM Computing Surveys, 54*(6), 1–35. doi:10.1145/3457607

Ministry of Innovation. (2023, December 17). *Israel's Policy on Artificial Intelligence Regulation and Ethics.* Ministry of Innovation, Science and Technology. https://www.gov.il/en/departments/policies/ai_2023#:~:text=Key%20Highlights%20of%20Israel's%20AI%20Policy%3A&text=Comprehensive%20approach%3A%20The%20AI%20Policy,safety%2C%20accountability%2C%20and%20privacy

Moor, J. (2006). The Dartmouth College Artificial Intelligence Conference: The Next Fifty Years. *AI Magazine, 27*(4), 87–87.

OECD. (2020). EXAMPLES OF AI NATIONAL POLICIES Report for the G20 Digital Economy Task Force. Organisation for Economic Co-operation and Development (OECD).

Ong, S., & Uddin, S. (2020). Data Science and Artificial Intelligence in Project Management: The Past, Present and Future. *The Journal of Modern Project Management, 7*(4).

Powers, D. (1998). The Total Turing Test and the Loebner Prize. *In D.M.W. Powers (ed.) NeMLaP3/CoNLL98 Workshop on Human Computer Conversation, ACL.*

Provost, F., & Fawcett, T. (2013). Data science and its relationship to big data and data-driven decision making. *Big Data, 1*(1), 51–59. doi:10.1089/big.2013.1508 PMID:27447038

SEP. (2014, August 19). *Philosophy of Statistics.* Stanford Encyclopedia of Philosophy. https://plato.stanford.edu/entries/statistics/#StaInd

Smaldone, F., Ippolito, A., Lagger, J., & Pellicano, M. (2022). Employability skills: Profiling data scientists in the digital labour market. *European Management Journal, 40*(5), 671–684. doi:10.1016/j.emj.2022.05.005

TEDxTalks. (2020, January 16). *Data Privacy and Consent | Fred Cate | TEDxIndianaUniversity* [Video]. YouTube. https://youtu.be/2iPDpV8ojHA

Thales. (2021, May 10). *BEYOND GDPR: DATA PROTECTION AROUND THE WORLD.* Thales. https://www.thalesgroup.com/en/markets/digital-identity-and-security/government/magazine/beyond-gdpr-data-protection-around-world

Turing, A. (2012). Computing Machinery and Intelligence (1950). In J. Copeland (Ed.), *A. Turing, The Essential Turing: Seminal Writings in Computing, Logic, Philosophy, Artificial Intelligence, and Artificial Life: Plus The Secrets of Enigma by B* (pp. 433–464).

Valinsky, J. (2019, April 11). *Amazon reportedly employs thousands of people to listen to your Alexa conversations*. CNN Business. https://edition.cnn.com/2019/04/11/tech/amazon-alexa-listening/index.html

Vesselinov, V., Alexandrov, B., & O'Malley, D. (2019). Non negative Tensor Factorization for Contaminant Source Identification. *Journal of Contaminant Hydrology, 220*, 66–97. doi:10.1016/j.jconhyd.2018.11.010 PMID:30528243

Wagstaff, K. (2013). *Massive Target credit card breach new step in security war with hackers: experts*. NBC News. https://www.nbcnews.com/technology/massive-target-credit-card-breach-new-step-security-war-hackers-2D11778083

Walmsley, J. (2021). Artificial intelligence and the value of transparency. *AI & Society, 36*(2), 585–595. https://doi.org/https://doi.org/10.1007/s00146-020-01066-z. doi:10.1007/s00146-020-01066-z

Wang, X. (2018). Overview of the research status on artificial neural networks. *Proceedings of the 2nd International Forum on Management, Education and Information Technology Application (IFMEITA 2017)*. Atlantis Press. 10.2991/ifmeita-17.2018.59

Wing, J. (2019). The Data Life Cycle. *Harvard Data Science Review, 1*(1), 6. doi:10.1162/99608f92.e26845b4

ADDITIONAL READINGS

Marcus, G., & Davis, E. (2019). *Rebooting AI: Building Artificial Intelligence We Can Trust*. Vintage.

Provost, F., & Fawcett, T. (2013). *Data Science for Business: What you need to know about data mining and data-analytic thinking*. O'Reilly Media, Inc.

Russell, S. J., & Norvig, P. (2010). *Artificial intelligence: A modern approach*.

Winston, P. H. (1984). *Artificial Intelligence*. Addison-Wesley Longman Publishing Co., Inc.

KEYWORDS AND DEFINITION

AI Winter: AI winter refers to the collapse of the Artificial Intelligence industry due to lack of funding, the disinterest of the public and stakeholders, etc.

Artificial Intelligence (AI): AI is the branch of Computer Science that deals with systems capable of solving problems and adapting to new environments by learning, understanding, and applying knowledge from past data.

Bias: Bias refers to the action or output of an AI system that discriminates against an individual, a group, or a thing due to the prejudiced data used for its training.

Big Data: Big data refers to the immense and complex raw data accumulating over time due to the internet and technology use.

Cookie: Cookies are small strings of data that a web server introduces into a user's computer to store the browsing activities and user information.

Data Science: Data Science is an interdisciplinary field that focuses on sourcing, managing, preparing, and analyzing raw data to extract meaningful insights that help find solutions to problems and in decision-making.

Data Scientist: A data scientist is the person responsible for sourcing, managing, preparing, and analyzing raw data to get insights for problem-solving and decision-making.

Machine Learning: Machine Learning is the field of Artificial Intelligence that deals with the study of developing algorithms that help AI machines learn and enhance from experience.

Neural Network: A Neural Network is an interconnected layered structure of multiple nodes or neurons, which enables the system to use information within these nodes or neurons simultaneously to analyze and process data.

Statistics: Statistics is a branch of mathematics that deals with finding insights from numerical data to make informed decisions.

ENDNOTES

[1] A fictional future where unpredictable and transformative technological advancement takes place, often involving the emergence of superintelligent machines.

[2] Superintelligence envisions a scenario where artificial intelligence surpasses human cognitive abilities, raising concerns about control, ethics, and societal impact. It is often linked to the concept of the technological singularity, where AI development becomes unpredictable and transformative.

[3] Initiated in 1991, this annual competition in the field of AI assessed the ability of AI programs to replicate human capabilities. However, it has been discontinued since 2020.

[4] *AI winter refers to the collapse of the AI industry due to lack of funding, the disinterest of the public and stakeholders, etc. The first AI winter occurred in the late 1960s to the 1980s, and the second from the late 1980s to the mid-1990s* (Toosi et al., 2021).

[5] Small strings of Data placed by a web server in the user's device often used to store information about the user's preferences, site interactions, or authentication credentials.

[6] A probabilistic graphical model using a directed acyclic graph to represent random variables and their conditional dependencies

Chapter 19
The Impact of 5G on the Future Development of the Healthcare Industry

Saurabh Srivastava

https://orcid.org/0000-0001-7654-0220

Department of Computer Science and Engineering (Data Science), Moradabad Institute of Technology, India

Harish Chandra Verma

ICAR-Central Institute for Subtropical Horticulture, Lucknow, India

Syed Adnan Ahaq

Advanced Computing Research Laboratory, Department of Computer Application, Integral University, Lucknow, India

Mohammad Faisal

Advanced Computing Research Laboratory, Department of Computer Application, Integral University, Lucknow, India

Tasneem Ahmed

Advanced Computing Research Laboratory, Department of Computer Application, Integral University, Lucknow, India

ABSTRACT

A 5G network can enable services like real-time remote patient monitoring and the distribution of huge files, including medical data for e-health systems. The internet of things (IoT), sensors, and other cutting-edge technologies will be used in the future to identify patients' illnesses and offer advice on how to treat them. The popularity of electric health care is increasing day by day, there are many applications available that can be used by the patient for routine checkups from the smartphone. Patients' private information is taken at the time of application downloads such as name, gender, and age is used by the application to increase its accuracy, as well as, the results of routine checkups are stored on the application's server (storage). The stored data can be used in different kinds of promotions. Hence, hackers are trying to steal information from users for their benefit and the IoT-based applications are not so reliable in terms of security.

DOI: 10.4018/979-8-3693-2964-1.ch019

INTRODUCTION

One of the most pressing issues in the modern world is medical health care. Concerns about the standard of treatment in healthcare are raised by inadequate infrastructure, subpar healthcare laws, and a lack of funding (Kumar et al., 2020). Medical healthcare is a major issue in developing nations, especially in rural regions, Rural healthcare systems face unique barriers such as the cost of caring for patients with chronic diseases, as well as the need for older people to receive care at home(Hamm et al., 2020). The healthcare industry is growing rapidly, and lots of applications are using networks to handle all types of data in different sizes and formats as a result (Rao, 2019.). The development of wireless telecommunication has enhanced medical health care in many ways, including easing patient struggles with remote health diagnoses that require traveling from remote locations to modern healthcare facilities and ensuring that patients receive proper medical care without having to pay exorbitant sums of money(Kumar et al., 2020). The healthcare sector is always on the lookout for cutting-edge technologies that will have a significant impact on the way healthcare is delivered. As technology advances globally, the healthcare sector also demands higher-quality networks, which is why 5G is essential to providing intelligent hospital care (Batool, 2022). The fifth generation of wireless networks is called 5G, and it was introduced by Germany in 2020 as a new mobile communication standard. Fast data throughput, low latency, and adequate coverage are key aspects of 5G networks. Fast Internet provides reliable connectivity to medical equipment and systems. Additionally, 5G enables instant downloads and communications between tablets and mobile devices used in smart healthcare systems. The 5G standard is considered to lead to the IoT and thus billions of networked end devices (Batool, 2022.; Hamm et al., 2020). The 5G is capable of activating some important features of smart applications of healthcare such as network slicing, where maintaining the performance of multiple network slices at once is difficult compared to the existing service assurances in legacy networks (Qureshi et al., 2021).

Data speed, latency, real-time multicasting, ad hoc peer-to-peer, and data encryption are the benefits that 5G has over current wireless network technologies (Le & Hsu, 2021; Mahmeen et al., 2021), and offers processing through accelerated edge data centers or edge clouds and artificial intelligence/machine learning (AI/ML) in addition to connection, which was already offered by earlier mobile technologies (Valcarenghi et al., 2022). The next generation of mobile cellular networks will be more functional and powerful than present wireless networks (Mahmeen et al., 2021). Digital technology has made enormous strides since the turn of the twenty-first century, and these changes are impacting the global healthcare system. Healthcare institutions are gradually and methodically switching from paper-based data to electronic records, ushering in a revolution in the sector (Chenthara et al., 2019). Health information is extremely sensitive and must be protected. The healthcare sector contributes significantly to the country's economy and provides a wide range of essential medical goods and services (Islam et al., 2023). The modern healthcare ecosystem faces many difficulties, including those related to infrastructure, connections, best utilization of resources, expertise, accuracy, data management, and real-time monitoring (, 2019). Network security and privacy issues, the expense of building and maintaining the network, and the impact of next-generation technology on human health are some of the potential drawbacks of the technology. The issues that services and applications are facing in the current environment are where modern technology must be ready to answer them, and they can be summarized in a few different ways (Srinivasu et al., 2022). Several use cases are considered, and classified by the following scenarios for cases related to location and mobility such as (i) Stable connection in a fixed, static environment, (ii)

A dynamically shifting environment with erratic connection, (iii) Local, within the hospital's physical plant, and (iv) Remote and distant from a hospital.

The health sector growing rapidly and utilizing modern technologies to overcome the difficulties of the traditional healthcare system. If technology is used in the right direction, different kinds of problems can be solved, but if it is used in the wrong direction, it creates a terrible situation. Today, many hackers are using modern technology for their selfish purposes which is causing harm to society. Scams happening in society can be stopped by using new technology and creating awareness. The impacts of 5G on the healthcare industry, cyber concerns, and attacks have been discussed in this study.

IMPACT OF 5G IN HEALTH CARE APPLICATIONS

Over the past two centuries, improvements in equipment and connections have made telemedicine viable. The telegraph was used during the Civil War in the 1860s to send signals about wounded soldiers to distant medical teams. William Enthoven, a physician from Leiden (Holland), made the first electrocardiogram in 1905 and successfully transmitted heart sounds between his laboratory and a hospital(Moglia et al., 2022). A very effective and adaptable method of data storage and sharing is made possible by electronic healthcare (E-health). E-health systems with 5G networks can allow services such as remote patient monitoring and transmission of large-scale health data files in real time. Since the communication path is open, system security and privacy are important considerations(Vedaei et al., 2020).

Healthcare institutions have deployed robotic technologies in several contexts. They have been thought of as aids that could assist in bridging the widening gap between the demand for and availability of healthcare. In the comfort of their own homes, they may assist a patient in maintaining their health and safety. Healthcare robots may carry out a variety of duties and are divided into two categories such as social robots and rehabilitation robots (Islam et al., 2023). Machine-type communication (MTC) is crucial to the mobile network society and has shown to be effective in helping mobile network operators make substantial sums of money. The MTC has emerged as the primary communication paradigm for several newly developing smart services, including public safety, healthcare, industrial automation, drones/robotics, utilities, and transportation (Mohammed et al., 2019). Significant advances in telecommunications, informatics, and wireless sensor networks have opened the way to the realization of ubiquitous intelligence, capable of making predictions with the help of the IoT (Shafique et al., 2020). Mixed Reality-Based Assistance has gained a lot of popularity in recent years to the point where it is now being used in a variety of industries and sectors, including video games, business, marketing, education, and healthcare. This technology is particularly influencing the way that healthcare is delivered today, accelerating diagnosis, lowering wait times, and enhancing personalization and results (Garcia et al., 2023).

CYBER SECURITY CONCERNS AND ATTACKS

Due to the growth of data in the current Big Data era, it has become necessary to outsource healthcare data to cloud servers. Despite the enormous benefits provided by the cloud, it also poses serious risks to the security and privacy of healthcare data (Chenthara et al., 2019). Security measures should be incorporated into the design of IoT devices, including risk assessment before the device is released for market use, and authentication measures should be built into the device (Chacko & Hayajneh, 2018). Infrastructure and coverage, healthcare concerns, patient data security, and privacy protection, 5G deployment with AI, blockchain, and IoT, verification, patient acceptance, and end-user training must also be taken into concern (Moglia et al., 2022). Some of the most important concerns that should be considered in the implementation and maintenance of smart applications of healthcare are shown in Figure 1.

Figure 1. Cyber security concerns of smart applications

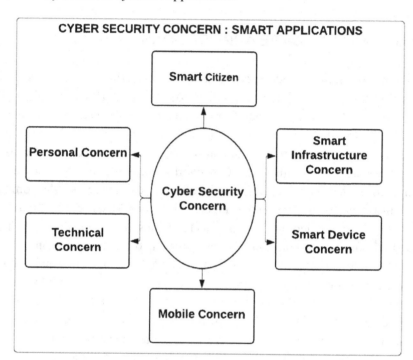

Cyber Security Concerns of Smart (Healthcare) Application

Personal Concern

A platform of services is offered by the smart application for people looking for and bringing up social issues. Concerns over cyber security are specific to each citizen and relate to their safety, communication, access to transit personal and professional information, banking details, etc. (Srivastava et al., 2023). User needs to know the importance of data before sharing it on different portals. Your sharing credentials could be the key for hackers to steal your details.

Technical Concern

Numerous applications offer a wide range of services that simplify human life. Everyone wants to increase their benefits from it, yet if a user makes a minor error, they risk hurting themselves. A hacker participates in a variety of activities to get unauthorized access to a user's system. Some technical instruments can be used for security, monitoring, or preventing unauthorized access such as Biometrics, Smart Grids, etc (Srivastava et al., 2023).

Mobile Concern

Mobile devices are necessary for connecting to the network architecture of smart apps, but they also introduce new security and privacy hazards for users by exposing sensitive data to dangers from the outside world. Privacy-preserving authentication (PPA) protocols for mobile services should be used as a potential cryptographic technique to provide authentication and privacy protection features for smart cities(Ismagilova et al., 2022).

Smart Device Concern

In IoT-based smart applications, human concerns, businesses, or industries make up the majority of the application data. To avoid obscuring, stealing, and altering, the data in these applications needs to be secure and protected. The safety of the communication channel is crucial. While these devices are the IoT's main focus, there are numerous problems with scalability, reaction time, and application availability (Singh et al., 2020).

Smart Infrastructure Concern

The IoT provides the network architecture required to gather and analyze data from scattered sensors and smart devices, which is crucial for the infrastructure of smart applications (Ismagilova et al., 2022). Confidentiality, integrity, and availability, sometimes known as the CIA trinity (i.e. confidentiality, integrity, and availability) are three crucial components of information technology component security, one of the key security components of IoT architecture (Singh et al., 2020).

The Most Papular Attacks on Smart (Healthcare) Applications

Information disclosure, Denial of Service (DoS) attacks, cloud malware injection attacks, man-in-the-middle cryptographic attacks, spoofing, and collusion assaults are a few examples of potential attacks on smart applications (Chenthara et al., 2019). The most popular attacks that are frequently used by hackers are mentioned below.

3.2.1. **Malicious Domains:** To find victims, hackers are making false and harmful websites. A large number of replicating domains confuse internet users, which makes them vulnerable to assault (Aa & Ahmad, 2021).

3.2.2. **Denial of service attack:** A distributed denial of service attack (DDoS) uses numerous hacked systems to assault a single system, resulting in a denial of service and the crash of that system. These attacks can be launched using a variety of techniques, including the employment of malicious botnets (Chacko & Hayajneh, 2018).

3.2.3. **Malware**: Cyber criminals are using websites of unknown companies to spread malware, spyware, and trojans into organizations and cell phones. Spam messages are used by hackers to trap their victims(Aa & Ahmad, 2021).

CONCLUSION

The health sector is growing rapidly due to the enhancement of technologies, the 5G of mobile networks helps to build applications that can respond quickly. In this study, different prospects of 5G in healthcare in the industry have been discussed, and observed that the 5G network infrastructure is intended to manage massive amounts of data. 5G enables a new form of network that delivers high peak data rates, ultra-low latency communication, and high user density. Electronic healthcare makes data storage and sharing a very efficient and flexible method possible. E-health services contribute to improving the long-term effects of treatments due to their collaborative nature. The 5G network could enable services such as real-time remote patient monitoring and the transmission of huge files, including medical data, to e-health systems.

REFERENCES

AaS.AhmadS. (2021). *Cybersecurity Threats and Attacks in Healthcare.* doi:10.13140/RG.2.2.17256.80647/2

BatoolI. (2022). *5G support in healthcare system.* doi:10.13140/RG.2.2.22073.85603

Chacko, A., & Hayajneh, T. (2018). Security and Privacy Issues with IoT in Healthcare. *EAI Endorsed Transactions on Pervasive Health and Technology*, 4(14), 155079. doi:10.4108/eai.13-7-2018.155079

Chenthara, S., Ahmed, K., Wang, H., & Whittaker, F. (2019). Security and Privacy-Preserving Challenges of e-Health Solutions in Cloud Computing. *IEEE Access : Practical Innovations, Open Solutions*, 7, 74361–74382. doi:10.1109/ACCESS.2019.2919982

Garcia, F. M., Moraleda, R., Schez-Sobrino, S., Monekosso, D. N., Vallejo, D., & Glez-Morcillo, C. (2023). Health-5G: A Mixed Reality-Based System for Remote Medical Assistance in Emergency Situations. *IEEE Access : Practical Innovations, Open Solutions*, 11, 59016–59032. doi:10.1109/ACCESS.2023.3285420

Hamm, S., Schleser, A. C., Hartig, J., Thomas, P., Zoesch, S., & Bulitta, C. (2020). 5G as enabler for Digital Healthcare. *Current Directions in Biomedical Engineering*, 6(3), 1–4. doi:10.1515/cdbme-2020-3001

Islam, M. A., Islam, M. A., Jacky, M. A. H., Al-Amin, M., Miah, M. S. U., Khan, M. M. I., & Hossain, M. I. (2023). Distributed Ledger Technology Based Integrated Healthcare Solution for Bangladesh. *IEEE Access : Practical Innovations, Open Solutions*, 11, 51527–51556. doi:10.1109/ACCESS.2023.3279724

Ismagilova, E., Hughes, L., Rana, N. P., & Dwivedi, Y. K. (2022). Security, Privacy and Risks Within Smart Cities: Literature Review and Development of a Smart City Interaction Framework. *Information Systems Frontiers*, 24(2), 393–414. doi:10.1007/s10796-020-10044-1 PMID:32837262

Kumar, A., Albreem, M. A., Gupta, M., Alsharif, M. H., & Kim, S. (2020). Future 5g network based smart hospitals: Hybrid detection technique for latency improvement. *IEEE Access : Practical Innovations, Open Solutions*, 8, 153240–153249. doi:10.1109/ACCESS.2020.3017625

Le, T. V., & Hsu, C. L. (2021). An Anonymous Key Distribution Scheme for Group Healthcare Services in 5G-Enabled Multi-Server Environments. *IEEE Access : Practical Innovations, Open Solutions*, 9, 53408–53422. doi:10.1109/ACCESS.2021.3070641

Mahmeen, M., Melconian, M. R., Haider, S., Friebe, M., & Pech, M. (2021). Next Generation 5G Mobile Health Network for User Interfacing in Radiology Workflows. *IEEE Access : Practical Innovations, Open Solutions*, 9, 102899–102907. doi:10.1109/ACCESS.2021.3097303

Moglia, A., Georgiou, K., Marinov, B., Georgiou, E., Berchiolli, R. N., Satava, R. M., & Cuschieri, A. (2022). 5G in Healthcare: From COVID-19 to Future Challenges. *IEEE Journal of Biomedical and Health Informatics*, 26(8), 4187–4196. doi:10.1109/JBHI.2022.3181205 PMID:35675255

Mohammed, N. A., Mansoor, A. M., & Ahmad, R. B. (2019). Mission-Critical Machine-Type Communication: An Overview and Perspectives towards 5G. In IEEE Access (Vol. 7, pp. 127198–127216). Institute of Electrical and Electronics Engineers Inc. doi:10.1109/ACCESS.2019.2894263

Qureshi, H. N., Manalastas, M., Zaidi, S. M. A., Imran, A., & Al Kalaa, M. O. (2021). Service Level Agreements for 5G and Beyond: Overview, Challenges and Enablers of 5G-Healthcare Systems. *IEEE Access : Practical Innovations, Open Solutions*, 9, 1044–1061. doi:10.1109/ACCESS.2020.3046927 PMID:35211361

Rao, K. (2019). *The Path to 5G for Health Care.*

Shafique, K., Khawaja, B. A., Sabir, F., Qazi, S., & Mustaqim, M. (2020). Internet of things (IoT) for next-generation smart systems: A review of current challenges, future trends and prospects for emerging 5G-IoT Scenarios. In *IEEE Access* (Vol. 8, pp. 23022–23040). Institute of Electrical and Electronics Engineers Inc. doi:10.1109/ACCESS.2020.2970118

Singh, D., Pati, B., Panigrahi, C. R., & Swagatika, S. (2020). Security Issues in IoT and their Countermeasures in Smart City Applications. *Advances in Intelligent Systems and Computing*, 1089, 301–313. doi:10.1007/978-981-15-1483-8_26

Srinivasu, P. N., Ijaz, M. F., Shafi, J., Wozniak, M., & Sujatha, R. (2022). *6G Driven Fast Computational Networking Framework for Healthcare Applications.* IEEE. doi:10.1109/ACCESS.2022.3203061

Srivastava, S., Ahmed, T., & Saxena, A. (2023). An Approach To Secure Iot Applications Of Smart City Using Blockchain Technology. In International Journal of Engineering Sciences & Emerging Technologies, 11(2).

Valcarenghi, L., Pacini, A., Borromeo, J. C., Fichera, S., Gagliardi, M., Amram, D., & Lionetti, V. (2022). Managing Physical Distancing Through 5G and Accelerated Edge Cloud. *IEEE Access : Practical Innovations, Open Solutions*, 10, 104169–104177. doi:10.1109/ACCESS.2022.3210262

Vedaei, S. S., Fotovvat, A., Mohebbian, M. R., Rahman, G. M. E., Wahid, K. A., Babyn, P., Marateb, H. R., Mansourian, M., & Sami, R. (2020). COVID-SAFE: An IoT-based system for automated health monitoring and surveillance in post-pandemic life. *IEEE Access : Practical Innovations, Open Solutions*, 8, 188538–188551. doi:10.1109/ACCESS.2020.3030194 PMID:34812362

Chapter 20
The Impact of Data Science and Participated Geographic Metadata on Improving Government Service Deliveries:
Prospects and Obstacles

Vivek Veeraiah

Sri Siddharth Institute of Technology, Sri Siddhartha Academy of Higher Education, India

Dharmesh Dhabliya

(iD) https://orcid.org/0000-0002-6340-2993

Vishwakarma Institute of Information Technology, India

Sukhvinder Singh Dari

(iD) https://orcid.org/0000-0002-6218-6600

Symbiosis Law School, Symbiosis International University, India

Jambi Ratna Raja Kumar

(iD) https://orcid.org/0000-0002-9870-7076

Genba Sopanrao Moze College of Engineering, India

Ritika Dhabliya

ResearcherConnect, India

Sabyasachi Pramanik

(iD) https://orcid.org/0000-0002-9431-8751

Haldia Institute of Technology, India

Ankur Gupta

(iD) https://orcid.org/0000-0002-4651-5830

Vaish College of Engineering, India

ABSTRACT

This chapter examined the profound influence of data science and volunteered geographic information (VGI) on the delivery of public services. Volunteered geographic information, being material created by users, has had a substantial impact on making geographic information accessible to everybody, enabling people to actively engage in the creation and management of data. The incorporation of VGI into government operations has introduced novel prospects for enhancing service provision in diverse sectors such as education, health, transportation, and waste management. In addition, data science has enhanced VGI by using sophisticated methodologies like artificial intelligence (AI), internet of things (IoT), big data, and blockchain, thereby transforming the whole framework of government service provision. Nevertheless, in order to effectively use VGI in public sector services, it is essential to tackle significant obstacles such as data accuracy, safeguarding, inclusiveness, technical framework, and specialized expertise.

DOI: 10.4018/979-8-3693-2964-1.ch020

INTRODUCTION

The integration of technology and data-driven methods has significantly revolutionized the provision of public services in recent years. The utilization of Volunteered Geographic Information (VGI) in conjunction with Data Science is a significant advancement in this domain. VGI, short for Volunteered Geographic Information, refers to location-based data that individuals willingly provide on various digital platforms, including social media, mobile applications, and crowdsourcing initiatives (Goodchild, 2007). Data science encompasses the methodologies and techniques used to derive insights and knowledge from vast databases (Brodie, 2019; Provost & Fawcett, 2013).

The use of VGI (Volunteered Geographic Information) and data science in the provision of public services has seen a substantial rise, owing to many factors. Currently, there is a significant increase in the collection and distribution of location-based data due to the widespread use of mobile devices and the availability of the Internet (Huang et al., 2021). In the present day, individuals have the ability to actively participate in the generation of spatial data, providing valuable perspectives on their requirements, choices, and encounters (Goodchild, 2007).Furthermore, conventional methods of gathering data incur significant costs and require a substantial amount of time. On the other hand, VGI offers a timely and cost-efficient alternative that allows public sector enterprises to get dependable information for decision-making processes (Ahmad et al., 2022).

The combination of Volunteered Geographic Information (VGI) with data science enables the effective analysis and interpretation of big datasets, facilitating informed and data-driven policy development (Arnaboldi & Azzone, 2020; Provost & Fawcett, 2013; Wong & C. Hinnant, 2022). Consequently, public service providers have the ability to enhance the distribution of resources, enhance the efficiency of their services, and meet the specific needs of the communities they serve. To evaluate poverty at the village level, one may integrate several data sources such as high-resolution imaging (HRI), point-of-interest (POI), OpenStreetMap (OSM), and digital surface model (DSM) data (Hu et al., 2022). In a similar vein, Ma et al. (2022) put out a method to assess the logic of the geographical distribution of public restrooms in urban functional areas. They achieved this by using POI big data and OSM. KUCUKALI et al., (2022) employed open-source geospatial data, such as OSM, to assess the ease of pedestrian access to crucial public services and facilities. In contrast, Abdulkarim et al., (2014) created a VGI application utilizing Google Street View to encourage individuals to contribute to the categorization of roof materials, thereby supporting energy efficiency initiatives.

The main purpose of this chapter is to analyze the use of donated geographic information and data science in the provision of public services. This study will examine the potential benefits and challenges of incorporating these two components in various aspects of the public domain. In order to accomplish this goal, the chapter is organized in the following manner: Section two offers a comprehensive examination of Volunteered Geographic Information (VGI), with a specific emphasis on its influence on the provision of public services. Section three explores the notion of data science and its potential to improve government services. The text examines the integration of Volunteered Geographic Information (VGI) with data science in four government sectors in section four. Section five is specifically focused on tackling the difficulties that emerge within this particular framework. Ultimately the last part functions as the closing comments of the chapter, along with providing prospective suggestions for more study.

Volunteered Geographic Information (VGI)

VGI may be defined as a kind of user-generated content that encompasses a diverse range of location-specific data, such as geotagged photographs, check-ins, reviews, and other contributions originating from a specific geographical region (Haklay, 2013; Kitchin et al., 2018). As individuals engage with their environment, they willingly collect and share these pieces of information. VGI is often generated by individuals using location-aware devices such as smartphones, where users gather geographic data in the form of text, images, or videos that are tagged with geographical information (Honarparvar et al., 2019). Subsequently, this data is disseminated via diverse online social media platforms, such as Twitter, so facilitating the establishment of Volunteered Geographic Information (VGI). OpenStreetMap (OSM) was founded in 2004 as a worldwide initiative with the goal of developing and managing an accessible and modifiable database and map of the whole globe (Grinberger et al., 2022). Regarded as a prominent illustration of Volunteered Geographic Information (VGI), it distinguishes itself via its substantial volunteer community and the huge collection of geographical data it has produced.

The core concept behind VGI is the empowerment of users to actively participate in the creation and management of geographic information, as opposed to being passive recipients of government datasets (Goodchild, 2007).Emerging digital platforms have emerged, enabling users to collectively contribute geographic information, resulting in the joint creation of digital representations and conceptualizations of nature (Calcagni et al., 2019). These systems allow users to connect with one other, working together to create and visualize geographical elements in the digital world.

In urban settings, the occurrence of catastrophic events, such as severe floods, has significant consequences that affect both the human population and vital infrastructure like road networks. These road networks are crucial for everyday living and, importantly, for emergency response systems. Volunteered Geographic Information (VGI) has the capability to provide immediate and up-to-date data changes, allowing for swift reactions to rapidly changing conditions or occurrences. This is particularly advantageous in the context of disaster management endeavors. The geo-located information provided by affected populations during catastrophes may be very beneficial and influential (Tzavella et al., 2018).

The incorporation of Volunteered Geographic Information (VGI) has the potential to enhance the recommendation process of Recommender Systems (RS) by integrating contextual information, as suggested by Honarparvar et al. (2019). Users have the ability to contribute valuable geographical data to RS by incorporating VGI, which offers valuable insights into their preferences and interests for specific regions and areas of interest. Utilizing Volunteered Geographic Information (VGI) may improve the accuracy and relevance of location-based recommendations provided by Remote Sensing (RS). RS may use Volunteered Geographic Information (VGI) to provide suggestions for neighboring healthcare facilities, educational institutions, or social welfare agencies. Moreover, VGI has the capability to enhance the ever-changing character of recommendations by promptly incorporating alterations in user preferences and interactions with their environment (Honarparvar et al., 2019). Therefore, Volunteered Geographic Information (VGI) may enhance the customization and timeliness of recommendation systems (RS).

Land cover maps may enhance government agencies' understanding of the current land use trends in metropolitan regions. Nevertheless, it is important to evaluate these maps in order to ensure the accuracy and reliability of the data, which is crucial for urban planning and development. Volunteered Geographic Information (VGI) may be used to facilitate the land cover validation procedures (Antoniou et al., 2016; Fonte et al., 2015; Olteanu-Raimond et al., 2020). Public service companies may use validated land cover maps to precisely identify optimal locations for new infrastructure projects, neighborhoods, commercial areas, and public facilities. By promoting the systematic expansion of urban areas and optimizing resource use, this enhances the citizens' quality of life.

Advantages of Volunteered Geographic Information (VGI) in Governmental Delivery of Services

Volunteered Geographic Information (VGI) has emerged as a valuable resource for governments and public service groups. The primary advantages of Volunteered Geographic Information (VGI) in relation to public services are outlined in Figure 1 and described below.

Figure 1. Benefits of VGI for government service delivery

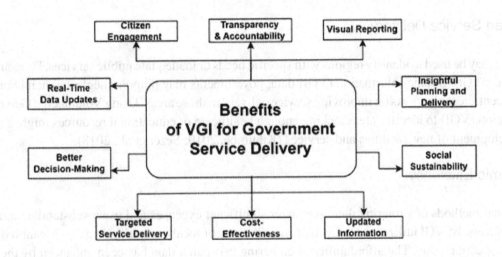

Public Participation

Public participation in civic affairs cultivates a feeling of empowerment among individuals, as they actively contribute to the development of public services. Government organizations may get valuable insights into the genuine requirements and preferences of the community by including individuals in the process of planning, designing, and evaluating services. Volunteered Geographic Information (VGI) may function as a significant instrument for involving individuals in public services (Güiza & Stuart, 2018; Sangiambut & Sieber, 2016). By engaging in this participation, it may be ensured that services are tailored to meet the distinct requirements of residents, so enhancing overall contentment and effectiveness.

Live Data Updates

The ability of VGI to provide real-time data updates is one of its most significant benefits. Users interact with their surroundings and willingly provide location-specific data via smartphones and other devices, allowing for immediate access to the data for analysis and decision-making purposes. The real-time capability of Volunteered Geographic Information (VGI) in public services may be particularly advantageous for emergency services, traffic management, and disaster response. It allows for prompt responses and informed decision-making (Annis & Nardi, 2019; Chen et al., 2016).

Enhanced Decision-Making

Volunteered Geographic Information (VGI) offers a diverse range of geospatial data that may assist different government sectors in making more informed choices. Public service companies may get valuable information about people' preferences, needs, and obstacles related to specific locations and services via the use of VGI data. This data-driven decision-making technique enables governments to develop policies that prioritize the needs of citizens and effectively allocate resources (Dias et al., 2019; Malik & Shaikh, 2019).

Focused Service Delivery

VGI data may be used to identify regions with specific needs or inadequate public services. By examining Volunteered Geographic Information (VGI) data, governments may pinpoint neglected neighborhoods and concentrate their efforts on improving service delivery in those areas. Using Volunteered Geographic Information (VGI) to identify areas lacking enough healthcare or educational resources might expedite the development of new facilities and services (Kihumbe, 2019; Sezer et al., 2018).

Economic Efficiency

Traditional methods of gathering data may incur significant expenses and use a substantial amount of time. Conversely, VGI utilizes the collective contributions of locals, resulting in a significant reduction in data collecting costs. The affordability of gathering geospatial data has been enhanced by the availability of extensive, collaboratively generated data via VGI platforms such as OpenStreetMap (OSM) (Grinberger et al., 2022).

Revised Data

VGI datasets often exhibit more currency and dynamism in comparison to conventional government datasets. Given the continuous influx of data from people, the information is consistently up-to-date and precise. Public service companies may consistently obtain the most current information using VGI data, allowing them to promptly respond to evolving situations (Honarparvar et al., 2019).

Social sustainability refers to the capacity of a society to meet the needs of its current and future members, while promoting social well-being, justice, and equity. It involves creating and maintaining social systems and

The social sustainability of VGI activities is enhanced by considering the needs of many groups, thereby promoting inclusivity. Vulnerable and disadvantaged communities have the opportunity to voice their concerns and actively engage in the development and implementation of public services using Volunteered Geographic Information (VGI). Consequently, the creation of services that are more inclusive and equitable may be facilitated (Bittner et al., 2016; Muzaffar et al., 2017).

Data Visualization

Geo-tagged photos and videos are specific types of Volunteered Geographic Information (VGI). Visual reporting utilizes geo-tagged photographs and videos obtained by modern smartphones and cameras to provide significant on-site data. Unmanned aerial vehicles equipped with cameras may assist rescue personnel in their endeavors and enhance the assessment of destruction in catastrophic situations. Geotagged images enhance the efficiency of delivering goods and services (To et al., 2015; Tsou et al., 2017).

Strategic Planning and Efficient Execution

Public sector firms may use location data to get insights into the geographical dispersion of various services and resources. Policymakers may effectively allocate resources by using this data to guide their decisions. It aids in identifying locations with specific needs for amenities such as healthcare facilities, educational facilities, transportation infrastructure, and more (Ahmad BinTouq, 2015).

Openness and Responsibility

Governments may use Volunteered Geographic Information (VGI) as a means to demonstrate openness by providing accessible data to the general public. Data science may also contribute to enhancing data quality, validating information, and preventing data exploitation, hence enhancing accountability in government services (Georgiadou et al., 2014).

DATA SCIENCE

Data science is a multidisciplinary domain that focuses on obtaining valuable information and insights from vast and diverse data sources (Subrahmanya et al., 2022). Data analytics encompasses several techniques for managing, examining, and interpreting data, including data mining, machine learning, statistical analysis, and data visualization. These techniques are used to facilitate decision-making processes (Asamoah et al., 2020). Data science may be used to provide public services by amalgamating data from diverse sources, such as administrative borders, surveys, social media, and VGI (Volunteered Geographic Information), in order to cultivate a more profound comprehension of societal patterns and citizen requirements. Data science enables governmental entities to get valuable insights from raw data, so promoting the formulation of well-informed policies and evidence-based decision-making (Gunapati, 2011). Integrating data analytics into public services enables governments and businesses to more effectively solve issues, anticipate people' demands, and allocate resources efficiently for maximum impact (Maffei et al., 2020).

In the field of governance, the introduction of data science technologies like artificial intelligence (AI), the Internet of Things (IoT), big data and behavioral/predictive analytics, and blockchain is about to bring about a significant and revolutionary period. The use of advanced technology has the potential to completely transform the way government services are provided by incorporating data-driven innovations (Engin & Treleaven, 2019; Mikhaylov et al., 2018). They can be employed in different fields, including public health development (R & B, 2022), providing public agricultural extension services (Namyenya et al., 2022), increasing awareness about climate change (Appelgren & Jönsson, 2021), creating policy innovation labs for digital governance of education (Williamson, 2015), and aiding in the planning and execution of public transportation (Keller et al., 2022).

The field of data science has initiated a significant and transformative movement in the public sector, leading to economic expansion and offering substantial prospects for improving the standard of public service provision (Manikam et al., 2019). Data science has the ability to establish a standard for the quality of public services by analyzing large amounts of data. This allows for the enhancement of services that prioritize the needs of citizens (Manikam et al., 2019). Data intelligence and analytics play a crucial role in the public sector by enabling decision-makers to make well-informed decisions, enhance efficiency, and promote positive outcomes (Di Vaio et al., 2022).The significant influence of big data analytics on the performance and efficiency of public organizations is clearly apparent, emphasizing its potential for transformation (Rogge et al., 2017; Wahdain et al., 2019).

Auditors in the public sector may use data analytics to maintain accountability and guarantee the appropriate administration of state funds (Novita & Indrany Nanda Ayu Anissa, 2022). The use of data analytics promotes transparency and enhances financial supervision in government bodies.The advancement of data science will have a substantial impact on the future duties and responsibilities of public sector organizations in many functional areas (Gamage, 2016).

Utilizing Data Science for Government Service Delivery

Data science applications have a broad scope of usefulness in the public sector. Below, a selection of these applications is summarized in Figure 2 and shown below.

Resource Planning May Be Optimized via the use of Predictive Analytics

Data science is widely used in the public sector for resource planning via the application of predictive analytics. Governmental entities may forecast future demand for a range of services, such as healthcare, education, and public safety, by analyzing historical data and trends. The Singaporean government use predictive analytics to strategically allocate resources and prepare for the healthcare needs of the aging population (Liming et al., 2015; Rangaswamy et al., 2021). By using this proactive approach, resources may be allocated more efficiently, resulting in enhanced service provision and decreased expenses.

Analyzing the Sentiment of Citizen Feedback

Extensive quantities of citizen comments collected via various channels, such as social media, surveys, and feedback forms, may be examined for sentiment analysis using data science techniques. By assessing the sentiment of individuals' thoughts and sentiments, governments may effectively gauge popular happiness, identify specific areas that need improvement, and promptly address public complaints (Muneeb & Chandler, 2021).

Transportation Optimization via Network Analysis

Network analysis is a crucial tool for enhancing urban transport networks by using data science approaches. Government entities may identify regions with significant traffic congestion, enhance public transportation pathways, and plan infrastructure development by examining traffic data. PTV Visum, a program developed by PTV Group, is used by the Barcelona Metropolitan Transport Authority (ATM) to conduct demand modeling and study different scenarios related to the provision of public transport services (Gomes, 2021).

Fraud Prevention via Anomaly Detection

Data science is crucial for detecting anomalies and fraudulent activities in government transactions. Government authorities have the ability to detect unusual trends and potential instances of fraud by examining financial data and transaction records. The Internal Revenue Service is using state-of-the-art technology to improve its capacity to detect tax fraud, concealed assets, money laundering, identity theft, and other types of noncompliance with greater effectiveness (Aprio, 2019; stahlesq.com, 2018).

Figure 2. Utilizing data science for government service delivery

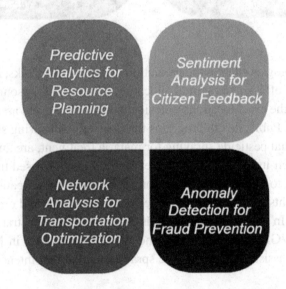

The synergy between Volunteered Geographic Information (VGI) and Data Science is being used to enhance government services.

The combination of data science and Volunteered Geographic Information (VGI) has the capacity to greatly improve government services via a range of methods. The possible uses of VGI and data science approaches are outlined in Figure 3 and further explained below.

Healthcare

Monitoring and Tracking of Diseases

Volunteered Geographic Information (VGI) may be used to provide up-to-date data on symptoms and locations, enabling the monitoring and analysis of disease transmission patterns. Analyzing this data

Figure 3. Potential benefits provided by volunteered geographic information (VGI) and the field of data science

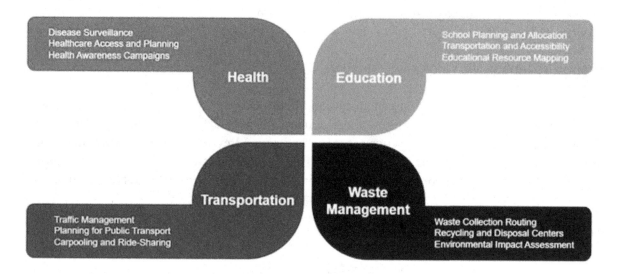

using data science methodologies enables the prediction of disease outbreaks, identification of high-risk areas, assessment of illness effect, and optimal allocation of healthcare resources. Precise cartographic representations that depict the exact locations of residences within a given area provide several benefits across multiple disciplines. Public health programs that focus on delivering services directly to households, such as indoor residual pesticide spraying for malaria treatment, are among the key areas where these maps demonstrate their usefulness. Sturrock et al., (2018) showcased the capability of ensemble machine learning methods to enhance the precision of building classification in VGI datasets. This research offers useful insights for targeted public health interventions and other uses, such as organizing fumigation campaigns. In a similar manner, Solís et al. (2018) demonstrated the use of Volunteered Geographic Information (VGI) to map building and road infrastructure in Mozambique and Kenya. This mapping was done to assist in an insecticide spray campaign that intended to reduce malaria and protect public health.

Rodriguez et al., (2021) conducted a comprehensive analysis to examine the influence of the Covid-19 pandemic on hospital services in key cities of Ecuador. They used advanced computational techniques and data from OSM to evaluate the spatial distribution and coverage of hospitals capable of treating Covid-19 patients. The study conducted by Qi et al. (2018) presented a new method for identifying and charting instances of food-borne illnesses by using Volunteered Geographic Information (VGI) datasets. Diggle (2005) emphasized the potential for corporations and governments to use data in order to enhance healthcare services for disadvantaged populations, hence guaranteeing fair and equal access to healthcare of superior quality.

Access to Healthcare and the Process of Planning for it

The use of VGI data enables the precise identification of healthcare institutions, identification of difficulties related to accessibility, and identification of gaps in services. These data may be evaluated utilizing the field of data science in order to strategically situate healthcare facilities and enhance healthcare accessibility for underprivileged groups. For example, Kihumbe (2019) illustrates the usefulness of Volunteered Geographic Information (VGI) in mapping regions with a high prevalence of HIV, whereas Mooney et al. (2013) evaluated the potential of OpenStreetMap (OSM) for widespread health computing applications. In addition, Malik & Shaikh (2019) examined the efficacy of VGI data in analyzing the spatial arrangement of healthcare institutions in Pakistan.

Ambulance services are crucial in the provision of healthcare services. The proper management and operation of ambulances is crucial, as it ensures optimal use of these costly and important assets. Ambulance services are essential for providing healthcare services. The importance of efficiently managing and running ambulances cannot be exaggerated, as it guarantees the optimal use of these costly and important assets. (Okyere et al., 2022) developed a VGI-based ambulance management system in Ghana.

Community asset mapping is essential in public health practices as it acknowledges a community's resources, needs, and successful techniques for urban health interventions. Kolak et al. (2020) introduced a novel participatory asset mapping system that utilizes Volunteered Geographic Information (VGI) in the field of urban health. The impact of air quality on public health has been significant for a considerable period of time. In order to successfully address air pollution in metropolitan areas, it is crucial to possess accurate and dependable data on air quality exposure. Gupta et al., (2018) provided useful insights on the systematic deployment of VGI air quality sensors to create precise air quality maps for urban monitoring.

Public Health Awareness Campaigns

Volunteered Geographic Information (VGI) data may be used to discern the population's specific information needs and apprehensions about their health. Data analytics may be used to identify patterns and develop targeted health awareness initiatives to effectively tackle certain health concerns. Noveck et al. (2017) have observed that the lack of public knowledge and misunderstandings considerably impede the efficiency of control efforts against mosquito-borne illnesses (MBDs). Dissemination of myths, rumors, and disinformation increases the likelihood of disease transmission within communities and hinders the effective mobilization of the public for preventative initiatives. Social media serves as a significant provider of Volunteered Geographic Information (VGI), and social media platforms are essential in bridging the information disparities within communities by promoting the interchange of information and raising awareness (Crowley et al., 2012). Noveck et al. (2017) proposed that sentiment analysis and network analysis of social media data obtained via crowdsourcing may be used to evaluate public sentiments on the dissemination of a particular illness, the efficacy of outreach initiatives, and potential areas for improvement.

Acquiring Knowledge and Skills Through Formal Instruction and Learning

School Planning and Allocation refers to the process of strategically organizing and distributing resources within an educational institution.

VGI data may be used to accurately identify areas characterized by high population density and limited access to educational services. Through the use of geographical data analysis, it is possible to identify areas that have limited access to high-quality educational opportunities. This information may then be used by governments and corporations to build and improve educational services for underprivileged populations. Data science methodologies may be used to examine the data and determine optimal sites for new schools or reallocate current resources, therefore promoting equitable access to top-notch education. Xie et al. (2019) observed that Volunteered Geographic Information (VGI) may be used for urban planning in regions where there is a lack of official building data. These projects have the potential to enhance social sustainability by reducing educational inequalities and empowering communities via education (Muzaffar et al., 2017).

Transportation and Accessibility

VGI data may be used to understand students' transportation needs. Data science may be used to create efficient school bus routes, identify areas with insufficient safe pedestrian walkways, and enhance transportation services, all with the aim of facilitating children' access to their schools. Sezer et al. (2018) performed a study that merged OpenStreetMap (OSM) data with demographic data in order to evaluate the level of accessibility to schools.

Educational Resource Mapping refers to the process of systematically identifying and categorizing educational resources available for use in teaching and learning.

Volunteered Geographic Information (VGI) may be used to compile comprehensive databases via the collective efforts of individuals who provide data about educational resources and facilities worldwide. Data science may evaluate this information to develop resource maps that facilitate data-driven decision-making for politicians and educators on resource allocation and improvement plans.In a study done by Filho et al. (2013), they tested the use of OpenStreetMap (OSM) in cartography disciplines to map cities. The findings of the study confirmed that OSM is very successful in improving the learning process. By using "local knowledge," participants were encouraged to actively participate in problem-oriented learning, focusing on real-life challenges within their communities. This method showcased the usefulness of OSM as a helpful instrument for integrating practical and pertinent activities into the learning process.

Means of Conveyance

Transportation Control

VGI data provides real-time information on accidents, road closures, and traffic congestion. These data may be evaluated using data science methodologies to predict traffic patterns, optimize signal timings, and provide alternative routes to enhance traffic flow. OpenStreetMap has shown to be a helpful data source for traffic simulations, offering information that is equivalent to, and in some instances, even better than the proprietary network models provided by government agencies (Rieck et al., 2015; Zilske et al., 2011).

Public Transport Planning

Public transportation is an essential and integral part of the transportation system, acting as a fundamental element for urban mobility. Nevertheless, the rapid development and urbanization have resulted in a significant increase in the need for public transit. Convenient and expeditious availability of both public and private amenities, such as educational institutions, recreational spaces, and retail establishments, plays a pivotal role in assessing the standard of living in metropolitan regions. Volunteered Geographic

Information (VGI) may provide insights about the demand for public transportation and the most popular transit routes. Data science may be used to design and improve public transport networks, which involves determining the most efficient routes and identifying areas with high levels of traffic.

Steiniger et al. (2016) developed a web-based framework to evaluate urban accessibility for transportation planning. The findings demonstrated that OSM data may be used to assess urban accessibility, albeit its comprehensiveness may differ across various regions.Tran et al., (2013) proposed a system for the synchronization of transport data between OpenStreetMap (OSM) and a General Transit Feed Specification (GTFS) dataset. Public transport providers were able to easily upload GTFS data to OSM and get valuable crowdsourced data in exchange. This framework facilitates the modification and enhancement of bus stop locations and facilities by using the existing GTFS data as a reference and leveraging online communities.

Shared Transportation and Ride-Sharing

Volunteered Geographic Information (VGI) may be used to advocate and improve initiatives for ride- and carpooling. Data science may be used to develop intuitive tools for facilitating carpooling arrangements and analyzing commuting trends in order to identify potential matches. Hariz et al. (2021) state that Volunteered Geographic Information (VGI) may improve transportation systems by maximizing passengers' happiness via the real-time interchange of data among the regional manager, public buses, vehicle ride-sharing services, and the riders themselves. In their study, Chao & Wang (2018) devised a model to evaluate the movement of bicycles by identifying optimal locations for docking stations. They accomplished this by using spatial information, including roads, walkways, bicycle lanes, and prominent landmarks from OSM, inside Taipei City.

Waste Management

Optimization of Waste Collection Routes

VGI data contains information on waste-generating trends and the specific locations of trash pickup sites. Data science has the potential to enhance garbage collection routes, hence reducing the environmental footprint and optimizing energy use. In their study, Dias et al. (2019) included Volunteered Geographic Information (VGI) into trash collection routes with the aim of increasing public participation, enhancing the collection rates of recyclable waste, and reducing logistical expenses. In a similar vein, Dowd et al. (2020) showcased the use of OpenStreetMap and Google OR-Tools in optimizing waste management collection routes, specifically tackling the challenges of mixed fleet routing difficulties. In addition, Felício et al. (2022) proposed an architecture that utilizes OpenStreetMap (OSM) for route planning in the specific context of solid waste collection.

Waste Management Facilities for Recycling and Disposing of Materials

VGI data may be used to cartographically represent recycling and disposal facilities and precisely identify areas without enough waste management infrastructures. Data analysis and data science may assist in strategically placing recycling plants in optimal locations and enhancing trash diversion rates. Inadequate planning has led to a significant volume of trash disposal, causing both economic and environmental concerns. In response to this issue, Thompson et al. (2013) proposed a solution for the distribution and movement of garbage containers. This method involves combining recyclable waste with other types of waste in the same containers and creating direct communication channels between residents and waste management authorities. In a similar vein, Mavakala et al. (2017) suggested investigating the use of crowdsourcing as a possible method for identifying, locating, and describing solid waste dumps.

Assessment of the Environmental Effects

Valuable information on illegal dumping and areas of high pollution may be obtained via Volunteered Geographic Information (VGI) data. Data science may be used to examine this data in order to ascertain the environmental impacts of inadequate waste management and to inform specific interventions aimed at mitigating environmental hazards. The use of spatial data may facilitate the implementation of sustainable waste reduction and management strategies, hence improving waste management policies (A. Singh, 2019).Singh et al. (2016) created a WebGIS platform that allows citizens to report, monitor, and visualize solid trash in metropolitan areas. A Geo-locator tool is developed to enable end-users to easily find the facility's locations for reporting and raising complaints.

Obstacles

Within the realm of public sector service delivery, Volunteered Geographic Information (VGI) offers many benefits, although it also presents certain obstacles that need thoughtful examination. This section examines notable VGI-related concerns, as emphasized and shown in Figure 4.

Data quality and validation refer to the process of ensuring the accuracy, consistency, and reliability of data. This involves verifying the integrity of the data and confirming that it meets certain standards and criteria.

Figure 4. Challenges of VGI

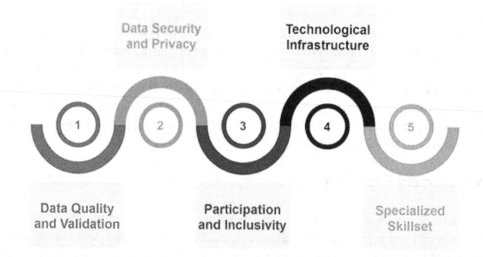

The dependability of VGI and crowdsourced data, which encompasses correctness, consistency, and completeness, might vary considerably. Therefore, it is crucial to use validation and verification processes to evaluate its quality. When it comes to delivering public services, using unreliable information might impede the efficient allocation of resources and decision-making processes (Ray & Bala, 2020). Inadequate data quality might result in suboptimal allocation of scarce resources and hinder the timely provision of assistance to vulnerable areas. Therefore, precise and verified data is crucial for making educated decisions and taking timely reaction measures.

Robust quality control techniques and data validation methods are essential in order to reduce the dissemination of erroneous information and the possible abuse of data. This includes the use of data cleansing, filtering, and cross-validation methodologies. Partnerships between academic institutions, organizations, and volunteers may be formed to guarantee the precision and dependability of data in diverse public service situations. Moreover, the use of Machine Learning methodologies may be applied to distinguish between spurious and genuine information and automate the identification and elimination of erroneous data (Ray & Bala, 2020).

Ensuring the Protection and Confidentiality of Data

Within the realm of public sector service delivery, the use of Volunteered Geographic Information (VGI) gives rise to apprehensions surrounding the safeguarding of data security and privacy. As the number of community members contributing data increases, it is crucial to protect their privacy and gain express agreement (Yingwei et al., 2020). Insufficient measures to protect data security and privacy may result in the exposure of personal information, undermining public trust and discouraging citizen involvement. Safeguarding privacy is essential for upholding the fundamental values of VGI, which include seizing opportunities, focusing on data, causing disruption, and promoting inclusion (B. T. Haworth et al., 2018).

The privacy issues faced by people may discourage them from sharing crucial information, so impeding the effectiveness of public sector operations. In order to guarantee the protection and confidentiality of data, it is necessary to develop rigorous protocols and rules. Utilizing data encryption techniques and adopting secure data storage protocols may effectively protect confidential information (Muthi Reddy et al., 2018). Moreover, it is essential to provide personnel engaged in VGI collecting with proper education about safe data management protocols.

Engagement and Accessibility

Facilitating the engagement of a wide range of community people in VGI programs might present difficulties. The inclusion of VGI activities may be limited by factors such as technological accessibility, literacy levels, language hurdles, and power imbalances within communities (B. T. Haworth et al., 2018). Limited engagement may lead to incomplete or prejudiced data, since it may neglect the needs and viewpoints of disadvantaged communities (B. T. Haworth et al., 2018). The participation of a diverse range of individuals improves the precision and pertinence of Volunteered Geographic Information (VGI) data, hence enhancing the efficacy of decision-making processes and focused actions.

It is important to make an active attempt to close these gaps and guarantee that underrepresented viewpoints are acknowledged and included in community-led initiatives (Bittner et al., 2016).One may arrange outreach programs to enhance knowledge and stimulate interest in VGI activities, particularly targeting vulnerable populations (B. Haworth et al., 2016). Offering instruction and assistance to persons lacking technology proficiency or digital literacy may significantly boost their involvement in VGI initiatives.

Technological infrastructure refers to the underlying framework of hardware, software, networks, and other technological components that support the functioning of a system or organization.

Restricted technological access may impede the maximum use of VGI in the realm of public sector service delivery (Li & Ulaganathan, 2017). Specific populations may have difficulties in accessing and effectively using the requisite platforms and tools for data input and analysis (Crawford & Finn, 2015). Inadequate technical infrastructure may limit the collection and use of VGI and data science approaches, resulting in delays in real-time monitoring, data transmission, and decision-making. The collection of crucial information may be hindered, which might obstruct the coordination of efficient reaction teams.

In order to surmount these challenges, it is essential to guarantee the presence of the necessary infrastructure and fair accessibility to technology and the Internet, particularly in disadvantaged regions. Creating software and platforms that are easy for users to use, can be used in several languages, and perform well with different hardware and networks helps reduce the negative effects of restricted technical infrastructure. Data science methodologies may also contribute to improving the transmission, storage, and processing of data to guarantee smooth information flow in public sector operations.

Expertise in a Certain Set of Skills

The lack of digital literacy and specialized skill sets among participants might impede the complete usage of Volunteered Geographic Information (VGI) in public sector delivery (Li & Ulaganathan, 2017). Insufficient skills and expertise might hinder people from successfully engaging in data collection, analysis, and interpretation (Sboui & Aissi, 2022). This limitation might undermine the precision and user-friendliness of the produced data. Additionally, the comprehensiveness and dependability of VGI may be constrained by a deficiency of specialized expertise and digital literacy, resulting in the gathering of inaccurate or insufficient data. Moreover, the examination and understanding of data might present difficulties, hindering the acquisition of key insights and informed decision-making in public sector endeavors.

In order to tackle this difficulty, it is essential to provide extensive training and technical assistance to people and communities involved in VGI initiatives (B. Haworth et al., 2016). Capacity-building initiatives may improve data collection techniques, data analysis approaches, and digital literacy proficiencies. Additionally, it is necessary to provide tools and platforms that are easy to use for prospective contributors of Volunteered Geographic Information (VGI), while also considering their varied backgrounds and talents. Data science has the potential to enhance data analysis tools, streamline data processing via automation, and provide valuable insights that may guide decision-making in the public sector.

CONCLUSION

The fields of data science and Volunteered Geographic Information (VGI) have emerged as transformative powers in the delivery of public services. VGI, a sort of user-generated content, democratizes geographic information by allowing users to actively contribute to data development and curation. Integrating Volunteered Geographic Information (VGI) into public sector operations might provide novel prospects for enhancing government services in several domains. Data science may enhance VGI by extracting relevant information from several data sources, including administrative borders, surveys, social media, and VGI itself. Data science facilitates the creation of policies based on data and enables decision-making based on evidence by using techniques such as data mining, machine learning, and statistical analysis. Moreover, contemporary technologies such as Artificial Intelligence (AI), Internet of Things (IoT), big data analytics and blockchain have the capability to fundamentally transform the provision of government services.

Volunteered Geographic Information (VGI) may provide immediate disease monitoring within the healthcare domain, hence improving the effectiveness of emergency response efforts. Integrating Volunteered Geographic Information (VGI) into recommender systems may enhance location-based suggestions for people, hence optimizing the delivery of public services. VGI's assistance in urban planning and land use validation enables a more efficient allocation of resources and the advancement of infrastructure development. In addition, VGI's citizen engagement program promotes empowerment and co-creation of public services, ensuring that they align with the authentic needs and preferences of the community. In order to fully harness the potential of VGI in public sector service provision, it is crucial to overcome many challenges such as data quality, security, inclusiveness, technical infrastructure, and specialized skill sets.

Future endeavors should prioritize on data quality control and validation methods for Volunteered Geographic Information (VGI) to further enhance government services. The facilitation of wider and more comprehensive citizen engagement will be achieved via the development of user-friendly tools and platforms, which will help to bridge the technical accessibility, divide and enhance digital literacy. Collaboration across organizations, volunteers, and academics may enhance the reliability and precision of Volunteered Geographic Information (VGI) in public sector applications. Furthermore, in order to effectively manage the ever-increasing amount of data and provide timely insights for prompt decision-making, it is essential to consistently enhance data science approaches. Governments should allocate resources towards data intelligence and analytics in order to enhance public service delivery, promote transparency, and strengthen financial monitoring. The use of data science will enhance efficiency, sustainability, and resource optimization in areas such as education, healthcare, transportation, and waste management.

REFERENCES

Abdulkarim, B., Kamberov, R., & Hay, G. J. (2014). Supporting Urban Energy Efficiency with Volunteered Roof Information and the Google Maps API. *Remote Sensing (Basel)*, 6(10), 9691–9711. Advance online publication. doi:10.3390/rs6109691

Ahmad, M., Khayal, M. S. H., & Tahir, A. (2022). Analysis of Factors Affecting Adoption of Volunteered Geographic Information in the Context of National Spatial Data Infrastructure. *ISPRS International Journal of Geo-Information*, 11(2), 120. doi:10.3390/ijgi11020120

Ahmad, B. (2015). The UAE Federal Government's E-Participation Roadmap: Developments in UAE Empowerment Initiatives With VGI/PGIS and Location Based Services (LBS). *Canadian Social Science*, 11(5). doi:10.3968/6919

Annis, A., & Nardi, F. (2019). Integrating VGI and 2D hydraulic models into a data assimilation framework for real time flood forecasting and mapping. *Geo-Spatial Information Science*, 22(4), 223–236. doi:10.1080/10095020.2019.1626135

Antoniou, V., Fonte, C. C., See, L., Estima, J., Arsanjani, J. J., Lupia, F., Minghini, M., Foody, G., & Fritz, S. (2016). Investigating the feasibility of geo-Tagged photographs as sources of land cover input data. *ISPRS International Journal of Geo-Information*, 5(5), 64. doi:10.3390/ijgi5050064

Appelgren, E., & Jönsson, A. M. (2021). Engaging Citizens for Climate Change—Challenges for Journalism. *Digital Journalism (Abingdon, England)*, 9(6), 755–772. doi:10.1080/21670811.2020.1827965

Aprio. (2019). *Artificial Intelligence to Help Resource-Strapped IRS More Efficiently Identify Tax Crimes.* https://www.aprio.com/artificial-intelligence-to-help-resource-strapped-irs-more-efficiency/

Arnaboldi, M., & Azzone, G. (2020). Data science in the design of public policies: Dispelling the obscurity in matching policy demand and data offer. *Heliyon*, 6(6), e04300. doi:10.1016/j.heliyon.2020.e04300 PMID:32637693

Asamoah, D. A., Doran, D., & Schiller, S. (2020). Interdisciplinarity in Data Science Pedagogy: A Foundational Design. *Journal of Computer Information Systems*, 60(4), 370–377. doi:10.1080/08874417.2018.1496803

Bittner, C., Michel, B., & Turk, C. (2016). Turning the spotlight on the crowd: Examining the participatory ethics and practices of crisis mapping. *ACME, 15*(1).

Brodie, M. L. (2019). What Is Data Science? In Applied Data Science: Lessons Learned for the Data-Driven Business. doi:10.1007/978-3-030-11821-1_8

Calcagni, F., Amorim Maia, A. T., Connolly, J. J. T., & Langemeyer, J. (2019). Digital co-construction of relational values: Understanding the role of social media for sustainability. *Sustainability Science*, 14(5), 1309–1321. doi:10.1007/s11625-019-00672-1

Chao, C. Y., & Wang, S. (2018). Using OpenStreetMap data for the location analysis on public bicycle stations – A case study on the Youbike system in downtown Taipei city. *Proceedings - 39th Asian Conference on Remote Sensing: Remote Sensing Enabling Prosperity, ACRS 2018*. IEEE.

Chen, X., Elmes, G., Ye, X., & Chang, J. (2016). Implementing a Real-Time Twitter-Based System for Resource Dispatch in Disaster Management. *GeoJournal*, 81(6), 863–873. doi:10.1007/s10708-016-9745-8

Crawford, K., & Finn, M. (2015). The limits of crisis data: Analytical and ethical challenges of using social and mobile data to understand disasters. *GeoJournal*, 80(4), 491–502. doi:10.1007/s10708-014-9597-z

Crowley, D. N., Breslin, J. G., Corcoran, P., & Young, K. (2012). Gamification of citizen sensing through mobile social reporting. *4th International IEEE Consumer Electronic Society - Games Innovation Conference, IGiC 2012*. IEEE. 10.1109/IGIC.2012.6329849

Di Vaio, A., Hassan, R., & Alavoine, C. (2022). Data intelligence and analytics: A bibliometric analysis of human–Artificial intelligence in public sector decision-making effectiveness. *Technological Forecasting and Social Change*, 174, 121201. doi:10.1016/j.techfore.2021.121201

Dias, V. E. C., Sperandio, V. G., & Lisboa-Filho, J. (2019). Routes generation for selective collection of urban waste using volunteered geographic information. *Iberian Conference on Information Systems and Technologies, CISTI, 2019-June*. IEEE. 10.23919/CISTI.2019.8760773

Diggle, P. (2005). Applied Spatial Statistics for Public Health Data. *Journal of the American Statistical Association, 100*(470), 702–703. Advance online publication. doi:10.1198/jasa.2005.s15

Dowd, M., Dixon, A., & Kinsella, B. (2020). Optimizing waste management collection routes in Urban Haiti: A collaboration between DataKind and SOIL. *ArXiv, 2012*.

Engin, Z., & Treleaven, P. (2019). Algorithmic Government: Automating Public Services and Supporting Civil Servants in using Data Science Technologies. *The Computer Journal, 62*(3), 448–460. Advance online publication. doi:10.1093/comjnl/bxy082

Felício, S., Hora, J., Ferreira, M. C., Abrantes, D., Costa, P. D., Dangelo, C., Silva, J., & Galvão, T. (2022). Handling OpenStreetMap georeferenced data for route planning. *Transportation Research Procedia, 62*, 189–196. doi:10.1016/j.trpro.2022.02.024

Filho, H. F., Leite, B. P., Pompermayer, G. A., Werneck, M. G., & Leyh, W. (2013). Teaching VGI as a strategy to promote the production of urban digital cartographic databases. *Joint Urban Remote Sensing Event 2013. JURSE, 2013*. doi:10.1109/JURSE#.2013.6550705

Fonte, C. C., Bastin, L., See, L., Foody, G., & Lupia, F. (2015). Usability of VGI for validation of land cover maps. *International Journal of Geographical Information Science, 29*(May), 1–23. doi:10.1080/13658816.2015.1018266

Gamage, P. (2016). New development: Leveraging 'big data' analytics in the public sector. *Public Money & Management, 36*(5), 385–390. doi:10.1080/09540962.2016.1194087

Georgiadou, Y., Lungo, J. H., & Richter, C. (2014). Citizen sensors or extreme publics? Transparency and accountability interventions on the mobile geoweb. *International Journal of Digital Earth, 7*(7), 516–533. doi:10.1080/17538947.2013.782073

Gomes, M. (2021). Barcelona improves public transport with modelling technology. *PTV Blog*. https://blog.ptvgroup.com/en/city-and-mobility/barcelona-improve-public-transport-modelling/

Goodchild, M. F. (2007). Citizens as sensors: The world of volunteered geography. In GeoJournal. doi:10.1007/s10708-007-9111-y

Grinberger, A. Y., Minghini, M., Juhász, L., Yeboah, G., & Mooney, P. (2022). OSM Science—The Academic Study of the OpenStreetMap Project, Data, Contributors, Community, and Applications. In ISPRS International Journal of Geo-Information, 11(4). doi:10.3390/ijgi11040230

Güiza, F., & Stuart, N. (2018). When citizens choose not to participate in volunteering geographic information to e-governance: a case study from Mexico. In GeoJournal. doi:10.1007/s10708-017-9820-9

Gunapati, S. (2011). Key Features for Designing a Dashboard. *Government Finance Review (0883-7856)*, 27.

Gupta, S., Pebesma, E., Degbelo, A., & Costa, A. C. (2018). Optimising citizen-driven air quality monitoring networks for cities. *ISPRS International Journal of Geo-Information*, *7*(12), 468. doi:10.3390/ijgi7120468

Haklay, M. (2013). Citizen science and volunteered geographic information: Overview and typology of participation. In Crowdsourcing Geographic Knowledge: Volunteered Geographic Information (VGI) in Theory and Practice. Springer. doi:10.1007/978-94-007-4587-2_7

Hariz, M. (2021). A dynamic mobility traffic model based on two modes of transport in smart cities. *Smart Cities*, *4*(1), 253–270. doi:10.3390/smartcities4010016

Haworth, B., Whittaker, J., & Bruce, E. (2016). Assessing the application and value of participatory mapping for community bushfire preparation. *Applied Geography (Sevenoaks, England)*, *76*, 115–127. doi:10.1016/j.apgeog.2016.09.019

Haworth, B. T., Bruce, E., Whittaker, J., & Read, R. (2018). The good, the bad, and the uncertain: Contributions of volunteered geographic information to community disaster resilience. In Frontiers in Earth Science (Vol. 6). doi:10.3389/feart.2018.00183

Honarparvar, S., Forouzandeh Jonaghani, R., Alesheikh, A. A., & Atazadeh, B. (2019). Improvement of a location-aware recommender system using volunteered geographic information. *Geocarto International*, *34*(13), 1496–1513. doi:10.1080/10106049.2018.1493155

Hu, S., Ge, Y., Liu, M., Ren, Z., & Zhang, X. (2022). Village-level poverty identification using machine learning, high-resolution images, and geospatial data. *International Journal of Applied Earth Observation and Geoinformation*, *107*, 102694. doi:10.1016/j.jag.2022.102694

Huang, H., Yao, X. A., Krisp, J. M., & Jiang, B. (2021). Analytics of location-based big data for smart cities: Opportunities, challenges, and future directions. *Computers, Environment and Urban Systems*, *90*, 101712. doi:10.1016/j.compenvurbsys.2021.101712

Keller, C., Glück, F., Gerlach, C. F., & Schlegel, T. (2022). Investigating the Potential of Data Science Methods for Sustainable Public Transport. *Sustainability (Basel)*, *14*(7), 4211. doi:10.3390/su14074211

Kihumbe. (2019). *Survey Digitization and Mapping for HIV Monitoring*. Kihumbe.

Kitchin, R., Lauriault, T. P., & Wilson, M. W. (2018). Understanding Spatial Media. In Understanding Spatial Media. doi:10.4135/9781526425850.n1

Kolak, M., Steptoe, M., Manprisio, H., Azu-Popow, L., Hinchy, M., Malana, G., & Maciejewski, R. (2020). *Extending Volunteered Geographic Information (VGI) with Geospatial Software as a Service: Participatory Asset Mapping Infrastructures for Urban Health*. Springer. doi:10.1007/978-3-030-19573-1_11

Kucukali, A., Pjeternikaj, R., Zeka, E., & Hysa, A. (2022). Evaluating the pedestrian accessibility to public services using open-source geospatial data and QGIS software. *Nova Geodesia*, *2*(2), 42. doi:10.55779/ng2242

Li, L., & Ulaganathan, M. N. (2017). Design and development of a crowdsourcing mobile app for disaster response. *International Conference on Geoinformatics, 2017-August*. IEEE. 10.1109/GEOINFORMATICS.2017.8090943

Liming, B., Gavino, A. I., Lee, P., Jungyoon, K., Na, L., Pink Pi, T. H., Xian, T. H., Buay, T. L., Xiaoping, T., Valera, A., Jia, E. Y., Wu, A., & Fox, M. S. (2015). SHINESeniors: Personalized services for active ageing-in-place. *2015 IEEE 1st International Smart Cities Conference, ISC2 2015*. IEEE. 10.1109/ISC2.2015.7366181

Ma, Q., Wang, L., Gong, X., & Li, K. (2022). Research on the Rationality of Public Toilets Spatial Layout based on the POI Data from the Perspective of Urban Functional Area. *Journal of Geo-Information Science*, 24(1). doi:10.12082/dqxxkx.2022.210331

Maffei, S., Leoni, F., & Villari, B. (2020). Data-driven anticipatory governance. Emerging scenarios in data for policy practices. *Policy Design and Practice*, 3(2), 123–134. doi:10.1080/25741292.2020.1763896

Malik, N. A., & Shaikh, M. A. (2019). Spatial distribution and accessibility to public sector tertiary care teaching hospitals: Case study from Pakistan. *Eastern Mediterranean Health Journal*, 25(6), 431–434. doi:10.26719/emhj.18.049 PMID:31469163

Manikam, S., Sahibudin, S., & Kasinathan, V. (2019). Business intelligence addressing service quality for big data analytics in public sector. *Indonesian Journal of Electrical Engineering and Computer Science*, 16(1), 491. doi:10.11591/ijeecs.v16.i1.pp491-499

Mavakala, B., Mulaji, C., Mpiana, P., Elongo, V., Otamonga, J.-P., Biey, E., Wildi, W., & Pote-Wembonyama, J. (2017). Citizen Sensing of Solid Waste Disposals : Crowdsourcing As Tool Supporting Waste Management in a. *Proceedings Sardinia 2017 / Sixteenth International Waste Management and Landfill Symposium.S. Margherita Di Pula, Cagliari, Italy - 2 - 6 October 2017 -*. 2017. IEEE. https://doi.org/https://archive-ouverte.unige.ch/unige:97650

Mikhaylov, S. J., Esteve, M., & Campion, A. (2018). Artificial intelligence for the public sector: Opportunities and challenges of cross-sector collaboration. *Philosophical Transactions. Series A, Mathematical, Physical, and Engineering Sciences*, 376(2128), 20170357. doi:10.1098/rsta.2017.0357 PMID:30082303

Muneeb, S., & Chandler, T. W. (2021). *Sentiment Analysis for COVID-19 National Vaccination Policy*. Pride. https://pide.org.pk/blog/sentiment-analysis-for-cov

Muthi Reddy, P., Manjula, S. H., & Venugopal, K. R. (2018). Secured Privacy Data using Multi Key Encryption in Cloud Storage. *Proceedings of 5th International Conference on Emerging Applications of Information Technology, EAIT 2018*. IEEE. 10.1109/EAIT.2018.8470399

Muzaffar, H. M., Tahir, A., Ali, A., Ahmad, M., & McArdle, G. (2017). Quality assessment of volunteered geographic information for educational planning. In *Volunteered Geographic Information and the Future of Geospatial Data*. IGI Global. doi:10.4018/978-1-5225-2446-5.ch005

Namyenya, A., Daum, T., Rwamigisa, P. B., & Birner, R. (2022). E-diary: A digital tool for strengthening accountability in agricultural extension. *Information Technology for Development*, 28(2), 319–345. doi:10.1080/02681102.2021.1875186

Noveck, B. S., Ayoub, R., Hermosilla, M., Marks, J., & Suwondo, P. (2017). *Smarter Crowdsourcing for Zika and Other Mosquito-Borne Diseases*.

Novita, N., & Indrany Nanda Ayu Anissa, A. (2022). The role of data analytics for detecting indications of fraud in the public sector. *International Journal of Research in Business and Social Science (2147-4478), 11*(7). doi:10.20525/ijrbs.v11i7.2113

Okyere, F., Minnich, T., Sproll, M., Mensah, E., Amartey, L., Otoo-Kwofie, C., & Brunn, A. (2022). Implementation of a low-cost Ambulance Management System. *Dgpf, 30*(March).

Olteanu-Raimond, A. M., See, L., Schultz, M., Foody, G., Riffler, M., Gasber, T., Jolivet, L., le Bris, A., Meneroux, Y., Liu, L., Poupée, M., & Gombert, M. (2020). Use of automated change detection and VGI sources for identifying and validating urban land use change. *Remote Sensing (Basel), 12*(7), 1186. doi:10.3390/rs12071186

Provost, F., & Fawcett, T. (2013). Data Science and its Relationship to Big Data and Data-Driven Decision Making. *Big Data, 1*(1), 51–59. doi:10.1089/big.2013.1508 PMID:27447038

Qi, Y., Guo, K., Zhang, C., Guo, D., & Zhi, Z. (2018). A VGI-based Foodborn Disease Report and Forecast System. *Proceedings of the 4th ACM SIGSPATIAL International Workshop on Safety and Resilience, EM-GIS 2018*. ACM. 10.1145/3284103.3284124

R, J., & B, S. (2022). Social Impacts of Data Science in Food, Housing And Medical Attention Linked to Public Service. *Technoarete Transactions on Advances in Data Science and Analytics, 1*(1). doi:10.36647/TTADSA/01.01.A004

Rangaswamy, E., Periyasamy, G., & Nawaz, N. (2021). A study on singapore's ageing population in the context of eldercare initiatives using machine learning algorithms. *Big Data and Cognitive Computing, 5*(4), 51. doi:10.3390/bdcc5040051

Ray, A., & Bala, P. K. (2020). Social media for improved process management in organizations during disasters. *Knowledge and Process Management, 27*(1), 63–74. doi:10.1002/kpm.1623

Rieck, D., Schünemann, B., & Radusch, I. (2015). Advanced Traffic Light Information in OpenStreetMap for Traffic Simulations. In Lecture Notes in Mobility. Springer. doi:10.1007/978-3-319-15024-6_2

Rodriguez, G., Torres, H., Fajardo, M., & Medina, J. (2021). Covid-19 in Ecuador: Radiography of Hospital Distribution Using Data Science. *ETCM 2021 - 5th Ecuador Technical Chapters Meeting*. IEEE. 10.1109/ETCM53643.2021.9590641

Rogge, N., Agasisti, T., & De Witte, K. (2017). Big data and the measurement of public organizations' performance and efficiency: The state-of-the-art. *Public Policy and Administration, 32*(4), 263–281. doi:10.1177/0952076716687355

Sangiambut, S., & Sieber, R. (2016). The V in VGI: Citizens or Civic Data Sources. *Urban Planning, 1*(2), 141–154. doi:10.17645/up.v1i2.644

Sboui, T., & Aissi, S. (2022). *A Risk-based Approach for Enhancing the Fitness of use of VGI*. IEEE., doi:10.1109/ACCESS.2022.3201022

Sezer, A., Deniz, M., & Topuz, M. (2018). Analysis of Accessibility of Schools in Usak City via Geographical Information Systems (GIS). *Tarih Kultur Ve Sanat Arastirmalari Dergisi-Journal Of History Culture And Art Research, 7*(5).

Singh, A. (2019). Remote sensing and GIS applications for municipal waste management. *Journal of Environmental Management, 243*, 22–29. doi:10.1016/j.jenvman.2019.05.017 PMID:31077867

Singh, Y. P., Singh, A. K., & Singh, R. P. (2016). *Web GIS based Framework for Citizen Reporting on Collection of Solid Waste and Mapping in GIS for Allahabad City. SAMRIDDHI : A Journal of Physical Sciences.* Engineering and Technology. doi:10.18090/samriddhi.v8i1.11405

Solís, P., McCusker, B., Menkiti, N., Cowan, N., & Blevins, C. (2018). Engaging global youth in participatory spatial data creation for the UN sustainable development goals: The case of open mapping for malaria prevention. *Applied Geography (Sevenoaks, England), 98*, 143–155. doi:10.1016/j.apgeog.2018.07.013

stahlesq.com. (2018). *How the IRS Uses Artificial Intelligence to Detect Tax Evaders.* Stahlesq. https://stahlesq.com/criminal-defense-law-blog/how-the-irs-uses-artificial-intelligence-to-detect-tax-evaders/

Steiniger, S., Poorazizi, M. E., Scott, D. R., Fuentes, C., & Crespo, R. (2016). Can we use OpenStreetMap POIs for the Evaluation of Urban Accessibility? *International Conference on GIScience Short Paper Proceedings, 1.* ACM. 10.21433/B31167F0678P

Sturrock, H. J. W., Woolheater, K., Bennett, A. F., Andrade-Pacheco, R., & Midekisa, A. (2018). Predicting residential structures from open source remotely enumerated data using machine learning. *PLoS One, 13*(9), e0204399. Advance online publication. doi:10.1371/journal.pone.0204399 PMID:30240429

Subrahmanya, S. V. G., Shetty, D. K., Patil, V., Hameed, B. M. Z., Paul, R., Smriti, K., Naik, N., & Somani, B. K. (2022). The role of data science in healthcare advancements: applications, benefits, and future prospects. In Irish Journal of Medical Science, 191(4). doi:10.1007/s11845-021-02730-z

Thompson, A. F., Afolayan, A. H., & Ibidunmoye, E. O. (2013). Application of geographic information system to solid waste management. *2013 Pan African International Conference on Information Science, Computing and Telecommunications, PACT 2013.* IEEE. 10.1109/SCAT.2013.7055110

To, H., Kim, S. H., & Shahabi, C. (2015). Effectively crowdsourcing the acquisition and analysis of visual data for disaster response. *Proceedings - 2015 IEEE International Conference on Big Data, IEEE Big Data 2015.* IEEE. 10.1109/BigData.2015.7363814

Tran, K., Barbeau, S., Hillsman, E., & Labrador, M. A. (2013). GO_Sync - A Framework to Synchronize Crowd-Sourced Mapping Contributors from Online Communities and Transit Agency Bus Stop Inventories. *International Journal of Intelligent Transportation Systems Research, 11*(2), 54–64. doi:10.1007/s13177-013-0056-x

Tsou, M. H., Jung, C. Te, Allen, C., Yang, J. A., Han, S. Y., Spitzberg, B. H., & Dozier, J. (2017). Building a real-time geo-targeted event observation (Geo) viewer for disaster management and situation awareness. *Lecture Notes in Geoinformation and Cartography.* doi:10.1007/978-3-319-57336-6_7

Tzavella, K., Fekete, A., & Fiedrich, F. (2018). Opportunities provided by geographic information systems and volunteered geographic information for a timely emergency response during flood events in Cologne, Germany. *Natural Hazards, 91.* doi:10.1007/s11069-017-3102-1

Wahdain, E. A., Baharudin, A. S., & Ahmad, M. N. (2019). Big data analytics in the malaysian public sector: The determinants of value creation. *Advances in Intelligent Systems and Computing, 843,* 139–150. doi:10.1007/978-3-319-99007-1_14

Williamson, B. (2015). Governing methods: Policy innovation labs, design and data science in the digital governance of education. *Journal of Educational Administration and History, 47*(3), 251–271. doi:10.1080/00220620.2015.1038693

Wong, W., & Hinnant, C. C. (2022). Competing perspectives on the Big Data revolution: A typology of applications in public policy. *Journal of Economic Policy Reform.* doi:10.1080/17487870.2022.2103701

Xie, X., Zhou, Y., Xu, Y., Hu, Y., & Wu, C. (2019). OpenStreetMap Data Quality Assessment via Deep Learning and Remote Sensing Imagery. *IEEE Access : Practical Innovations, Open Solutions, 7,* 176884–176895. doi:10.1109/ACCESS.2019.2957825

Yingwei, Y., Dawei, M., & Hongchao, F. (2020). A research framework for the application of volunteered geographic information in post-disaster recovery monitoring. *Tropical Geography, 40*(2). doi:10.13284/j.cnki.rddl.003239

Zilske, M., Neumann, A., & Nagel, K. (2011). OpenStreetMap For Traffic Simulation. M. Schmidt, G. Gartner *(Eds.), Proceedings of the 1st European State of the Map – OpenStreetMap Conference, No. 11-10.* Springer.

Chapter 21
The Proposed Framework of View–Dependent Data Integration Architecture

Pradeep Kumar

 https://orcid.org/0000-0002-0206-9808

Teerthanker Mahaveer University, India

Madhurendra Kumar

Ministry of Electronics and Information Technology, CDAC, India

Rajeev Kumar

 https://orcid.org/0000-0002-4141-1282

Moradabad Institute of Technology, India

ABSTRACT

In this chapter the authors have proposed framework that have we are using various techniques to overcome the above mentioned challenges and proving the exact identification and extraction of WQI by classifying them according to their domain. The framework has been represented in system level design as a high level view of the view dependent data integration system and the architectural framework of the system design. Being a multi-database-oriented system, it is scalable to structure as well as unstructured data source. The wrapper and mediator module of the system is used to map the web-query-interface to the global schema of the integrated web query interfaces. In this the authors have implemented high level view of the system modeling from the end users' point of view; and the operational framework design also been represented.

INTRODUCTION

Deep web provides huge amount of domain-specific content like medical data, houses on sale data, e-commerce and science data etc. All these data stored on deep web can be accessed with the help of various HTML forms named Web Query Interface. This information stored in deep web can be fetched

DOI: 10.4018/979-8-3693-2964-1.ch021

from multiple database servers but one at a time that makes the information accessing process inefficient. To overcome this, a different approach is used by integrating Web Query Interface (IWQI) that is able to perform the query operation on multiple database servers at a time and acts as an independent entry point. In the proposed architecture we have given an innovative method for integrating web forms along with IWQI based on VDIS (View-based Data Integration System) and linked data. We have proposed a novel and alternative solution for combining various WQI into single IWQI for a specific domain for the advanced registrations based systems for enhanced security and reliability. Our proposed method starts with an integrated system by wrapping of various domain-wise WQI till the development of single integrated WQI.

The deep web, also called invisible web, hidden web is a part of World Wide Web. Its content can't directly index by common search engines like Yahoo, Google or Bing. Deep web acts like an umbrella for various parts of internet that are not fully accessible through the mentioned search engines. In the working process of crawling a web, every search engine uses search robots to add new contents to find any appropriate search engine index. The range of deep web is unknown to everyone, but according to the estimates by experts that 1% of all content over the web is indexed and crawled by search engines. The content that is available for access through search engines is known as surface web. As the content stored in it, deep web includes private data over various social media sites, emails, chat messages, bank statements etc. Search Robots are capable to search the paid sites like news articles or educational sites those requires some kind of subscription and for the open access sites like YouTube, Netflix. The terms deep web and dark web are uses similarly sometimes, still they are having some differences. Dark web is a part of larger web and the content stored in it is accessible via any search engine like Google.

Dark web is less reputed as it involves many illegal activities like malware attacks, cyber attacks, black market of stolen credit card, etc. In contrast of this, deep web contains legal content like academic journals and research databases. Although deep web can't be indexed by standard search engines, still, to access deep web is safe and easily accessible by many users. Gmail, LinkedIn sign in are some of the examples of deep web sites. The data on deep web contains personal information, that's why they are having restrictive access. The value of deep web content becomes enormous if the most iconic commodity of current information age is actually information. By having these facts, Bright Planet shows various facts and findings in a study. The study is actually based on collected data between 13 and 20 March 2000. The findings include-

- Deep web is having a huge public information dataset. There available information is 400-500 times larger than the data in commonly defined in www.
- Deep web is having 7500 Tb of information in its database. On contrary surface web is having only 19 Tb of information.
- 550 billion individual documents are stored in deep web that are far more than surface web.
- In current time near about 200,000 deep websites exists all over the web.
- On an average, data traffic received by the deep websites is 50% more than the data traffic received by the surface web.
- Content available on deep web is highly connected and relevant to any information related to market, need or domain.
- Deep web provides quality content 1000- 2000 times greater than surface web.

The content stored in deep web database can be accessed directly by URLs or through specific IP addresses if the website is open source kind of and it may require security access to fetch public-web pages. The deep web content is stored behind HTTP forms and it is useful for various common platforms like online banking, emailing, paywalls, etc. Figure 1 is showing the step-wise sequence of various phases involved in deep web query interface.

Phase 1: Query to Deep Websites

There is millions of data is stored on the web and to access the content with more optimization there are various query interfaces are being used. Queries are helpful in getting the proper and specific content from the web instantly. Two approaches are used to accomplish this task. First is to access the content from the deep web and add the index of the content to standard search engines. Second is by integrating search capabilities by various sources and supports integrated access to the content. This phase of deep web architecture is having the queries that are working in fetching and manipulating the data stored in deep web. Queries written in various languages like Oracle, SQL, etc are used to access the content in deep web integration approach.

Phase 2: Web Form Identification, Filling, and Submission

Deep web working nature is quite different in comparison to surface web. In this phase the data sent by the web forms from previous phase is received by the web forms classifiers for managing the data storage standards. Further, that classified data became available for search through any appropriate and automatic form filling and submission web page. The web forms that are being used to access or store data to a particular location are identified for various security aspects, so that the data stored in web page can be submitted to the deep web server for further processing.

Phase 3: Displaying Results

After the successful completion of previous phases an appropriate result is displayed to the user. In this phase actually the data received by the previous phase gets stored in the form of response pages and then the results are displayed to the user on his/ her browser.

VIEW DEPENDENT DATA INTEGRATION ARCHITECTURE: THE PROPOSED FRAMEWORK

The deep web is a dynamic web that uses web query interface methods to retrieve its content from various data sources like database or traditional file system. Web query interface is nothing but a HTML form that is beyond the search mechanism of various standard search engines like Google, Yahoo, Bing, etc. Deep web is huge in size and called as larger than surface web 500 times and continuously it is growing its size at an exponential speed. For the purpose of connecting end-user with deep web database WQI are introduced. They are performing as an intermediate between the user and deep web database through various communication mechanisms. The working culture of deep web query interface can be discussed in multiple sequential steps that may include the following-

Figure 1. Stages involved in deep web processing

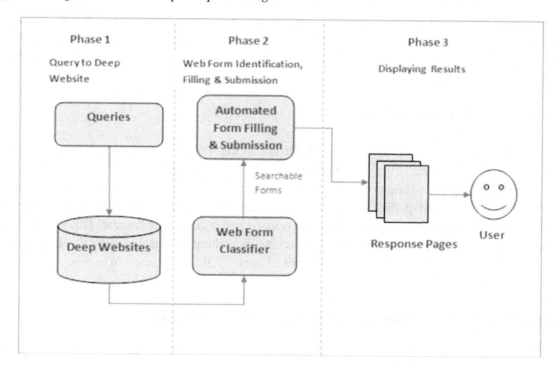

Submission of Query by User- It is the very first and basic step for the execution of any WQI. In this step a user submits his/ her query to the WQI having the consideration of semantics for every element involved in WQI. The information related to the knowledge of metadata of every element and the organizational structure of WQI is also considered here.

Sending Query to Domain-Specific Web Query Interface- There can be several WQIs available for any given domain for ex- medical information. To retrieve the information from the deep web essentially requires filling of every WQI, sending, gathering, integrating and inspecting the results. It became a time consuming task due to its complexity in nature. To resolve this issue, an integrated WQI represents a complete set of multiple but related WQI for a given specific domain that when receives any domain-specific query; it directly and automatically transforms and submit the query to an individual WQIs (Single WQI for each database).

Results Generated by Integrated Web Query Interface- after developing a domain-specific integrated web query interface the results are generated according to the nature of the query to fetch or manipulate data to/ from multiple deep web servers. The generated results are represented through specific HTML web forms.

Various challenges that occurs while developing an automatic integrated web query interface through combining multiple independent WQIs for a given domain may contain many challenges-

- Different WQI's elements like labels, text boxes, check boxes, radio buttons that are used in multiple domains for representing the same concept. For ex- in medical domain, the label, Age can be represented by selection button or textboxes.

- In different WQIs various labels are used for the description of same concept. For ex- Town, City, State can be referred as Location.
- Here, an ambiguous searching criterion exists. For ex- the word "tenure" can refer to choice of renting or buy something.

By considering various schema matching concepts and techniques, various researchers represented the integration issue in WQI. In their work, they have considered various interface matching issues like- A domain containing large set of source, finding mapping between source and query interface attributes called semantic correspondence. An important component that plays a vital role to integrate multiple WQIs is named as Interface matching that involves majorly three main tasks that are Schema Extraction, Schema Matching and Interface Modeling.

Schema Extraction- A WQI renders in the form of multiple HTML forms. HTML form is basically consists of multiple visual representations of attributes. Although, in WQI the form doesn't directly specify attribute to attribute or attribute to label relationship. Schema extraction is used to infer the structural aspects of WQI from its visual representation.

Schema Matching- It is required to determine the groups accurately by mapping various attributes among different WQIs. Mapping is of two types: Simple mapping and Complex mapping. Simple mapping represents the semantic correspondence of one-to-one among two attributes while in Complex mapping includes semantics correspondence of one-to-many among multiple attributes among various another WQI.

Interface Modeling- WQI consists of various multiple attributes by combining related attributes with each other to form a group. The groups that define close relations with each other can further form a group having subgroups. In WQIs, attributes and attribute are intuitively ordered.

For creating integrated Web Queries, the proposed framework represents an innovative method, that is based on linked data techniques. The data that can be published over internet according to the rules and standards of www is called Linked data. Linked data follows open W3C standards like SPARQL and RDF for the construction of meaningful information. Our proposed method is capable in integration of WQI and flexible to apply on any domain. By using linked data approach it became easy in data exploration, retrieval, comparison of various labels retrieved from different web pages. By using this approach the analysis of WQI is also became easy through its structured format and it provides an organized and reliable identification of particular labels like range or dates. The language, SPARQL is used to access largest possible count of elements included in each web form like- textboxes, drop down lists, check boxes, etc. The presented work is based on an ontology that supports every interested domain by the use of semantic information using linked data techniques. The proposed architecture is basically considers a set of common standards to identify WQI elements and to map them in a structures W3C standard formats. The requirement to represent WQI as linked data is the initialization of various relationships among different W3C string based elements to obtain the similarity which further follows the integration process. The SPARQL language is used to query the linked data. Major contribution of the proposed architecture can be listed as follows-

- A new method for integration of WQI using linked data.
- An ontology that is domain-independent to map various WQI schemas.

- To determine the relationships and relevancy of each individual WQI element for creating Integrated WQI.

Design Oriented Methodology

The complete methodology for a model for integrated web query interface is proposed in this section. The methodology is divided in two categories according to its levels of complexity. Firstly, the high level schema design is represented through figure- 1.2 it is showing the brief process of web query interface integration, next an architectural diagram is shown in figure- 1.3 to show the deep architectural workflow of the model. The high level design diagram is having various levels and the working of each level is described significantly. The detailed architectural approach is described by the global schema in terms of local information among various independent OR-WQIs in this framework. Architectural diagram shows the broad working architecture by initially starting with multiple WQIs generated from various source up-to the generation of web query interface integration. The complete description of each level is discussed further.

High Level Schema Design

The proposed framework initially starts with High Level Schema Design model architecture (Figure. 1.2). We have used multi-databases systems and federated system technologies as an inspiration in development of our method. Our work is providing single point access of all heterogeneous source data. The proposed method is having four major modules that are source, mediator, wrapper and application module. The complete description of their work abilities can be explained as follows-

Source Module

This module is one of the primary building blocks of High Level Schema Design model. It is constructed by the structured data like web pages, text files, databases, and XML files. At the source module the structured linked data is collected from various data sources to perform the extraction operation on them. All the resources that a source module uses to get the linked data can be described more precisely as follows-

❖ **Web Pages**

According to the user and business requirements website architecture is utilized by creating a logical layout of website. It basically defines the various components that will make a complete website and each component or website as a whole will provide services. Here, some factors that are the parts of website architecture are mentioned-

- **Technical aspects** like server, memory, interface for communication etc can be considered as the technological requirements of any website architecture.
- **Functional Constraints** like processes, services, various types of services that a website will provide.

- **Graphics & Visual advents** that describes the user interface, graphical elements like color, text styles, buttons, textboxes, etc.
- **Authenticity parameters** tell how the website architecture will follow the security constraints by ensuring secure control on data access while performing various transactions.

While talking about website architecture it became essential to discuss the roles of various prime components in development of any particular website. Web Page is one of the major components that make a website meaningful. A web page is a document in website that is developed in hypertext format to show the relevant information to the user with the help of any web-browser. A website is known as a collection of multiple inter-related WebPages.

Web pages also referred as a Page can be considered as the pages in a book. It is a document that exist on www (World Wide Web) containing multiple HTML elements to render on different browsers on various web browsers like Mozilla, Chrome, Internet Explorer, Firefox, etc. The web pages are contained in multiple web servers that can be accessed by a unique address called URL. To access any web page its URL can be entered and the page will be loaded and start appearing on user's browser. A web page can have images, text, and links for the navigation, videos and other graphics to make the web environment more interactive for the user.

❖ XML File

XML stands for Extensible Markup Language is a kind of markup language that defines various rules for document encoding in a format that can provide the readability for both human and machines. The XML 1.0 specifications of W3C and many other relevant specifications, and other open standards combinedly define XML. The prime objective of developing XML files is to achieve simplicity, usability over internet, interoperability among multiple platforms, etc. It provides textual formatting of data involving strong support through Unicode for various languages that are known to human. XML is a language that is widely used for representing many arbitrary data structures those are used in several web services. For the processing of XML data many programmers have developed many APIs (Application Programming Interfaces). To aid the definition of various XML-based languages several schema systems are created.

XML works as an independent tool beyond the limitations of software and hardware for sorting, transporting and manipulating data. XML document is a self-descriptive language and it is very much similar to HTML but having some variations between both. XML is a W3C recommended language that is able to transfer the data from one database server to another in simple steps. The data XML document contains is derived from the characters Unicode repository.

Only a small number of control characters are excluded to be there as the supported content. Any character may appear within XML document content that is defined by Unicode formats. XML document also include various facilities related to identification and encoding of Unicode characters that are used to make up the whole data document. The Unicode data is having the ability to encode the characters in the form of bytes and can be easily store and traverse from one location to another in a variety of transmission ways. Unicode characters itself contains a number of other formats like UTF-8, UTF-16, etc. Many other text encoding formats are also available to predate Unicode characters like ASCII, ISO/IEC 8859 character formats. All the characters stored in these repositories are just a subset of Unicode character set.

❖ Text File

A text file also called flat file is a kind of computer file that contains the information about the electronic text. A text file is always getting the storage in computer's file system. In some operating systems like MS-DOS, the file system doesn't know the file size in the form of bytes. To represent the end of the file in those operating systems, an end of the file marker is used to show the end of the file. But, in modern OS, text file doesn't contain any kind of end of the file marker because the file system of such OS records the file size in bytes. To transfer the files in an online environment any web application needs a File Transfer Protocol as this protocol supports file transfer of any size, type. In the proposed work at source module also we have used a FTP protocol to transfer the files to the upper level module that is the wrapper module. The files from source module can be uploaded and downloaded using a FTP application along with some FTP commands.

❖ Databases

Actually a database is nothing but an organized and well maintained collection of electronic data for the easy access and manipulation in a computer system. In the proposed architecture we have used database as one of the resources from which the data is transferring to the upper level. Database requires various design and modeling techniques to store and organize the data in a structured form as sometimes the database may have the complex nature. Database is managed through appropriate database management software and that software interacts with user, application or database itself directly.

In our proposed architecture database is also working as a kind of source from which the data is accessed and transferred to the upper levels. Our model is able to fetch the data from any kind of database source and ably transfers it to the wrapper module. To fetch the data from the database various kinds of query languages are used like SQL-Server, Oracle, DB2, MongoDB, Informix, PostgreSQL etc. We have proposed to use the most suitable query language to fetch the data from the database. Database is mainly classified in four components and it provides various functions and services. The grouping of data in database management system can be discussed as follows-

Data Definition- It contains the definitions related to the creation, removal and modification of data.
Updating Data- It is the category that includes insertion, deletion and the modification of data.
Data Retrieval- To retrieve the existing data in a form similar to that is stored in the database or a new form can be created for the accessed data from the database.
Administration- This grouping can have the constraints related to registering, monitoring, security, and integrity and concurrency control.

Wrapper module

It is used to map the content of source module in a form of schema for solving the heterogeneity among various sources. The wrapper module maps the (WQI-OR), 1 to 1 WQI into ontological representation where each element of WQI represents linked data resources. This module receives the data from various data sources in source module. The data received by this module can be in the form of any text file, XML file, through the WebPages or directly from any database server. Wrapper module is able to generate the local schema for each data received from various sources.

If we talk about local schema then it became mandatory to understand the schema first as it initially inherits from the database schema. The database schema is a formal language representation of data. It also refers as the blueprint for the database construction. By formally defining database schema we can say that it is a set of rules or formulas that are known as Integrity constraints applied on a particular database to ensure the compatibility among other schemas. So, in wrapper module the local schema is generated in the form of a formal language to define the data along with the integrity constraints for the data collected from various sources.

To represent the logical view of whole database, schema is used. It can be represented as the skeleton type of structure of any database. In wrapper module of proposed architecture, local schema is generating for every data source to show how the data is organized in that data source and also to know about how they are related with each other. It prepares all the constraints that a data must hold when it is used with various applications. In our module, local schema is defining the set of integrity constraints to make data valid and consistent.

Mediator module

It is basically for generating Global schema. By using schema information this module generates the schema to represent wrapped data globally. Query to this module extracts all the results from various sources and show them in the final module. This module receives the input data in the form of local schema that contains the information about the data collected from various sources.

By receiving multiple local schema, the mediator module combines the set of rules collecting from different local schema into one by applying certain other integrity constraints and that merged schema is known as Global schema. Mediator module stores the entire information of data along with the version history of all local schema based on a particular strategy, provides numerous compatibility settings and allows development of global schema according to the compatible configuration settings.

Global schema defines all the entities available in various local schemas and also defines the relationship among them. It contains the complete and whole descriptive details of database. Global schema contains the information in two major forms, that is described as follows-

Physical Database Schema- This schema is concerned with the actual storage of data and the various forms in which data is stored like –indices and files etc. It describes how the data will get store as secondary storage.

Logical Database Schema- This type of schema defines how the constraints are to be applied on the data. The major components of logical schema can be integrity constraints, tables and views.

By having local schema information at mediator level, global schema needs a schema registry in a separate distributed layer that the proposed model is using as its storage mechanism. It got done by assigning a global unique id to each schema, providing a durable and reliable backend.

Application Module

This module is working as the representation of output generated by mediator module in the form of various applications like-Web portal or Desktop application (Figure 2). Web portal is nothing but a special kind of website that is able to bring the information from various sources like- online forms, email, search engine, etc and converts them in a uniform way.

Web portal in application module has a dedicated area for the information received by the global schema from the mediator module. For this web portal is using the application programming interface of search engine to permit the users to search the data from our application content. Desktop application is a kind of software program that can run only on stand-alone computer to perform a specific task. In our proposed framework, the results can also be displayed on a particular desktop application. They receive the information about the data in the form of global schema and represents it accordingly.

Figure 2. High level schema design of view dependent data integration system

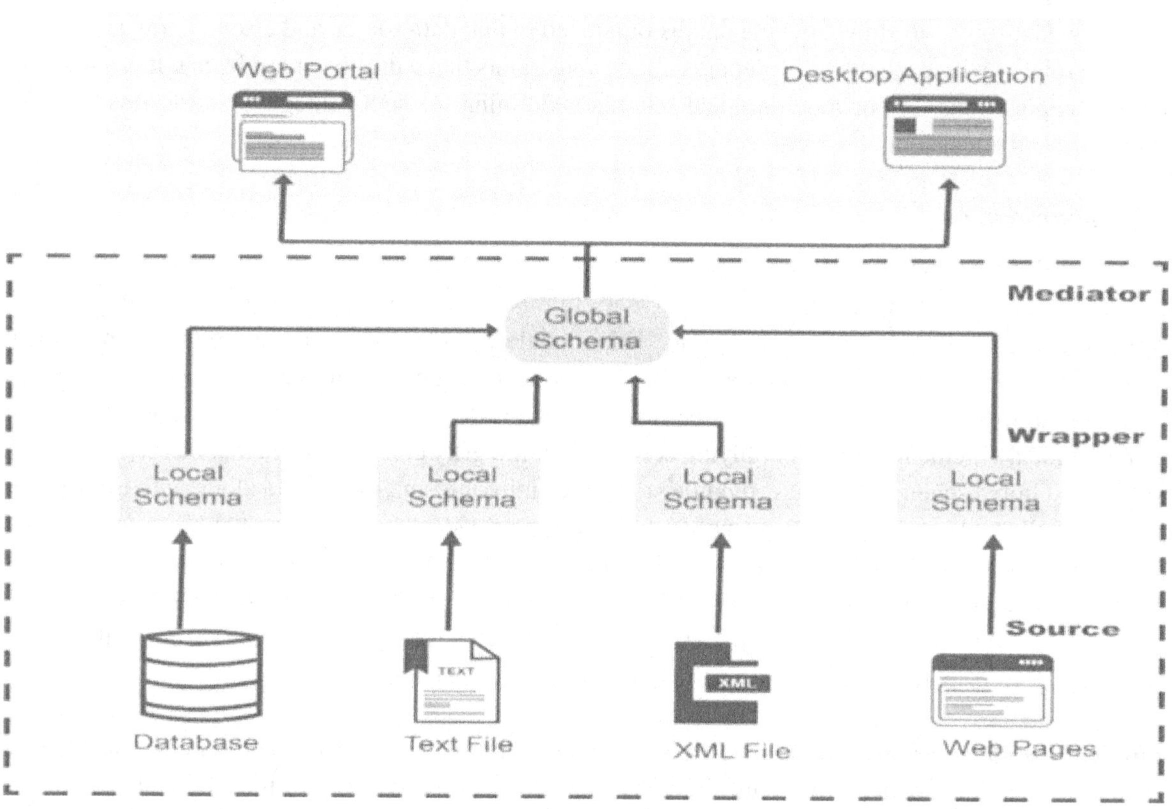

Architectural Framework

Generally, an architecture framework describes the establishment of common working culture for the creating, analyzing, and interpreting by using architectural behavior and description for a particular domain. The Architectural framework of proposed model is representing the explanatory working procedure of the implemented model. Here at the lowest level in Figure 3 we are having the source module which is having the collection of multiple web query interface that are generated by various kinds of sources shown in High level schema design diagram like- Text file, XML file, database servers, and web pages. Various WQIs are identified and the fruitful information is extracted from them and it is passed to the upper level i.e. the wrapper module. This module is performing the task to map each web query interface with a particular related ontological representation.

By having such kind of one-to-one relationship among the WQI and OR it is able represent linked data sources in the form of WQI. Further, the global mapping and view mapping is applied to the each OR-WQIs in wrapper module. By having the related information about the data in global schema and in its components like tables, views, procedures, etc. the integrated web query interface is developed at the next module, i.e. the mediator module. Further, the detailed description of proposed architecture framework can be discussed by exploring the working of each module at various levels separately.

Source

As it is described in figure 1.2, the source module is having various data sources and these data sources are able to generate their own data. The role of source module is to develop web query interface by each of this data source. A web query interface provides some basic features as it provides an exact way to browse various tables into the actual database along with other information like database schema and the field description of the database. In source module web query interface is providing a pattern to query the tables for fetching, manipulating, and storing the data in the database.

It collects and shows the results to the virtual storage space of the database. Further, the identification and the extraction techniques are applied through web query interfaces as it is a crucial task to perform the mapping through the WQI in comparison to schema matching process. Schema contains the structural information about a given database that may help in identification and extraction of dependencies among various attributes. Such information can't be provided by WQIs. For the mentioned reason the local schema is generated at the wrapper module so that the mapping of OR-WQI can be done easily. WQIs provide the best approximation of hidden schemas in database. The web query interface involves some challenges of deep web integration as follows-

- Identifying query interface
- Matching the appropriate query interface
- Classification of deep web sources domain-wise

In our proposed framework we are using various techniques to overcome the above mentioned challenges and proving the exact identification and extraction of WQI by classifying them according to their domain.

Wrapper

The wrapper module is working as mapping of received WQI with some ontological representation of each. Generally, an agreed, detailed and shared model for a certain problem is represented by ontology. By having a domain-specific ontology, it is possible to define a semantic model of combined data by having its domain knowledge. To define various links and relationships among different semantic knowledge, ontology is used. Thus we can say that we can use ontology to apply some kind of data search strategies to the database. It is not possible by the end-user to write complex search requests to fetch the desired results.

To resolve this issue we have focused on a knowledge representation techniques by having interactive queries through ontologies to accelerate the process of searching the database from the traditional queries. The wrapper module of proposed framework is discussing the mapping of ontological representation for each web query interface in order to achieve the highly effective search mechanism by considering the aspects of ontological modeling. This module is generating the one-to-one mapping of each WQI with OR by having certain aspects into consideration those are as follows-

- Ontology developed from local schema
- Representing ontological knowledge by its domain knowledge
- Translating domain knowledge in the form of relational queries
- Translating OR-WQI into global schema and view mapping

Mediator

This is the final phase of developing integrated web query interface in our proposed architectural framework. This phase is modeling the integration process in a number of sequential steps. Initially, it classifies the multiple OR-WQI according to their domain knowledge. Next, the web query interface is transformed into machine specific format called extraction to allow the analysis on data.

Further, similar semantic elements are analyzed among various OR-WQIs. By performing all these tasks successfully, the system will be able to combine distributed WQIs into single integrated web query interface.

In the proposed architecture we specially concentrated the integration of web query interface that aims to search the contents in a deep web with more knowledge in multiple domains. We have completed the task in multiple sequential steps starting from the data collection from various sources, identification, extraction, classification, mapping and matching local schema to global schema and views. Finally, the integrated web query interface generation is done in the last step of the proposed architecture.

User and Service Oriented Diagram

High Level Schema

The proposed framework is being represented as a layered architecture. The complete black-box view or the user oriented high level schema of the framework contains three layers, namely- First, User Interface, Second System Level Processing & Service Composition, Third Mediator and Wrapper, Last Multi Database Sources.

Below given Figure 4 showing User Oriented High-Level / Black Box Schema of the Proposed Framework.

User Interface- User-Interface is the entry and gateway point of the interaction of the proposed framework. As discussed above, user-interface is having both type of standard ways (Web-Portal and Desktop Application) of using the system. Towards the solution of the problem statement, web portal based system is to enable usage of the system to the wide public era, rather the desktop application is majorly providing ways to administrators for the administration kind of service enablement.

System Level Processing & Service Composition- Queries fired from user interface either by web-portal or desktop application, Service Composer performs authentication services and then forward extracted queries to further processing based on semantic web-query processing. Finally, after final results computation it will respond to the user interface with the desired / processed results.

Mediator and Wrapper- The wrapper module is working as mapping of received web-query with some ontological representation of each. Generally, an agreed, detailed and shared model for a certain

Figure 3. Architectural framework

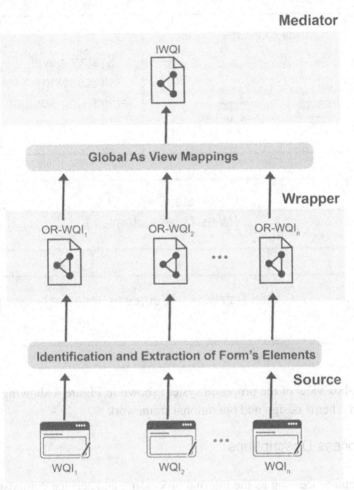

problem is represented by ontology. By having a domain-specific ontology, it is possible to define a semantic model of combined data by having its domain knowledge.

On the other hand, Mediator is the final phase of developing integrated web query interface in our proposed architectural framework. This phase is modeling the integration process in a number of sequential steps. Initially, it classifies the multiple OR-WQI according to their domain knowledge. Next, the web query interface is transformed into machine specific format called extraction to allow the analysis on data. Further, similar semantic elements are analyzed among various OR-WQIs. By performing all these tasks successfully, the system will be able to combine distributed WQIs into single integrated web query interface.

Multi-Database Sources / Schemas- At very bottom layer of the proposed system's black-box layered design is the databases sources with different facets. Different facets databases sources providing the view and scalability towards multi-schematic architectural design.

Figure 4. User oriented high-level / black box schema of the proposed framework

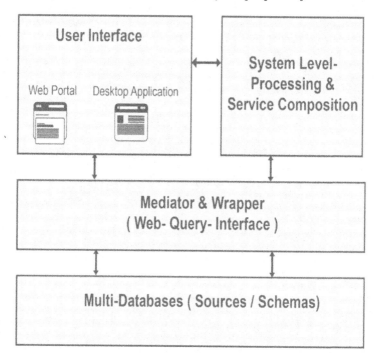

Hence, the black-box view of the proposed system shown in Figure 4 showing the overall layout of the proposed system schema design and operational framework.

Services and Process Descriptions

User level services and processing as the internal processing and service composition of the proposed framework shown above are being detailed here in Figure 5.

User interface of the proposed system is divided in two main interaction ways Web-Portal and Desktop-Application based, and two main services containers- Student Services and Management Services. Student services contains following main service modules.

Registration Services- Registration Services enables new users to register with the system towards the system usage. Even the registration services are also having pre-authentication phase before registration to authenticate a valid candidate to be registered. After registration system will automatically composed student dashboard with the standard modules compositions.

Announcement Services- Announcement and notification services are also similarly essential to always be in afresh of the system. The student dashboard will fetch newly fed information and compose all as new announcements to the students.

Programs and Courses Services- Program and courses services are enabling students to enrollment, and participate towards a particular and registered for program / courses. All academic and related services would be listed here in this module as sab modules.

Semantic Search- Search is again essential towards utilizing the system and relocation of the services. In the proposed framework semantic search is being composed for the semantic enabled response towards a fired query for the fast, reliable and meaningful search facility.

The second main service container in user interface module is Management Services which contains following main services inside.

Admin Services- Very first service list in the proposed framework would be the Admin Service that will be responsible for all the way administration of in and outside the system. All add/edit/update services and candidates can be performed by the administrator.

Authentication Services- All authentication processing / services and controlled modules also been handled by the announced administrators of the system.

Indexing and Cataloging Services- All indexing and catalogs of the major services and changes can be controlled by management services panel too.

Component Management Services- Even the user dashboard will be composed by system itself as default services dashboard but from management services panel various services for a particular user can be composed or disposed with a proper authentication of administration account.

After user interface, the next component is System Level Processing and Service Composition. The detailed functionality of this component is being described below.

Service Composer- The main component of this container is service composer at very first it will authorize a fired query and the query generator source using the standard authorization process of
- CIA-Confidentiality, Integrity and Authentication. After successful authentication it will deliver and evaluate the given query, a pre-fetcher will have fetched the results and store the logs. Pre-Fetcher is responsible for logs recording and dumping the dump-able items. Service composer will compose all the desired queries results towards the user interface as response.

Rest two components of the framework Mediator & Wrapper and Multi-Databases Sources already been detailed in the first half of this chapter.

Implementation Aspects

As far as the implementation aspects, we will be discussing detailed implementation in the next chapter, for now it will be implemented as a web based environment to test the basic and essential functionalities of the proposed framework.

In name of implementation the basic flow of the system to be simulated and tested for the optimized service delivery. The proposed framework will be tested against following-

- The cohesion of the modules and formation of the system as single system image
- Working of the each module and intermediate results as the input to the next phases in well refined manner
- Overall performance of the complete system in terms of optimized service delivery

Figure 5. User level detailed operational architecture of the system

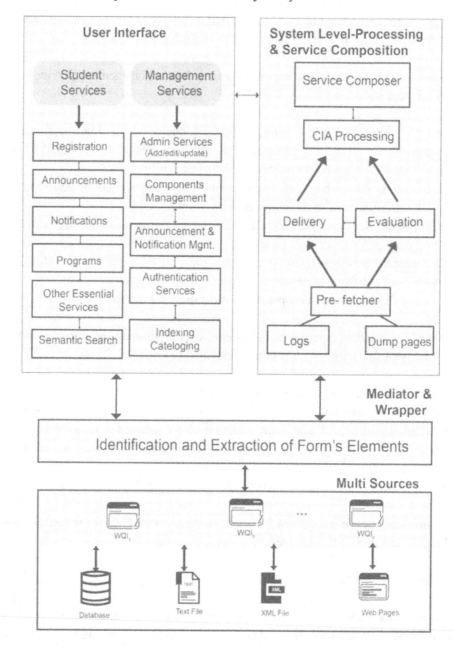

CHAPTER SUMMARY

In this chapter the proposed framework as the main contribution, View Dependent Data Integration System has been proposed and presented as the solution towards advanced, reliable and view oriented registration system with multi-databases sources support. The framework has been represented in system level design as high level view of the View Dependent Data Integration System and the architectural framework of the system design. Being multi-database oriented system it is scalable to structure as well as unstructured data source. The wrapper and mediator module of the system used to map the web-query-interface to the global schema of the integrated web query interfaces.

Secondly the complete system has been represented as the user end usage, the black-box i.e. again high level view of the system modeling from the end users point of view, and the operational framework design also been represented.

REFERENCES

Barbosa, L., Nguyen, H., Nguyen, T., Pinnamaneni, R., & Freire, J. (2010). Creating and exploring web form repositories. *Proc. ACM SIGMOD Int.Conf. Manage.* ACM. 10.1145/1807167.1807311

Begum, A., & Kumar, R. (2022). *Design an Archetype to Predict the impact of diet and lifestyle interventions in autoimmune diseases using Deep Learning and Artificial Intelligence.* Research Square. doi:10.21203/rs.3.rs-1405206/v1

Bizer, C., Vidal, M.-E., & Skaf-Molli, H. (2018). Linked Open Data. Springer.

H. M. M. Castro, V. S. Sosa, V., & Maganda, M. (2018). Automatic construction of vertical search tools for the deep web. *IEEE Latin Amer. Trans., 16*(2), 574_584.

Chang, K., He, B., Li, C., Patel, M., & Zhang, Z. (2004). Structured databases on the web: Observations and implications. *ACM SIGMOD Rec.* ACM.

Gosain, M. S., Aggarwal, N., & Kumar, R. (2023). A Study of 5G and Edge Computing Integration with IoT- A Review. *2023 International Conference on Computational Intelligence and Sustainable Engineering Solutions (CISES)*, Greater Noida, India. 10.1109/CISES58720.2023.10183438

Gupta, N., Sharma, H., Kumar, S., Kumar, A., & Kumar, R. (2022). *A Comparative Study of Implementing Agile Methodology and Scrum Framework for Software Development.* 2022 11th International Conference on System Modeling & Advancement in Research Trends (SMART), Moradabad, India. 10.1109/SMART55829.2022.10047477

He, B., & Chang, K. C.-C. (2003). Statistical schema matching across web query interfaces. *ACM SIGMOD Int. Conf. Manage. Data.* ACM. 10.1145/872757.872784

He, Z., Hong, J., & Bell, D. (2008). Schema matching across query interfaces on the deep web. A. Gray, K. Jeffery, J. Shao, (eds.). Sharing Data, Information and Knowledge. Berlin, Germany: Springer. doi:10.1007/978-3-540-70504-8_6

Jaiswal, A., & Kumar, R. (2022). *Breast cancer diagnosis using Stochastic Self-Organizing Map and Enlarge C4.5.* Multimed Tools Appl. doi:10.1007/s11042-022-14265-1

Jaiswal, A., & Kumar, R. (2023). Breast Cancer Prediction Using Greedy Optimization and Enlarge C4.5. In S. Maurya, S. K. Peddoju, B. Ahmad, & I. Chihi (Eds.), *Cyber Technologies and Emerging Sciences. Lecture Notes in Networks and Systems* (Vol. 467). Springer. doi:10.1007/978-981-19-2538-2_4

Jou, C. (2016). *Deep web query interface integration based on incremental schema matching and merging.* Proc. 3rd Multidisciplinary Int. Social Netw. Conf. Social Inform., Data Sci., New York, NY, USA. 10.1145/2955129.2955170

Kumar, A., Tewari, N. & Kumar, R. (2022) A comparative study of various techniques of image segmentation for the identification of hand gesture used to guide the slide show navigation. *Multimed Tools Appl (2022).* doi:10.1007/s11042-022-12203-9

Kumar, P., Kumar, M., Singh, K. B., Tripathi, A. R., & Kumar, A. (2021, December). Blockchain Security Detection Condition Light Module. In *2021 10th International Conference on System Modeling & Advancement in Research Trends (SMART)* (pp. 363-367). IEEE. 10.1109/SMART52563.2021.9676302

Kumar, R., & Kumar, R. (2022, May). Kumar, Sandeep. (2022). Intelligent Model to Image Enrichment for Strong Night-Vision Surveillance Cameras in Future Generation. *Multimedia Tools and Applications, 81*(12), 16335–16351. doi:10.1007/s11042-022-12496-w

Landers, T., & Rosenberg, R. L. (1986). An Overview. Artech House.

Bergman, M. (2001). White paper: The deep web: Surfacing hidden value. *J. Electron., 7*(1).

Madhavan, J., Cohen, S., Dong, X. L., Alon Halevy, Y., Jeffery, R. S., Ko, D., & Yu, C. (2007). *Web-scale data integration: You can afford to pay as you go.* Proc. 3rd Biennial Conf. Innov. Data Syst. Res., Asilomar, CA, USA. www.cidrdb.org

Mahammad, A. B., & Kumar, R. (2022). Machine Learning Approach to Predict Asthma Prevalence with Decision Trees. *2022 2nd International Conference on Technological Advancements in Computational Sciences (ICTACS),* (pp. 263-267). IEEE. 10.1109/ICTACS56270.2022.9988210

Mahammad, A. B., & Kumar, R. (2022). Design a Linear Classification model with Support Vector Machine Algorithm on Autoimmune Disease data. *2022 3rd International Conference on Intelligent Engineering and Management (ICIEM),* (pp. 164-169). IEEE. 10.1109/ICIEM54221.2022.9853182

Mahammad, A. B., & Kumar, R. (2023). Scalable and Security Framework to Secure and Maintain Healthcare Data using Blockchain Technology. *2023 International Conference on Computational Intelligence and Sustainable Engineering Solutions (CISES),* Greater Noida, India. 10.1109/CISES58720.2023.10183494

Mohtashim Mian, S., & Kumar, R. (2023). Deep Learning for Performance Enhancement Robust Underwater Acoustic Communication Network. In S. Maurya, S. K. Peddoju, B. Ahmad, & I. Chihi (Eds.), *Cyber Technologies and Emerging Sciences. Lecture Notes in Networks and Systems* (Vol. 467). Springer. doi:10.1007/978-981-19-2538-2_24

Sakr, S., Wylot, M., Mutharaju, R., Le Phuoc, D., & Fundulaki, I. (2018). *Linked Data_Storing, Querying, Reasoning.* Springer.

Shah, P. K., Pandey, R. P., & Kumar, R. (2016, November). Vector quantization with codebook and index compression. In *2016 International Conference System Modeling & Advancement in Research Trends (SMART)* (pp. 49-52). IEEE.

Sharma, N., Chakraborty, C., & Kumar, R. (2022). (2022) Optimized multimedia data through computationally intelligent algorithms. *Multimedia Systems*. doi:10.1007/s00530-022-00918-6

Wu, W., Doan, A., & Yu, C. (2009). Modeling and Extracting Deep-Web Query Interfaces. Berlin, Germany: Springer.

Chapter 22
Toward a More Ethical Future of Artificial Intelligence and Data Science

Wasswa Shafik

https://orcid.org/0000-0002-9320-3186

School of Digital Science, Universiti Brunei Darussalam, Brunei & Dig Connectivity Research Laboratory (DCRLab), Kampala, Uganda

ABSTRACT

Examining the ethical aspects of artificial intelligence (AI) and data science (DS) recognizes their impressive progress in innovation while emphasizing the pressing necessity to tackle intricate ethical dilemmas. The chapter provides a detailed framework for navigating the changing environment, beginning with an examination of the increasing ethical challenges. The study highlights transparency, fairness, and responsibility as crucial for cultivating confidence in AI systems. The chapter emphasizes the urgent requirement to address problems such as algorithmic bias and privacy breaches with strong mitigation techniques. Furthermore, it promotes flexible policies that strike a balance between innovation and ethical safeguards. The examination of societal effects, particularly on various socioeconomic groups, economies, and cultures, is conducted thoroughly, with a focus on equity and the protection of individual rights. Finally, to proactively tackle future ethical challenges in technology, it is advisable to employ proactive solutions such as implementing AI ethics by design.

INTRODUCTION

The development of Artificial Intelligence and data science (AIDS) has undergone significant changes throughout history, dating back to the mid-20th century when fundamental ideas of AI were initially formulated (Gil, 2017). Since its early phases, which were marked by lofty speculations of machine intelligence, to the present day, where AI systems are present in all aspects of our lives, this path has been filled with significant developments (Khan et al., 2022). Concurrently, the discussion on the moral consequences of AI and data science has become more prominent as these technologies have grown rapidly.

DOI: 10.4018/979-8-3693-2964-1.ch022

The historical story is a complex interweaving of technological advancement and ethical introspection, which is crucial for comprehending the present terrain (Shaban-Nejad et al., 2018).

Ethical considerations in AIDS have followed a complex trajectory, developing alongside technological advancements. At first, ethical considerations were not given much importance since they were overwhelmed by the fascination with scientific advancements (Wu et al., 2021). As AI systems advanced and became more prevalent in several fields, ethical considerations became more prominent. The progress of these factors reflects the acknowledgment of their inherent significance, going beyond mere theoretical discussion to actual application in the creation and deployment of AI systems (Ronmi et al., 2023).

At present, the field of AI and data science is characterized by complex problems and ethical dilemmas. These obstacles include a wide range of issues, from algorithmic prejudice and unclear decision-making processes to privacy infringements and the growing socioeconomic gap (Vinod & Prabaharan, 2020). The widespread presence of AI technology has intensified these challenges, underscoring the urgency for prompt and thorough response. The rapid advancement of technology has created ethical challenges that require a thorough reassessment of the moral principles that guide these groundbreaking innovations (Atov et al., 2020). It is crucial to comprehend and tackle these difficulties to guide AIDS toward a trajectory that is ethically conscientious, as presented by the AI branches in Figure 1.

Figure 1. Artificial intelligence branches

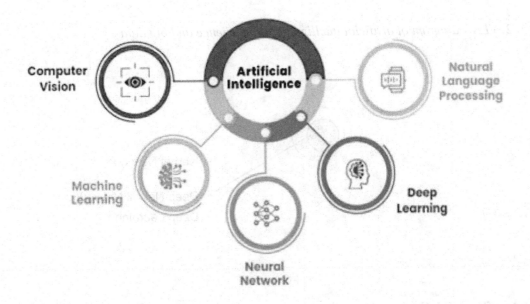

Algorithmic biases are significant obstacles in the field of AIDS, as they highlight the unintentional reproduction of societal preconceptions in machine learning algorithms. These prejudices are shown in judgments relating to hiring, lending, and criminal justice, perpetuating inequality and discrimination (Jain, 2023; Shafik, 2024a). Furthermore, the lack of transparency in AI algorithms intensifies these problems, impeding the capacity to assign responsibility and worsening ethical dilemmas. Concurrently, the issue of data privacy has become increasingly important due to the rapid expansion of data gathering and usage (Kumar et al., 2023). The utilization of personal data elicits ethical concerns, leading to the examination of issues related to consent, control, and security.

Ethical concerns involve not just technical difficulties but can have significant effects on society. The democratization of AI technologies, however encouraging, magnifies socioeconomic gaps, potentially worsening the digital divide (Choi et al., 2023a; Shafik, 2024a). Furthermore, the ethical dilemma of the displacement of jobs by AI technology has significant implications for society, necessitating ethical deliberation in restructuring labor markets and guaranteeing fair chances. Aside from ethical dilemmas, the opaqueness of AI decision-making processes gives rise to questions regarding accountability and transparency (Chkoniya, 2021). The absence of interpretability in AI systems hinders the understanding of their judgments, making it difficult to correct errors or biases and undermining the establishment of trust among stakeholders (Maxwell et al., 2021).

Effectively navigating complex ethical situations necessitates employing a comprehensive strategy that combines technical advancement with thoughtful self-reflection on moral principles as Figure 2 presents the Venn diagram intersecting Big Data and AIDS. To tackle these difficulties, it is essential for a wide range of individuals and groups, including technologists, legislators, ethicists, and society, to work together in a coordinated manner (Shafik, 2023a; AstraZeneca, 2022). Gaining insight into the historical backdrop and progression of ethical considerations in AI and data science establishes the basis for effectively tackling present-day challenges. It stresses the importance of adopting an ethically aware strategy to guide the development of these groundbreaking technologies toward a fair and accountable future (Malik et al., 2023).

Figure 2. VENN diagram of artificial intelligence, data science and big data

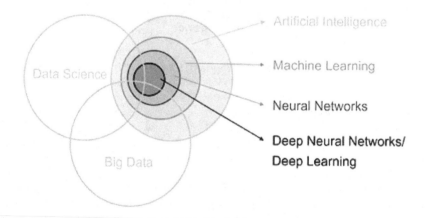

Chapter Contribution

This chapter provides the following contributions:

- Accentuate the increasing ethical challenges in AI and data science, which require a well-organized ethical framework for responsible development and implementation.
- Rationalize the fundamental role of ethical concepts such as transparency, fairness, and accountability in establishing trust and dependability in AI systems.

- Explore obstacles, including algorithmic bias and privacy concerns, while highlighting the significance of strong mitigation measures to tackle these problems properly.
- Examines the existing regulatory framework, identifying deficiencies and promoting flexible policies that reconcile technological advancement with the protection of ethical principles.
- Analyzes the extensive societal effects of AI and data science, with a specific emphasis on how ethical considerations might influence a fairer and more advantageous technology environment for many communities.
- Stars the development of moral systems and suggests the incorporation of forthcoming technologies, for example, AI ethics by design, as a proactive strategy for upcoming obstacles.
- Summarizes the crucial significance of ethics in directing the progress and implementation of AI and data science, emphasizing the ongoing requirement for ethical self-reflection and adjustment in technical advancements.

Chapter Organization

Section 2 presents identified ethical principles in AI and Data Science. Section 3 illustrates the Ethical Challenges and Mitigation Strategies. Section 4 demonstrates the Regulatory and Policy Framework. Section 5 presents the Impact on Society. Section 6 pertains to Future Directions. Finally, Section 7 presents the lessons learned from the chapter and the chapter conclusion.

ETHICAL PRINCIPLES IN ARTIFICIAL INTELLIGENCE AND DATA SCIENCE

Ethical principles act as fundamental principles that guide the creation, implementation, and utilization of AIDS. Transparency, justice, accountability, and privacy are key values that play a crucial role in establishing the ethical framework in applications as presented.

Transparency Within Artificial Intelligence and Data Science

Transparent AI systems offer transparency by clearly elucidating their internal mechanisms, so ensuring that users and relevant stakeholders comprehend the underlying logic behind the decisions made by these systems (Jordan & Mitchell, 2015). The transparency of AI models allows for accountability and simplifies the process of identifying and correcting biases or errors. Furthermore, it enables humans to make well-informed decisions and cultivates a feeling of influence over outcomes driven by artificial intelligence (Wan et al., 2022; Shafik, 2023b). Transparency in AI decision-making, by offering clear and understandable explanations, fosters user trust and confidence. This is essential for the general acceptance and ethical implementation of these technologies. Figure 3 presents the DS life cycle from business problem understanding through data visualization.

Fairness

Fairness in AI examines the necessity of providing equal treatment and results to all persons, irrespective of their history or demographic attributes. Ethical AI systems prioritize equity by actively working to eradicate prejudices and guaranteeing that AI applications do not perpetuate or worsen societal inequities

Figure 3. Data science life cycle

(Sagi et al., 2020; Shafik et al., 2023). To ensure fairness, it is essential to conduct thorough evaluations to identify biases, using algorithms that are designed to promote justice and consistently monitor and address any inequalities that may arise. The dedication to fairness not only maintains ethical norms but also aids in constructing more comprehensive and equitable AI systems, promoting public trust and credibility (Gujral et al., 2019).

Accountability

Accountability in AIDS stresses the significance of being responsible and having proper supervision at every stage of AI systems' development and operation. Ensuring ethical AIDS necessitates an unambiguous assignment of accountability for the decisions and behaviors executed by AI models (Naudé et al., 2022). Implementing accountability measures guarantees that in the event of errors or biases being detected, there are established channels for seeking redress, rectification, and gaining insights from these occurrences. Furthermore, it encourages developers and organizations to prioritize ethical issues in their AI projects, thus promoting responsible behavior (Choi et al., 2021; Shafik, 2023c). A responsible AI ecosystem fosters trust among stakeholders by showcasing a dedication to ethical behavior and prudent advancement.

Privacy

Privacy is of utmost significance in the field of AIDS, especially when it comes to the management and use of personal data. Ethical AI frameworks provide priority to upholding user privacy by protecting sensitive data and guaranteeing adherence to privacy standards (Mikhaylov et al., 2018). This entails adopting strong data security measures, acquiring informed consent, and reducing data gathering to the level required for the intended objectives. Privacy-preserving methodologies, such as anonymization or differential privacy, are utilized to achieve a harmonious equilibrium between fostering innovation and safeguarding the private rights of individuals (Drobot, 2020). Adhering to privacy rules not only protects user data but also promotes trust and confidence in AIDS systems, which is crucial for maintaining user engagement and societal acceptance of these technologies.

Integrity

Integrity in AIDS pertains to the ethical obligation of upholding honesty, dependability, and ethical behavior over the whole lifespan of AI. It entails guaranteeing that AIDS systems function in accordance with ethical norms and societal principles (Sbailò et al., 2022; Jun et al., 2021). This principle underscores the need to prevent deceit, guarantee the integrity of data, and uphold the dependability and resilience of AI systems (Villarreal-Torres et al., 2023). Maintaining integrity necessitates being open and honest about the capabilities and limitations of AIDS, refraining from deceitful behaviors, and cultivating a culture of ethical accountability among organizations involved in the development and implementation of AI technology. AIDS systems promote trust, credibility, and ethical reliability by upholding integrity, leading to increased acceptance and good impact on society (Brzezinski & Krzeminska, 2023).

Beneficence

Generosity involves the ethical principle of optimizing the advantages of AIDS while limiting potential negative consequences. The objective of ethical AI systems is to give priority to the welfare of society by guaranteeing that the results and uses of AI technologies have a positive impact on individuals and communities (Górriz et al., 2020; Alnssyan et al., 2023). This approach necessitates meticulous deliberation of the potential ramifications, both deliberate and accidental, of AIDS systems on different stakeholders. Ethical AI development entails taking proactive steps to minimize risks, emphasize safety, and improve society's well-being by responsibly deploying and continuously evaluating the societal consequences of AIDS (Stracener et al., 2019).

ETHICAL CHALLENGES AND MITIGATION STRATEGIES

The ethical hurdles in AIDS are diverse and intricate, posing intricate dilemmas that necessitate meticulous examination and solutions, as illustrated.

Ethical Challenges of Artificial Intelligence and Data Science

There are notable ethical dilemmas as presented.

Algorithmic Bias

The obstacle comes from the utilization of biased or faulty data inputs during the training of machine learning models, resulting in unintentional discrimination. Biased algorithms have a far-reaching impact, as they reinforce past preconceptions in decision-making processes across many disciplines (Popescu, 2018). Biased algorithms have the potential to perpetuate preexisting disparities in industries such as employment, banking, criminal justice, and healthcare. The difficulty lies in recognizing and correcting biases that are deeply embedded in AI models. Effective management of AI requires the implementation of intricate and ongoing measures to detect, mitigate, and avoid issues spanning the whole lifecycle of the technology (Thuraisingham, 2020). To tackle algorithmic bias effectively, it is crucial to adopt a comprehensive strategy that includes a wide range of datasets, powerful bias detection algorithms, and ethical principles to guarantee fairness in decision-making powered by AIDS (Cao, 2023a).

Privacy Concerns

The consequential impact encompasses the possibility of confidentiality breaches, unauthorized access, and the inappropriate use of sensitive information. These violations of privacy undermine user confidence and present substantial ethical dilemmas (Choi et al., 2022a). Stringent safeguards are required to maintain the delicate balance between utilizing data for innovation and protecting individual privacy. It is crucial to implement thorough methods that protect privacy and comply with changing regulations (Choi et al., 2022b). Achieving this equilibrium is crucial for establishing and maintaining confidence among users and stakeholders while also facilitating the appropriate and ethical utilization of data-driven technology.

Lack of Transparency

AI systems that lack transparency obstruct the understanding of decision-making processes, making it difficult to explain outcomes or correct errors efficiently. The lack of transparency in AI systems impedes accountability, hence obstructing the capacity to identify biases and correct erroneous results (Choi et al., 2023b). The development of explainable AI systems is crucial, as it involves finding a middle ground between sophisticated models and the requirement for understandability. Promoting openness is crucial for cultivating user trust and confidence in AI technology, as well as allowing stakeholders to comprehend, examine, and enhance AI decision-making procedures (Kadam et al., 2020).

Accountability Gap

The ethical difficulty of establishing accountability for AI-generated decisions is significant due to the intricate nature of responsibility in the development and deployment of AI. The decentralized nature of accountability among many stakeholders complicates the attribution of responsibility for negative results or mistakes (Raschka et al., 2020). The difficulty lies in creating strong procedures that assign responsibility and ensure accountability across the AI ecosystem. It is crucial to establish clear and precise definitions of the roles and duties of developers, organizations, and regulatory agencies to address this lack of accountability effectively (Octavian Dumitru et al., 2019). Strong frameworks for accountability are essential to guarantee the implementation of corrective measures and the acquisition of knowledge from mistakes, all while promoting ethical behavior in AIDS (Leung et al., 2021).

Ethical Decision-Making

The field of ethical decision-making in AI deals with instances where robots are presented with ethical issues, such as autonomous vehicles having to make life-or-death judgments or AI systems making crucial choices in healthcare (Mainali & Park, 2023). To resolve these challenges, it is necessary to synchronize technological capabilities with ethical considerations. It is imperative to create frameworks that include ethical principles in the process of designing and implementing AI systems (Han & Liu, 2019). Adopting ethical principles from the start and fostering collaborations across different fields are crucial in developing guidelines and frameworks that address the intricate ethical dilemmas in AIDS (Schwaller et al., 2022).

Societal Impact and Bias Amplification

Insufficiently built AI systems have the potential to strengthen and worsen preexisting societal biases and disparities unintentionally. In fields like recruitment, where AI-powered technologies support decision-making, biased algorithms have the potential to perpetuate past biases against specific demographic groups (Allen et al., 2021). Furthermore, within the healthcare sector, AI diagnostics that exhibit bias have the potential to disproportionately impact underprivileged communities, exacerbating existing inequalities in their ability to obtain high-quality healthcare services. This difficulty necessitates the use of proactive strategies to identify and address biases, as well as actively minimize their magnification within AI systems (Hameed et al., 2021). To tackle the societal impact and amplification of bias, it is crucial to make a focused endeavor in developing AI models that are more inclusive and equitable. These models should actively strive to reduce societal biases and inequities rather than perpetuating them (Hilal et al., 2022).

Explainability and Decision Interpretability

AI systems frequently generate results without offering intelligible justifications for their choices, impeding users' comprehension and confidence in such systems. AI judgments, particularly those that have substantial effects on persons or society, must be interpretable and explainable in order to meet ethical requirements (Cao, 2023b). To do this, it is necessary to devise approaches that make intricate AI processes comprehensible without losing their efficiency or effectiveness as some insights are presented in Figure 4. Setting criteria for decision interpretability guarantees that AI systems are transparent, responsible, and able to justify their judgments, promoting trust and ethical assurance in their implementation (Peña-Guerrero et al., 2021).

Mitigation Strategies of Artificial Intelligence and Data Science

The task of making AI judgments more interpretable and accessible to stakeholders while maintaining the subtleties of the underlying technology is a big challenge that involves finding the right balance between complexity and clarity, among others.

Algorithmic Bias Mitigation

To tackle algorithmic bias, a comprehensive strategy is required, starting with the careful selection of diverse and representative datasets throughout the training of AI models. This technique seeks to alleviate

Figure 4. XAI and No XAI decision phases

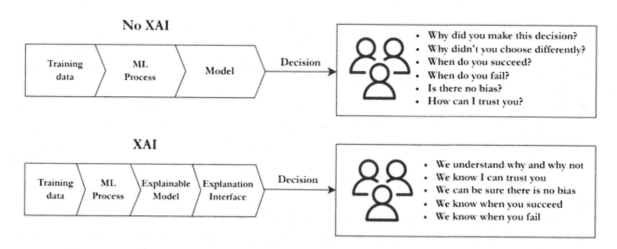

biases by ensuring that the data utilized to train AI systems contains a diverse range of demographics, hence reducing the likelihood of reproducing historical preconceptions (Cui et al., 2021). Incorporating algorithmic techniques like fairness-aware algorithms or debiasing methods throughout the model construction phase enables the detection and correction of biases. Consistent audits and ongoing refining of models additionally enhance the development of fair and equitable AI systems (Saura et al., 2022).

Privacy-Preserving Technologies

The application of privacy-preserving technology presents a viable resolution to the ethical dilemmas associated with data privacy in AIDS. Methods such as differential privacy, federated learning, and homomorphic encryption facilitate the extraction of valuable insights from sensitive data while safeguarding individual privacy (Ray et al., 2021). Differential privacy, for example, incorporates random perturbations to safeguard individual data points, guaranteeing that the addition or removal of any data does not have a substantial effect on the result. Federated learning facilitates the training of AI models across distributed devices, eliminating the need for centralized data aggregation. It ensures that data remains on users' devices and that only model updates are shared (Ward et al., 2021). Homomorphic encryption enables the execution of computations on encrypted data while maintaining data confidentiality without the need for decryption.

Explainable Artificial Intelligence

Explainable Artificial Intelligence[1] (XAI) approaches aim to clarify the complex processes of AI models, enabling stakeholders to comprehend and explain their conclusions, as Figure 5 demonstrates. Interpretable models, attention mechanisms, and model-agnostic techniques provide a valuable understanding of how AI algorithms reach specific results, improving transparency without sacrificing model performance (Nishimwe et al., 2022). These strategies enable stakeholders to authenticate and comprehend AI-generated outputs, offering vital insights into the reasoning behind decisions. Through the incorporation of

Figure 5. Potential stakeholders of explainable Artificial Intelligence (XAIR)

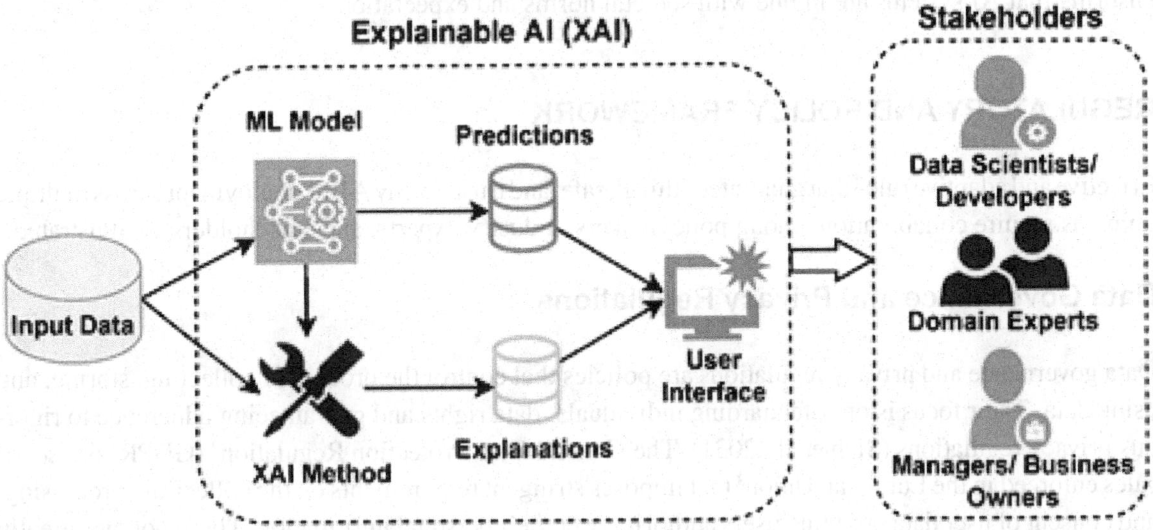

XAI methodologies, companies, and developers strive to cultivate confidence, augment responsibility, and expedite the implementation of more ethical AI systems in diverse fields (Trujillo-Cabezas, 2020).

Accountability Frameworks

Implementing strong accountability frameworks is essential for addressing the issue of attributing responsibility for results generated by artificial intelligence. These frameworks provide distinct boundaries of accountability among the parties engaged in the creation, implementation, and oversight of AI (Ong & Uddin, 2020). Clearly defined duties and responsibilities guarantee that individuals can be held accountable in situations involving errors, biases, or negative outcomes. Organizations and regulatory agencies should work together to create clear principles and standards for ethical behavior in AI, with a focus on fostering transparency and accountability (González-Méijome et al., 2022). Transparent accountability systems promote a culture of ethical responsibility among developers and organizations, encouraging compliance with ethical norms and enabling appropriate corrective measures to be implemented as needed (Borrego-Díaz & Galán-Páez, 2022b).

Ethical Artificial Intelligence Design Principles

Incorporating ethical considerations into the fundamental design principles of AI systems is crucial for actively tackling ethical challenges. Ethical AI design involves incorporating concepts like fairness, transparency, accountability, and societal effect into the AI development process from the very beginning (Gruson et al., 2019). This method guarantees that ethical considerations are fundamental to AI design decisions, directing the development of responsible and fair AI systems. Developers and organizations strive to reduce ethical risks and increase social trust in AI technology by integrating ethical frameworks and principles into every stage of the AIDS lifecycle (Young et al., 2023). This proactive strategy promotes the responsible implementation of AI by methodically addressing ethical considerations and ensuring that AI systems are in line with societal norms and expectations.

REGULATORY AND POLICY FRAMEWORK

Effective and adaptive rules that guarantee ethical, safe, and trustworthy AIDS deployment across multiple domains require collaboration among policymakers, industry experts, and stakeholders, as illustrated.

Data Governance and Privacy Regulations

Data governance and privacy regulations are policies that control the process of collecting, storing, and using data. Their focus is on safeguarding individuals' data rights and guaranteeing adherence to rigorous privacy regulations (Siah et al., 2021). The General Data Protection Regulation[2] (GDPR) is a set of rules enforced in the European Union[3] that imposes stringent requirements on the collection, processing, and consent of user data, granting users authority over their personal information. These policies greatly enhance user confidence by protecting sensitive data, promoting ethical data management practices, and granting individuals authority over their information (Aguilar-Esteva et al., 2023). Failure to comply with regulations results in significant penalties, motivating firms to prioritize data protection as a vital component of their operations.

Ethical Guidelines and Standards

Ethical norms and standards establish principles that direct the responsible development of AIDS, with a focus on ensuring justice, openness, and responsibility in AI systems. The ethically aligned design and AI ethics guidelines by the Institute of Electrical and Electronics Engineers[4] (IEEE) provide extensive frameworks that guide developers in implementing ethical norms in AI (Borrego-Díaz & Galán-Páez, 2022a). These recommendations serve as a moral compass for developers and organizations, promoting ethical decision-making in all aspects of AI development and implementation. By ensuring that AI systems are in accordance with social values, they encourage confidence and acceptance among users, creating a climate that is favorable for ethically responsible AIDS advancements (Sohail, 2023).

Algorithmic Accountability and Transparency

Enforce regulations that require AI systems to be transparent and provide explanations for their judgments, allowing users to comprehend the underlying logic behind AI-generated outcomes. The AI Impact Assessment[5] (AIIA) framework in the United Kingdom requires enterprises to assess the consequences and responsibility of their AI systems (Cruz-Jesus et al., 2020). Requiring algorithmic openness improves user confidence and responsibility by enabling the identification and correction of biases. Users experience heightened assurance when AI systems offer comprehensible justifications for their judgments, resulting in enhanced confidence in the dependability of these systems (An et al., 2023).

Safety and Security Regulations

Regulations aimed at safeguarding the security, dependability, and robustness of AI systems in the face of possible malfunctions and cybersecurity risks. ISO/IEC 27001[6] establishes criteria for managing information security, encompassing AI systems as well. ISO/IEC 27001 is a globally recognized standard for the effective management of information security (Marr, 2020). The standard was initially co-published by the International Organization for Standardization and the International Electrotechnical Commission[7] in 2005. It was subsequently amended in 2013 and, most recently, in 2022 (Yang et al., 2022). Safety and security laws help to minimize potential dangers, guaranteeing the consistent functioning of AI systems while decreasing the chances of data breaches or system failures. Adherence to these regulations is crucial to cultivating user confidence and trust in AI technologies (Gil, 2017).

Sector-Specific Regulations

Customized legislation that specifically targets the unique issues and ethical concerns present in industries such as healthcare, banking, or transportation. Food and Drug Administration[8] (FDA) rules ensure the safety and effectiveness of AI-based medical devices within healthcare applications. Industry-specific rules ensure that AI systems are deployed ethically and safely in specialized sectors by enforcing conformity with industry norms (Shaban-Nejad et al., 2018). These policies safeguard users and promote industry norms by ensuring that AI apps and data science adhere to precise ethical and safety criteria.

International Collaboration and Standards

The aim is to work together to create worldwide guidelines and regulations for governing artificial intelligence, with the goal of fostering consistency and compatibility. The Global Partnership on AI[9] (GPAI) promotes international collaboration on policies and ethics related to artificial intelligence (Wu et al., 2021). Thus, international collaboration promotes the development of common principles and standards, helping the harmonization of AI rules across different countries. It fosters worldwide adherence to moral principles and collaboration, guaranteeing uniform and compatible systems that surpass geographical limits (Vinod & Prabaharan, 2020).

Regulatory Adaptability and Innovation

Adaptable frameworks that can accommodate fast-paced technological breakthroughs and emerging difficulties in AIDS while ensuring that legislation remains up-to-date and supportive of innovation. Regulatory sandboxes provide the controlled experimentation of cutting-edge AI technologies inside a set of regulations (Kumar et al., 2023). Regulatory flexibility fosters innovation by creating a setting that allows for regulated experimentation while upholding ethical and legal norms. It promotes responsible innovation and ensures that legislation is continuously updated to align with the fast-changing AI environment (Choi et al., 2023a). The ability to adapt guarantees that rules stay up to date, creating an environment that is favorable for both technological progress and the ethical development of AI.

Consumer Protection Regulations

The primary objective of this legislation is to protect customers from fraudulent acts, promote transparency, and establish channels for seeking redress in products and services driven by artificial intelligence (Chkoniya, 2021). Consumer protection[10] laws in many jurisdictions require explicit disclosures and safeguards against unjust practices in AI-powered marketing, services, or products. Consumer protection standards enhance user confidence by imposing responsibility on AI developers and organizations to conduct interactions that are equitable and transparent (AstraZeneca, 2022). They offer users channels for addressing AI-related complaints, guaranteeing the ethical and responsible implementation of AI that prioritizes the well-being of consumers and DS applications.

Continual Evaluation and Updating of Regulations

Frameworks that require regular assessments and revisions of current legislation to keep up with technical improvements and changing ethical issues. Regulatory authorities are utilizing agile methodologies to develop AI policies and regulations (Malik et al., 2023). They are actively seeking feedback and promptly adjusting their approaches to address emerging difficulties in AIDS. Ongoing assessment and adjustment of regulations guarantee their pertinence and efficacy in a constantly changing technological environment (Wan et al., 2022). This methodology enables prompt reactions to evolving ethical issues and technological breakthroughs, ensuring a harmonious equilibrium between promoting innovation and adhering to ethical principles in the field of AIDS (Gujral et al., 2019). Consistent updates also indicate a dedication to maintaining regulations that are in line with the changing requirements and difficulties of AI implementation, guaranteeing long-term user trust and confidence.

ARTIFICIAL INTELLIGENCE AND DATA SCIENCE IMPACT ON SOCIETY

The revolutionary potential of AIDS offers both advantageous prospects and complex obstacles. Responsible deployment and ethical considerations are essential for mitigating potential hazards and ensuring that these technologies serve society while stimulating innovation and breakthroughs.

Automation and Labor Shifts

AIDS-powered automation is revolutionizing various industries by enhancing human abilities and fundamentally changing the way work is done. Automation optimizes operations but simultaneously affects the job market. Automating repetitive operations has the potential to result in job displacement in certain industries (Choi et al., 2021). However, emerging sectors such as AI development, data analytics, and robots present fresh prospects, underscoring the importance of acquiring new skills and adjusting to the changing employment landscape. It is crucial to implement strategies that prioritize retraining and education to minimize possible labor disruptions and take advantage of the benefits of AI-driven efficiencies while managing changes in the workforce (Drobot, 2020).

Healthcare Advancements

The application of AIDS in healthcare is transforming the industry by utilizing extensive databases to tailor therapies, forecast diseases, and improve diagnostic capabilities. Machine learning algorithms examine patient data to facilitate early disease identification and provide personalized therapy recommendations (Sbailò et al., 2022). Predictive analytics can assist in the identification of health concerns, optimization of therapies, and potentially lower healthcare expenses. Incorporating these technologies into healthcare systems necessitates thorough testing, ethical deliberations, and adherence to regulatory standards to guarantee their dependability, precision, and adherence to medical ethics (Villarreal-Torres et al., 2023). This ultimately enhances patient outcomes and revolutionizes healthcare provision.

Ethical and Bias Challenges

The use of data in AI systems raises ethical concerns, namely around algorithmic biases and privacy preservation. Biased datasets have the potential to sustain societal disparities, resulting in biased judgments in crucial domains like employment or criminal justice (Górriz et al., 2020). To tackle these difficulties, it is necessary to implement comprehensive strategies such as collecting varied and representative data, conducting audits to ensure algorithmic fairness, and establishing ethical norms in the creation of AI (Thuraisingham, 2020). Ensuring a proper equilibrium between innovation and ethical responsibility becomes of utmost importance, requiring the implementation of transparent and accountable AI systems that reduce biases and prioritize fairness and equity.

Education and Accessibility

AI-powered personalized learning customizes instructional content to meet the specific needs of each student, hence boosting the quality of their learning experiences. In addition, AI enables the use of accessibility tools such as speech recognition, language translation, and assistive technologies, promoting inclusive education for individuals with impairments (Choi et al., 2022b). Incorporating AI into educational institutions necessitates ethical deliberations, measures to protect data privacy, and continuous assessment to guarantee effectiveness and prevent exacerbating inequalities. Incorporating AI into education allows institutions to customize learning experiences, foster diversity, and empower a wide range of learners, potentially transforming conventional educational frameworks (Kadam et al., 2020).

Economic Growth and Innovation

AIDS drives economic growth by extracting valuable insights from extensive information, promoting innovation, and improving decision-making in various industries. These technologies enable firms to make decisions based on data, improve operations, and create new products and services through innovation (Khan et al., 2022). However, to fully enjoy the advantages of AI, it is necessary to make investments in infrastructure, enhance the skills of the workforce, and establish legal frameworks that strike a balance between innovation and ethical concerns. Adopting AI promotes a culture of creativity, increases efficiency, and improves competitiveness in a quickly changing global market, ultimately leading to economic growth and progress in various industries (Gil, 2017).

Privacy and Security Challenges

The exponential growth of AIDS gives rise to significant issues surrounding the protection of data privacy and cybersecurity. Given the comprehensive gathering and examination of vast datasets, safeguarding personal information becomes of utmost importance (Atov et al., 2020). Ensuring the protection of data from breaches, unauthorized access, or misuse becomes a primary focus. Essential components include strong encryption techniques, strict data access restrictions, and adherence to privacy rules like the Health Insurance Portability and Accountability Act[11] (HIPAA) (Chkoniya, 2021). Achieving a balance between the usefulness of data and the protection of privacy necessitates the adoption of responsible data management methods, clear disclosure of data usage, and continuous evaluation of data security measures (Gujral et al., 2019). It is essential to address these difficulties to build confidence among users, preserve the accuracy of data, and protect against the ethical and legal consequences of data breaches.

Environmental Sustainability

The utilization of AIDS is crucial in advancing environmental sustainability through the optimization of resource allocation, accurate prediction of climatic trends, and effective monitoring of ecological transformations (Mikhaylov et al., 2018). These technologies facilitate climate modeling, scrutinize environmental data, and formulate strategies for sustainable practices in industries such as energy, agriculture, and transportation. Machine learning algorithms aid in minimizing energy use, forecasting weather patterns for disaster management, and tracking changes in biodiversity (Sbailò et al., 2022). Collaboration among technology specialists, environmental scientists, and policymakers is essential for leveraging AI to achieve environmental sustainability. This collaboration aims to create AI-driven solutions that effectively tackle ecological concerns. Utilizing AI's predictive skills and data analytics aids in making well-informed decisions, which promotes the development of a sustainable and resilient environment for future generations (Górriz et al., 2020).

CURRENT REGULATORY FRAMEWORK AND ADAPTABLE POLICIES

Within this section, we illustrate the current regulatory framework and adaptable policies that reconcile technological advancements with the safeguarding of ethical principles.

General Data Protection Regulation

The General Data Protection Regulation[12] (GDPR) is a crucial regulatory framework established in the European Union (EU) to protect the privacy of persons' data. The GDPR applies the principles of transparency, fairness, and legitimate processing of personal data to regulate the ethical development and deployment of AI (Gujral et al., 2019). Organizations that operate within the European Union or handle data belonging to European Union people are required to adhere to the GDPR. This regulation influences their procedures regarding the gathering, handling, and utilization of data for AI applications (Thuraisingham, 2020). The GDPR has a far-reaching influence that extends beyond conventional data protection, penetrating the ethical aspects of AI development.

AI Ethics Guidelines by International Organizations

Several international organizations, including the Organization for Economic Cooperation and Development[13] (OECD) and the United Nations Educational, Scientific, and Cultural Organization[14] (UNESCO), have produced guidelines for the development of artificial intelligence that comply with ethical principles (Drobot, 2020). When it comes to the process of implementing artificial intelligence systems, these guidelines emphasize transparency, accountability, and fairness. Although these principles are not legally enforceable, they have a substantial impact on discussions regarding the ethics of artificial intelligence and contribute to the formation of laws and practices on a worldwide scale (Thuraisingham, 2020). This is the case even though they are not legally enforceable.

Algorithmic Impact Assessments

Because it is a framework that requires enterprises to review the societal ramifications of their artificial intelligence algorithms, the concept of algorithmic impact assessments is gaining traction as a framework. Emphasizing fairness and avoiding biases, these evaluations are designed to make certain that artificial intelligence systems do not mistakenly discriminate or cause harm to individuals (Drobot, 2020). The influence of such evaluations is significant, and they encourage the development of AI that is responsible and ethical. This is especially true in fields where skewed results may have far-reaching consequences.

National Legislation and Initiatives

Legislation that specifically targets artificial intelligence and data ethics is currently being actively considered or adopted in a few countries throughout the world. A wide variety of subjects are covered by these regulations, such as the security of data, the transparency of algorithmic processes, accountability, and the appropriate application of artificial intelligence among others (Choi et al., 2022b). Legal frameworks are a way by which states want to regulate and supervise the ethical aspects of artificial intelligence development and deployment within their borders. This is accomplished through the construction of legal frameworks.

Industry Self-Regulation

To govern their artificial intelligence operations, technology corporations, and industry associations are increasingly implementing ethical principles and self-regulatory norms. When it comes to the development and implementation of artificial intelligence, these objectives include commitments to justice, openness, and appropriate data handling. While the self-regulation of the industry does not have the force of law behind it, it does reflect a proactive approach toward ethical considerations and has the potential to greatly influence public perception and trust in artificial intelligence technologies (Gil, 2017). A culture of ethical responsibility is being fostered inside the technology sector because of the impact, which goes beyond the mere fulfillment of legal responsibilities.

Government Initiatives and AI Ethics Offices

Several countries have formed AI ethics offices or task forces, particularly for the aim of resolving ethical problems considering the quickly changing environment of artificial intelligence ethics. These offices or task forces are intended to address ethical concerns (Borrego-Díaz & Galán-Páez, 2022a). Several organizations are currently striving to develop and amend policies to guarantee that artificial intelligence technology is utilized responsibly. A comprehensive and adaptable regulatory system that navigates the problems given by technological breakthroughs while respecting ethical standards is the goal of these projects (Vinod & Prabaharan, 2020). These projects strive to create such an environment by fostering collaboration between government agencies, industry stakeholders, and the public. The goal of these projects is to create such an environment.

FUTURE DIRECTIONS OF ARTIFICIAL INTELLIGENCE AND DATA SCIENCE

The future of AIDS has significant potential but also requires responsible development, ethical deliberation, and collaboration to fully utilize its capacity for social benefit while minimizing potential hazards and obstacles.

Artificial Intelligence in Personalization and Augmentation

Future breakthroughs in AI will drive hyper-personalization, leading to a revolution in industries through the customization of services, content, and goods according to individual tastes. Artificial intelligence algorithms will process extensive datasets, enabling the provision of tailored recommendations and services, thus boosting user experiences (Gil, 2017). The progress in artificial intelligence will facilitate enhanced cooperation between people and machines. Artificial intelligence tools will enhance human talents by assisting in decision-making, improving creativity, and streamlining operations. This partnership will lead to enhanced efficiency and effectiveness in multiple fields, using the powers of AI while also harnessing human intuition and innovation (Khan et al., 2022).

Artificial Intelligence Ethics and Responsible AI Development

The future of AI will see a heightened emphasis on strong ethical frameworks. These frameworks will tackle prejudices, guarantee openness, and emphasize the welfare of society in the implementation of AI, promoting trust and responsibility (Wu et al., 2021). In the upcoming years, comprehensive governance structures and regulatory frameworks for AIDS will be constructed. These frameworks will address new ethical issues, guaranteeing responsible development, deployment, and use of AI systems, thus protecting against potential threats and fostering ethical behavior (Vinod & Prabaharan, 2020).

Artificial Intelligence in Healthcare and Biotechnology

The progress in AIDS will enhance precision medicine through the utilization of individual health data and genetic information. This will enable the provision of individualized treatment regimens and the prediction of disease patterns, leading to a substantial enhancement in patient care (Jain, 2023). AI-powered technologies will enhance drug discovery procedures by evaluating extensive datasets, discovering viable drug candidates, expediting development, and optimizing treatment regimens. This has the potential to revolutionize pharmaceutical research and development (Maxwell et al., 2021).

Artificial Intelligence and Sustainability

AIDS will have a crucial role in forecasting and alleviating the effects of climate change by examining environmental data in climate modeling and sustainability solutions. These technologies will enable the implementation of sustainable practices in various industries, thereby assisting in efforts to adapt to and mitigate climate change (Gujral et al., 2019). In the future, AI will play a crucial role in pushing innovation in environmentally friendly technology. It will focus on optimizing energy use and bringing forth revolutionary eco-friendly solutions. The utilization of AI-driven analysis would facilitate the development of sustainable practices and eco-friendly technologies, thereby making a significant contribution towards a more environmentally conscious and sustainable future (Mikhaylov et al., 2018).

Artificial Intelligence-driven Automation and Industry 4.0

The utilization of AI-driven automation will persist in revolutionizing several sectors through the improvement of efficiency and optimization of workflows. The implementation of advanced robotics and automation will optimize manufacturing processes, logistics, and diverse industries, fundamentally transforming existing workflows (Villarreal-Torres et al., 2023). Industry 4.0 refers to the merging of AIDS with the Internet of Things (IoT), resulting in the development of intelligent and networked systems. This integration will facilitate predictive maintenance, the creation of smart cities, and the advancement of autonomous cars (Cao, 2023a). This merger will stimulate innovation by optimizing operations across several industries and establishing interconnected ecosystems.

Artificial Intelligence in Education and Skills Development

The future progress of AI will revolutionize education by providing customized learning experiences that cater to the specific demands of each student. Artificial intelligence-powered technologies will modify educational content, promoting various learning styles and improving involvement (Choi et al., 2022b). AI-driven tools will assist in enhancing and retraining the workforce for new job positions in the AI-dominated economy. These technologies will have a vital role in equipping individuals with changing employment environments and providing customized learning routes to match the needs of the future workforce (Kadam et al., 2020).

Advancements in Natural Language Processing and Artificial Intelligence Interfaces

The future will see notable progress in conversational AI, with breakthroughs in natural language processing and generation. This will allow for more intuitive and human-like interactions between humans and machines (Leung et al., 2021). Conversational AI has the potential to completely transform user interfaces by providing effortless communication and customized interactions. AI Interfaces: Advancing AI interfaces will improve user experiences by incorporating voice commands, gesture recognition, and immersive technology (Han & Liu, 2019). These interfaces will enhance user experience and facilitate interaction with technology, enabling access to AI-driven features.

Artificial Intelligence in Ethical and Social Implications

The future will prioritize promoting the extensive implementation of ethical AI techniques and concepts across all businesses. This entails incorporating ethical considerations into AI systems, guaranteeing fairness, transparency, and accountability at every stage of their development and implementation (Saura et al., 2022). Social Impact Assessment: There will be an increased focus on completing thorough social impact evaluations prior to implementing AI technologies. These assessments will examine possible social consequences, such as unemployment or prejudices, to take proactive steps to reduce negative effects and promote fair outcomes (Ong & Uddin, 2020).

Artificial Intelligence Collaboration and Interdisciplinary Integration

The future progress of AI will promote greater cooperation among different fields, facilitating the integration of several disciplines. This partnership will bring together specialists from several fields, such as psychology, ethics, sociology, and technology, to tackle intricate problems and create comprehensive AI solutions (Borrego-Díaz & Galán-Páez, 2022b). The integration of AI with diverse disciplines will result in novel applications and solutions. Combining AI with neuroscience has the potential to result in significant advancements in brain-computer interfaces. Similarly, the integration of AI with environmental science can fundamentally transform ecological conservation efforts (Borrego-Díaz & Galán-Páez, 2022a). The collaboration between different disciplines will lead to new and innovative findings and progress in other fields, utilizing the potential of artificial intelligence in different situations.

LESSONS LEARNED THE CONCLUSION

These lessons learned work as fundamental principles to manage the intricate, developing, and influential fields of AIDS safely and ethically, as summarized.

- The research highlights the utmost significance of ethical factors in the field of AIDS. It is crucial to incorporate fairness, transparency, accountability, and societal well-being into the development, deployment, and regulation of AI systems.
- The field of AIDS is constantly changing and developing. It is imperative to uphold a mentality of perpetual learning and adjustment to stay updated on evolving technology, ethical frameworks, and regulatory modifications.
- Understanding the importance of interdisciplinary teamwork is crucial. The potential of AI surpasses conventional limitations, and incorporating knowledge from several disciplines amplifies creativity and the ability to solve problems.
- The advancement of AI necessitates the integration of a conscientious approach. Ensuring a harmonious integration of technological progress with ethical concerns and the potential impact on society is essential for establishing long-term viability and fostering trust.
- Prior to using AI technology, it is crucial to conduct thorough impact evaluations. Early identification of potential societal, ethical, and economic ramifications enables the implementation of proactive mitigation techniques to tackle difficulties effectively.

- By prioritizing design that focuses on the requirements of users and promotes inclusion, we can ensure that AI technologies are accessible to everyone and do not worsen existing societal inequalities.
- Regulators and politicians must consistently modify and elucidate regulatory frameworks to stay abreast of technological progress. Regulatory flexibility and clear compliance criteria are crucial for promoting innovation while guaranteeing the ethical deployment of AI.
- It is imperative for all parties involved to actively promote education and raise awareness of the ethical considerations of AIDS. By providing users, developers, politicians, and organizations with the necessary tools and knowledge, they are enabled to make well-informed decisions and actively participate in the development of AI in a responsible manner.
- Prioritizing international cooperation and the establishment of common standards in the administration of artificial intelligence promotes the development of shared principles and standards, which in turn enhances the unity and ethicality of the AI ecosystem.
- The future of AIDS is intrinsically unpredictable. By accepting and embracing this lack of certainty and implementing a systematic and repetitive approach to development and policymaking, it becomes possible to be flexible and responsive to the ever-changing ethical and technological environments.

CONCLUSION

Examining the complex terrain of AIDS uncovered vital observations across various aspects. Ethical considerations have developed as fundamental principles that require prioritizing to ensure fairness, transparency, and societal well-being in the development and implementation of AI. The study shed light on ethical problems, highlighting the necessity of tackling biases, safeguarding privacy, and navigating ethical dilemmas to foster responsible AI. Regulatory frameworks have emerged as crucial instruments, requiring ongoing adjustment to promote innovation while maintaining ethical norms. Likewise, the analysis of AI's influence on society has highlighted its significant capacity to revolutionize healthcare, environmental initiatives, and education while also emphasizing the importance of ethical and responsible implementations. The future progress of AI and Data Science relies on interdisciplinary collaboration, user-centric design, and worldwide cooperation to ensure the widespread adoption of ethical AI. The lessons learned highlighted the continuous requirement for ethical concerns, flexibility in rules, collaboration across different fields, and innovation centered around user needs to navigate the ever-changing AI world properly. As AI becomes more prevalent in various industries, the importance of combining innovation and ethical responsibility is being emphasized. The future development of AI and Data Science depends not just on scientific progress but also on a unified, ethical, and inclusive strategy to create a future where AI responsibly enhances human talents for the overall advantage of society.

REFERENCES

Aguilar-Esteva, V., Acosta-Banda, A., Carreño Aguilera, R., & Patiño Ortiz, M. (2023). Sustainable Social Development through the Use of Artificial Intelligence and Data Science in Education during the COVID Emergency: A Systematic Review Using PRISMA. In Sustainability (Switzerland), 15(8). doi:10.3390/su15086498

Allen, B., Agarwal, S., Coombs, L., Wald, C., & Dreyer, K. (2021). 2020 ACR Data Science Institute Artificial Intelligence Survey. *Journal of the American College of Radiology*, 18(8), 1153–1159. doi:10.1016/j.jacr.2021.04.002 PMID:33891859

Alnssyan, B., Ahmad, Z., Malela-Majika, J. C., Seong, J. T., & Shafik, W. (2023). On the identifiability and statistical features of a new distributional approach with reliability applications. *AIP Advances*, 13(12), 125211. doi:10.1063/5.0178555

An, L., Grimm, V., Bai, Y., Sullivan, A., Turner, B. L. II, Malleson, N., Heppenstall, A., Vincenot, C., Robinson, D., Ye, X., Liu, J., Lindkvist, E., & Tang, W. (2023). Modeling agent decision and behavior in the light of data science and artificial intelligence. *Environmental Modelling & Software*, 166, 105713. Advance online publication. doi:10.1016/j.envsoft.2023.105713

AstraZeneca. (2022). *Data Science & Artificial Intelligence: Unlocking new science insights*. AstraZeneca.

Atov, I., Chen, K. C., Kamal, A., & Louta, M. (2020). Data Science and Artificial Intelligence for Communications. In IEEE Communications Magazine, 58(10). doi:10.1109/MCOM.2020.9247523

Borrego-Díaz, J., & Galán-Páez, J. (2022a). Explainable Artificial Intelligence in Data Science. *Minds and Machines*, 32(3), 485–531. doi:10.1007/s11023-022-09603-z

Borrego-Díaz, J., & Galán-Páez, J. (2022b). Explainable Artificial Intelligence in Data Science: From Foundational Issues Towards Socio-technical Considerations. *Minds and Machines*, 32(3), 485–531. doi:10.1007/s11023-022-09603-z

Brzezinski, M., & Krzeminska, I. (2023). The strategies for innovating with virtual reality and artificial intelligence: A literature review. *Technium*, 8, 72–83. doi:10.47577/technium.v8i.8671

Cao, L. (2023a). Trans-AI/DS: Transformative, transdisciplinary and translational artificial intelligence and data science. *International Journal of Data Science and Analytics*, 15(2), 119–132. doi:10.1007/s41060-023-00383-y

Cao, L. (2023b). Trans-AI/DS: Transformative, transdisciplinary and translational artificial intelligence and data science. *International Journal of Data Science and Analytics*, 15(2), 119–132. doi:10.1007/s41060-023-00383-y

Chkoniya, V. (2021). Handbook of research on applied data science and artificial intelligence in business and industry. In Handbook of Research on Applied Data Science and Artificial Intelligence in Business and Industry. doi:10.4018/978-1-7998-6985-6

Choi, Y., Kamal, A., & Louta, M. (2023b). Series Editorial: Artificial Intelligence and Data Science for Communications. In IEEE Communications Magazine, 61(6). doi:10.1109/MCOM.2023.10155727

Choi, Y., Kamal, A. E., & Louta, M. (2021). Series Editorial: Artificial Intelligence and Data Science for Communications. In IEEE Communications Magazine, 59(11). doi:10.1109/MCOM.2021.9665429

Choi, Y., Kamal, A. E., & Louta, M. (2022a). Series Editorial: Artificial Intelligence and Data Science for Communications. In IEEE Communications Magazine, 60(11). doi:10.1109/MCOM.2022.9946952

Choi, Y., Kamal, A. E., & Louta, M. (2022b). Series Editorial: Artificial Intelligence and Data Science for Communications. In IEEE Communications Magazine, 60(7). doi:10.1109/MCOM.2022.9831135

Choi, Y., Kamal, A. E., & Louta, M. (2023a). Series Editorial: Artificial Intelligence and Data Science for Communications. In IEEE Communications Magazine, 61(3). doi:10.1109/MCOM.2023.10080874

Cruz-Jesus, F., Castelli, M., Oliveira, T., Mendes, R., Nunes, C., Sa-Velho, M., & Rosa-Louro, A. (2020). Using artificial intelligence methods to assess academic achievement in public high schools of a European Union country. *Heliyon*, 6(6), e04081. doi:10.1016/j.heliyon.2020.e04081 PMID:32551378

Cui, X., Li, W., & Gu, C. (2021). Big Data of Food Science and Artificial Intelligence Technology. *Journal of Chinese Institute of Food Science and Technology*, 21(2). doi:10.16429/j.1009-7848.2021.02.001

Drobot, A. T. (2020). Industrial Transformation and the Digital Revolution: A Focus on Artificial Intelligence, Data Science and Data Engineering. *2020 ITU Kaleidoscope. Industry-Driven Digital Transformation, ITU K, 2020*, 1–11. doi:10.23919/ITUK50268.2020.9303221

Gil, Y. (2017). Thoughtful artificial intelligence: Forging a new partnership for data science and scientific discovery. *Data Science*, 1(1–2), 119–129. doi:10.3233/DS-170011

González-Méijome, J. M., Piñero, D. P., & Villa-Collar, C. (2022). Upcoming Special Issue: "Artificial Intelligence, Data Science and E-health in Vision Research and Clinical Activity.". *Journal of Optometry*, 15(1), 1–2. doi:10.1016/j.optom.2021.11.003 PMID:34933741

Górriz, J. M., Ramírez, J., Ortíz, A., Martínez-Murcia, F. J., Segovia, F., Suckling, J., Leming, M., Zhang, Y. D., Álvarez-Sánchez, J. R., Bologna, G., Bonomini, P., Casado, F. E., Charte, D., Charte, F., Contreras, R., Cuesta-Infante, A., Duro, R. J., Fernández-Caballero, A., Fernández-Jover, E., & Ferrández, J. M. (2020). Artificial intelligence within the interplay between natural and artificial computation: Advances in data science, trends and applications. *Neurocomputing*, 410, 237–270. doi:10.1016/j.neucom.2020.05.078

Gruson, D., Helleputte, T., Rousseau, P., & Gruson, D. (2019). Data science, artificial intelligence, and machine learning: Opportunities for laboratory medicine and the value of positive regulation. In Clinical Biochemistry (Vol. 69). doi:10.1016/j.clinbiochem.2019.04.013

Gujral, G., Shivarama, J., & Mariappan, M. (2019). Artificial Intelligence (Ai) and Data Science for Developing Intelligent Health Informatics Systems. *Proceedings of the Ntional Conference on AI in HI & VR, SHSS-TISS Mumbai, January*. IEEE.

Hameed, B. M. Z., Prerepa, G., Patil, V., Shekhar, P., Zahid Raza, S., Karimi, H., Paul, R., Naik, N., Modi, S., Vigneswaran, G., Prasad Rai, B., Chłosta, P., & Somani, B. K. (2021). Engineering and clinical use of artificial intelligence (AI) with machine learning and data science advancements: radiology leading the way for future. In Therapeutic Advances in Urology (Vol. 13). doi:10.1177/17562872211044880

Han, H., & Liu, W. (2019). The coming era of artificial intelligence in biological data science. In BMC Bioinformatics (Vol. 20). doi:10.1186/s12859-019-3225-3

Hilal, A. M., Alsolai, H., Al-Wesabi, F. N., Al-Hagery, M. A., Hamza, M. A., & Al Duhayyim, M. (2022). Artificial intelligence based optimal functional link neural network for financial data Science. *Computers, Materials & Continua, 70*(3). doi:10.32604/cmc.2022.021522

Jain, T. (2023). Applications of Artificial Intelligence & Machine Learning: A study on Automotive Industry. *Interantional Journal of Scientific Research In Engineering And Management, 07*(07). doi:10.55041/IJSREM23445

Jordan, M. I., & Mitchell, T. M. (2015). Machine learning: Trends, perspectives, and prospects. In Science, 349(6245). doi:10.1126/science.aaa8415

Jun, Y., Craig, A., Shafik, W., & Sharif, L. (2021). Artificial intelligence application in cybersecurity and cyberdefense. *Wireless Communications and Mobile Computing, 2021*, 1–10. doi:10.1155/2021/3329581

Kadam, M. S., Krutika Kajulkar, M., & Ovhal, M. A. (2020). Importance of Human-Machine Interface in Artificial Intelligence and Data Science. [IJERT]. *International Journal of Engineering Research & Technology (Ahmedabad), 8*(5).

Khan, C., Blount, D., Parham, J., Holmberg, J., Hamilton, P., Charlton, C., Christiansen, F., Johnston, D., Rayment, W., Dawson, S., Vermeulen, E., Rowntree, V., Groch, K., Levenson, J. J., & Bogucki, R. (2022). Artificial intelligence for right whale photo identification: From data science competition to worldwide collaboration. *Mammalian Biology, 102*(3), 1025–1042. doi:10.1007/s42991-022-00253-3

Kumar, K., George, K. S., Bhatt, D., & Paul, O. P. (2023). A Brief Study on the Fake Review Detection methods on Ecommerce Websites using Machine Learning, Artificial Intelligence, and Data Science. *International Journal for Research in Applied Science and Engineering Technology, 11*(9), 721–726. doi:10.22214/ijraset.2023.52743

Leung, C. K., Pazdor, A. G. M., & Souza, J. (2021). Explainable Artificial Intelligence for Data Science on Customer Churn. *2021 IEEE 8th International Conference on Data Science and Advanced Analytics, DSAA 2021*. IEEE. 10.1109/DSAA53316.2021.9564166

Mainali, S., & Park, S. (2023). Artificial Intelligence and Big Data Science in Neurocritical Care. In Critical Care Clinics, 39(1). doi:10.1016/j.ccc.2022.07.008

Malik, M., Gahlawat, V. K., Mor, R. S., Agnihotri, S., Panghal, A., Rahul, K., & Emanuel, N. (2023). Artificial Intelligence and Data Science in Food Processing Industry. In EAI/Springer Innovations in Communication and Computing. Springer. doi:10.1007/978-3-031-19711-6_11

Marr, B. (2020). Coronavirus: How Artificial Intelligence, Data Science And Technology Is Used To Fight The Pandemic. *Forbes*.

Maxwell, D., Meyer, S., & Bolch, C. (2021). DataStory™: An interactive sequential art approach for data science and artificial intelligence learning experiences. *Innovación Educativa (México, D.F.), 3*(1), 8. doi:10.1186/s42862-021-00015-x

Mikhaylov, S. J., Esteve, M., & Campion, A. (2018). Artificial intelligence for the public sector: Opportunities and challenges of cross-sector collaboration. *Philosophical Transactions. Series A, Mathematical, Physical, and Engineering Sciences, 376*(2128), 20170357. doi:10.1098/rsta.2017.0357 PMID:30082303

Naudé, W., Bray, A., & Lee, C. (2022). Crowdsourcing Artificial Intelligence in Africa: Analysis of a Data Science Contest. SSRN *Electronic Journal.* doi:10.2139/ssrn.4076351

Nishimwe, A., Ruranga, C., Musanabaganwa, C., Mugeni, R., Semakula, M., Nzabanita, J., Kabano, I., Uwimana, A., Utumatwishima, J. N., Kabakambira, J. D., Uwineza, A., Halvorsen, L., Descamps, F., Houghtaling, J., Burke, B., Bahati, O., Bizimana, C., Jansen, S., Twizere, C., & Twagirumukiza, M. (2022). Leveraging artificial intelligence and data science techniques in harmonizing, sharing, accessing and analyzing SARS-COV-2/COVID-19 data in Rwanda (LAISDAR Project): Study design and rationale. *BMC Medical Informatics and Decision Making, 22*(1), 214. doi:10.1186/s12911-022-01965-9 PMID:35962355

Octavian Dumitru, C., Schwarz, G., Castel, F., Lorenzo, J., & Datcu, M. (2019). Artificial Intelligence Data Science Methodology for Earth Observation. In Advanced Analytics and Artificial Intelligence Applications. InTech Open. doi:10.5772/intechopen.86886

Ong, S., & Uddin, S. (2020). Data science and artificial intelligence in project management: The past, present and future. *Journal of Modern Project Management, 7*(4). doi:10.19255/JMPM02202

Peña-Guerrero, J., Nguewa, P. A., & García-Sosa, A. T. (2021). Machine learning, artificial intelligence, and data science breaking into drug design and neglected diseases. In Wiley Interdisciplinary Reviews: Computational Molecular Science, 11(5). doi:10.1002/wcms.1513

Popescu, C.-C. (2018). Improvements in business operations and customer experience through data science and Artificial Intelligence. *Proceedings of the International Conference on Business Excellence, 12*(1). 10.2478/picbe-2018-0072

Raschka, S., Patterson, J., & Nolet, C. (2020). Machine learning in python: Main developments and technology trends in data science, machine learning, and artificial intelligence. In Information (Switzerland), 11(4). doi:10.3390/info11040193

Ray, R., Agar, Z., Dutta, P., Ganguly, S., Sah, P., & Roy, D. (2021). MenGO: A Novel Cloud-Based Digital Healthcare Platform for Andrology Powered By Artificial Intelligence, Data Science & Analytics, Bio-Informatics And Blockchain. *Biomedical Sciences Instrumentation, 57*(4), 476–485. doi:10.34107/KSZV7781.10476

Ronmi, A. E., Prasad, R., & Raphael, B. A. (2023). How can artificial intelligence and data science algorithms predict life expectancy - An empirical investigation spanning 193 countries. *International Journal of Information Management Data Insights, 3*(1), 100168. doi:10.1016/j.jjimei.2023.100168

Sagi, T., Lehahn, Y., & Bar, K. (2020). Artificial intelligence for ocean science data integration: Current state, gaps, and way forward. *Elementa, 8*, 21. doi:10.1525/elementa.418

Saura, J. R., Ribeiro-Soriano, D., & Palacios-Marqués, D. (2022). Assessing behavioral data science privacy issues in government artificial intelligence deployment. *Government Information Quarterly*, *39*(4), 101679. doi:10.1016/j.giq.2022.101679

Sbailò, L., Fekete, Á., Ghiringhelli, L. M., & Scheffler, M. (2022). The NOMAD Artificial-Intelligence Toolkit: Turning materials-science data into knowledge and understanding. *npj Computational Materials*, *8*(1), 250. doi:10.1038/s41524-022-00935-z

Schwaller, P., Vaucher, A. C., Laplaza, R., Bunne, C., Krause, A., Corminboeuf, C., & Laino, T. (2022). Machine intelligence for chemical reaction space. In Wiley Interdisciplinary Reviews: Computational Molecular Science, 12(5). doi:10.1002/wcms.1604

Shaban-Nejad, A., Michalowski, M., & Buckeridge, D. L. (2018). Health intelligence: how artificial intelligence transforms population and personalized health. In npj Digital Medicine. doi:10.1038/s41746-018-0058-9

Shafik, W. (2023a). Cyber Security Perspectives in Public Spaces: Drone Case Study. In Handbook of Research on Cybersecurity Risk in Contemporary Business Systems (pp. 79-97). IGI Global. doi:10.4018/978-1-6684-7207-1.ch004

Shafik, W. (2023b). Making Cities Smarter: IoT and SDN Applications, Challenges, and Future Trends. In Opportunities and Challenges of Industrial IoT in 5G and 6G Networks (pp. 73-94). IGI Global. doi:10.4018/978-1-7998-9266-3.ch004

Shafik, W. (2023c). A Comprehensive Cybersecurity Framework for Present and Future Global Information Technology Organizations. In *Effective Cybersecurity Operations for Enterprise-Wide Systems* (pp. 56–79). IGI Global. doi:10.4018/978-1-6684-9018-1.ch002

Shafik, W. (2024a). Introduction to ChatGPT. In *Advanced Applications of Generative AI and Natural Language Processing Models* (pp. 1–25). IGI Global. doi:10.4018/979-8-3693-0502-7.ch001

Shafik, W. (2024b). Predicting Future Cybercrime Trends in the Metaverse Era. In Forecasting Cyber Crimes in the Age of the Metaverse (pp. 78-113). IGI Global. doi:10.4018/979-8-3693-0220-0.ch005

Shafik, W., Matinkhah, S. M., & Shokoor, F. (2023). Cybersecurity in unmanned aerial vehicles: A review. *International Journal on Smart Sensing and Intelligent Systems*, *16*(1), 20230012. doi:10.2478/ijssis-2023-0012

Siah, K. W., Kelley, N. W., Ballerstedt, S., Holzhauer, B., Lyu, T., Mettler, D., Sun, S., Wandel, S., Zhong, Y., Zhou, B., Pan, S., Zhou, Y., & Lo, A. W. (2021). Predicting drug approvals: The Novartis data science and artificial intelligence challenge. *Patterns (New York, N.Y.)*, *2*(8), 100312. doi:10.1016/j.patter.2021.100312 PMID:34430930

Sohail, A. (2023). Genetic Algorithms in the Fields of Artificial Intelligence and Data Sciences. In Annals of Data Science, 10(4). doi:10.1007/s40745-021-00354-9

Stracener, C., Samelson, Q., MacKie, J., Ihaza, M., Laplante, P. A., & Amaba, B. (2019). The Internet of Things Grows Artificial Intelligence and Data Sciences. *IT Professional*, *21*(3), 55–62. doi:10.1109/MITP.2019.2912729

Thuraisingham, B. (2020). Artificial Intelligence and Data Science Governance: Roles and Responsibilities at the C-Level and the Board. *Proceedings - 2020 IEEE 21st International Conference on Information Reuse and Integration for Data Science, IRI 2020*. IEEE. 10.1109/IRI49571.2020.00052

Trujillo-Cabezas, R. (2020). Integrating Foresight, Artificial Intelligence and Data Science to Develop Dynamic Futures Analysis. *Journal of Information Systems Engineering & Management, 5*(3), em0120. doi:10.29333/jisem/8428

Villarreal-Torres, H., Ángeles-Morales, J., Cano-Mejía, J., Mejía-Murillo, C., Flores-Reyes, G., Cruz-Cruz, O., Marín-Rodriguez, W., Andrade-Girón, D., Carreño-Cisneros, E., & Boscán-Carroz, M. C. (2023). Development of a Classification Model for Predicting Student Payment Behavior Using Artificial Intelligence and Data Science Techniques. *EAI Endorsed Transactions on Scalable Information Systems, 10*(5). doi:10.4108/eetsis.3489

Vinod, D. N., & Prabaharan, S. R. S. (2020). Data science and the role of Artificial Intelligence in achieving the fast diagnosis of Covid-19. *Chaos, Solitons, and Fractals, 140*, 110182. doi:10.1016/j.chaos.2020.110182 PMID:32834658

Wan, W. Y., Tsimplis, M., Siau, K. L., Yue, W. T., Nah, F. F. H., & Yu, G. M. (2022). Legal and Regulatory Issues on Artificial Intelligence, Machine Learning, Data Science, and Big Data. Lecture Notes in Computer Science (Including Subseries Lecture Notes in Artificial Intelligence and Lecture Notes in Bioinformatics), 13518 LNCS. Springer. doi:10.1007/978-3-031-21707-4_40

Ward, T. M., Mascagni, P., Madani, A., Padoy, N., Perretta, S., & Hashimoto, D. A. (2021). Surgical data science and artificial intelligence for surgical education. In Journal of Surgical Oncology, 124(2). doi:10.1002/jso.26496

Wu, Y., Zhang, Z., Kou, G., Zhang, H., Chao, X., Li, C. C., Dong, Y., & Herrera, F. (2021). Distributed linguistic representations in decision making: Taxonomy, key elements and applications, and challenges in data science and explainable artificial intelligence. In Information Fusion (Vol. 65). doi:10.1016/j.inffus.2020.08.018

Young, E., Wajcman, J., & Sprejer, L. (2023). Mind the gender gap: Inequalities in the emergent professions of artificial intelligence (AI) and data science. *New Technology, Work and Employment, 38*(3), 391–414. doi:10.1111/ntwe.12278

ENDNOTES

1   https://www.ibm.com/topics/explainable-ai
2   https://gdpr-info.eu/
3   https://european-union.europa.eu/
4   https://www.ieee.org/
5   https://bipartisanpolicy.org/blog/impact-assessments-for-ai/
6   https://en.wikipedia.org/wiki/ISO/IEC_27001
7   https://www.iso.org/
8   https://www.fda.gov/

[9] https://gpai.ai/
[10] https://en.wikipedia.org/wiki/Consumer_protection
[11] https://www.cdc.gov/phlp/publications/topic/hipaa.html
[12] https://gdpr-info.eu/
[13] https://www.oecd.org/
[14] https://www.unesco.org/en

Compilation of References

Chen, M., Zhang, Z., Wang, T., & Backe, M. (2021). *When Machine Unlearning Jeopardizes Privacy*. ACM.

Aaronson, D., Faber, J., Hartley, D., Mazumder, B., & Sharkey, P. (2021). The long-run effects of the 1930s HOLC "redlining" maps on place-based measures of economic opportunity and socioeconomic success. *Regional Science and Urban Economics*, *86*, 103622. doi:10.1016/j.regsciurbeco.2020.103622

AaS.AhmadS. (2021). *Cybersecurity Threats and Attacks in Healthcare*. doi:10.13140/RG.2.2.17256.80647/2

Abbasi-Sureshjani, S., Raumanns, R., Michels, B., Schouten, G., & Cheplygina, V. (2020). Risk of Training Diagnostic Algorithms on Data with Demographic Bias. *Interpretable and Annotation-Efficient Learning for Medical Image Computing: Third International Workshop, iMIMIC 2020*. Springer International Publishing.

Abd El Kader, I., Xu, G., Shuai, Z., Saminu, S., Javaid, I., Ahmad, I. S., & Kamhi, S. (2021). Brain Tumor Detection and Classification on MR Images by a Deep Wavelet Auto-Encoder Model. *Diagnostics (Basel)*, *11*(9), 1589. doi:10.3390/diagnostics11091589 PMID:34573931

Abdulkarim, B., Kamberov, R., & Hay, G. J. (2014). Supporting Urban Energy Efficiency with Volunteered Roof Information and the Google Maps API. *Remote Sensing (Basel)*, *6*(10), 9691–9711. Advance online publication. doi:10.3390/rs6109691

Abdulwahid, A. H., Pattnaik, M., Palav, M. R., Babu, S. T., Manoharan, G., & Selvi, G. P. (2023, April). Library Management System Using Artificial Intelligence. In *2023 Eighth International Conference on Science Technology Engineering and Mathematics (ICONSTEM)* (pp. 1-7). IEEE.

Abrokwah-Larbi, K., & Awuku-Larbi, Y. (2023). The impact of artificial intelligence in marketing on the performance of business organizations: evidence from SMEs in an emerging economy. *Journal of Entrepreneurship in Emerging Economies*.

Achar, A. (2017). *Artificial intelligence for early literacy in India: A case study of Pratham*. Retrieved from https://files.eric.ed.gov/fulltext/ED496345.pdf

Acikkar, M., & Akay, M. F. (2009). Support vector machines for predicting the admission decision of a candidate to the School of Physical Education and Sports at Cukurova University. *Expert Systems with Applications*, *36*(3), 7228–7233. doi:10.1016/j.eswa.2008.09.007

Adadi, A. (2021). A survey on data-efficient algorithms in big data era. *Journal of Big Data*, *8*(1), 24. doi:10.1186/s40537-021-00419-9

Adams, A., Adelfio, A., Barnes, B., Berlien, R., Branco, D., Coogan, A., Garson, L., Ramirez, N., Stansbury, N., Stewart, J., Worman, G., Butler, P. J., & Brown, D. (2023). Risk-Based Monitoring in Clinical Trials: 2021 Update. *Therapeutic Innovation & Regulatory Science*, *57*(3), 529–537. doi:10.1007/s43441-022-00496-9 PMID:36622566

Adel, T., Ghahramani, Z., & Weller, A. (2018, July). Discovering interpretable representations for both deep generative and discriminative models. In *International Conference on Machine Learning* (pp. 50-59). PMLR.

Agarwal, A., Gans, J., & Goldfarb, A. (2016, December 21). The Obama Administration's Roadmap for AI Policy. *Harvard Business Review*. https://hbr.org/2016/12/the-obama-administrations-roadmap-for-ai-policy

Agbo, C., Mahmoud, Q., & Eklund, J. (2019). Blockchain Technology in Healthcare: A Systematic Review. *Health Care*, 7(2), 56. doi:10.3390/healthcare7020056 PMID:30987333

Agostini, M., Pucciarelli, S., Enzo, M. V., Del Bianco, P., Briarava, M., Bedin, C., Maretto, I., Friso, M. L., Lonardi, S., Mescoli, C., Toppan, P., Urso, E., & Nitti, D. (2011). Circulating Cell-Free DNA: A Promising Marker of Pathologic Tumor Response in Rectal Cancer Patients Receiving Preoperative Chemoradiotherapy. *Annals of Surgical Oncology*, 18(9), 2461–2468. doi:10.1245/s10434-011-1638-y PMID:21416156

Aguilar-Esteva, V., Acosta-Banda, A., Carreño Aguilera, R., & Patiño Ortiz, M. (2023). Sustainable Social Development through the Use of Artificial Intelligence and Data Science in Education during the COVID Emergency: A Systematic Review Using PRISMA. In Sustainability (Switzerland), 15(8). doi:10.3390/su15086498

Aguilar, J., Garces-Jimenez, A., R-Moreno, M. D., & García, R. (2021). A systematic literature review on using artificial intelligence in energy self-management in smart buildings. *Renewable & Sustainable Energy Reviews*, 151, 111530. doi:10.1016/j.rser.2021.111530

Ahasan, R., & Hossain, M. M. (2021). Leveraging GIS and spatial analysis for informed decision-making in COVID-19 pandemic. *Health Policy and Technology*, 10(1), 7–9. Advance online publication. doi:10.1016/j.hlpt.2020.11.009 PMID:33318916

Ahmad, B. (2015). The UAE Federal Government's E-Participation Roadmap: Developments in UAE Empowerment Initiatives With VGI/PGIS and Location Based Services (LBS). *Canadian Social Science*, 11(5). doi:10.3968/6919

Ahmad, H.A., Hanandeh, R., Alazzawi, F.R., Al-Daradkah, A., ElDmrat, A.T., Ghaith, Y.M., & Darawsheh, S.R. (2023). The effects of big data, artificial intelligence, and business intelligence on e-learning and business performance: Evidence from Jordanian telecommunication firms. *International Journal of Data and Network Science*.

Ahmad, M. (2023a). Exploring the Potential of Spatial Data for Enhancing Higher Education Learners' Learning Outcomes. In K. Barua, N. Radwan, V. Singh, & R. Figueiredo (Eds.), *Design and Implementation of Higher Education Learners' Learning Outcomes (HELLO)* (pp. 128–145). IGI Global. doi:10.4018/978-1-6684-9472-1.ch008

Ahmad, M. (2023b). Leveraging Social Media Geographic Information for Smart Governance and Policy Making:Opportunities and Challenges. In C. Chavadi & D. Thangam (Eds.), *Global Perspectives on Social Media Usage Within Governments* (pp. 192–213). IGI Global. doi:10.4018/978-1-6684-7450-1.ch013

Ahmad, M. (2023c). Spatial Data as a Catalyst to Drive Entrepreneurial Growth and Sustainable Development. In N. Yaw Asabere, G. K. Gyimah, A. Acakpovi, & F. Plockey (Eds.), *Technological Innovation Driving Sustainable Entrepreneurial Growth in Developing Nations* (pp. 79–104). IGI Global. doi:10.4018/978-1-6684-9843-9.ch004

Ahmad, M. (2023d). Unlocking the Power of Spatial Big Data for Sustainable Development:From Capacity Building to Food Security and Food Traceability. In J. M. Falcó, B. M. Lajara, E. S. García, & L. A. M. Tudela (Eds.), *Crafting a Sustainable Future Through Education and Sustainable Development* (pp. 204–218). IGI Global. doi:10.4018/978-1-6684-9601-5.ch010

Ahmad, M., Khayal, M. S. H., & Tahir, A. (2022). Analysis of Factors Affecting Adoption of Volunteered Geographic Information in the Context of National Spatial Data Infrastructure. *ISPRS International Journal of Geo-Information*, 11(2), 120. doi:10.3390/ijgi11020120

Aïvodji, U., Arai, H., Fortineau, O., Gambs, S., Hara, S., & Tapp, A. (2019). Fairwashing: The risk of rationalization. *Proceedings of the International Conference on Machine Learning*, (pp. 161-170). IEEE.

Ajzen, I. (1985). *From intentions to actions: A theory of planned behavior. Action control.* Springer.

Ajzen, I. (1991). The theory of planned behavior. *Organizational Behavior and Human Decision Processes*, *50*(2), 179–211. doi:10.1016/0749-5978(91)90020-T

Ajzen, I. (2002). Perceived behavioral control, self-efficacy, locus of control, and the theory of planned behavior 1. *Journal of Applied Social Psychology*, *32*(4), 665–683. doi:10.1111/j.1559-1816.2002.tb00236.x

Akgun, S., & Greenhow, C. (2021). Artificial Intelligence in Education: Addressing Ethical Challenges in K-12 Settings. *AI and Ethics*, 1–10. PMID:34790956

Akgun, S., & Greenhow, C. (2022). Artificial Intelligence (AI) in Education: Addressing Societal and Ethical Challenges in K-12 Settings. *Proceedings of the 16th International Conference of the Learning Sciences-ICLS 2022*, (pp. 1373-1376). IEEE.

Akgun, S., & Greenhow, C. (2022). Artificial intelligence in education: Addressing ethical challenges in K-12 settings. *AI and Ethics*, *2*(3), 431–440. doi:10.1007/s43681-021-00096-7 PMID:34790956

Akshara Foundation. (n.d.). *Home*. Akshara Foundation. https://akshara.org.in/

Akter, S., McCarthy, G., Sajib, S., Michael, K., Dwivedi, Y. K., D'Ambra, J., & Shen, K. N. (2021). Algorithmic bias in data-driven innovation in the age of AI. *International Journal of Information Management*, *60*, 102387. doi:10.1016/j.ijinfomgt.2021.102387

Alblooshi, M.A., Mohamed, A.M., & Yusr, M.M. (2023). Moderating Role of Artificial Intelligence Between Leadership Skills and Business Continuity. *International Journal of Professional Business Review*.

Albuquerque, R., Koskinen, Y., & Zhang, C. (2019). Corporate social responsibility and firm risk: Theory and empirical evidence. *Management Science*, *65*(10), 4451–4469. doi:10.1287/mnsc.2018.3043

Aldaghri, N. (2021). Coded Machine Unlearning. *IEEE Access : Practical Innovations, Open Solutions*.

Aler Tubella, A., Mora-Cantallops, M., & Nieves, J. C. (2024). How to teach responsible AI in Higher Education: Challenges and opportunities. *Ethics and Information Technology*, *26*(1), 3. doi:10.1007/s10676-023-09733-7

Alexandron, G., Yoo, L. Y., Ruipérez-Valiente, J. A., Lee, S., & Pritchard, D. E. (2019). Are MOOC learning analytics results trustworthy? With fake learners, they might not be! *International Journal of Artificial Intelligence in Education*, *29*(4), 484–506. doi:10.1007/s40593-019-00183-1

Alexopoulos, G. S. (2023). Artificial Intelligence in Geriatric Psychiatry Through the Lens of Contemporary Philosophy. *The American Journal of Geriatric Psychiatry*. Advance online publication. doi:10.1016/j.jagp.2023.09.006 PMID:37813788

Aliakbari Sani, S., Khorram, A., Jaffari, A., & Ebrahimi, G. (2021). Development of processing map for InX-750 superalloy using hyperbolic sinus equation and ANN model. *Rare Metals*, *40*(12), 3598–3607. doi:10.1007/s12598-018-1043-9

Allen, B., Agarwal, S., Coombs, L., Wald, C., & Dreyer, K. (2021). 2020 ACR Data Science Institute Artificial Intelligence Survey. *Journal of the American College of Radiology*, *18*(8), 1153–1159. doi:10.1016/j.jacr.2021.04.002 PMID:33891859

Allioui, H., & Mourdi, Y. (2023). Unleashing the potential of AI: Investigating cutting-edge technologies that are transforming businesses. [IJCEDS]. *International Journal of Computer Engineering and Data Science*, *3*(2), 1–12.

Al-Mahairah, M. S., Manoharan, G., Singh, J., & Krishna, S. H. (2022). *Principles of Management*. Book Rivers.

Alnssyan, B., Ahmad, Z., Malela-Majika, J. C., Seong, J. T., & Shafik, W. (2023). On the identifiability and statistical features of a new distributional approach with reliability applications. *AIP Advances*, *13*(12), 125211. doi:10.1063/5.0178555

Alzamil, H., Aloraini, K., AlAgeel, R., Ghanim, A., Alsaaran, R., Alsomali, N., Albahlal, R. A., & Alnuaim, L. (2020). Disparity among endocrinologists and gynaecologists in the diagnosis of polycystic ovarian syndrome. *Sultan Qaboos University Medical Journal*, *20*(3), 323. doi:10.18295/squmj.2020.20.03.012 PMID:33110648

Amifor, J. (2016). Political Advertising Design in Nigeria, 1960 – 2007. *International Journal of Arts and Humanities.*, *4*(2), 149–163.

Anadioti, E., Musharbash, L., Blatz, M. B., Papavasiliou, G., & Kamposiora, P. (2020). 3D printed complete removable dental prostheses: A narrative review. *BMC Oral Health*, *20*(1), 343. doi:10.1186/s12903-020-01328-8 PMID:33246466

Ananth, M. & Mathioudakis, M. (2021). *Certifiable Machine Un- learning for Linear Models*. arXiv:2106.15093v3 (cs.LG).

André, Q., Carmon, Z., Wertenbroch, K., Crum, A., Frank, D., Goldstein, W., Huber, J., Van Boven, L., Weber, B., & Yang, H. (2018). Consumer choice and autonomy in the age of artificial intelligence and big data. *Customer Needs and Solutions*, *5*(1), 28–37. doi:10.1007/s40547-017-0085-8

Andrushia, A. D., Neebha, T. M., Patricia, A. T., Sagayam, K. M., & Pramanik, S. (2023). Capsule Network based Disease Classification for Vitis Vinifera Leaves. *Neural Computing & Applications*. doi:10.1007/s00521-023-09058-y

Angwin, J., Larson, J., Mattu, S., & Kirchner, L. (2016). *Machine Bias*. ProPublica. https://www.propublica.org/article/machine-bias-risk-assessments-in-criminal-sentencing.

Anitha, C., Komala, C., Vivekanand, C. V., Lalitha, S., & Boopathi, S. (2023). Artificial Intelligence driven security model for Internet of Medical Things (IoMT). *IEEE Explore*, (pp. 1–7). IEEE.

An, L., Grimm, V., Bai, Y., Sullivan, A., Turner, B. L. II, Malleson, N., Heppenstall, A., Vincenot, C., Robinson, D., Ye, X., Liu, J., Lindkvist, E., & Tang, W. (2023). Modeling agent decision and behavior in the light of data science and artificial intelligence. *Environmental Modelling & Software*, *166*, 105713. Advance online publication. doi:10.1016/j.envsoft.2023.105713

Annis, A., & Nardi, F. (2019). Integrating VGI and 2D hydraulic models into a data assimilation framework for real time flood forecasting and mapping. *Geo-Spatial Information Science*, *22*(4), 223–236. doi:10.1080/10095020.2019.1626135

Antoniou, V., Fonte, C. C., See, L., Estima, J., Arsanjani, J. J., Lupia, F., Minghini, M., Foody, G., & Fritz, S. (2016). Investigating the feasibility of geo-Tagged photographs as sources of land cover input data. *ISPRS International Journal of Geo-Information*, *5*(5), 64. doi:10.3390/ijgi5050064

Antunes, P., Sapateiro, C., Zurita, G., & Baloian, N. (2010). Integrating spatial data and decision models in an e-planning tool. Lecture Notes in Computer Science (Including Subseries Lecture Notes in Artificial Intelligence and Lecture Notes in Bioinformatics), 6257 LNCS. Springer. doi:10.1007/978-3-642-15714-1_8

Anu, S. (2017). Literature Review and Challenges of Data Mining Techniques for Social Network Analysis. *journal Advances in Computational Sciences and Technology, 10*(5)

Anu, S. (2019). Hybrid Neuro-Fuzzy Classification Algorithm for Social Network. International *Journal of Engineering and Advanced Technology, 8*(6).

Appelgren, E., & Jönsson, A. M. (2021). Engaging Citizens for Climate Change—Challenges for Journalism. *Digital Journalism (Abingdon, England)*, *9*(6), 755–772. doi:10.1080/21670811.2020.1827965

Aprio. (2019). *Artificial Intelligence to Help Resource-Strapped IRS More Efficiently Identify Tax Crimes.* https://www.aprio.com/artificial-intelligence-to-help-resource-strapped-irs-more-efficiency/

Apte, A., Ingole, V., Lele, P., Marsh, A., Bhattacharjee, T., Hirve, S., Campbell, H., Nair, H., Chan, S., & Juvekar, S. (2019). Ethical considerations in the use of GPS-based movement tracking in health research-Lessons from a care-seeking study in rural west India. *Journal of Global Health*, 9(1), 010323. Advance online publication. doi:10.7189/jogh.09.010323 PMID:31275566

Ariful, I. (2020). Brain Tumor Detection from MRI Images using Image Processing. *International Journal of Innovative Technology and Exploring Engineering (IJITEE)*, 9(8).

Arnaboldi, M., & Azzone, G. (2020). Data science in the design of public policies: Dispelling the obscurity in matching policy demand and data offer. *Heliyon*, 6(6), e04300. doi:10.1016/j.heliyon.2020.e04300 PMID:32637693

Arroyo, I., Woolf, B. P., Burleson, W., Muldner, K., Rai, D., & Tai, M. (2014). A multimedia adaptive tutoring system for mathematics that addresses cognition, metacognition and affect. *International Journal of Artificial Intelligence in Education*, 24(4), 387–426. doi:10.1007/s40593-014-0023-y

Asamoah, D. A., Doran, D., & Schiller, S. (2020). Interdisciplinarity in Data Science Pedagogy: A Foundational Design. *Journal of Computer Information Systems*, 60(4), 370–377. doi:10.1080/08874417.2018.1496803

Assefa, S. A., Dervovic, D., Mahfouz, M., Tillman, R. E., Reddy, P., & Veloso, M. (2020, October). Generating synthetic data in finance: opportunities, challenges and pitfalls. In *Proceedings of the First ACM International Conference on AI in Finance* (pp. 1-8). ACM. 10.1145/3383455.3422554

AstraZeneca. (2022). *Data Science & Artificial Intelligence: Unlocking new science insights.* AstraZeneca.

Athanasopoulou, K., Daneva, G. N., Adamopoulos, P. G., & Scorilas, A. (2022). Artificial intelligence: The milestone in modern biomedical research. *BioMedInformatics*, 2(4), 727–744. doi:10.3390/biomedinformatics2040049

Atkin, C., & Heald, G. (1976). Effects of Political Advertising. *Public Opinion Quarterly*, 40(2), 216–228. doi:10.1086/268289

Atov, I., Chen, K. C., Kamal, A., & Louta, M. (2020). Data Science and Artificial Intelligence for Communications. In *IEEE Communications Magazine*, 58(10). doi:10.1109/MCOM.2020.9247523

Ayush, K. (2021). *Fast yet effective Machine Unlearning.* arXiv:2111.08947v2 (cs.LG).

Ayush, S., Acharya, J., Kamath G., & Suresh, A. (2021). Remember What You Want to Forget: Algorithms for Machine Unlearning. *35th Conference on Neural Information Processing Sys- tems(NIPS)*. IEEE.

Azevedo, R., & Aleven, V. (Eds.). (2013). *International handbook of metacognition and learning technologies.* Springer. doi:10.1007/978-1-4419-5546-3

Babu, B. S., Kamalakannan, J., Meenatchi, N., Karthik, S., & Boopathi, S. (2022). Economic impacts and reliability evaluation of battery by adopting Electric Vehicle. *IEEE Explore*, (pp. 1–6). IEEE.

Bahadure, N. B., Ray, A. K., & Thethi, H. P. (2017). Image Analysis for MRI Based Brain Tumor Detection and Feature Extraction Using Biologically Inspired BWT and SVM. *International Journal of Biomedical Imaging*, 2017, 1–12. doi:10.1155/2017/9749108 PMID:28367213

Baker, T., Smith, L., & Anissa, N. (2019). *Educ-AI-tion rebooted? Exploring the future of artificial intelligence in schools and colleges.* NESTA. https://media.nesta.org.uk/documents/Future_of_AI_and_education_v5_WEB.pdf

Baker, R. S., & Hawn, A. (2021). Algorithmic bias in education. *International Journal of Artificial Intelligence in Education*, 1–41.

Baker, R., & Hawn, A. (2021). Algorithmic Bias in Education. *International Journal of Artificial Intelligence in Education*, 1–41.

Bandura, A. (1986). *Social foundation of thought and action: A social-cognitive view.*

Bandura, A. (1991). Social cognitive theory of self-regulation. *Organizational Behavior and Human Decision Processes*, *50*(2), 248–287. doi:10.1016/0749-5978(91)90022-L

Barbosa, L., Nguyen, H., Nguyen, T., Pinnamaneni, R., & Freire, J. (2010). Creating and exploring web form repositories. *Proc. ACM SIGMOD Int.Conf. Manage.* ACM. 10.1145/1807167.1807311

Barocas, S., Hardt, M., & Narayanan, A. (2017). Fairness in machine learning. *Nips tutorial, 1,* 2017.

Baroody, R. A. (2004). *Sensory experiences and the development of mathematical reasoning.* RoutledgeFalmer. https://link.springer.com/article/10.1007/s10763-004-3224-2

Bates, T., Cobo, C., Mariño, O., & Wheeler, S. (2020). Can artificial intelligence transform higher education? *International Journal of Educational Technology in Higher Education*, *17*(1), 42. doi:10.1186/s41239-020-00218-x

BatoolI. (2022). *5G support in healthcare system.* doi:10.13140/RG.2.2.22073.85603

Bautista, Y. J. P., Theran, C., Aló, R., & Lima, V. (2023, October). Health Disparities Through Generative AI Models: A Comparison Study Using a Domain Specific Large Language Model. In *Proceedings of the Future Technologies Conference* (pp. 220-232). Cham: Springer Nature Switzerland. 10.1007/978-3-031-47454-5_17

Beauchamp, T. L., & Childress, J. F. (1995). Principles of Biomedical Ethics. Oxford University Press, 6.

BegumA.KumarR. (2022). Design an Archetype to Predict the impact of diet and lifestyle interventions in autoimmune diseases using Deep Learning and Artificial Intelligence. Research Square. doi:10.21203/rs.3.rs-1405206/v1

Belenguer, L. (2022). AI bias: Exploring discriminatory algorithmic decision-making models and the application of possible machine-centric solutions adapted from the pharmaceutical industry. *AI and Ethics*, *2*(4), 771–787. doi:10.1007/s43681-022-00138-8 PMID:35194591

Bellamy, R. K., Dey, K., Hind, M., Hoffman, S. C., Houde, S., Kannan, K., Lohia, P., Martino, J., Mehta, S., Mojsilovic, A., Nagar, S., Ramamurthy, K. N., Richards, J., Saha, D., Sattigeri, P., Singh, M., Varshney, K. R., & Zhang, Y. (2019). AI Fairness 360: An extensible toolkit for detecting and mitigating algorithmic bias. *IBM Journal of Research and Development*, *63*(4/5), 4–1. doi:10.1147/JRD.2019.2942287

Benabed, A. (2023). *Artificial Intelligence's Relevance for Energy Optimization, Companies and Business Internationalization.* New Trends in Sustainable Business and Consumption.

Benke, I., Feine, J., Venable, J. R., & Maedche, A. (2020). On implementing ethical principles in design science research. *AIS Transactions on Human-Computer Interaction*, *12*(4), 206–227. doi:10.17705/1thci.00136

Berelson, B. (1969). The Debates in the Light of Research: A Survey of Surveys. In S. Kraus (Ed.), *The Great Debates, Bloomington, Indiana University Press, 1962; Jay Blumler and Denis McQuail, Television in Politics.* University of Chicago Press.

Berente, N., Gu, B., Recker, J., & Santhanam, R. (2021). Managing artificial intelligence. *Management Information Systems Quarterly*, *45*(3).

Bergman, M. (2001). White paper: The deep web: Surfacing hidden value. *J. Electron., 7*(1).

Berman, G., de la Rosa, S., Accone, T., & Associates. (2018). Ethical considerations when using geospatial technologies for evidence generation. In *Innocenti Discussion Papers: Vol. no. 2018-02*. UNICEF Office of Research - Innocenti, Florence. https://www.unicef-irc.org/publications/971-ethical-considerations-when-using-geospatial-technologies-for-evidence-generation.html

Bhakri, S. & Shri, D. (2021). *Artificial Intelligence(Ai): Applications And Implications (Ai) For Indian Economy*. Research Gate. doi:10.1729/Journal.27139

Bharadiya, J. P. (2023). Machine learning and AI in business intelligence: Trends and opportunities. [IJC]. *International Journal of Computer*, *48*(1), 123–134.

Bhatia, S. N., & Ingber, D. E. (2014). Microfluidic organs-on-chips. *Nature Biotechnology*, *32*(8), 760–772. doi:10.1038/nbt.2989 PMID:25093883

Bhattacharya, S. (2018). Dimensions of Political Marketing: A Study from the Indian Perspective. *International Journal of Current Advanced Research*, *7*, 11138–11143. doi:10.24327/ijcar.2018.11143.1920

Bhattacharyya, D. (2011). *Brain Tumor Detection Using MRI Image Analysis*. Springer-Verlag Berlin Heidelberg. doi:10.1007/978-3-642-20998-7_38

Bird, S., Dudík, M., Edgar, R., Horn, B., Lutz, R., Milan, V., & Walker, K. (2020). Fairlearn: A toolkit for assessing and improving fairness in AI. *Microsoft, Tech. Rep. MSR-TR-2020-32*.

Biswas, G., Segedy, J. R., & Bunchongchit, K. (2016). From design to implementation to practice a learning by teaching system: Betty's Brain. *International Journal of Artificial Intelligence in Education*, *26*(1), 350–364. doi:10.1007/s40593-015-0057-9

Bittner, C., Michel, B., & Turk, C. (2016). Turning the spotlight on the crowd: Examining the participatory ethics and practices of crisis mapping. *ACME*, *15*(1).

Bizer, C., Vidal, M.-E., & Skaf-Molli, H. (2018). Linked Open Data. Springer.

Black, P., & Wiliam, D. (1998). *Assessment and classroom learning*. Guilford Publications. doi:10.1080/0969595980050102

Bodenstedt, S., Wagner, M., Müller-Stich, B. P., Weitz, J., & Speidel, S. (2020). Artificial intelligence-assisted surgery: Potential and challenges. *Visceral Medicine*, *36*(6), 450–455. doi:10.1159/000511351 PMID:33447600

Boehmke, B. C., & Greenwell, B. M. (2019). *Interpretable Machine Learning*. Hands-On Machine Learning with R. doi:10.1201/9780367816377-16

Bohr, A., & Memarzadeh, K. (2020). Chapter 2 - The rise of artificial intelligence in healthcare applications. A. Bohr & K. Memarzadeh, (eds) Artificial Intelligence in Healthcare. Academic Press. doi:10.1016/B978-0-12-818438-7.00002-2

Boje, D. (2011). Narrative Methods for Organizational & Communication Research. In Narrative Methods for Organizational & Communication Research. Sage. doi:10.4135/9781849209496

Bolander, T. (2019). What do we lose when machines take the decisions? *The Journal of Management and Governance*, *23*(4), 849–867. doi:10.1007/s10997-019-09493-x

Bongard, A. (2019). Automating talent acquisition: Smart recruitment, predictive hiring algorithms, and the data-driven nature of artificial intelligence. *Psychosociological Issues in Human Resource Management*, *7*(1), 36–41.

Boopathi, S. (2013). *Experimental study and multi-objective optimization of near-dry wire-cut electrical discharge machining process* [PhD Thesis]. http://hdl.handle.net/10603/16933

Boopathi, S. (2022c). Cryogenically treated and untreated stainless steel grade 317 in sustainable wire electrical discharge machining process: A comparative study. *Environmental Science and Pollution Research*, 1–10. Springer.

Boopathi, S., Kumar, P. K. S., Meena, R. S., Sudhakar, M., & Associates. (2023). Sustainable Developments of Modern Soil-Less Agro-Cultivation Systems: Aquaponic Culture. In Human Agro-Energy Optimization for Business and Industry (pp. 69–87). IGI Global.

Boopathi, S. (2022a). An extensive review on sustainable developments of dry and near-dry electrical discharge machining processes. *ASME: Journal of Manufacturing Science and Engineering*, *144*(5), 050801–1.

Boopathi, S. (2022b). An investigation on gas emission concentration and relative emission rate of the near-dry wire-cut electrical discharge machining process. *Environmental Science and Pollution Research International*, *29*(57), 86237–86246. doi:10.1007/s11356-021-17658-1 PMID:34837614

Boopathi, S., Alqahtani, A. S., Mubarakali, A., & Panchatcharam, P. (2023). Sustainable developments in near-dry electrical discharge machining process using sunflower oil-mist dielectric fluid. *Environmental Science and Pollution Research International*, 1–20. doi:10.1007/s11356-023-27494-0 PMID:37199846

Boopathi, S., & Davim, J. P. (2023a). Applications of Nanoparticles in Various Manufacturing Processes. In *Sustainable Utilization of Nanoparticles and Nanofluids in Engineering Applications* (pp. 1–31). IGI Global. doi:10.4018/978-1-6684-9135-5.ch001

Boopathi, S., & Davim, J. P. (2023b). *Sustainable Utilization of Nanoparticles and Nanofluids in Engineering Applications*. IGI Global. doi:10.4018/978-1-6684-9135-5

Boopathi, S., & Kanike, U. K. (2023). Applications of Artificial Intelligent and Machine Learning Techniques in Image Processing. In *Handbook of Research on Thrust Technologies' Effect on Image Processing* (pp. 151–173). IGI Global. doi:10.4018/978-1-6684-8618-4.ch010

Boopathi, S., & Sivakumar, K. (2013). Experimental investigation and parameter optimization of near-dry wire-cut electrical discharge machining using multi-objective evolutionary algorithm. *International Journal of Advanced Manufacturing Technology*, *67*(9–12), 2639–2655. doi:10.1007/s00170-012-4680-4

Boopathi, S., Sureshkumar, M., & Sathiskumar, S. (2022). Parametric Optimization of LPG Refrigeration System Using Artificial Bee Colony Algorithm. *International Conference on Recent Advances in Mechanical Engineering Research and Development*, (pp. 97–105). IEEE.

Boopathi, S., Umareddy, M., & Elangovan, M. (2023). Applications of Nano-Cutting Fluids in Advanced Machining Processes. In *Sustainable Utilization of Nanoparticles and Nanofluids in Engineering Applications* (pp. 211–234). IGI Global. doi:10.4018/978-1-6684-9135-5.ch009

Borrego-Díaz, J., & Galán-Páez, J. (2022a). Explainable Artificial Intelligence in Data Science. *Minds and Machines*, *32*(3), 485–531. doi:10.1007/s11023-022-09603-z

Bostrom, N., & Yudkowsky, E. (2014). The ethics of artificial intelligence. In The Cambridge Handbook of Artificial Intelligence (pp. 316-334). Cambridge Handbook. doi:10.1017/CBO9781139046855.020

Brei, V. A. (2020). Machine learning in marketing: Overview, learning strategies, applications, and future developments. *Foundations and Trends® in Marketing*, *14*(3), 173-236.

Brey, P. A. E. (2012). Anticipatory Ethics for Emerging Technologies. *NanoEthics*, *6*(1), 1–13. doi:10.1007/s11569-012-0141-7

Briel, M., Elger, B. S., McLennan, S., Schandelmaier, S., von Elm, E., & Satalkar, P. (2021). Exploring reasons for recruitment failure in clinical trials: A qualitative study with clinical trial stakeholders in Switzerland, Germany, and Canada. *Trials*, *22*(1), 844. doi:10.1186/s13063-021-05818-0 PMID:34823582

Brock, D. W. (2019). Ethical Issues in the Use of Cost Effectiveness Analysis for the Prioritisation of Health Care Resources. In The Ethics of Public Health. Taylor & Francis. doi:10.4324/9781315239927-28

Brodie, M. L. (2019). What Is Data Science? In Applied Data Science: Lessons Learned for the Data-Driven Business. doi:10.1007/978-3-030-11821-1_8

Brophy, J., & Lowd, D. (2021). Machine Unlearning for Random Forests. arXiv:2009.05567v2(cs.LG).

Brundage, M., Avin, S., Wang, J., Belfield, H., Krueger, G., Hadfield, G., & Anderljung, M. (2020). Toward trustworthy AI development: mechanisms for supporting verifiable claims. *arXiv preprint arXiv:2004.07213*.

Brynjolfsson, E., Li, D., & Raymond, L. R. (2023). Generative AI at work (No. w31161). National Bureau of Economic Research.

Brynjolfsson, E., & Mcafee, A. (2017). Artificial intelligence, for real. *Harvard Business Review*, *1*, 1–31.

Brzezinski, M., & Krzeminska, I. (2023). The strategies for innovating with virtual reality and artificial intelligence: A literature review. *Technium*, *8*, 72–83. doi:10.47577/technium.v8i.8671

Buallay, A., Fadel, S. M., Al-Ajmi, J. Y., & Saudagaran, S. (2020). Sustainability reporting and performance of MENA banks: Is there a trade-off? *Measuring Business Excellence*, *24*(2), 197–221. doi:10.1108/MBE-09-2018-0078

Buchanan, B. G. (1988). Artificial intelligence as an experimental science. In J. H. Fetzer (Ed.), *Aspects of artificial intelligence* (pp. 209–250). Kluwer Academic Publishers. doi:10.1007/978-94-009-2699-8_8

Bugaj, M., Kliestik, T., & Lăzăroiu, G. (2023). Generative Artificial Intelligence-based Diagnostic Algorithms in Disease Risk Detection, Personalized and Targeted Healthcare Procedures, and Patient Care Safety and Quality. *Contemporary Readings in Law and Social Justice*, *15*(1), 9–26. doi:10.22381/CRLSJ15120231

Bugaj, M., Kliestik, T., & Lăzăroiu, G. (2023). Generative Artificial Intelligence-based Diagnostic Pagano, S., Holzapfel, S., Kappenschneider, T., Meyer, M., Maderbacher, G., Grifka, J., & Holzapfel, D. E. (2023). Arthrosis diagnosis and treatment recommendations in clinical practice: An exploratory investigation with the generative AI model GPT-4. *Journal of Orthopaedics and Traumatology*, *24*(1), 61. PMID:38015298

Bughin, J., Hazan, E., & Sree Ramaswamy, P. (2017). Artificial intelligence the next digital frontier. Research Gate.

Bull, S., & Kay, J. (2016). SMILI☺: A framework for interfaces to learning data in open learner models, learning analytics and related fields. *International Journal of Artificial Intelligence in Education*, *26*(1), 293–331. doi:10.1007/s40593-015-0090-8

Buolamwini, J. (2018). Gender Shades: Intersectional Accuracy Disparities in Commercial Gender Classification. In Proceedings of Machine Learning Research (Vol. 81). ACM.

Burkhardt, G., Boy, F., Doneddu, D., & Hajli, N. (2022). Privacy Behaviour: A Model for Online Informed Consent. *Journal of Business Ethics*, *186*(1), 237–255. doi:10.1007/s10551-022-05202-1

Burton, R. R., & Brown, J. S. (1979). An investigation of computer coaching for informal learning activities. *International Journal of Man-Machine Studies*, *11*(1), 5–24. doi:10.1016/S0020-7373(79)80003-6

Cai, Y. (1995). Using a self-organizing artificial neural network model to distinguish the onset period of citrus canker disease. *Journal of Plant Pathology*, (01), 43–46.

Calcagni, F., Amorim Maia, A. T., Connolly, J. J. T., & Langemeyer, J. (2019). Digital co-construction of relational values: Understanding the role of social media for sustainability. *Sustainability Science, 14*(5), 1309–1321. doi:10.1007/s11625-019-00672-1

Cao, L. (2017). Data Science: A Comprehensive Overview. *ACM Computing Surveys (CSUR), 50*(3), 1-42.

Cao, L. (2017). Data science: A comprehensive overview. *ACM Computing Surveys, 50*(3), 1–42. doi:10.1145/3076253

Cao, L. (2023a). Trans-AI/DS: Transformative, transdisciplinary and translational artificial intelligence and data science. *International Journal of Data Science and Analytics, 15*(2), 119–132. doi:10.1007/s41060-023-00383-y

Cao, Y., & Yang, J. (2015). Towards making systems forget with machine unlearning. *Proc. IEEE Symposium*. IEEE. 10.1109/SP.2015.35

Carbonell, J. R. (1970). AI in CAI: An artificial-intelligence approach to computer-assisted instruction. *IEEE Transactions on Man-Machine Systems, 11*(4), 190–202. doi:10.1109/TMMS.1970.299942

Carney, M. (2021). *Value(s): Building a Better World for All*. Public Affairs.

Carter, S. M., Rogers, W., Win, K. T., Frazer, H., Richards, B., & Houssami, N. (2020). The ethical, legal and social implications of using artificial intelligence systems in breast cancer care. *The Breast, 49*, 25–32. doi:10.1016/j.breast.2019.10.001 PMID:31677530

Castello-Sirvent, F., García Felix, V., & Canos-Daros, L. (2023). AI In Higher Education: New Ethical Challenges For Students And Teachers. *EDULEARN23 Proceedings* (pp. 4463-4470). Research Gate.

Cate, F., & Mayer-Schönberger, V. (2013). Notice and consent in a world of Big Data. *International Data Privacy Law, 3*(2), 67-73.

Cath, C. (2018). Governing artificial intelligence: Ethical, legal and technical opportunities and challenges. *Philosophical Transactions. Series A, Mathematical, Physical, and Engineering Sciences, 376*(2133), 20180080. doi:10.1098/rsta.2018.0080 PMID:30322996

Cath, C., Wachter, S., Mittelstadt, B., Taddeo, M., & Floridi, L. (2018). Artificial intelligence and the 'good society': The US, EU, and UK approach. *Science and Engineering Ethics, 24*, 505–528. PMID:28353045

Cavazos, J. G., Phillips, P. J., Castillo, C. D., & O'Toole, A. J. (2020). Accuracy comparison across face recognition algorithms: Where are we on measuring race bias? *IEEE Transactions on Biometrics, Behavior, and Identity Science, 3*(1), 101–111. doi:10.1109/TBIOM.2020.3027269 PMID:33585821

Cen, Li, & Shi. (2007). Research on the Recognition of Cucumber Anthracnose and Brown Spot Based on Color Statistical Features of Color Images. *Journal of Horticulture, 34*(6), 124–124.

Chacko, A., & Hayajneh, T. (2018). Security and Privacy Issues with IoT in Healthcare. *EAI Endorsed Transactions on Pervasive Health and Technology, 4*(14), 155079. doi:10.4108/eai.13-7-2018.155079

Chandgude, V., & Kawade, B. (2023). Role of Artificial Intelligence and Machine Learning in Decision Making for Business Growth. International Journal of Advanced Research in Science.

Chang, K., He, B., Li, C., Patel, M., & Zhang, Z. (2004). Structured databases on the web: Observations and implications. *ACM SIGMOD Rec*. ACM.

Chankoson, T., Chen, F., Wang, Z., Wang, M., & Sukpasjaroen, K. (2023). Knowledge Mapping for the Study of Artificial Intelligence in Education Research: Literature Reviews. *Journal of Intelligence Studies in Business*.

Chao, C. Y., & Wang, S. (2018). Using OpenStreetMap data for the location analysis on public bicycle stations – A case study on the Youbike system in downtown Taipei city. *Proceedings - 39th Asian Conference on Remote Sensing: Remote Sensing Enabling Prosperity, ACRS 2018*. IEEE.

Chatila, R., Firth-Butterfield, K., & Havens, J. C. (2018). *Ethically aligned design: A vision for prioritizing human well-being with autonomous and intelligent systems version 2*. University of southern California Los Angeles.

Chattopadhyay, A., & Maitra, M. (2022). MRI-based brain tumour image detection using CNN based deep learning method. *Neuroscience Informatics (Online)*, 2(4), 100060. doi:10.1016/j.neuri.2022.100060

Chauhan, N. R., Shukla, R. K., Sengar, A. S., & Gupta, A. (2022, December). Classification of Nutritional Deficiencies in Cabbage Leave Using Random Forest. In *2022 11th International Conference on System Modeling & Advancement in Research Trends (SMART)* (pp. 1314-1319). IEEE. 10.1109/SMART55829.2022.10047282

Chawla, R. N., & Goyal, P. (2022). Emerging trends in digital transformation: A bibliometric analysis. *Benchmarking*, 29(4), 1069–1112. doi:10.1108/BIJ-01-2021-0009

Chen, D., & Qi, E. Y. (2020). Innovative highlights of clinical drug trial design. *Translational Research; the Journal of Laboratory and Clinical Medicine*, 224, 71–77. doi:10.1016/j.trsl.2020.05.007 PMID:32504825

Chen, M., & Decary, M. (2020, January). Artificial intelligence in healthcare: An essential guide for health leaders. *Healthcare Management Forum*, 33(1), 10–18. doi:10.1177/0840470419873123 PMID:31550922

Chenthara, S., Ahmed, K., Wang, H., & Whittaker, F. (2019). Security and Privacy-Preserving Challenges of e-Health Solutions in Cloud Computing. *IEEE Access : Practical Innovations, Open Solutions*, 7, 74361–74382. doi:10.1109/ACCESS.2019.2919982

Chen, X., Elmes, G., Ye, X., & Chang, J. (2016). Implementing a Real-Time Twitter-Based System for Resource Dispatch in Disaster Management. *GeoJournal*, 81(6), 863–873. doi:10.1007/s10708-016-9745-8

Chheang, V., Marquez-Hernandez, R., Patel, M., Rajasekaran, D., Sharmin, S., Caulfield, G., . . . Barmaki, R. L. (2023). Towards anatomy education with generative AI-based virtual assistants in immersive virtual reality environments. *arXiv preprint arXiv:2306.17278*.

Chiarello, F., Belingheri, P., & Fantoni, G. (2021). Data science for engineering design: State of the art and future directions. *Computers in Industry*, 129, 103447. doi:10.1016/j.compind.2021.103447

Chkoniya, V. (2021). Handbook of research on applied data science and artificial intelligence in business and industry. In Handbook of Research on Applied Data Science and Artificial Intelligence in Business and Industry. doi:10.4018/978-1-7998-6985-6

Choi, Y., Kamal, A. E., & Louta, M. (2021). Series Editorial: Artificial Intelligence and Data Science for Communications. In IEEE Communications Magazine, 59(11). doi:10.1109/MCOM.2021.9665429

Choi, Y., Kamal, A. E., & Louta, M. (2022a). Series Editorial: Artificial Intelligence and Data Science for Communications. In IEEE Communications Magazine, 60(11). doi:10.1109/MCOM.2022.9946952

Choi, Y., Kamal, A. E., & Louta, M. (2022b). Series Editorial: Artificial Intelligence and Data Science for Communications. In IEEE Communications Magazine, 60(7). doi:10.1109/MCOM.2022.9831135

Choi, Y., Kamal, A. E., & Louta, M. (2023a). Series Editorial: Artificial Intelligence and Data Science for Communications. In IEEE Communications Magazine, 61(3). doi:10.1109/MCOM.2023.10080874

Choi, Y., Kamal, A., & Louta, M. (2023b). Series Editorial: Artificial Intelligence and Data Science for Communications. In IEEE Communications Magazine, 61(6). doi:10.1109/MCOM.2023.10155727

Chong, S., Rahman, A., & Narayan, A. K. (2022). Guest editorial: Accounting in transition: influence of technology, sustainability and diversity. *Pacific Accounting Review*, *34*(4), 517–525. doi:10.1108/PAR-07-2022-210

Chui, M., Manyika, J., & Miremadi, M. (2018). What AI can and can't do (yet) for your business. *McKinsey Quarterly*, *1*(97-108), 1.

Clancey, W. J. (1979). Tutoring rules for guiding a case method dialogue. *International Journal of Man-Machine Studies*, *11*(1), 25–50. doi:10.1016/S0020-7373(79)80004-8

Clandinin, D. J., Caine, V., Lessard, S., & Huber, J. (2016). Engaging in narrative inquiries with children and youth. In Engaging in Narrative Inquiries with Children and Youth. Springer. doi:10.4324/9781315545370

Clark, M., & Severn, M. (2023). Artificial Intelligence in Prehospital Emergency Health Care. *Canadian Journal of Health Technologies*, *3*(8). doi:10.51731/cjht.2023.712 PMID:37934833

Cockburn, I. M., Henderson, R., & Stern, S. (2018). The impact of artificial intelligence on innovation: An exploratory analysis. In *The economics of artificial intelligence: An agenda* (pp. 115–146). University of Chicago Press.

Collins, A., & Halverson, J. (Eds.). (2019). *Rethinking the Future of Learning: Augmented Reality and Mixed Reality in Education*. Vanderbilt Center for Teaching. https://cft. vanderbilt.edu/

Collins, F. S., & Varmus, H. (2015). A new initiative on precision medicine. *The New England Journal of Medicine*, *372*(9), 793–795. doi:10.1056/NEJMp1500523 PMID:25635347

Comendador, B. E. V., Rabago, L. W., & Tanguilig, B. T. (2016, August). An educational model based on Knowledge Discovery in Databases (KDD) to predict learner's behavior using classification techniques. In *2016 IEEE International Conference on Signal Processing, Communications and Computing (ICSPCC)* (pp. 1-6). IEEE. 10.1109/ICSPCC.2016.7753623

Cooke, L. (2020). Ethics in information technology. *ISACA Journal*, *6*, 69–106. doi:10.1017/CBO9781107445666.007

Copeland, B. (2000). The turing test*. *Minds and Machines*, *10*(4), 519–539. doi:10.1023/A:1011285919106

Cordeschi, R. (2007). AI TURNS FIFTY: REVISITING ITS ORIGINS. *Applied Artificial Intelligence*, *21*(4-5), 259–279. doi:10.1080/08839510701252304

Coston, A., Mishler, A., Kennedy, E. H., & Chouldechova, A. (2020, January). Counterfactual risk assessments, evaluation, and fairness. In *Proceedings of the 2020 conference on fairness, accountability, and transparency* (pp. 582-593). ACM. 10.1145/3351095.3372851

Crawford, K., & Finn, M. (2015). The limits of crisis data: Analytical and ethical challenges of using social and mobile data to understand disasters. *GeoJournal*, *80*(4), 491–502. doi:10.1007/s10708-014-9597-z

Crofts, P., & van Rijswijk, H. (2020). Negotiating 'evil': Google, project maven and the corporate form. *Law, technology and humans, 2*(1), 75-90.

Croskerry, P. (2003). The importance of cognitive errors in diagnosis and strategies to minimize them. In Academic Medicine, 78(8). doi:10.1097/00001888-200308000-00003

Crowley, D. N., Breslin, J. G., Corcoran, P., & Young, K. (2012). Gamification of citizen sensing through mobile social reporting. *4th International IEEE Consumer Electronic Society - Games Innovation Conference, IGiC 2012*. IEEE. 10.1109/IGIC.2012.6329849

Cruz-Jesus, F., Castelli, M., Oliveira, T., Mendes, R., Nunes, C., Sa-Velho, M., & Rosa-Louro, A. (2020). Using artificial intelligence methods to assess academic achievement in public high schools of a European Union country. *Heliyon*, *6*(6), e04081. doi:10.1016/j.heliyon.2020.e04081 PMID:32551378

Cui, M., & Zhang, D. Y. (2021). Artificial intelligence and computational pathology. *Laboratory Investigation*, *101*(4), 412–422. doi:10.1038/s41374-020-00514-0 PMID:33454724

Cui, X., Li, W., & Gu, C. (2021). Big Data of Food Science and Artificial Intelligence Technology. *Journal of Chinese Institute of Food Science and Technology*, *21*(2). doi:10.16429/j.1009-7848.2021.02.001

Cummiskey, D. (1990). Kantian consequentialism. *Ethics*, *100*(3), 586–615. doi:10.1086/293212

d'Amato, C., Fernandez, M., Tamma, V., Lecue, F., Cudré-Mauroux, P., Sequeda, J., & Heflin, J. (Eds.). (2017). *The Semantic Web–ISWC 2017: 16th International Semantic Web Conference*, (Vol. 10587). Springer.

Dastile, X., Celik, T., & Potsane, M. (2020). Statistical and machine learning models in credit scoring: A systematic literature survey. *Applied Soft Computing*, *91*, 106263. doi:10.1016/j.asoc.2020.106263

Dastin, J. (2022). Amazon scraps secret AI recruiting tool that showed bias against women. In K. Martin, Ethics of Data and Analytics. (pp. 296-299). Auerbach Publications. doi:10.1201/9781003278290-44

Datta, A., Tschantz, M. C., & Datta, A. (2015). Automated Experiments on Ad Privacy Settings – A Tale of Opacity, Choice, and Discrimination. *Proceedings on Privacy Enhancing Technologies. Privacy Enhancing Technologies Symposium*, *1*(1), 92–112. doi:10.1515/popets-2015-0007

Dave, B., Patel, S., Shivani, R., Purohit, S., & Chaudhury, B. (2023). Synthetic data generation using generative adversarial network for tokamak plasma current quench experiments. *Contributions to Plasma Physics*, *63*(5-6), e202200051. doi:10.1002/ctpp.202200051

Davenport, T. H. (2018). From analytics to artificial intelligence. *Journal of Business Analytics*, *1*(2), 73–80. doi:10.1080/2573234X.2018.1543535

Davenport, T. H., & Ronanki, R. (2018). Artificial intelligence for the real world. *Harvard Business Review*, *96*(1), 108–116.

Davenport, T., Guha, A., Grewal, D., & Bressgott, T. (2020). How artificial intelligence will change the future of marketing. *Journal of the Academy of Marketing Science*, *48*(1), 24–42. doi:10.1007/s11747-019-00696-0

Davenport, T., & Patil, D. (2012). Data Scientist. *Harvard Business Review*, *90*(5), 70–76. PMID:23074866

Davies, B. (2020). The right not to know and the obligation to know. *Journal of Medical Ethics*, *46*(5), 300–303. doi:10.1136/medethics-2019-106009

Dawson, J., Jr. (2007). The Essential Turing: Seminal Writings in Computing, Logic, Philosophy, Artificial Intelligence, and Artificial Life plus The Secrets of Enigma, by Alan M. Turing (author) and B. Jack Copeland (editor). The Review of Modern Logic, 10(32), 179-181.

De Freitas, J., Uğuralp, A. K., Oğuz-Uğuralp, Z., & Puntoni, S. (2022). Chatbots and Mental Health: Insights into the Safety of Generative AI. *Journal of Consumer Psychology*.

De Jong, B. C., Gaye, B. M., Luyten, J., Van Buitenen, B., André, E., Meehan, C. J., O'Siochain, C., Tomsu, K., Urbain, J., Grietens, K. P., Njue, M., Pinxten, W., Gehre, F., Nyan, O., Buvé, A., Roca, A., Ravinetto, R., & Antonio, M. (2019). Ethical considerations for movement mapping to identify disease transmission hotspots. *Emerging Infectious Diseases*, *25*(7). doi:10.3201/eid2507.181421 PMID:31211938 de Melo, C. M., Torralba, A., Guibas, L., DiCarlo, J., Chellappa, R., & Hodgins, J. (2022). Next-generation deep learning based on simulators and synthetic data. *Trends in Cognitive Sciences*, 26(2), 174–187. doi:10.1016/j.tics.2021.11.008 PMID:34955426

Delichatsios, S., & Sonuyi, T. (2005). Get to Know Google… Because They Know You. *MIT, Ethics and Law on the Electronic Frontier, 6*(14).

Deljoo & Keshtgari. (2011). Wireless Sensor Network Solution for Precision Agriculture Based on Zigbee Technology. *Wireless Sensor Network, 4*(1).

Demšar, U., Harris, P., Brunsdon, C., Fotheringham, A. S., & McLoone, S. (2013). Principal Component Analysis on Spatial Data: An Overview. *Annals of the Association of American Geographers*, 103(1), 106–128. doi:10.1080/0004 5608.2012.689236

Deviprasad, S., Madhumithaa, N., Vikas, I. W., Yadav, A., & Manoharan, G. (2023). The Machine Learning-Based Task Automation Framework for Human Resource Management in MNC Companies. *Engineering Proceedings*, 59(1), 63.

Dhamodaran, S., Ahamad, S., Ramesh, J. V. N., Sathappan, S., Namdev, A., Kanse, R. R., & Pramanik, S. (2023). *Fire Detection System Utilizing an Aggregate Technique in UAV and Cloud Computing, Thrust Technologies' Effect on Image Processing*. IGI Global.

Dhanda, U., & Shrotryia, V. K. (2021). Corporate sustainability: The new organizational reality. *Qualitative Research in Organizations and Management*, 16(3/4), 464–487. doi:10.1108/QROM-01-2020-1886

Dhar, V. (2013). Data science and prediction. *Communications of the ACM*, 56(12), 64–73. doi:10.1145/2500499

Di Vaio, A., Hassan, R., & Alavoine, C. (2022). Data intelligence and analytics: A bibliometric analysis of human–Artificial intelligence in public sector decision-making effectiveness. *Technological Forecasting and Social Change*, 174, 121201. doi:10.1016/j.techfore.2021.121201

Dias, V. E. C., Sperandio, V. G., & Lisboa-Filho, J. (2019). Routes generation for selective collection of urban waste using volunteered geographic information. *Iberian Conference on Information Systems and Technologies, CISTI, 2019-June*. IEEE. 10.23919/CISTI.2019.8760773

Diggle, P. (2005). Applied Spatial Statistics for Public Health Data. *Journal of the American Statistical Association*, 100(470), 702–703. Advance online publication. doi:10.1198/jasa.2005.s15

Division, F. N. (2020). Artificial Intelligence (AI) Policies in India- A Status Paper. Future Networks (FN) Division, Telecommunication Engineering Center Janpath, New Delhi.

Dixit, S. K., & Singh, A. K. (2022). Predicting electric vehicle (EV) buyers in India: A machine learning approach. *The Review of Socionetwork Strategies*, 16(2), 221–238. doi:10.1007/s12626-022-00109-9 PMID:35600566

Dolšak, B., & Novak, M. (2011). Intelligent decision support for structural design analysis. *Advanced Engineering Informatics*, 25(2), 330–340. doi:10.1016/j.aei.2010.11.001

Domakonda, V. K., Farooq, S., Chinthamreddy, S., Puviarasi, R., Sudhakar, M., & Boopathi, S. (2022). Sustainable Developments of Hybrid Floating Solar Power Plants: Photovoltaic System. In Human Agro-Energy Optimization for Business and Industry (pp. 148–167). IGI Global.

Donoho, D. (2017). 50 years of data science. *Journal of Computational and Graphical Statistics*, 26(4), 745–766. doi:10.1080/10618600.2017.1384734

Doshi-Velez, F., Kortz, M., Budish, R., Bavitz, C., Gershman, S., O'Brien, D., & Wood, A. (2017). *Accountability of AI under the law: The role of explanation.* arXiv preprint arXiv:1711.01134.

Doss, C., Mondschein, J., Shu, D., Wolfson, T., Kopecky, D., Fitton-Kane, V. A., Bush, L., & Tucker, C. (2023). Deepfakes and scientific knowledge dissemination. *Scientific Reports*, *13*(1), 13429. doi:10.1038/s41598-023-39944-3 PMID:37596384

Dougherty, J., & Ilyankou, I. (2021). Hands-on data visualization. In O'Reilly Media, Inc.

Dowd, M., Dixon, A., & Kinsella, B. (2020). Optimizing waste management collection routes in Urban Haiti: A collaboration between DataKind and SOIL. *ArXiv, 2012.*

Drachsler, H., & Greller, W. (2016). Privacy and analytics: It's a DELICATE issue a checklist for trusted learning analytics. In S. Dawson, H. Drachsler, & C. P. Rosé (Eds.), *Enhancing impact: Convergence of communities for grounding, implementation, and validation* (pp. 89–98). ACM. doi:10.1145/2883851.2883893

Drobot, A. T. (2020). Industrial Transformation and the Digital Revolution: A Focus on Artificial Intelligence, Data Science and Data Engineering. *2020 ITU Kaleidoscope. Industry-Driven Digital Transformation, ITU K, 2020*, 1–11. doi:10.23919/ITUK50268.2020.9303221

du Boulay, B. (2016). Artificial intelligence as an effective classroom assistant. *IEEE Intelligent Systems*, *31*(6), 76–81. doi:10.1109/MIS.2016.93

Duan, Y., Edwards, J. S., & Dwivedi, Y. K. (2019). Artificial Intelligence for Decision Making in the Era of Big Data–Evolution, Challenges, and Research Agenda. *International Journal of Information Management*, *48*, 63–71. doi:10.1016/j.ijinfomgt.2019.01.021

Dunn, A. G., Shih, I., Ayre, J., & Spallek, H. (2023). What generative AI means for trust in health communications. *Journal of Communication in Healthcare*, *16*(4), 385–388. doi:10.1080/17538068.2023.2277489 PMID:37921509

Durai, S., Krishnaveni, K., & Manoharan, G. (2022, May). Designing entrepreneurial performance metric (EPM) framework for entrepreneurs owning small and medium manufacturing units (SME) in coimbatore. In AIP Conference Proceedings (Vol. 2418, No. 1). AIP Publishing.

Durai, S., Krishnaveni, K., & Manoharan, G. (2022, May). Leveraging HR metrics for effective recruitment & selection process in IT industries in Chennai and Coimbatore, Tamil Nadu. In AIP Conference Proceedings (Vol. 2418, No. 1). AIP Publishing.

Durai, S., Krishnaveni, K., & Manoharan, G. (2022, May). Metric based performance management of employees–A case study of MSME unit in Coimbatore, Tamil Nadu. In AIP Conference Proceedings (Vol. 2418, No. 1). AIP Publishing.

Dweck, C. S. (2002). Beliefs that make smart people dumb. In R. J. Sternberg (Ed.), *Why smart people can be so stupid* (pp. 24–41). Yale University Press.

Dwijendra, N. (2022). Machine learning time series models for tea pest Helopeltis infestation in India. *Webology*, *19*(2). Available at: https://www.webology.org/abstract.php?id=1625

Dwijendra, N., Pandey, A. K., & Dwivedi, A. D. (2023). Examining the emotional tone in politically polarized speeches in India: An in-depth analysis of two contrasting perspectives. *SOUTH INDIA JOURNAL OF SOCIAL SCIENCES, 21*(2), 125-136. https://journal.sijss.com/index.php/home/article/view/65

Dwivedi, D. N., & Mahanty, G. (2021). A text mining-based approach for accessing AI risk incidents. *International Conference on Artificial Intelligence.* AI Foundation Trust, India.

Dwivedi, D. N., Mahanty, G., & Pathak, Y. K. (2023). AI applications for financial risk management. In M. Irfan, M. Elmogy, M. Shabri Abd. Majid, & S. El-Sappagh (Eds.), The Impact of AI Innovation on Financial Sectors in the Era of Industry 5.0 (pp. 17-31). IGI Global. doi:10.4018/979-8-3693-0082-4.ch002

Dwivedi, D. N. (2022). Benchmarking of traditional and advanced machine learning modelling techniques for forecasting. In *Visualization Techniques for Climate Change with Machine Learning and Artificial Intelligence* (pp. x–x). Elsevier. doi:10.1016/B978-0-323-99714-0.00017-0

Dwivedi, D. N. (2022). Benchmarking of traditional and deep learning time series modelling techniques for prediction of rainfall in United Arab Emirates. *8th International conference on Time Series and Forecasting (ITISE 2022)*.

Dwivedi, D. N. (2024). The use of artificial intelligence in supply chain management and logistics. In D. Sharma, B. Bhardwaj, & M. Dhiman (Eds.), *Leveraging AI and Emotional Intelligence in Contemporary Business Organizations* (pp. 306–313). IGI Global. doi:10.4018/979-8-3693-1902-4.ch018

Dwivedi, D. N., & Anand, A. (2021). Trade heterogeneity in the EU: Insights from the emergence of COVID-19 using time series clustering. *Zeszyty Naukowe Uniwersytetu Ekonomicznego w Krakowie*, 3(993), 9–26. doi:10.15678/ZNUEK.2021.0993.0301

Dwivedi, D. N., & Gupta, A. (2022). Artificial intelligence-driven power demand estimation and short-, medium-, and long-term forecasting. In *Artificial Intelligence for Renewable Energy Systems* (pp. 231–242). Woodhead Publishing. doi:10.1016/B978-0-323-90396-7.00013-4

Dwivedi, D. N., & Mahanty, G. (2023). Human creativity vs. machine creativity: Innovations and challenges. In Z. Fields (Ed.), *Multidisciplinary Approaches in AI, Creativity, Innovation, and Green Collaboration* (pp. 19–28). IGI Global. doi:10.4018/978-1-6684-6366-6.ch002

Dwivedi, D. N., & Mahanty, G. (2024). AI-powered employee experience: Strategies and best practices. In M. Rafiq, M. Farrukh, R. Mushtaq, & O. Dastane (Eds.), *Exploring the Intersection of AI and Human Resources Management* (pp. 166–181). IGI Global. doi:10.4018/979-8-3693-0039-8.ch009

Dwivedi, D. N., Mahanty, G., & Vemareddy, A. (2022). How responsible is AI?: Identification of key public concerns using sentiment analysis and topic modeling. [IJIRR]. *International Journal of Information Retrieval Research*, 12(1), 1–14. doi:10.4018/IJIRR.298646

Dwivedi, D. N., Tadoori, G., & Batra, S. (2023). Impact of women leadership and ESG ratings in organizations: A time series segmentation study. *Academy of Strategic Management Journal*, 22(S3), 1–6.

Dwivedi, D., Batra, S., & Pathak, Y. K. (2023). A machine learning-based approach to identify key drivers for improving corporate's ESG ratings. *Journal of Law and Sustainable Development*, 11(1), e0242. doi:10.37497/sdgs.v11i1.242

Dwivedi, D., Kapur, P. N., & Kapur, N. N. (2023). Machine learning time series models for tea pest looper infestation in Assam, India. In A. Sharma, N. Chanderwal, & R. Khan (Eds.), *Convergence of Cloud Computing, AI, and Agricultural Science* (pp. 280–289). IGI Global. doi:10.4018/979-8-3693-0200-2.ch014

Eitel-Porter, R. (2021). Beyond the promise: Implementing ethical AI. *AI and Ethics*, 1(1), 73–80. doi:10.1007/s43681-020-00011-6

Electronic Privacy Information Center. (2016). *Children's Online Privacy Protection Act (COPPA)*. EPIC. https://epic.org/privacy/kids/

Ellili, N. O. D. (2022). Impact of ESG disclosure and financial reporting quality on investment efficiency. *Corporate Governance: An International Journal of Business in Society*.

Elliott, D., Husbands, S., Hamdy, F. C., Holmberg, L., & Donovan, J. L. (2017). Understanding and Improving Recruitment to Randomised Controlled Trials: Qualitative Research Approaches. *European Urology*, *72*(5), 789–798. doi:10.1016/j.eururo.2017.04.036 PMID:28578829

Elwood, S., & Leszczynski, A. (2013). New spatial media, new knowledge politics. *Transactions of the Institute of British Geographers*, *38*(4), 544–559. doi:10.1111/j.1475-5661.2012.00543.x

Enayat, U. (2021). Machine Unlearning via Algorithmic Stability. *Proceedings of 34th Conference on Learning Theory*. IEEE.

Engin, Z., & Treleaven, P. (2019). Algorithmic Government: Automating Public Services and Supporting Civil Servants in using Data Science Technologies. *The Computer Journal*, *62*(3), 448–460. Advance online publication. doi:10.1093/comjnl/bxy082

Enholm, I. M., Papagiannidis, E., Mikalef, P., & Krogstie, J. (2022). Artificial Intelligence and business value: A literature review. *Information Systems Frontiers*, *24*(8), 1709–1734. doi:10.1007/s10796-021-10186-w

Epstein, A. (2008). *What can schools do to promote the involvement of parents?* American Education Consulting Firm. https://www.aecf.org/resources/ parental-involvement-in-education

European Commission. (2020). *White Paper on Artificial Intelligence: A European approach to excellence and trust* (COM(2020) 65 final). EC. https://ec.europa.eu/info/publications/white-paper-artificial-intelligence-european-approach-excellence-and-trust_en

EY. (2023). *Artificial intelligence ESG stakes, Discussion paper*. Assets. https://assets.ey.com/content/dam/ey-sites/ey-com/en_ca/topics/ai/ey-artificial-intelligence-esg-stakes-discussion-paper.pdf

Faden, R. R., & Beauchamp, T. L. (1986). *A history and theory of informed consent*. Oxford University Press.

Fairbairn, G. J. (2002). Ethics, empathy and storytelling in professional development. *Learning in Health and Social Care*, *1*(1), 22–32. doi:10.1046/j.1473-6861.2002.00004.x

Falk, S., & van Wynsberghe, A. (2023). Challenging AI for Sustainability: What ought it mean? *AI and Ethics*. doi:10.1007/s43681-023-00323-3

FDA. (2020). *Postmarket Drug Safety Information for Patients and Providers*. FDA. https://www.fda.gov/drugs/drug-safety-and-availability/postmarket-drug-safety-information-patients-and-providers

Feeney, M., Rajabifard, A., & Williamson, I. P. (2001). Spatial Data Infrastructure Frameworks to Support Decision-Making for Sustainable Development. *5th Global Spatial Data Infrastucture Conference*. ACM.

Felício, S., Hora, J., Ferreira, M. C., Abrantes, D., Costa, P. D., Dangelo, C., Silva, J., & Galvão, T. (2022). Handling OpenStreetMap georeferenced data for route planning. *Transportation Research Procedia*, *62*, 189–196. doi:10.1016/j.trpro.2022.02.024

Feng, Y. (2021). Comprehensive evaluation of quality indicators for different varieties of kiwifruit based on principal component analysis and cluster analysis. *Jiangsu Agricultural Science and Technology Xue*, *49*(22), 180–185.

Feuerriegel, S., Dolata, M., & Schwabe, G. (2020). Fair AI: Challenges and opportunities. *Business & Information Systems Engineering*, *62*(4), 379–384. doi:10.1007/s12599-020-00650-3

Figueira, A., & Vaz, B. (2022). Survey on synthetic data generation, evaluation methods and GANs. *Mathematics*, *10*(15), 2733. doi:10.3390/math10152733

Filho, H. F., Leite, B. P., Pompermayer, G. A., Werneck, M. G., & Leyh, W. (2013). Teaching VGI as a strategy to promote the production of urban digital cartographic databases. *Joint Urban Remote Sensing Event 2013. JURSE, 2013.* doi:10.1109/JURSE#.2013.6550705

Firouzi, F., Farahani, B., Barzegari, M., & Daneshmand, M. (2020). AI-driven data monetization: The other face of data in IoT-based smart and connected health. *IEEE Internet of Things Journal, 9*(8), 5581–5599. doi:10.1109/JIOT.2020.3027971

Firth, K. (2023). The future of the workforce: How human-AI collaboration will redefine the industry. *Forbes.* https://www.forbes.com/ sites/forbestechcouncil/2023/05/04/the-future-of-the-workforce-\ \how-human-ai-collaboration-will-redefine-the-industry

Fischer, M. M., Scholten, H. J., & Unwin, D. (2019). Geographic information systems, spatial data analysis and spatial modelling: an introduction. In Spatial Analytical Perspectives on GIS. Springer. doi:10.1201/9780203739051-1

Fisher, M., Fradley, M., Flohr, P., Rouhani, B., & Simi, F. (2021). Ethical considerations for remote sensing and open data in relation to the endangered archaeology in the Middle East and North Africa project. *Archaeological Prospection, 28*(3), 279–292. doi:10.1002/arp.1816

Fisher, W. R. (1984). Narration as a human communication paradigm: The case of public moral argument. *Communication Monographs, 51*(1), 1–22. doi:10.1080/03637758409390180

Florance, P. (2006). GIS collection development within an academic library. *Library Trends, 55*(2), 222–234. Advance online publication. doi:10.1353/lib.2006.0057

Floridi, L., & Cowls, J. (2022). A unified framework of five principles for AI in society. *Machine learning and the city: Applications in architecture and urban design*, 535-545.

Floridi, L., Cowls, J., Beltrametti, M., Chatila, R., Chazerand, P., Dignum, V., & Vayena, E. (2021). An ethical framework for a good AI society: Opportunities, risks, principles, and recommendations. *Ethics, governance, and policies in artificial intelligence*, 19-39.

Floridi, L., & Cowls, J. (2019). A unified framework of five principles for AI in society. *Harvard Data Science Review, 1*(1), 1–13. doi:10.1162/99608f92.8cd550d1

Floridi, L., & Taddeo, M. (2016). What is data ethics? *Philosophical Transactions. Series A, Mathematical, Physical, and Engineering Sciences, 374*(2083), 20160360. doi:10.1098/rsta.2016.0360 PMID:28336805

Fonte, C. C., Bastin, L., See, L., Foody, G., & Lupia, F. (2015). Usability of VGI for validation of land cover maps. *International Journal of Geographical Information Science, 29*(May), 1–23. doi:10.1080/13658816.2015.1018266

Fowziya, S., Sivaranjani, S., Devi, N. L., Boopathi, S., Thakur, S., & Sailaja, J. M. (2023). Influences of nano-green lubricants in the friction-stir process of TiAlN coated alloys. *Materials Today: Proceedings.* doi:10.1016/j.matpr.2023.06.446

Fox, E., Tarzian, A. J., Danis, M., & Duke, C. C. (2021). Ethics Consultation in United States Hospitals: Assessment of Training Needs. *The Journal of Clinical Ethics, 32*(3), 247–255. doi:10.1086/JCE2021323247 PMID:34339396

Fox, S., Farr-Jones, S., Sopchak, L., Boggs, A., Nicely, H. W., Khoury, R., & Biros, M. (2006). High-Throughput Screening: Update on Practices and Success. *SLAS Discovery, 11*(7), 864–869. doi:10.1177/1087057106292473 PMID:16973922

Friedman, B., Felten, E., & Millett, L. I. (2000). Informed consent online: A conceptual model and design principles. In: *University of Washington Computer Science & Engineering Technical Report 00–12-2*. University of Washington.

Gabriel, I. (2022). Toward a theory of justice for artificial intelligence. *Daedalus, 151*(2), 218–231. doi:10.1162/daed_a_01911

Galinsky, A. D., Maddux, W. W., Gilin, D., & White, J. B. (2008). Why It Pays to Get Inside the Head of Your Opponent. *Psychological Science, 19*(4), 378–384. doi:10.1111/j.1467-9280.2008.02096.x PMID:18399891

Gamage, P. (2016). New development: Leveraging 'big data' analytics in the public sector. *Public Money & Management, 36*(5), 385–390. doi:10.1080/09540962.2016.1194087

Gao, X., Ma, X., Wang, J., Sun, Y., Li, B., Ji, S., Cheng, P., & Chen, J. (2022). *Verifi: Towards verifiable federated unlearning.* arXiv preprint arXiv:2205.12709.

Garbayo, E., Pascual-Gil, S., Rodríguez-Nogales, C., Saludas, L., Estella-Hermoso de Mendoza, A., & Blanco-Prieto, M. J. (2020). Nanomedicine and drug delivery systems in cancer and regenerative medicine. *Wiley Interdisciplinary Reviews. Nanomedicine and Nanobiotechnology, 12*(5), e1637. doi:10.1002/wnan.1637 PMID:32351045

Garcia, F. M., Moraleda, R., Schez-Sobrino, S., Monekosso, D. N., Vallejo, D., & Glez-Morcillo, C. (2023). Health-5G: A Mixed Reality-Based System for Remote Medical Assistance in Emergency Situations. *IEEE Access : Practical Innovations, Open Solutions, 11*, 59016–59032. doi:10.1109/ACCESS.2023.3285420

Garcia-Jimeno, C., & Yildirim, P. (2015). *Matching pennies on the campaign trail: An empirical study of senate elections and media coverage* (University of Pennsylvania working paper). University of Pennsylvania.

García-Tadeo, D. A., Peram, D. R., Kumar, K. S., Vives, L., Sharma, T., & Manoharan, G. (2022). Comparing the impact of Internet of Things and cloud computing on organisational behavior: A survey. *Materials Today: Proceedings, 51*, 2281–2285. doi:10.1016/j.matpr.2021.11.399

Garg, S., Goldwasser, S., & Vasudevan, P. N. (2020). Formalizing data deletion in the context of the right to be forgotten. *Proc. Annu. Int. Conf. Theory Appl. Cryptograph. Techn.* Cham, Switzerland: Springer. 10.1007/978-3-030-45724-2_13

Garzcarek, U., & Steuer, D. (2019). Approaching Ethical Guidelines for Data Scientists. In N. I. Bauer, *Applications in Statistical Computing. Studies in Classification, Data Analysis, and Knowledge Organization.* Springer.

Gee, J. (2013). *What video games have to teach us about learning and literacy.* Routledge. https://blog.ufes.br/kyriafinardi/files/2017/10/ What-Video-Games-Have-to-Teach-us-About-Learning-and-Literacy-2003. -ilovepdf-compressed.pdf

Geetha, M., Pavan, K. B., & Sunitha, P. A. (2022). *COVID-19: A special review of MSMEs for sustaining entrepreneurship.* Trueline Academic and Research Centre.

Geetha, M., & Sunitha, P. A. (2022). *Work-Life Balance among Women in the Private Higher Education Industry during COVID-19: A Path to Organisational Sustainability.* Building Resilient Organizations.

General Teaching Council Scotland. (2012). *Code of Professionalism and Conduct.* GTCS. https://www.gtcs.org.uk/regulation/copac.aspx

George, D. (2015). Brain Tumor Detection Using Shape features and Machine Learning Algorithms. *International Journal of Scientific & Engineering Research, 6*(12), 454-459.

Georgiadou, Y., Lungo, J. H., & Richter, C. (2014). Citizen sensors or extreme publics? Transparency and accountability interventions on the mobile geoweb. *International Journal of Digital Earth, 7*(7), 516–533. doi:10.1080/17538947.2013.782073

Getting Smart. (2023). *Home.* Getting Smart. https://www.gettingsmart.com/2023/10/25/ one-year-into-the-ai-revolution-and-most-schools-are-still\ \-seeking-direction

Gil, Y. (2017). Thoughtful artificial intelligence: Forging a new partnership for data science and scientific discovery. *Data Science, 1*(1–2), 119–129. doi:10.3233/DS-170011

Gokila, B. P. (2021). Brain tumor detection from MRI images using deep learning techniques. *IOP Conf. Series: Materials Science and Engineering*. IOP Science. 10.1088/1757-899X/1055/1/012115

Golatkar, A., Achille, A., & Soatto, S. (2020). Eternal sunshine of the spotless net: Selective forgetting in deep networks. *Proc. IEEE/CVF Conf. Comput. Vis. Pattern Recognit*. IEEE. 10.1109/CVPR42600.2020.00932

Goldstein, I. P. (1979). The genetic graph: A representation for the evolution of procedural knowledge. *International Journal of Man-Machine Studies*, *11*(1), 51–78. doi:10.1016/S0020-7373(79)80005-X

Gomes, M. (2021). Barcelona improves public transport with modelling technology. *PTV Blog*. https://blog.ptvgroup.com/en/city-and-mobility/barcelona-improve-public-transport-modelling/

González-Gonzalo, C., Thee, E., Klaver, C., Lee, A., Schlingemann, R., Tufail, Verbraak, F., & Sánchez, C. (2022). Trustworthy AI: Closing the gap between development and integration of AI systems in ophthalmic practice. *Progress in Retinal and Eye Research, 90*. doi:10.1016/j.preteyeres.2021.101034

González-Méijome, J. M., Piñero, D. P., & Villa-Collar, C. (2022). Upcoming Special Issue: "Artificial Intelligence, Data Science and E-health in Vision Research and Clinical Activity.". *Journal of Optometry*, *15*(1), 1–2. doi:10.1016/j.optom.2021.11.003 PMID:34933741

González-Sendino, R., Serrano, E., Bajo, J., & Novais, P. (2023). A Review of Bias and Fairness in Artificial Intelligence. *International Journal of Interactive Multimedia and Artificial Intelligence*.

Goodchild, M. F. (2007). Citizens as sensors: The world of volunteered geography. In GeoJournal. doi:10.1007/s10708-007-9111-y

Goodfellow, I., Bengio, Y., & Courville, A. (2016). *Deep Learning*. MIT Press. https://www.deeplearningbook.org/

Gopinathan, R., & Manoharan, G. (2022). Work-Life Balance Practices and Organizational Performance for Achieving Superior Performance. *The Changing Role of Human Resource Management in the Global Competitive Environment*, 217.

Górriz, J. M., Ramírez, J., Ortíz, A., Martínez-Murcia, F. J., Segovia, F., Suckling, J., Leming, M., Zhang, Y. D., Álvarez-Sánchez, J. R., Bologna, G., Bonomini, P., Casado, F. E., Charte, D., Charte, F., Contreras, R., Cuesta-Infante, A., Duro, R. J., Fernández-Caballero, A., Fernández-Jover, E., & Ferrández, J. M. (2020). Artificial intelligence within the interplay between natural and artificial computation: Advances in data science, trends and applications. *Neurocomputing*, *410*, 237–270. doi:10.1016/j.neucom.2020.05.078

Gosain, M. S., Aggarwal, N., & Kumar, R. (2023). A Study of 5G and Edge Computing Integration with IoT- A Review. *2023 International Conference on Computational Intelligence and Sustainable Engineering Solutions (CISES)*, Greater Noida, India. 10.1109/CISES58720.2023.10183438

Gowri, N. V., Dwivedi, J. N., Krishnaveni, K., Boopathi, S., Palaniappan, M., & Medikondu, N. R. (2023). Experimental investigation and multi-objective optimization of eco-friendly near-dry electrical discharge machining of shape memory alloy using Cu/SiC/Gr composite electrode. *Environmental Science and Pollution Research International*, *30*(49), 1–19. doi:10.1007/s11356-023-26983-6 PMID:37126160

Grant, C., & Sonesh, L. (2019). Skillset diversity: Why it matters and how to harness it. Skill Up Powered by Sertifier. *Medium*. https://medium.com/skill-up-powered-by-sertifier/the-importance-of-skillset-diversity-in-the-modern-workplace-4a9d581dd72

Greene, J. D. (2015). Beyond Point-and-Shoot Morality: Why Cognitive (Neuro)Science Matters for Ethics. *Law and Ethics of Human Rights*, *9*(2), 141–172. doi:10.1515/lehr-2015-0011

Greening, D. W., & Gray, B. (1994). Testing a model of organizational response to social and political issues. *Academy of Management Journal, 37*(3), 467–498. doi:10.2307/256697

Grinberger, A. Y., Minghini, M., Juhász, L., Yeboah, G., & Mooney, P. (2022). OSM Science—The Academic Study of the OpenStreetMap Project, Data, Contributors, Community, and Applications. In ISPRS International Journal of Geo-Information, 11(4). doi:10.3390/ijgi11040230

Gross, J. J. (2015). Emotion Regulation: Current Status and Future Prospects. *Psychological Inquiry, 26*(1), 1–26. doi: 10.1080/1047840X.2014.940781

Grupac, M., Zauskova, A., & Nica, E. (2023). Generative Artificial Intelligence-based Treatment Planning in Clinical Decision-Making, in Precision Medicine, and in Personalized Healthcare. *Contemporary Readings in Law and Social Justice, 15*(1).

Gruson, D., Helleputte, T., Rousseau, P., & Gruson, D. (2019). Data science, artificial intelligence, and machine learning: Opportunities for laboratory medicine and the value of positive regulation. In Clinical Biochemistry (Vol. 69). doi:10.1016/j.clinbiochem.2019.04.013

Guardian. (2018). Fitness tracking app Strava gives away location of secret US army bases. *The Guardian*. https://www.theguardian.com/world/2018/jan/28/fitness-tracking-app-gives-away-location-of-secret-us-army-bases

Güiza, F., & Stuart, N. (2018). When citizens choose not to participate in volunteering geographic information to e-governance: a case study from Mexico. In GeoJournal. doi:10.1007/s10708-017-9820-9

Gujral, G., Shivarama, J., & Mariappan, M. (2019). Artificial Intelligence (Ai) and Data Science for Developing Intelligent Health Informatics Systems. *Proceedings of the Ntional Conference on AI in HI & VR, SHSS-TISS Mumbai, January*. IEEE.

Gulati, N., Sethi, A., Mahesh, D., Makkar, S., Manoharan, G., Megaladevi, M., & Palanivel, R. (2022, May). Implementation of blockchain in supply chain. In AIP Conference Proceedings (Vol. 2418, No. 1). AIP Publishing. doi:10.1063/5.0083008

Gunapati, S. (2011). Key Features for Designing a Dashboard. *Government Finance Review (0883-7856), 27*.

Gunasekaran, K., & Boopathi, S. (2023). Artificial Intelligence in Water Treatments and Water Resource Assessments. In *Artificial Intelligence Applications in Water Treatment and Water Resource Management* (pp. 71–98). IGI Global. doi:10.4018/978-1-6684-6791-6.ch004

Guo, K., Yang, Z., Yu, C.-H., & Buehler, M. J. (2021). Artificial intelligence and machine learning in design of mechanical materials. *Materials Horizons, 8*(4), 1153–1172. doi:10.1039/D0MH01451F PMID:34821909

Gupta, A., Dwivedi, D. N., & Jain, A. (2021). Threshold fine-tuning of money laundering scenarios through multi-dimensional optimization techniques. *Journal of Money Laundering Control*. doi:10.1108/JMLC-12-2020-0138

Gupta, A., Dwivedi, D. N., & Shah, J. (2023). Applying Artificial Intelligence on Investigation. In: Artificial Intelligence Applications in Banking and Financial Services. Future of Business and Finance. Springer, Singapore. doi:10.1007/978-981-99-2571-1_9

Gupta, A., Dwivedi, D. N., & Shah, J. (2023). Applying Machine Learning for Effective Customer Risk Assessment. In: Artificial Intelligence Applications in Banking and Financial Services. Future of Business and Finance. Springer, Singapore. doi:10.1007/978-981-99-2571-1_6

Gupta, A., Dwivedi, D. N., & Shah, J. (2023). Artificial Intelligence-Driven Effective Financial Transaction Monitoring. In: Artificial Intelligence Applications in Banking and Financial Services. Future of Business and Finance. Springer, Singapore. doi:10.1007/978-981-99-2571-1_7

Gupta, A., Dwivedi, D. N., & Shah, J. (2023). Data Organization for an FCC Unit. In: Artificial Intelligence Applications in Banking and Financial Services. Future of Business and Finance. Springer, Singapore. doi:10.1007/978-981-99-2571-1_4

Gupta, A., Dwivedi, D. N., & Shah, J. (2023). Ethical Challenges for AI-Based Applications. In: Artificial Intelligence Applications in Banking and Financial Services. Future of Business and Finance. Springer, Singapore. doi:10.1007/978-981-99-2571-1_10

Gupta, A., Dwivedi, D. N., & Shah, J. (2023). Financial Crimes Management and Control in Financial Institutions. In: Artificial Intelligence Applications in Banking and Financial Services. Future of Business and Finance. Springer, Singapore. doi:10.1007/978-981-99-2571-1_2

Gupta, A., Dwivedi, D. N., & Shah, J. (2023). Machine Learning-Driven Alert Optimization. In: Artificial Intelligence Applications in Banking and Financial Services. Future of Business and Finance. Springer, Singapore. doi:10.1007/978-981-99-2571-1_8

Gupta, A., Dwivedi, D. N., & Shah, J. (2023). Overview of Money Laundering. In: Artificial Intelligence Applications in Banking and Financial Services. Future of Business and Finance. Springer, Singapore. doi:10.1007/978-981-99-2571-1_1

Gupta, A., Dwivedi, D. N., & Shah, J. (2023). Overview of Technology Solutions. In: Artificial Intelligence Applications in Banking and Financial Services. Future of Business and Finance. Springer, Singapore. doi:10.1007/978-981-99-2571-1_3

Gupta, A., Dwivedi, D. N., & Shah, J. (2023). Planning for AI in Financial Crimes. In: Artificial Intelligence Applications in Banking and Financial Services. Future of Business and Finance. Springer, Singapore. doi:10.1007/978-981-99-2571-1_5

Gupta, A., Dwivedi, D. N., & Shah, J. (2023). Setting up a Best-In-Class AI-Driven Financial Crime Control Unit (FCCU). In: Artificial Intelligence Applications in Banking and Financial Services. Future of Business and Finance. Springer, Singapore. doi:10.1007/978-981-99-2571-1_11

Gupta, A., Dwivedi, D. N., Shah, J., & Jain, A. (2021). Data quality issues leading to suboptimal machine learning for money laundering models. *Journal of Money Laundering Control.* doi:10.1108/JMLC-05-2021-0049

Gupta, A., Shukla, R. K., Bhola, A., & Sengar, A. S. (2021, December). Comparative Analysis of Supervised Learning Techniques of Machine Learning for Software Defect Prediction. In *2021 10th International Conference on System Modeling & Advancement in Research Trends (SMART)* (pp. 406-409). IEEE. 10.1109/SMART52563.2021.9676307

Gupta, N., Sharma, H., Kumar, S., Kumar, A., & Kumar, R. (2022). *A Comparative Study of Implementing Agile Methodology and Scrum Framework for Software Development.* 2022 11th International Conference on System Modeling & Advancement in Research Trends (SMART), Moradabad, India. 10.1109/SMART55829.2022.10047477

Gupta, M., Akiri, C., Aryal, K., Parker, E., & Praharaj, L. (2023). From chatgpt to threatgpt: Impact of generative ai in cybersecurity and privacy. *IEEE Access: Practical Innovations, Open Solutions, 11,* 80218–80245. doi:10.1109/ACCESS.2023.3300381

Gupta, S., Pebesma, E., Degbelo, A., & Costa, A. C. (2018). Optimising citizen-driven air quality monitoring networks for cities. *ISPRS International Journal of Geo-Information, 7*(12), 468. doi:10.3390/ijgi7120468

H. M. M. Castro, V. S. Sosa, V., & Maganda, M. (2018). Automatic construction of vertical search tools for the deep web. *IEEE Latin Amer. Trans., 16*(2), 574_584.

Haefner, N., Wincent, J., Parida, V., & Gassmann, O. (2021). Artificial intelligence and innovation management: A review, framework, and research agenda. *Technological Forecasting and Social Change, 162,* 120392. doi:10.1016/j.techfore.2020.120392

Hagendorff, T. (2020). The Ethics of AI Ethics: An Evaluation of Guidelines. *Minds and Machines, 30*(1), 99–120. doi:10.1007/s11023-020-09517-8

Haklay, M. (2013). Citizen science and volunteered geographic information: Overview and typology of participation. In Crowdsourcing Geographic Knowledge: Volunteered Geographic Information (VGI) in Theory and Practice. Springer. doi:10.1007/978-94-007-4587-2_7

Haley, D. F., Matthews, S. A., Cooper, H. L. F., Haardörfer, R., Adimora, A. A., Wingood, G. M., & Kramer, M. R. (2016). Confidentiality considerations for use of social-spatial data on the social determinants of health: Sexual and reproductive health case study. *Social Science & Medicine, 166*, 49–56. doi:10.1016/j.socscimed.2016.08.009 PMID:27542102

Hameed, B. M. Z., Prerepa, G., Patil, V., Shekhar, P., Zahid Raza, S., Karimi, H., Paul, R., Naik, N., Modi, S., Vigneswaran, G., Prasad Rai, B., Chłosta, P., & Somani, B. K. (2021). Engineering and clinical use of artificial intelligence (AI) with machine learning and data science advancements: radiology leading the way for future. In Therapeutic Advances in Urology (Vol. 13). doi:10.1177/17562872211044880

Hamm, S., Schleser, A. C., Hartig, J., Thomas, P., Zoesch, S., & Bulitta, C. (2020). 5G as enabler for Digital Healthcare. *Current Directions in Biomedical Engineering, 6*(3), 1–4. doi:10.1515/cdbme-2020-3001

Han, B., Nawaz, S., Buchanan, G., & McKay, D. (2023). Ethical and Pedagogical Impacts of AI in Education. *International Conference on Artificial Intelligence in Education*, (pp. 667-673). Springer. 10.1007/978-3-031-36272-9_54

Han, H., & Liu, W. (2019). The coming era of artificial intelligence in biological data science. In BMC Bioinformatics (Vol. 20). doi:10.1186/s12859-019-3225-3

Hancock, J. T., Naaman, M., & Levy, K. (2020). AI-mediated communication: Definition, research agenda, and ethical considerations. *Journal of Computer-Mediated Communication, 25*(1), 89–100. doi:10.1093/jcmc/zmz022

Hanson, R. (2016). The age of. In *Work, love, and life when robots rule the Earth*. Oxford University Press.

Hanumanthakari, S., Gift, M. M., Kanimozhi, K., Bhavani, M. D., Bamane, K. D., & Boopathi, S. (2023). Biomining Method to Extract Metal Components Using Computer-Printed Circuit Board E-Waste. In *Handbook of Research on Safe Disposal Methods of Municipal Solid Wastes for a Sustainable Environment* (pp. 123–141). IGI Global. doi:10.4018/978-1-6684-8117-2.ch010

Hao, K. (2020). The UK exam debacle reminds us that algorithms can't fix broken systems. *The MIT Technology Review.* https://www.technologyreview.com/2020/08/20/1007502/uk-exam-algorithm-cant-fix-broken-system/

Hao, X., & Demir, E. (2023). Artificial intelligence in supply chain decision-making: an environmental, social, and governance triggering and technological inhibiting protocol. *Journal of Modelling in Management.* doi:10.1108/JM2-01-2023-0009

Haokun, Y. (2021). Image Segmentation of Pitaya Disease Based on Genetic Algorithm and Otsu Algorithm . *Journal of Physics: Conference Series, 1955*(1).

Haque, N. (2023). Artificial intelligence and geriatric medicine: New possibilities and consequences. *Journal of the American Geriatrics Society, 71*(6), 2028–2031. doi:10.1111/jgs.18334 PMID:36930059

Hargreaves, A., & Avelino, A. (2022). *Humanized learning: Rethinking education in a globalized age*. Routledge.

Haribalaji, V., Boopathi, S., & Asif, M. M. (2021). Optimization of friction stir welding process to join dissimilar AA2014 and AA7075 aluminum alloys. *Materials Today: Proceedings, 50*, 2227–2234. doi:10.1016/j.matpr.2021.09.499

Hariz, M. (2021). A dynamic mobility traffic model based on two modes of transport in smart cities. *Smart Cities, 4*(1), 253–270. doi:10.3390/smartcities4010016

Harrer, S., Shah, P., Antony, B., & Hu, J. (2019). Artificial Intelligence for Clinical Trial Design. *Trends in Pharmacological Sciences, 40*(8), 577–591. doi:10.1016/j.tips.2019.05.005 PMID:31326235

Harrison, G., Hanson, J., Jacinto, C., Ramirez, J., & Ur, B. (2020, January). An empirical study on the perceived fairness of realistic, imperfect machine learning models. In *Proceedings of the 2020 conference on fairness, accountability, and transparency* (pp. 392-402). ACM. 10.1145/3351095.3372831

Hartley, J. R. (1973). The design and evaluation of an adaptive teaching system. *International Journal of Man-Machine Studies, 5*(3), 421–436. doi:10.1016/S0020-7373(73)80029-X

Hassani, H., & Silva, E. S. (2023). The Role of ChatGPT in Data Science: How AI-Assisted Conversational Interfaces Are Revolutionizing the Field. *Big Data Cognitive Computing, 7*(62).

Hassani, B. K. (2021). Societal bias reinforcement through machine learning: A credit scoring perspective. *AI and Ethics, 1*(3), 239–247. doi:10.1007/s43681-020-00026-z

Hassan, W. M., Aldoseri, D. T., Saeed, M. M., Khder, M. A., & Ali, B. (2022). Utilization of Artificial Intelligence and Robotics Technology in Business. *2022 ASU International Conference in Emerging Technologies for Sustainability and Intelligent Systems (ICETSIS),* (pp. 443-449). IEEE. 10.1109/ICETSIS55481.2022.9888895

Hasselgren, C., & Oprea, T. I. (2024). Artificial Intelligence for Drug Discovery: Are We There Yet? *Annual Review of Pharmacology and Toxicology, 64*(1), 64. doi:10.1146/annurev-pharmtox-040323-040828 PMID:37738505

Hastings, J. (2024). Preventing harm from non-conscious bias in medical generative AI. *The Lancet. Digital Health, 6*(1), e2–e3. doi:10.1016/S2589-7500(23)00246-7 PMID:38123253

Hattie, J. (2008). *Visible learning: The key to improving education.* Routledge. doi:10.4324/9780203887332

Haworth, B. T., Bruce, E., Whittaker, J., & Read, R. (2018). The good, the bad, and the uncertain: Contributions of volunteered geographic information to community disaster resilience. In Frontiers in Earth Science (Vol. 6). doi:10.3389/feart.2018.00183

Haworth, B., Whittaker, J., & Bruce, E. (2016). Assessing the application and value of participatory mapping for community bushfire preparation. *Applied Geography (Sevenoaks, England), 76*, 115–127. doi:10.1016/j.apgeog.2016.09.019

Hayward, R. (2023). Generative Artificial Intelligence-driven Healthcare Systems in Medical Imaging Analysis, in Clinical Decision Support, and in Patient Engagement and Monitoring. *Contemporary Readings in Law and Social Justice, 15*(1), 63–80. doi:10.22381/CRLSJ15120234

He, B., & Chang, K. C.-C. (2003). Statistical schema matching across web query interfaces. *ACM SIGMOD Int. Conf. Manage. Data.* ACM. 10.1145/872757.872784

He, Y. (2015). *DEEP- OBLIVIATE.* arXiv:2105.06209(cs.LG).

He, Z., Hong, J., & Bell, D. (2008). Schema matching across query interfaces on the deep web. A. Gray, K. Jeffery, J. Shao, (eds.). Sharing Data, Information and Knowledge. Berlin, Germany: Springer. doi:10.1007/978-3-540-70504-8_6

Hegde, S. K., & Mundada, M. R. (2022). Hybrid generative regression-based deep intelligence to predict the risk of chronic disease. *International Journal of Intelligent Computing and Cybernetics, 15*(1), 144–164. doi:10.1108/IJICC-06-2021-0103

Hehman, E., & Xie, S. Y. (2021). Doing Better Data Visualization. *Advances in Methods and Practices in Psychological Science, 4*(4). doi:10.1177/25152459211045334

Hendler, J. (2008). Avoiding another AI winter. *IEEE Intelligent Systems, 23*(2), 2–4. doi:10.1109/MIS.2008.20

Henriques, R., Antunes, C., & Madeira, S. C. (2015). Generative modeling of repositories of health records for predictive tasks. *Data Mining and Knowledge Discovery, 29*(4), 999–1032. doi:10.1007/s10618-014-0385-7

Hernandez, J. (2009). Redlining revisited: Mortgage lending patterns in Sacramento 1930-2004. *International Journal of Urban and Regional Research, 33*(2), 291–313. doi:10.1111/j.1468-2427.2009.00873.x

Hilal, A. M., Alsolai, H., Al-Wesabi, F. N., Al-Hagery, M. A., Hamza, M. A., & Al Duhayyim, M. (2022). Artificial intelligence based optimal functional link neural network for financial data Science. *Computers, Materials & Continua, 70*(3). doi:10.32604/cmc.2022.021522

Hinton, G. E. & Osidero, S., & The, Y.-W. (2006). A fast learning algorithm for deep belief nets. *Neural Calculation, 18*(7), 1527-1554.

Hinton, G. E., & Salakhutdinov, R. R. (2006). Reducing the dimensionality of data with neural networks. *Science, 313*(5786), 504–507. doi:10.1126/science.1127647 PMID:16873662

Hirsch, P. B. (2021). Footprints in the cloud: The hidden cost of IT infrastructure. *The Journal of Business Strategy, 43*(1), 65–68. doi:10.1108/JBS-11-2021-0175

Hobday, A. J., Hartog, J. R., Manderson, J. P., Mills, K. E., Oliver, M. J., Pershing, A. J., Siedlecki, S., & Browman, H. (2019). Ethical considerations and unanticipated consequences associated with ecological forecasting for marine resources. *ICES Journal of Marine Science, 76*(5), fsy210. doi:10.1093/icesjms/fsy210

Holmes, W. (2021). Ethics of AI in education: Towards a community-wide framework. *International Journal of Artificial Intelligence in Education*, 1–23.

Holstein, K., McLaren, B. M., & Aleven, V. (2018). Student learning benefits of a mixed-reality teacher awareness tool in ai-enhanced classrooms. In C. P. Rosé, R. Martínez-Maldonado, H. U. Hoppe, R. Luckin, M. Mavrikis, K. Porayska-Pomsta, B. McLaren, & B. du Boulay (Eds.), *Artificial intelligence in education: 19th international conference,* (pp. 154–168). Cham: Springer. 10.1007/978-3-319-93843-1_12

Holtje, J. (2011). *The Power of Storytelling: Captivate, Convince, or Convert Any Business Audience UsingStories from Top CEOs*. Penguin.

Holzinger, A., Haibe-Kains, B., & Jurisica, I. (2019). Why imaging data alone is not enough: AI-based integration of imaging, omics, and clinical data. *European Journal of Nuclear Medicine and Molecular Imaging, 46*(13), 2722–2730. doi:10.1007/s00259-019-04382-9 PMID:31203421

Honarparvar, S., Forouzandeh Jonaghani, R., Alesheikh, A. A., & Atazadeh, B. (2019). Improvement of a location-aware recommender system using volunteered geographic information. *Geocarto International, 34*(13), 1496–1513. doi:10.1080/10106049.2018.1493155

Honey, M., & Osborne, A. (2017). *The future of jobs*. Oxford University Press.

Huafang, L., Hanbo, Y., & Xinyuan, W. (2020). Development and application of a "diatomaceous earth filter" type soil moisture sensor. *Water Resources and Hydropower Technology*.

Huang, H., Yao, X. A., Krisp, J. M., & Jiang, B. (2021). Analytics of location-based big data for smart cities: Opportunities, challenges, and future directions. *Computers, Environment and Urban Systems, 90*, 101712. doi:10.1016/j.compenvurbsys.2021.101712

Huang, J., Neill, L., Wittbrodt, M., Melnick, D., Klug, M., Thompson, M., Bailitz, J., Loftus, T., Malik, S., Phull, A., Weston, V., Heller, J. A., & Etemadi, M. (2023). Generative artificial intelligence for chest radiograph interpretation in the emergency department. *JAMA Network Open*, *6*(10), e2336100–e2336100. doi:10.1001/jamanetworkopen.2023.36100 PMID:37796505

Huang, Y., & Fu, J. (2019). Review on application of artificial intelligence in civil engineering. *Computer Modeling in Engineering & Sciences*, *121*(3), 845–875. doi:10.32604/cmes.2019.07653

Hughes, A. C., Orr, M. C., Ma, K., Costello, M. J., Waller, J., Provoost, P., Yang, Q., Zhu, C., & Qiao, H. (2021). Sampling biases shape our view of the natural world. *Ecography*, *44*(9), 1259–1269. doi:10.1111/ecog.05926

Hu, S., Ge, Y., Liu, M., Ren, Z., & Zhang, X. (2022). Village-level poverty identification using machine learning, high-resolution images, and geospatial data. *International Journal of Applied Earth Observation and Geoinformation*, *107*, 102694. doi:10.1016/j.jag.2022.102694

Hussain, Z., & Srimathy, G. (2023). *IoT and AI Integration for Enhanced Efficiency and Sustainability*.

Hu, X., Zheng, Z., Chen, D., Zhang, X., & Sun, J. (2022). Processing, assessing, and enhancing the Waymo autonomous vehicle open dataset for driving behavior research. *Transportation Research Part C, Emerging Technologies*, *134*, 103490. doi:10.1016/j.trc.2021.103490

Hyysalo, S., Savolainen, K., Pirinen, A., Mattelmäki, T., Hietanen, P., & Virta, M. (2023). *Design types in diversified city administration: The case City of Helsinki*. Design Journal., doi:10.1080/14606925.2023.2181886

Ingle, R. B., Senthil, T. S., Swathi, S., Muralidharan, N., Mahendran, G., & Boopathi, S. (2023). Sustainability and Optimization of Green and Lean Manufacturing Processes Using Machine Learning Techniques. IGI Global. doi:10.4018/978-1-6684-8238-4.ch012

International Association of Privacy Professionals. (2023). *Government receives more than 20K submissions on India's proposed DPDPB*. IAPP.

Isgut, M., Rao, M., Yang, C., Subrahmanyam, V., Rida, P. C. G., & Aneja, R. (2018). Application of Combination High-Throughput Phenotypic Screening and Target Identification Methods for the Discovery of Natural Product-Based Combination Drugs. *Medicinal Research Reviews*, *38*(2), 504–524. doi:10.1002/med.21444 PMID:28510271

Islam, M. A., Islam, M. A., Jacky, M. A. H., Al-Amin, M., Miah, M. S. U., Khan, M. M. I., & Hossain, M. I. (2023). Distributed Ledger Technology Based Integrated Healthcare Solution for Bangladesh. *IEEE Access : Practical Innovations, Open Solutions*, *11*, 51527–51556. doi:10.1109/ACCESS.2023.3279724

Ismagilova, E., Hughes, L., Rana, N. P., & Dwivedi, Y. K. (2022). Security, Privacy and Risks Within Smart Cities: Literature Review and Development of a Smart City Interaction Framework. *Information Systems Frontiers*, *24*(2), 393–414. doi:10.1007/s10796-020-10044-1 PMID:32837262

Iwaya, L. H., Fischer-Hübner, S., Åhlfeldt, R.-M., & Martucci, L. A. (2018). *mHealth: A Privacy Threat Analysis for Public Health Surveillance Systems*. 2018 IEEE 31st International Symposium on Computer-Based Medical Systems (CBMS), Karlstad, Sweden. 10.1109/CBMS.2018.00015

Iwaya, L. H., Li, J., Fischer-Hübner, S., Åhlfeldt, R. M., & Martucci, L. A. (2019, August 21). E-Consent for Data Privacy: Consent Management for Mobile Health Technologies in Public Health Surveys and Disease Surveillance. *Studies in Health Technology and Informatics*, *264*, 1223–1227. doi:10.3233/SHTI190421 PMID:31438120

Izzo, Z., & Smart, M. A. (2021). Ap- proximate Data Deletion from Machine Learning Models", San Diego, California, USA. *Proceedings of the 24th International Conference on Artificial Intelligence and Statistics(AISTATS)*. IEEE.

Jaber, T. A. (2022). Artificial intelligence in computer networks. *Periodicals of Engineering and Natural Sciences*, *10*(1), 309–322. doi:10.21533/pen.v10i1.2616

Jack, S. (2017, November 21). *Google - powerful and responsible?* BBC. https://www.bbc.com/news/business-42060091

Jackson, M. (2020). Good Financial Practice and Clinical Research Coordinator Responsibilities. *Seminars in Oncology Nursing*, *36*(2), 150999. doi:10.1016/j.soncn.2020.150999 PMID:32253048

Jaichandran, R., Krishna, S. H., Madhavi, G. M., Mohammed, S., Raj, K. B., & Manoharan, G. (2023, January). Fuzzy Evaluation Method on the Financing Efficiency of Small and Medium-Sized Enterprises. In *2023 International Conference on Artificial Intelligence and Knowledge Discovery in Concurrent Engineering (ICECONF)* (pp. 1-7). IEEE.

Jain, A., Gupta, A., Sengar, A. S., Shukla, R. K., & Jain, A. (2021, December). Application of Deep Learning for Image Sequence Classification. In *2021 10th International Conference on System Modeling & Advancement in Research Trends (SMART)* (pp. 280-284). IEEE. 10.1109/SMART52563.2021.9676200

Jain, T. (2023). Applications of Artificial Intelligence & Machine Learning: A study on Automotive Industry. *Interantional Journal of Scientific Research In Engineering And Management*, *07*(07). doi:10.55041/IJSREM23445

Jain, V., Rastogi, M., Ramesh, J. V. N., Chauhan, A., Agarwal, P., Pramanik, S., & Gupta, A. (2023). FinTech and Artificial Intelligence in Relationship Banking and Computer Technology. In K. Saini, A. Mummoorthy, R. Chandrika, N. S. Gowri Ganesh, & I. G. I. Global (Eds.), *AI, IoT, and Blockchain Breakthroughs in E-Governance*. doi:10.4018/978-1-6684-7697-0.ch011

Jaiswal, A., & Kumar, R. (2022). *Breast cancer diagnosis using Stochastic Self-Organizing Map and Enlarge C4.5*. Multimed Tools Appl. doi:10.1007/s11042-022-14265-1

Jaiswal, A., & Kumar, R. (2023). Breast Cancer Prediction Using Greedy Optimization and Enlarge C4.5. In S. Maurya, S. K. Peddoju, B. Ahmad, & I. Chihi (Eds.), *Cyber Technologies and Emerging Sciences. Lecture Notes in Networks and Systems* (Vol. 467). Springer. doi:10.1007/978-981-19-2538-2_4

Jeevanantham, Y. A., Saravanan, A., Vanitha, V., Boopathi, S., & Kumar, D. P. (2022). Implementation of Internet-of-Things (IoT) in Soil Irrigation System. *IEEE Explore*, (pp. 1–5). IEEE.

Jivet, I., Wong, J., Scheffel, M., Specht, M., & Drachsler, H. (2021). Quantum of choice: How learners' feedback monitoring decisions, goals and self-regulated learning skills are related. In M. Scheffel, N. Dowell, S. Joksimovic, & G. Siemens (Eds.), *The impact we make: The contributions of learning analytics to learning* (pp. 416–427). ACM. doi:10.1145/3448139.3448179

Johnson, C. E. Jr, Stout, J. H., & Walter, A. C. (2020). Profound Change: The Evolution of ESG. *Business Lawyer*, *75*, 2567–2608.

Johnson, W. L. (2019). Data-driven development and evaluation of Enskill English. *International Journal of Artificial Intelligence in Education*, *29*(3), 425–457. doi:10.1007/s40593-019-00182-2

Jordan, M. I. (2019). Artificial intelligence—The revolution hasn't happened yet. *Harvard Data Science Review*, *1*(1), 1–9.

Jordan, M. I., & Mitchell, T. M. (2015). Machine learning:Trends, perspectives,and prospects. *Science*, *349*(6245), 255–260. doi:10.1126/science.aaa8415 PMID:26185243

Joshi, A., & Tiwari, H. (2023). No. 10. An Overview of Python Libraries for Data Science: Manuscript Received: 20 March 2023, Accepted: 12 May 2023, Published: 15 September 2023. *Journal of Engineering Technology and Applied Physics*, *5*(2), 85–90. doi:10.33093/jetap.2023.5.2.10

Jou, C. (2016). *Deep web query interface integration based on incremental schema matching and merging.* Proc. 3rd Multidisciplinary Int. Social Netw. Conf. Social Inform., Data Sci., New York, NY, USA. 10.1145/2955129.2955170

Juergens, C., & Redecker, A. P. (2023). Basic Geo-Spatial Data Literacy Education for Economic Applications. *KN - Journal of Cartography and Geographic Information, 73*(2), 1–13. doi:10.1007/s42489-023-00135-9 PMID:37361712

Jung, M., Lim, C., Lee, C., Kim, S., & Kim, J. (2021). Human dietitians vs. Artificial intelligence: Which diet design do you prefer for your children? *The Journal of Allergy and Clinical Immunology, 147*(2), AB117. doi:10.1016/j.jaci.2020.12.430

Junxiao, W. (2022). *Federated Unlearning via Class-discriminative Pruning.* Hong Kong Polytechnic University, Dalian University of Technology.

Jun, Y., Craig, A., Shafik, W., & Sharif, L. (2021). Artificial intelligence application in cybersecurity and cyberdefense. *Wireless Communications and Mobile Computing, 2021*, 1–10. doi:10.1155/2021/3329581

Jurado-Camino, M. T., Chushig-Muzo, D., Soguero-Ruiz, C., de Miguel-Bohoyo, P., & Mora-Jiménez, I. (2023). On the Use of Generative Adversarial Networks to Predict Health Status Among Chronic Patients. In HEALTHINF (pp. 167-178). doi:10.5220/0011690500003414

Kadam, M. S., Krutika Kajulkar, M., & Ovhal, M. A. (2020). Importance of Human-Machine Interface in Artificial Intelligence and Data Science. [IJERT]. *International Journal of Engineering Research & Technology (Ahmedabad), 8*(5).

Kalidas, R., Boopathi, S., Sivakumar, K., & Mohankumar, P. (2012). Optimization of Machining Parameters of WEDM Process Based On the Taguchi Method. *IJEST, 6*(1).

Kamruzzaman, M. M. (2022). Impact of social media on geopolitics and economic growth: Mitigating the risks by developing artificial intelligence and cognitive computing tools. *Computational Intelligence and Neuroscience, 2022*, 2022. doi:10.1155/2022/7988894 PMID:35602647

Kapp, K. (2014). *Learning in the age of disruption: How personalized learning and microlearning are changing education.* John Wiley & Sons. https://mitpress.mit.edu/ 9780262542210/the-next-age-of-disruption/

Kar, A. K., Choudhary, S. K., & Singh, V. K. (2022). How can artificial intelligence impact sustainability: A systematic literature review. *Journal of Cleaner Production, 376*, 134120. doi:10.1016/j.jclepro.2022.134120

Kassymova, G. K. (2023). Ethical Problems of Digitalization and Artificial Intelligence in Education: A Global Perspective. *Journal of Pharmaceutical Negative Results*, 21502161.

Katare, D., Kourtellis, N., Park, S., Perino, D., Janssen, M., & Ding, A. (2022). Bias Detection and Generalization in AI Algorithms on Edge for Autonomous Driving. *Proceedings of the IEEE International Conference on Edge Computing.* IEEE. 10.1109/SEC54971.2022.00050

Kaur, A. (2019). *Relevance of artificial intelligence in politics.* VirtuInterpress. . doi:10.22495/ncpr_24

Kaushik, D.P. (2022). Role and Application of Artificial Intelligence in Business Analytics: A Critical Evaluation. *International Journal for Global Academic & Scientific Research.*

Kavitha, C. R., Varalatchoumy, M., Mithuna, H. R., Bharathi, K., Geethalakshmi, N. M., & Boopathi, S. (2023). Energy Monitoring and Control in the Smart Grid: Integrated Intelligent IoT and ANFIS. In M. Arshad (Ed.), (pp. 290–316). Advances in Bioinformatics and Biomedical Engineering. IGI Global. doi:10.4018/978-1-6684-6577-6.ch014

Keller, C., Glück, F., Gerlach, C. F., & Schlegel, T. (2022). Investigating the Potential of Data Science Methods for Sustainable Public Transport. *Sustainability (Basel), 14*(7), 4211. doi:10.3390/su14074211

Kenthapadi, K., Lakkaraju, H., & Rajani, N. (2023, August). Generative AI Meets Responsible AI: Practical Challenges and Opportunities. In *Proceedings of the 29th ACM SIGKDD Conference on Knowledge Discovery and Data Mining* (pp. 5805-5806). ACM. 10.1145/3580305.3599557

Ke, R., Taoxiong, L., Di, Z., & Fei, H. (2022). Hierarchical data licensing mechanisms in the data market. *Journal of Industrial Engineering and Engineering Management, 36*(6). doi:10.13587/j.cnki.jieem.2022.06.002

Kerr, K. (2020). *Ethical Considerations When Using Artificial Intelligence-Based Assistive Technologies in Education.* Open Educational Berta. https:// openeducationalberta.ca/educationaltechnologyethics/chapter/ ethical-considerations-when-using-artificial-intelligence-based-\ \assistive-technologies-in-education

Keserwani, H., PT, R., PR, J., Manoharan, G., Mane, P., & Gupta, S. K. (2021). Effect Of Employee Empowerment On Job Satisfaction In Manufacturing Industry. *Turkish Online Journal of Qualitative Inquiry, 12*(3).

Kettering, A. H., Baker, R. S., & Mathews, M. M. (2020). *Personalized learning with adaptive systems: From theory to practice.* Cambridge University Press.

Khanan, A., Abdullah, S., Mohamed, A., Mehmood, A., & Ariffin, K. (2019). Big Data Security and Privacy Concerns: A Review. *In Smart Technologies and Innovation for a Sustainable Future: Proceedings of the 1st American University in the Emirates International Research Conference.* Dubai UAE: Springer International Publishing.

Khan, C., Blount, D., Parham, J., Holmberg, J., Hamilton, P., Charlton, C., Christiansen, F., Johnston, D., Rayment, W., Dawson, S., Vermeulen, E., Rowntree, V., Groch, K., Levenson, J. J., & Bogucki, R. (2022). Artificial intelligence for right whale photo identification: From data science competition to worldwide collaboration. *Mammalian Biology, 102*(3), 1025–1042. doi:10.1007/s42991-022-00253-3

Khandare, N. B., Nikam, V. B., Banerjee, B., & Kiwelekar, A. (2022). *Spatial Data Infrastructure for Suitable Land Identification for Government Projects.* Springer. doi:10.1007/978-981-19-0725-8_7

Khanh, P. T., Ngoc, T. T. H., & Pramanik, S. (2024). AI-Decision Support System: Engineering, Geology, Climate, and Socioeconomic Aspects' Implications on Machine Learning. Using Traditional Design Methods to Enhance AI-Driven Decision Making. IGI Global.

Khonturaev, S. I. (2023). The Evolution Of Artificial Intelligence: A Comprehensive Exploration For Higher Education. *Research for Development, 2*(11), 700–706.

Khozin, S., & Coravos, A. (2019). Decentralized Trials in the Age of Real-World Evidence and Inclusivity in Clinical Investigations. *Clinical Pharmacology and Therapeutics, 106*(1), 25–27. doi:10.1002/cpt.1441 PMID:31013350

Kibriya, H. (2020). *A Novel Approach for Brain Tumor Classification Using an Ensemble of Deep and Hand-Crafted Features.* MDPI.

Kietzmann, J., Paschen, J., & Treen, E. (2018). Artificial Intelligence in Advertising: How Marketers Can Leverage Artificial Intelligence Along the Consumer Journey. *Journal of Advertising Research, 58*, 263-267. . doi:10.2501/JAR-2018-035

Kihumbe. (2019). *Survey Digitization and Mapping for HIV Monitoring.* Kihumbe.

Kiley, R., Peatfield, T., Hansen, J., & Reddington, F. (2017). Data Sharing from Clinical Trials—A Research Funder's Perspective. *The New England Journal of Medicine, 377*(20), 1990–1992. doi:10.1056/NEJMsb1708278 PMID:29141170

Kim, B., Rudin, C., & Shah, J. A. (2014). The bayesian case model: A gen- erative approach for case-based reasoning and prototype classification. *Advances in Neural Information Processing Systems.*

King, D. R., Nanda, G., Stoddard, J., Dempsey, A., Hergert, S., Shore, J. H., & Torous, J. (2023). An Introduction to Generative Artificial Intelligence in Mental Health Care: Considerations and Guidance. *Current Psychiatry Reports*, *25*(12), 1–8. doi:10.1007/s11920-023-01477-x PMID:38032442

Kitchin, R. (2014b). The Data Revolution: Big Data, Open Data, Data Infrastructures & Their Consequences. In The Data Revolution: Big Data, Open Data, Data Infrastructures & Their Consequences. Springer. doi:10.4135/9781473909472

Kitchin, R., Lauriault, T. P., & Wilson, M. W. (2018). Understanding Spatial Media. In Understanding Spatial Media. doi:10.4135/9781526425850.n1

Kitchin, R. (2014a). Big Data, new epistemologies and paradigm shifts. *Big Data & Society*, *1*(1). doi:10.1177/2053951714528481

Kitchin, R., & Lauriault, T. P. (2014). *Towards critical data studies : Charting and unpacking data assemblages and their work (preprint)*. Geoweb and Big Data.

Kitsios, F., & Kamariotou, M. (2021). Artificial Intelligence and Business Strategy towards Digital Transformation: A Research Agenda. *Sustainability (Basel)*, *13*(4), 2025. doi:10.3390/su13042025

Koedinger, K. R., & Aleven, V. (2016). An interview reflection on "Intelligent Tutoring Goes to School in the Big City". *International Journal of Artificial Intelligence in Education*, *16*(1), 13–24. doi:10.1007/s40593-015-0082-8

Kohn, M. S., Sun, J., Knoop, S., Shabo, A., Carmeli, B., Sow, D., & Rapp, W. (2014). IBM's health analytics and clinical decision support. *Yearbook of Medical Informatics*, *23*(01), 154–162. doi:10.15265/IY-2014-0002 PMID:25123736

Kolak, M., Steptoe, M., Manprisio, H., Azu-Popow, L., Hinchy, M., Malana, G., & Maciejewski, R. (2020). *Extending Volunteered Geographic Information (VGI) with Geospatial Software as a Service: Participatory Asset Mapping Infrastructures for Urban Health*. Springer. doi:10.1007/978-3-030-19573-1_11

Kolluri, S., Lin, J., Liu, R., Zhang, Y., & Zhang, W. (2022). Machine Learning and Artificial Intelligence in Pharmaceutical Research and Development: A Review. *The AAPS Journal*, *24*(1), 19. doi:10.1208/s12248-021-00644-3 PMID:34984579

Kordon, A. (2020). *Applying Data Science How to Create Value with Artificial Intelligence*. Springer International Publishing. doi:10.1007/978-3-030-36375-8

Kordzadeh, N., & Ghasemaghaei, M. (2022). Algorithmic bias: Review, synthesis, and future research directions. *European Journal of Information Systems*, *31*(3), 388–409. doi:10.1080/0960085X.2021.1927212

Kose, U. (2015). On the intersection of artificial intelligence and distance education. In U. Kose & D. Koc (Eds.), *Artificial intelligence applications in distance education* (pp. 1–11). IGI Global. doi:10.4018/978-1-4666-6276-6.ch001

Kose, U., & Koc, D. (2015). *Artificial intelligence applications in distance education*. IGI Global. doi:10.4018/978-1-4666-6276-6

Koshariya, A. K., Kalaiyarasi, D., Jovith, A. A., Sivakami, T., Hasan, D. S., & Boopathi, S. (2023). AI-Enabled IoT and WSN-Integrated Smart Agriculture System. In *Artificial Intelligence Tools and Technologies for Smart Farming and Agriculture Practices* (pp. 200–218). IGI Global. doi:10.4018/978-1-6684-8516-3.ch011

Kousa, P., & Niemi, H. (2023). AI ethics and learning: EdTech companies' challenges and solutions. *Interactive Learning Environments*, *31*(10), 6735–6746. doi:10.1080/10494820.2022.2043908

Krishna, S. H., & Manoharan, G. (2022). Making the Link between Work-Life Balance Practices and Organizational Performance in the Hospitality Industry. *The Changing Role of Human Resource Management in the Global Competitive Environment*, 201.

Krishnamoorthy, R. (2021). Environmental, Social, and Governance (ESG) Investing: Doing Good to Do Well. *Open Journal of Social Sciences*, *9*(7), 189–197. doi:10.4236/jss.2021.97013

Kshetri, N., Dwivedi, Y. K., Davenport, T. H., & Panteli, N. (2023). Generative artificial intelligence in marketing: Applications, opportunities, challenges, and research agenda. *International Journal of Information Management*, 102716. doi:10.1016/j.ijinfomgt.2023.102716

Kucukali, A., Pjeternikaj, R., Zeka, E., & Hysa, A. (2022). Evaluating the pedestrian accessibility to public services using open-source geospatial data and QGIS software. *Nova Geodesia*, *2*(2), 42. doi:10.55779/ng2242

Kumar, A., Tewari, N. & Kumar, R. (2022) A comparative study of various techniques of image segmentation for the identification of hand gesture used to guide the slide show navigation. *Multimed Tools Appl.*. doi:10.1007/s11042-022-12203-9

Kumar, P., Kumar, M., Singh, K. B., Tripathi, A. R., & Kumar, A. (2021, December). Blockchain Security Detection Condition Light Module. In *2021 10th International Conference on System Modeling & Advancement in Research Trends (SMART)* (pp. 363-367). IEEE. 10.1109/SMART52563.2021.9676302

Kumar, S., Das, D., & Agarwal, A. (2016, March). A novel method for identification and performance improvement of Blurred and Noisy Images using modified facial deblur inference (FADEIN) algorithms. In *2016 IEEE Students' Conference on Electrical, Electronics and Computer Science (SCEECS)* (pp. 1-7). IEEE.

Kumar, A., Albreem, M. A., Gupta, M., Alsharif, M. H., & Kim, S. (2020). Future 5g network based smart hospitals: Hybrid detection technique for latency improvement. *IEEE Access : Practical Innovations, Open Solutions*, *8*, 153240–153249. doi:10.1109/ACCESS.2020.3017625

Kumara, V., Mohanaprakash, T., Fairooz, S., Jamal, K., Babu, T., & Sampath, B. (2023). Experimental Study on a Reliable Smart Hydroponics System. In *Human Agro-Energy Optimization for Business and Industry* (pp. 27–45). IGI Global. doi:10.4018/978-1-6684-4118-3.ch002

Kumar, K., George, K. S., Bhatt, D., & Paul, O. P. (2023). A Brief Study on the Fake Review Detection methods on Ecommerce Websites using Machine Learning, Artificial Intelligence, and Data Science. *International Journal for Research in Applied Science and Engineering Technology*, *11*(9), 721–726. doi:10.22214/ijraset.2023.52743

Kumar, P. R., Meenakshi, S., Shalini, S., Devi, S. R., & Boopathi, S. (2023). Soil Quality Prediction in Context Learning Approaches Using Deep Learning and Blockchain for Smart Agriculture. In R. Kumar, A. B. Abdul Hamid, & N. I. Binti Ya'akub (Eds.), (pp. 1–26). Advances in Computational Intelligence and Robotics. IGI Global. doi:10.4018/978-1-6684-9151-5.ch001

Kumar, R. (2022). Intelligent Model to Image Enrichment for Strong Night-Vision Surveillance Cameras in Future Generation. *Multimedia Tools and Applications*. doi:10.1007/s11042-022- 12496

Kumar, R., & Kumar, R. (2022, May). Kumar, Sandeep. (2022). Intelligent Model to Image Enrichment for Strong Night-Vision Surveillance Cameras in Future Generation. *Multimedia Tools and Applications*, *81*(12), 16335–16351. doi:10.1007/s11042-022-12496-w

Kumar, S., Kumar, P., Ahmad, S. S., Jayasaradadevi, P., Rajeyyagari, S., Manoharan, G., & Radhakrishnan, V. (2024). A Novel Approach for IoT-Based Cloud Computing Technology and its Impact on Business Entrepreneurship. *International Journal of Intelligent Systems and Applications in Engineering*, *12*(10s), 624–628.

Kwasniewska, A., & Szankin, M. (2022). Can AI See Bias in X-ray Images? *International Journal of New Developments in Imaging*.

La Torre, M., Sabelfeld, S., Blomkvist, M., Tarquinio, L., & Dumay, J. (2028). Harmonising non-financial reporting regulation in Europe: Practical forces and projections for future research. *Meditari Accountancy Research, 26*(4), 598-621.

Lam, J., Brinkman, W. P., & Bruijnes, M. (2021). *Generative algorithms to improve mental health issue detection.*

Landers, T., & Rosenberg, R. L. (1986). An Overview. Artech House.

Landrum, G., Tosco, P., & Kelley, B. Sriniker, Gedeck, NadineSchneider, Vianello, R., Ric, Dalke, A., Cole, B., AlexanderSavelyev, Swain, M., Turk, S., N, D., Vaucher, A., Kawashima, E., Wójcikowski, M., Probst, D., Godin, G., & Doliath, G. (2020). *rdkit/rdkit: 2020_03_1 (Q1 2020) Release* (Release_2020_03_1) [Computer software]. Zenodo. doi:10.5281/ZENODO.3732262

Larsson, S. (2019). Artificial Intelligence as a Normative Societal Challenge: Bias, Responsibility, and Transparency. Banakar, Dahlstrand & Ryberg-Welander (Eds.), Festschrift for Håkan Hydén. Lund: Juristförlaget.

Larsson, S., & Heintz, F. (2020). Transparency in artificial intelligence. *Internet Policy Review, 9*(2). doi:10.14763/2020.2.1469

Lawrence, T. B., & Suddaby, R. (2006). Institutions and institutional work. In The SAGE Handbook of Organization Studies. Sage. doi:10.4135/9781848608030.n7

Lawrence, J., Rasche, A., & Kenny, K. (2018). Sustainability as Opportunity: Unilever's Sustainable Living Plan. In G. Lenssen & N. Smith (Eds.), *Managing Sustainable Business* (pp. 435–455). Springer. doi:10.1007/978-94-024-1144-7_21

Lee, Y.-J., & Zhang, X. T. (2019). AI-Generated Corporate Environmental Data: An Event Study with Predictive Power. In J. J. Choi & B. Ozkan (Eds.), Disruptive Innovation in Business and Finance in the Digital World. Emerald Publishing Limited. doi:10.1108/S1569-376720190000020009

Lee, C., Kim, S., Kim, J., Lim, C., & Jung, M. (2022). Challenges of diet planning for children using artificial intelligence. *Nutrition Research and Practice, 16*(6), 801–812. doi:10.4162/nrp.2022.16.6.801 PMID:36467765

Lee, H. Y., Tseng, H. Y., Mao, Q., Huang, J. B., Lu, Y. D., Singh, M., & Yang, M. H. (2020). Drit++: Diverse image-to-image translation via disentangled representations. *International Journal of Computer Vision, 128*(10-11), 2402–2417. doi:10.1007/s11263-019-01284-z

Lescano, A. R., Blazes, D. L., Montano, S. M., Moran, Z., Naquira, C., Ramirez, E., Lie, R., Martin, G. J., Lescano, A. G., & Zunt, J. R. (2008). Research ethics training in Peru: A case study. *PLoS One, 3*(9), e3274. doi:10.1371/journal.pone.0003274 PMID:18818763

Le, T. V., & Hsu, C. L. (2021). An Anonymous Key Distribution Scheme for Group Healthcare Services in 5G-Enabled Multi-Server Environments. *IEEE Access : Practical Innovations, Open Solutions, 9*, 53408–53422. doi:10.1109/ACCESS.2021.3070641

Leung, C. K., Pazdor, A. G. M., & Souza, J. (2021). Explainable Artificial Intelligence for Data Science on Customer Churn. *2021 IEEE 8th International Conference on Data Science and Advanced Analytics, DSAA 2021*. IEEE. 10.1109/DSAA53316.2021.9564166

Li, L., He, D., & Wang, M. (2021). Research on Plant Leaf Image Recognition Based on Improved LBP Algorithm. *Computer Engineering and Applications, 57*(19), 228-34.

Li, Y., Chen, C., Zheng, X., & Zhang, J. (2023). *Federated unlearning via active forgetting.* arXiv preprint arXiv:2307.03363

Li, A. P., Bode, C., & Sakai, Y. (2004). A novel in vitro system, the integrated discrete multiple organ cell culture (Id-MOC) system, for the evaluation of human drug toxicity: Comparative cytotoxicity of tamoxifen towards normal human cells from five major organs and MCF-7 adenocarcinoma breast cancer cells. *Chemico-Biological Interactions, 150*(1), 129–136. doi:10.1016/j.cbi.2004.09.010 PMID:15522266

Li, H., Yu, L., & He, W. (2019). The impact of GDPR on global technology development. *Journal of Global Information Technology Management, 22*(1), 1–6. doi:10.1080/1097198X.2019.1569186

Li, L., & Ulaganathan, M. N. (2017). Design and development of a crowdsourcing mobile app for disaster response. *International Conference on Geoinformatics, 2017-August.* IEEE. 10.1109/GEOINFORMATICS.2017.8090943

Li, L., & Vakanski, A. (2018). Generative adversarial networks for generation and classification of physical rehabilitation movement episodes. *International Journal of Machine Learning and Computing, 8*(5), 428. PMID:30344962

Liming, B., Gavino, A. I., Lee, P., Jungyoon, K., Na, L., Pink Pi, T. H., Xian, T. H., Buay, T. L., Xiaoping, T., Valera, A., Jia, E. Y., Wu, A., & Fox, M. S. (2015). SHINESeniors: Personalized services for active ageing-in-place. *2015 IEEE 1st International Smart Cities Conference, ISC2 2015.* IEEE. 10.1109/ISC2.2015.7366181

Linardatos, P., Papastefanopoulos, V., & Kotsiantis, S. (2020). Explainable ai: A review of machine learning interpretability methods. *Entropy (Basel, Switzerland), 23*(1), 18. doi:10.3390/e23010018 PMID:33375658

Litvina, E., Adams, A., Barth, A., Bruchez, M., Carson, J., Chung, J. E., Dupre, K. B., Frank, L. M., Gates, K. M., Harris, K. M., Joo, H., William Lichtman, J., Ramos, K. M., Sejnowski, T., Trimmer, J. S., White, S., & Koroshetz, W. (2019). BRAIN Initiative: Cutting-Edge Tools and Resources for the Community. *The Journal of Neuroscience : The Official Journal of the Society for Neuroscience, 39*(42), 8275–8284. doi:10.1523/JNEUROSCI.1169-19.2019 PMID:31619497

Liu, S., Tomizuka, M., & Ulsoy, A. (2004). *Challenges and opportunities in the engineering of intelligent systems. Proc. of the 4th International Workshop on Structural Control*, New York.

LiuY.MaZ.LiuJ.PhilipY. (2021). *Learn to Forget: Ma- chine Unlearning via Neuron Masking.* arXiv:2003.10933v3 [cs.LG].

Li, Y., Zhang, L., & Liu, Z. (2018). Multi-objective de novo drug design with conditional graph generative model. *Journal of Cheminformatics, 10*(1), 33. doi:10.1186/s13321-018-0287-6 PMID:30043127

Llorens, A., Tzovara, A., Bellier, L., Bhaya-Grossman, I., Bidet-Caulet, A., Chang, W. K., Cross, Z. R., Dominguez-Faus, R., Flinker, A., Fonken, Y., Gorenstein, M. A., Holdgraf, C., Hoy, C. W., Ivanova, M. V., Jimenez, R. T., Jun, S., Kam, J. W. Y., Kidd, C., Marcelle, E., ... Dronkers, N. F. (2021). Gender bias in academia: A lifetime problem that needs solutions. *Neuron, 109*(13), 2047–2074. doi:10.1016/j.neuron.2021.06.002 PMID:34237278

Lombardi, M. (2021). Introduction to the columns of the Internet of Things Technology magazine. *Internet of Things Technology, 11*(12), 114.

Lourenço, I., Branco, M., Curto, J., & Eugénio, T. (2012). How does the market value corporate sustainability performance? *Journal of Business Ethics, 108*(4), 417–428. doi:10.1007/s10551-011-1102-8

Lourens, M., Raman, R., Vanitha, P., Singh, R., Manoharan, G., & Tiwari, M. (2022, December). Agile Technology and Artificial Intelligent Systems in Business Development. In *2022 5th International Conference on Contemporary Computing and Informatics (IC3I)* (pp. 1602-1607). IEEE. 10.1109/IC3I56241.2022.10073410

Lourens, M., Sharma, S., Pulugu, R., Gehlot, A., Manoharan, G., & Kapila, D. (2023, May). Machine learning-based predictive analytics and big data in the automotive sector. In *2023 3rd International Conference on Advance Computing and Innovative Technologies in Engineering (ICACITE)* (pp. 1043-1048). IEEE. 10.1109/ICACITE57410.2023.10182665

Lu, A. J., Marcu, G., Ackerman, M. S., & Dillahunt, T. R. (2021). *Coding bias in the use of behavior management technologies: Uncovering socio-technical consequences of data-driven surveillance in classrooms.* Paper presented at the DIS'21: Conference on Designing Interactive Systems, Virtual Event, USA. https://static1.squarespace.com/static/5ebb1d874617b44f913c6d4b/t/609afa7f6ca7b40f39e55106/1620769435258/lu_dis21.pdf

Lu, K., Mardziel, P., Wu, F., Amancharla, P., & Datta, A. (2020). Gender bias in neural natural language processing. *Logic, Language, and Security: Essays Dedicated to Andre Scedrov on the Occasion of His 65th Birthday*, 189-202.

Lu, B. (2022). Analysis on Innovation Path of Business Administration Based on Artificial Intelligence. *Mathematical Problems in Engineering*, 2022, 1–7. doi:10.1155/2022/6790836

Luxton-Earl, C. (2020). *The future of artificial intelligence in education: Promises and perils.* John Wiley & Sons.

MacEachren, A. M. (1992). Visualizing Uncertain Information. *Cartographic Perspectives*, *13*(13), 10–19. doi:10.14714/CP13.1000

Macklin, R. (2003). Bioethics, vulnerability, and protection. *Bioethics*, *17*(5–6), 472–486. doi:10.1111/1467-8519.00362 PMID:14959716

MacphersonM.GasperiniA.BoscoM. (2021). Artificial Intelligence and FinTech Technologies for ESG Data and Analysis. SSRN. https://ssrn.com/abstract=3790774 or doi:10.2139/ssrn.3790774

Madaio, M., Stark, L., Vaughan, J. W., & Wallach, H. (2020). Co-Designing Checklists to Understand Organizational Challenges and Opportunities around Fairness in AI. *In Proceedings of the 2020 CHI conference on human factors in computing systems.* Springer.

Madaio, M. A., Stark, L., Wortman Vaughan, J., & Wallach, H. (2020, April). Co-designing checklists to understand organizational challenges and opportunities around fairness in AI. In *Proceedings of the 2020 CHI conference on human factors in computing systems* (pp. 1-14). ACM. 10.1145/3313831.3376445

Madhavan, J., Cohen, S., Dong, X. L., Alon Halevy, Y., Jeffery, R. S., Ko, D., & Yu, C. (2007). *Web-scale data integration: You can afford to pay as you go.* Proc. 3rd Biennial Conf. Innov. Data Syst. Res., Asilomar, CA, USA. www.cidrdb.org

Maffei, S., Leoni, F., & Villari, B. (2020). Data-driven anticipatory governance. Emerging scenarios in data for policy practices. *Policy Design and Practice*, *3*(2), 123–134. doi:10.1080/25741292.2020.1763896

Maguluri, L. P., Ananth, J., Hariram, S., Geetha, C., Bhaskar, A., & Boopathi, S. (2023). Smart Vehicle-Emissions Monitoring System Using Internet of Things (IoT). In Handbook of Research on Safe Disposal Methods of Municipal Solid Wastes for a Sustainable Environment (pp. 191–211). IGI Global.

Mahammad, A. B., & Kumar, R. (2022). Design a Linear Classification model with Support Vector Machine Algorithm on Autoimmune Disease data. *2022 3rd International Conference on Intelligent Engineering and Management (ICIEM)*, (pp. 164-169). IEEE. 10.1109/ICIEM54221.2022.9853182

Mahammad, A. B., & Kumar, R. (2022). Machine Learning Approach to Predict Asthma Prevalence with Decision Trees. *2022 2nd International Conference on Technological Advancements in Computational Sciences (ICTACS)*, (pp. 263-267). IEEE. 10.1109/ICTACS56270.2022.9988210

Mahammad, B., & Kumar, R. (2023). Scalable and Security Framework to Secure and Maintain Healthcare Data using Blockchain Technology. *2023 International Conference on Computational Intelligence and Sustainable Engineering Solutions (CISES)*, Greater Noida, India. 10.1109/CISES58720.2023.10183494

Maheswari, B. U., Imambi, S. S., Hasan, D., Meenakshi, S., Pratheep, V., & Boopathi, S. (2023). Internet of Things and Machine Learning-Integrated Smart Robotics. In Global Perspectives on Robotics and Autonomous Systems: Development and Applications (pp. 240–258). IGI Global. doi:10.4018/978-1-6684-7791-5.ch010

Mahmeen, M., Melconian, M. R., Haider, S., Friebe, M., & Pech, M. (2021). Next Generation 5G Mobile Health Network for User Interfacing in Radiology Workflows. *IEEE Access : Practical Innovations, Open Solutions*, 9, 102899–102907. doi:10.1109/ACCESS.2021.3097303

Mainali, S., & Park, S. (2023). Artificial Intelligence and Big Data Science in Neurocritical Care. In Critical Care Clinics, 39(1). doi:10.1016/j.ccc.2022.07.008

Makridakis, S. (2017). The forthcoming Artificial Intelligence (AI) revolution: Its impact on society and firms. *Futures*, 90, 46–60. doi:10.1016/j.futures.2017.03.006

Malik, M., Gahlawat, V. K., Mor, R. S., Agnihotri, S., Panghal, A., Rahul, K., & Emanuel, N. (2023). Artificial Intelligence and Data Science in Food Processing Industry. In EAI/Springer Innovations in Communication and Computing. Springer. doi:10.1007/978-3-031-19711-6_11

Malik, N. A., & Shaikh, M. A. (2019). Spatial distribution and accessibility to public sector tertiary care teaching hospitals: Case study from Pakistan. *Eastern Mediterranean Health Journal*, 25(6), 431–434. doi:10.26719/emhj.18.049 PMID:31469163

Mallick, S., & Bhadra, S. (2023). CDGCN: Conditional de novo Drug Generative Model Using Graph Convolution Networks. In H. Tang (Ed.), *Research in Computational Molecular Biology* (Vol. 13976, pp. 104–119). Springer Nature Switzerland. doi:10.1007/978-3-031-29119-7_7

Manikam, S., Sahibudin, S., & Kasinathan, V. (2019). Business intelligence addressing service quality for big data analytics in public sector. *Indonesian Journal of Electrical Engineering and Computer Science*, 16(1), 491. doi:10.11591/ijeecs.v16.i1.pp491-499

Manoharan, G., & Ashtikar, S. P. (2022). The Relationship between Job-Related Factors on Work Life Balance and Job Satisfaction. *The Changing Role of Human Resource Management in the Global Competitive Environment*, 169.

Manoharan, G., & Narayanan, S. (2021). A research study to investigate the feasibility of digital marketing strategies in advertising. *PalArch's Journal of Archaeology of Egypt/Egyptology, 18*(09), 450-456.

Manoharan, G., Durai, S., & Rajesh, G. A. (2022, May). Emotional intelligence: A comparison of male and female doctors in the workplace. In AIP Conference Proceedings (Vol. 2418, No. 1). AIP Publishing. doi:10.1063/5.0081816

Manoharan, G., Durai, S., & Rajesh, G. A. (2022, May). Identifying performance indicators and metrics for performance measurement of the workforce is the need of the hour: A case of a retail garment store in Coimbatore. In AIP Conference Proceedings (Vol. 2418, No. 1). AIP Publishing. doi:10.1063/5.0081821

Manoharan, G., Durai, S., Ashtikar, S. P., & Kumari, N. (2024). Artificial Intelligence in Marketing Applications. In Artificial Intelligence for Business (pp. 40-70). Productivity Press.

Manoharan, G., Durai, S., Rajesh, G. A., & Ashtikar, S. P. (2023). Credit and Risk Analysis in the Financial and Banking Sectors: An Investigation. In Artificial Intelligence for Capital Markets (pp. 57-72). Chapman and Hall/CRC.

Manoharan, G., Ashtikar, S. P., Smitha, V., Sundaramoorthi, S., & Krishna, I. M. (2023). Work-Life Balance Perceptions of Women in the IT and ITeS Sectors in Kerala: A Research Study. *Journal of Pharmaceutical Negative Results*, 3363–3375.

Manoharan, G., Durai, S., Rajesh, G. A., Razak, A., Rao, C. B., & Ashtikar, S. P. (2023). A study of postgraduate students' perceptions of key components in ICCC to be used in artificial intelligence-based smart cities. In *Artificial Intelligence and Machine Learning in Smart City Planning* (pp. 117–133). Elsevier. doi:10.1016/B978-0-323-99503-0.00003-X

Manoharan, G., Durai, S., Rajesh, G. A., Razak, A., Rao, C. B., & Ashtikar, S. P. (2023). A study on the perceptions of officials on their duties and responsibilities at various levels of the organizational structure in order to accomplish artificial intelligence-based smart city implementation. In *Artificial Intelligence and Machine Learning in Smart City Planning* (pp. 1–10). Elsevier. doi:10.1016/B978-0-323-99503-0.00007-7

Manoharan, G., Durai, S., Rajesh, G. A., Razak, A., Rao, C. B., & Ashtikar, S. P. (2023). An investigation into the effectiveness of smart city projects by identifying the framework for measuring performance. In *Artificial Intelligence and Machine Learning in Smart City Planning* (pp. 71–84). Elsevier. doi:10.1016/B978-0-323-99503-0.00004-1

Manoharan, K. G., Nehru, J. A., & Balasubramanian, S. (2021). *Artificial Intelligence and IoT*. Springer. doi:10.1007/978-981-33-6400-4

Ma, Q., Wang, L., Gong, X., & Li, K. (2022). Research on the Rationality of Public Toilets Spatial Layout based on the POI Data from the Perspective of Urban Functional Area. *Journal of Geo-Information Science*, 24(1). doi:10.12082/dqxxkx.2022.210331

Marković, M. G., Debeljak, S., & Kadoić, N. (2019). Preparing students for the era of the General Data Protection Regulation (GDPR). TEM Journal: Technology, Education, Management. *Informatics (MDPI)*, 8, 150–156. doi:10.18421/TEM81-21

Marr, B. (2020). Coronavirus: How Artificial Intelligence, Data Science And Technology Is Used To Fight The Pandemic. *Forbes*.

Martinho, A., Kroesen, M., & Chorus, C. (2020). *An Empirical Approach to Capture Moral Uncertainty in AI*. 101–101. doi:10.1145/3375627.3375805

Martinho, A., Herber, N., Kroesen, M., & Chorus, C. (2021). Ethical issues in focus by the autonomous vehicles industry. *Transport Reviews*, 41(5), 556–577. doi:10.1080/01441647.2020.1862355

Masevski, S., & Stojanovski, S. (2023). Artificial Intelligence: Geopolitical Tool Of Modern Countries. *44VOL. XXIII*.

Maslej, N., Fattorini, L., Brynjolfsson, E., Etchemendy, J., Ligett, K., Lyons, T., & Perrault, R. (2023). *The AI Index 2023 Annual Report*. AI Index Steering Committee, Institute for Human-Centered AI, Stanford University, Stanford, CA.

Mavakala, B., Mulaji, C., Mpiana, P., Elongo, V., Otamonga, J.-P., Biey, E., Wildi, W., & Pote-Wembonyama, J. (2017). Citizen Sensing of Solid Waste Disposals : Crowdsourcing As Tool Supporting Waste Management in a. *Proceedings Sardinia 2017 / Sixteenth International Waste Management and Landfill Symposium.S. Margherita Di Pula, Cagliari, Italy - 2 - 6 October 2017 -. 2017*. IEEE. https://doi.org/https://archive-ouverte.unige.ch/unige:97650

Maxwell, D., Meyer, S., & Bolch, C. (2021). DataStory™: An interactive sequential art approach for data science and artificial intelligence learning experiences. *Innovación Educativa (México, D.F.)*, 3(1), 8. doi:10.1186/s42862-021-00015-x

Mayer, R. C., Davis, J. H., & Schoorman, F. D. (1995). An Integrative Model Of Organizational Trust. *Academy of Management Review*, 20(3), 709. doi:10.2307/258792

Mayo, C. S., Matuszak, M. M., Schipper, M. J., Jolly, S., Hayman, J. A., & Ten Haken, R. K. (2017). Big Data in Designing Clinical Trials: Opportunities and Challenges. *Frontiers in Oncology*, 7, 187. doi:10.3389/fonc.2017.00187 PMID:28913177

Ma, Z., Liu, Y., Liu, X., Liu, J., Ma, J., & Ren, K. (2022). Learn to forget: Machine unlearning via neuron masking. *IEEE Transactions on Dependable and Secure Computing*.

McDonnell, M.-H., & Cobb, J. (2020). Take a Stand or Keep Your Seat: Board Turnover after Social Movement Boycotts. *Academy of Management Journal*, *63*(4), 1028–1053. doi:10.5465/amj.2017.0890

McKee, R. (1997). *Story: style, structure, substance, and the principles of screenwriting*. Harper Collins.

Meenaakumari, M., Jayasuriya, P., Dhanraj, N., Sharma, S., Manoharan, G., & Tiwari, M. (2022, December). Loan Eligibility Prediction using Machine Learning based on Personal Information. In *2022 5th International Conference on Contemporary Computing and Informatics (IC3I)* (pp. 1383-1387). IEEE. 10.1109/IC3I56241.2022.10073318

Mehrabi, N., Morstatter, F., Saxena, N., Lerman, K., & Galstyan, A. (2021). A survey on bias and fairness in machine learning. *ACM Computing Surveys*, *54*(6), 1–35. doi:10.1145/3457607

Mehr, H., Ash, H., & Fellow, D. (2017). Artificial intelligence for citizen services and government. *Ash Cent. Democr. Gov. Innov. Harvard Kennedy Sch*, (August), 1–12.

Meng, C., Trinh, L., Xu, N., Enouen, J., & Liu, Y. (2022). Interpretability and fairness evaluation of deep learning models on MIMIC-IV dataset. *Scientific Reports*, *12*(1), 7166. doi:10.1038/s41598-022-11012-2 PMID:35504931

Mennella, C., Maniscalco, U., De Pietro, G., & Esposito, M. (2023). Generating a novel synthetic dataset for rehabilitation exercises using pose-guided conditioned diffusion models: A quantitative and qualitative evaluation. *Computers in Biology and Medicine*, *167*, 107665. doi:10.1016/j.compbiomed.2023.107665 PMID:37925908

Meyrink, G. (1915). *Dr Golem*. Kurt Wolff.

Mikhaylov, S. J., Esteve, M., & Campion, A. (2018). Artificial intelligence for the public sector: Opportunities and challenges of cross-sector collaboration. *Philosophical Transactions. Series A, Mathematical, Physical, and Engineering Sciences*, *376*(2128), 20170357. doi:10.1098/rsta.2017.0357 PMID:30082303

Miller, M. L. (1979). A structured planning and debugging environment for elementary programming. *International Journal of Man-Machine Studies*, *11*(1), 79–95. doi:10.1016/S0020-7373(79)80006-1

Miller, T. (2019). Explanation in artificial intelligence: Insights from the social sciences. *Artificial Intelligence*, *267*, 1–38. doi:10.1016/j.artint.2018.07.007

Ministry of Innovation. (2023, December 17). *Israel's Policy on Artificial Intelligence Regulation and Ethics*. Ministry of Innovation, Science and Technology. https://www.gov.il/en/departments/policies/ai_2023#:~:text=Key%20Highlights%20of%20Israel's%20AI%20Policy%3A&text=Comprehensive%20approach%3A%20The%20AI%20Policy,safety%2C%20accountability%2C%20and%20privacy

Mishra, J. L., Allen, D. K., & Pearman, A. D. (2013). Information use, support and decision making in complex, uncertain environments. *Proceedings of the ASIST Annual Meeting*, *50*(1). IEEE. 10.1002/meet.14505001045

Mishra, S., & Tripathi, A. R. (2021). AI business model: An integrative business approach. *Journal of Innovation and Entrepreneurship*, *10*(1), 18. doi:10.1186/s13731-021-00157-5

Mithilesh, K., & Choubey, M. (2019). *Political Adverting In India*. Research Gate. doi:10.13140/RG.2.2.16402.30406

Mitrofanov, S. (2021). Tree retraining in the decision tree learning algorithm. *IOP Conference Series. Materials Science and Engineering*, *1047*(1).

Moglia, A., Georgiou, K., Marinov, B., Georgiou, E., Berchiolli, R. N., Satava, R. M., & Cuschieri, A. (2022). 5G in Healthcare: From COVID-19 to Future Challenges. *IEEE Journal of Biomedical and Health Informatics, 26*(8), 4187–4196. doi:10.1109/JBHI.2022.3181205 PMID:35675255

Mohai, P., & Saha, R. (2006). Reassessing racial and socioeconomic disparities in environmental justice research. *Demography, 43*(2), 383–399. doi:10.1353/dem.2006.0017 PMID:16889134

Mohammed, N. A., Mansoor, A. M., & Ahmad, R. B. (2019). Mission-Critical Machine-Type Communication: An Overview and Perspectives towards 5G. In IEEE Access (Vol. 7, pp. 127198–127216). Institute of Electrical and Electronics Engineers Inc. doi:10.1109/ACCESS.2019.2894263

Mohanty, A., Venkateswaran, N., Ranjit, P., Tripathi, M. A., & Boopathi, S. (2023). Innovative Strategy for Profitable Automobile Industries: Working Capital Management. In Handbook of Research on Designing Sustainable Supply Chains to Achieve a Circular Economy (pp. 412–428). IGI Global.

Möhring, H.-C., Müller, M., Krieger, J., Multhoff, J., Plagge, C., de Wit, J., & Misch, S. (2020). Intelligent lightweight structures for hybrid machine tools. *Production Engineering, 14*(5-6), 583–600. doi:10.1007/s11740-020-00988-3

Mohtashim Mian, S., & Kumar, R. (2023). Deep Learning for Performance Enhancement Robust Underwater Acoustic Communication Network. In S. Maurya, S. K. Peddoju, B. Ahmad, & I. Chihi (Eds.), *Cyber Technologies and Emerging Sciences. Lecture Notes in Networks and Systems* (Vol. 467). Springer. doi:10.1007/978-981-19-2538-2_24

Moon, J. A. (2010). Using story: In higher education and professional development. In Using Story: In Higher Education and Professional Development. Springer. doi:10.4324/9780203847718

Moor, J. (2006). The Dartmouth College Artificial Intelligence Conference: The Next Fifty Years. *AI Magazine, 27*(4), 87–87.

Moor, J. (2006). The Dartmouth College Artificial Intelligence conference: The next fifty years. *AI Magazine, 27*(4), 87–91. doi:10.1609/aimag.v27i4.1911

Morley, J., Floridi, L., Kinsey, L., & Elhalal, A. (2020). From what to how: An initial review of publicly available AI ethics tools, methods and research to translate principles into practices. *Science and Engineering Ethics, 26*(4), 2141–2168. doi:10.1007/s11948-019-00165-5 PMID:31828533

Moro-Visconti, R. (2022). *Augmented Corporate Valuation: From Digital Networking to ESG Compliance*. Palgrave Macmillan. doi:10.1007/978-3-030-97117-5

Muhammad, J. (2018). Detection of Brain Tumor based on Features Fusion and Machine Learning. *Journal of Ambient Intelligence and Humanized Computing Online Publication*.

Müller, V. (2023). Ethics of Artificial Intelligence and Robotics. E. N. Zalta & U. Nodelman (eds.), *The Stanford Encyclopedia of Philosophy*. Stanford Press. <https://plato.stanford.edu/archives/fall2023/entries/ethics-ai/>

Muneeb, S., & Chandler, T. W. (2021). *Sentiment Analysis for COVID-19 National Vaccination Policy*. Pride. https://pide.org.pk/blog/sentiment-analysis-for-cov

Munoko, I., Brown-Liburd, H. L., & Vasarhelyi, M. (2020). The ethical implications of using artificial intelligence in auditing. *Journal of Business Ethics, 167*(2), 209–234. doi:10.1007/s10551-019-04407-1

Musalamadugu, T. S., & Kannan, H. (2023). Generative AI for medical imaging analysis and applications. *Future Medicine AI*, (0), FMAI5.

Musleh Al-Sartawi, A. M., Hussainey, K., & Razzaque, A. (2022). The role of artificial intelligence in sustainable finance. *Journal of Sustainable Finance & Investment*, 1–6. doi:10.1080/20430795.2022.2057405

Musleh Al-Sartawi, A. M., Razzaque, A., & Kamal, M. M. (Eds.). (2021). *Artificial Intelligence Systems and the Internet of Things in the Digital Era. EAMMIS 2021. Lecture Notes in Networks and Systems* (Vol. 239). Springer.

Muthi Reddy, P., Manjula, S. H., & Venugopal, K. R. (2018). Secured Privacy Data using Multi Key Encryption in Cloud Storage. *Proceedings of 5th International Conference on Emerging Applications of Information Technology, EAIT 2018*. IEEE. 10.1109/EAIT.2018.8470399

Muzaffar, H. M., Tahir, A., Ali, A., Ahmad, M., & McArdle, G. (2017). Quality assessment of volunteered geographic information for educational planning. In *Volunteered Geographic Information and the Future of Geospatial Data*. IGI Global. doi:10.4018/978-1-5225-2446-5.ch005

Myilsamy, S., & Sampath, B. (2017). Grey Relational Optimization of Powder Mixed Near-Dry Wire Cut Electrical Discharge Machining of Inconel 718 Alloy. *Asian Journal of Research in Social Sciences and Humanities*, *7*(3), 18–25. doi:10.5958/2249-7315.2017.00157.5

Nadeem, A., Marjanovic, O., & Abedin, B. (2022). Gender bias in AI-based decision-making systems: A systematic literature review. *AJIS. Australasian Journal of Information Systems*, *26*, 26. doi:10.3127/ajis.v26i0.3835

Nadikattu, R. (2016). The Emerging Role Of Artificial Intelligence In *Modern Society, 4*, 906-911.

Nalbalwar, R. (2014). Detection of Brain Tumor by using ANN. *International Journal of Research in Advent Technology*.

Namyenya, A., Daum, T., Rwamigisa, P. B., & Birner, R. (2022). E-diary: A digital tool for strengthening accountability in agricultural extension. *Information Technology for Development*, *28*(2), 319–345. doi:10.1080/02681102.2021.1875186

Naqvi, A. (Ed.). (2021). *Artificial intelligence for asset management and investment: a strategic perspective*. Wiley. doi:10.1002/9781119601838

Nath, R. (2020). Face Detection and Recognition Using Machine Learning Algorithm. *Sambodhi (UGC Care Journal)*, *43*(3).

Naudé, W., Bray, A., & Lee, C. (2022). Crowdsourcing Artificial Intelligence in Africa: Analysis of a Data Science Contest. SSRN *Electronic Journal*. doi:10.2139/ssrn.4076351

Naudé, W. (2020). Artificial Intelligence against [An early review.]. *COVID*, 19.

Neel, S., Roth, A., & Sharifi-Malvajerdi, S. (2019). Descent-to-delete: Gradint based methods for machine unlearning.

Newman-Griffis, D., Rauchberg, J., Alharbi, R., Hickman, L., & Hochheiser, H. (2022). Definition drives design: Disability models and mechanisms of bias in AI technologies. *First Monday*.

Newman, P. A., Guta, A., & Black, T. (2021). Ethical Considerations for Qualitative Research Methods During the COVID-19 Pandemic and Other Emergency Situations: Navigating the Virtual Field. *International Journal of Qualitative Methods*, 20. doi:10.1177/16094069211047823

Ngayua, E. N., He, J., & Agyei-Boahene, K. (2021). Applying advanced technologies to improve clinical trials: A systematic mapping study. *Scientometrics*, *126*(2), 1217–1238. doi:10.1007/s11192-020-03774-1 PMID:33250544

Ng, C. K. (2023). Generative adversarial network (generative artificial intelligence) in pediatric radiology: A systematic review. *Children (Basel, Switzerland)*, *10*(8), 1372. doi:10.3390/children10081372 PMID:37628371

Nguyen, A., Ngo, H., Hong, Y., Dang, B., & Nguyen, B. (2023). Ethical principles for artificial intelligence in education. *Education and Information Technologies*, 28(4), 42214241. doi:10.1007/s10639-022-11316-w PMID:36254344

Nishanth, J., Deshmukh, M. A., Kushwah, R., Kushwaha, K. K., Balaji, S., & Sampath, B. (2023). Particle Swarm Optimization of Hybrid Renewable Energy Systems. In *Intelligent Engineering Applications and Applied Sciences for Sustainability* (pp. 291–308). IGI Global. doi:10.4018/979-8-3693-0044-2.ch016

Nishant, R., Kennedy, M., & Corbett, J. (2020). Artificial Intelligence for Sustainability: Challenges, Opportunities, and a Research Agenda. *International Journal of Information Management*, 53, 102104. doi:10.1016/j.ijinfomgt.2020.102104

Nishimwe, A., Ruranga, C., Musanabaganwa, C., Mugeni, R., Semakula, M., Nzabanita, J., Kabano, I., Uwimana, A., Utumatwishima, J. N., Kabakambira, J. D., Uwineza, A., Halvorsen, L., Descamps, F., Houghtaling, J., Burke, B., Bahati, O., Bizimana, C., Jansen, S., Twizere, C., & Twagirumukiza, M. (2022). Leveraging artificial intelligence and data science techniques in harmonizing, sharing, accessing and analyzing SARS-COV-2/COVID-19 data in Rwanda (LAISDAR Project): Study design and rationale. *BMC Medical Informatics and Decision Making*, 22(1), 214. doi:10.1186/s12911-022-01965-9 PMID:35962355

Norori, N., Hu, Q., Aellen, F. M., Faraci, F. D., & Tzovara, A. (2021). Addressing bias in big data and AI for health care: A call for open science. *Patterns (New York, N.Y.)*, 2(10), 100347. doi:10.1016/j.patter.2021.100347 PMID:34693373

Nova, K. (2023). Generative AI in healthcare: Advancements in electronic health records, facilitating medical languages, and personalized patient care. *Journal of Advanced Analytics in Healthcare Management*, 7(1), 115–131.

Noveck, B. S., Ayoub, R., Hermosilla, M., Marks, J., & Suwondo, P. (2017). *Smarter Crowdsourcing for Zika and Other Mosquito-Borne Diseases*.

Novita, N., & Indrany Nanda Ayu Anissa, A. (2022). The role of data analytics for detecting indications of fraud in the public sector. *International Journal of Research in Business and Social Science (2147- 4478)*, 11(7). doi:10.20525/ijrbs.v11i7.2113

NSPCC. (2023). *Keeping children safe online*. NSPCC. https://learning.nspcc.org.uk/safeguardingchild-protection/

Ntoutsi, E., Fafalios, P., Gadiraju, U., Iosifidis, V., Nejdl, W., Vidal, M. E., Ruggieri, S., Turini, F., Papadopoulos, S., Krasanakis, E., Kompatsiaris, I., Kinder-Kurlanda, K., Wagner, C., Karimi, F., Fernandez, M., Alani, H., Berendt, B., Kruegel, T., Heinze, C., & Staab, S. (2020). Bias in data-driven artificial intelligence systems—An introductory survey. *Wiley Interdisciplinary Reviews. Data Mining and Knowledge Discovery*, 10(3), e1356. doi:10.1002/widm.1356

Nuzzi, O. (2020). What It's Like to Get Doxed for Taking a Bike Ride. *NY Mag*. Https://Nymag.Com/. https://nymag.com/intelligencer/2020/06/what-its-like-to-get-doxed-for-taking-a-bike-ride.html

O'Reilly-Shah, V. N., Gentry, K. R., Walters, A. M., Zivot, J., Anderson, C. T., & Tighe, P. J. (2020). Bias and ethical considerations in machine learning and the automation of perioperative risk assessment. *British Journal of Anaesthesia*, 125(6), 843–846. doi:10.1016/j.bja.2020.07.040 PMID:32838979

O'Shea, T. (1979). A self-improving quadratic tutor. *International Journal of Man-Machine Studies*, 11(1), 97–124. doi:10.1016/S0020-7373(79)80007-3

O'Sullivan, M. E., Considine, E. C., O'Riordan, M., Marnane, W., Rennie, J., & Boylan, G. (2021). Challenges of Developing Robust AI for Intrapartum Fetal Heart Rate Monitoring. *Frontiers in Artificial Intelligence*, 4, 4. doi:10.3389/frai.2021.765210 PMID:34765970

Obermeyer, Z., & Emanuel, E. J. (2016). Predicting the Future—Big Data, Machine Learning, and Clinical Medicine. *The New England Journal of Medicine*, 375(13), 1216–1219. doi:10.1056/NEJMp1606181 PMID:27682033

Octavian Dumitru, C., Schwarz, G., Castel, F., Lorenzo, J., & Datcu, M. (2019). Artificial Intelligence Data Science Methodology for Earth Observation. In Advanced Analytics and Artificial Intelligence Applications. InTech Open. doi:10.5772/intechopen.86886

OECD. (2020). EXAMPLES OF AI NATIONAL POLICIES Report for the G20 Digital Economy Task Force. Organisation for Economic Co-operation and Development (OECD).

Okyere, F., Minnich, T., Sproll, M., Mensah, E., Amartey, L., Otoo-Kwofie, C., & Brunn, A. (2022). Implementation of a low-cost Ambulance Management System. *Dgpf, 30*(March).

Olteanu-Raimond, A. M., See, L., Schultz, M., Foody, G., Riffler, M., Gasber, T., Jolivet, L., le Bris, A., Meneroux, Y., Liu, L., Poupée, M., & Gombert, M. (2020). Use of automated change detection and VGI sources for identifying and validating urban land use change. *Remote Sensing (Basel), 12*(7), 1186. doi:10.3390/rs12071186

Oneto, L., & Chiappa, S. (2020). *Fairness in Machine Learning*. ArXiv, abs/2012.15816.

Ong, S., & Uddin, S. (2020). Data science and artificial intelligence in project management: The past, present and future. *Journal of Modern Project Management, 7*(4). doi:10.19255/JMPM02202

Ong, S., & Uddin, S. (2020). Data science and artificial intelligence in project management: The past, present and future. *The Journal of Modern Project Management, 7*(4).

Ong, S., & Uddin, S. (2020). Data Science and Artificial Intelligence in Project Management: The Past, Present and Future. *The Journal of Modern Project Management, 7*(4).

Osório, L. A., Silva, E., & Mackay, R. E. (2021). A Review of Biomaterials and Scaffold Fabrication for Organ-on-a-Chip (OOAC) Systems. *Bioengineering (Basel, Switzerland), 8*(8), 113. doi:10.3390/bioengineering8080113 PMID:34436116

Ostrom, E. (2009). A general framework for analyzing sustainability of social-ecological systems. In Science (Vol. 325, Issue 5939). doi:10.1126/science.1172133

Othman, S. (2021). A Low Cost Platform for Environmental Smart Farming Monitoring System Based on IoT and UAVs. *Sustainability, 13*(11).

Özyurt, F. (2019). *Brain tumor detection based on Convolutional Neural Network with neutrosophic expert maximum fuzzy sure entropy*. Elsevier.

Pagano, S., Holzapfel, S., Kappenschneider, T., Meyer, M., Maderbacher, G., Grifka, J., & Holzapfel, D. E. (2023). Arthrosis diagnosis and treatment recommendations in clinical practice: An exploratory investigation with the generative AI model GPT-4. *Journal of Orthopaedics and Traumatology, 24*(1), 61. doi:10.1186/s10195-023-00740-4 PMID:38015298

Paladugu, P. S., Ong, J., Nelson, N., Kamran, S. A., Waisberg, E., Zaman, N., Kumar, R., Dias, R. D., Lee, A. G., & Tavakkoli, A. (2023). Generative adversarial networks in medicine: Important considerations for this emerging innovation in artificial intelligence. *Annals of Biomedical Engineering, 51*(10), 2130–2142. doi:10.1007/s10439-023-03304-z PMID:37488468

Pane, J. F., Griffin, B. A., McCaffrey, D. F., & Karam, R. (2014). Effectiveness of Cognitive Tutor Algebra I at scale. *Educational Evaluation and Policy Analysis, 36*(2), 127–144. doi:10.3102/0162373713507480

Pan, Y., & Zhang, L. (2021). Roles of artificial intelligence in construction engineering and management: A critical review and future trends. *Automation in Construction, 122*, 103517. doi:10.1016/j.autcon.2020.103517

Parker, D. M., Pine, S. G., & Ernst, Z. W. (2019). Privacy and Informed Consent for Research in the Age of Big Data. *Penn State Law Review, 123*(3), 4. https://elibrary.law.psu.edu/pslr/vol123/iss3/4

Pathak, S., & Patra, R. (2015). *Evolution of Political Campaign in India., 2*, 55–59.

Peach, R. L., Yaliraki, S. N., Lefevre, D., & Barahona, M. (2019). Data-driven unsupervised clustering of online learner behaviour. *npj Science of Learning, 4*, 14. doi:10.1038/s41539-019-0054-0

Pedro, F., Subosa, M., Rivas, A., & Valverde, P. (2019). *Artificial intelligence in education: Challenges and opportunities for sustainable development.* Unesco.

Pekrun, R. (2014). Emotions and learning. Retrieved from https://www.ibe.unesco.org/en/document/emotions-and-learning-educational-practices-24

Peña-Guerrero, J., Nguewa, P. A., & García-Sosa, A. T. (2021). Machine learning, artificial intelligence, and data science breaking into drug design and neglected diseases. In Wiley Interdisciplinary Reviews: Computational Molecular Science, 11(5). doi:10.1002/wcms.1513

Peng, J., Jury, E. C., Dönnes, P., & Ciurtin, C. (2021). Machine learning techniques for personalised medicine approaches in immune-mediated chronic inflammatory diseases: Applications and challenges. *Frontiers in Pharmacology, 12*, 720694. doi:10.3389/fphar.2021.720694 PMID:34658859

Perry, B. J., Guo, Y., & Mahmoud, H. N. (2022). Automated site-specific assessment of steel structures through integrating machine learning and fracture mechanics. *Automation in Construction, 133*, 104022. doi:10.1016/j.autcon.2021.104022

Persad, G., Wertheimer, A., & Emanuel, E. J. (2009). Principles for allocation of scarce medical interventions. In The Lancet (Vol. 373, Issue 9661). doi:10.1016/S0140-6736(09)60137-9

Pessach, D., & Shmueli, E. (2023). Algorithmic fairness. In *Machine Learning for Data Science Handbook: Data Mining and Knowledge Discovery Handbook* (pp. 867–886). Springer International Publishing. doi:10.1007/978-3-031-24628-9_37

Peters, D., Vold, K., Robinson, D., & Calvo, R. A. (2020). Responsible AI—Two frameworks for ethical design practice. *IEEE Transactions on Technology and Society, 1*(1), 34–47. doi:10.1109/TTS.2020.2974991

Pierce, F J. (2007). Regional and on arm wireless sensor networks for agricultural Systems in Eastern Washington. *Computers and Electronics in Agriculture, 61*(1).

Pinaya, W. H., Graham, M. S., Kerfoot, E., Tudosiu, P. D., Dafflon, J., Fernandez, V., & Cardoso, M. J. (2023). Generative ai for medical imaging: extending the monai framework. *arXiv preprint arXiv:2307.15208.*

Pink, D. H. (2006). *A whole new mind: Why right-brainers will rule the future.* Penguin.

Ponduri, S. B., Ahmad, S. S., Ravisankar, P., Thakur, D. J., Chawla, K., Chary, D. T., & Sharma, S. (2024). A Study on Recent Trends of Technology and its Impact on Business and Hotel Industry. *Migration Letters : An International Journal of Migration Studies, 21*(S1), 801–806.

Popescu, C.-C. (2018). Improvements in business operations and customer experience through data science and Artificial Intelligence. *Proceedings of the International Conference on Business Excellence, 12*(1). 10.2478/picbe-2018-0072

Porro, C. & Bierce, K. (2018). *AI for good: what CSR professionals should know.* CECP.

Powers, D. (1998). The Total Turing Test and the Loebner Prize. *In D.M.W. Powers (ed.) NeMLaP3/CoNLL98 Workshop on Human Computer Conversation, ACL.*

Pozzi, F. A., & Dwivedi, D. (2023). ESG and IoT: Ensuring Sustainability and Social Responsibility in the Digital Age. In S. Tiwari, F. Ortiz-Rodríguez, S. Mishra, E. Vakaj, & K. Kotecha (Eds.), *Artificial Intelligence: Towards Sustainable Intelligence. AI4S 2023. Communications in Computer and Information Science* (Vol. 1907). Springer. doi:10.1007/978-3-031-47997-7_2

Prakash, V., Moore, M., & Yáñez-Muñoz, R. J. (2016). Current Progress in Therapeutic Gene Editing for Monogenic Diseases. *Molecular Therapy*, 24(3), 465–474. doi:10.1038/mt.2016.5 PMID:26765770

Pramanik, S. (2023). A Novel Data Hiding Locating Approach in Image Steganography. *Multimedia Tools and Applications*. doi:10.1007/s11042-023-16762-3

Prinsloo, P., & Slade, S. (2016). Big data, higher education and learning analytics: Beyond justice, towards an ethics of care. In K. D. Ben (Ed.), *Big data and learning analytics in higher education: Current theory and practice* (pp. 109–124). Springer.

Provost, F., & Fawcett, T. (2013). Data science and its relationship to big data and data-driven decision making. *Big Data*, 1(1), 51–59. doi:10.1089/big.2013.1508 PMID:27447038

Pushpakom, S., Iorio, F., Eyers, P. A., Escott, K. J., Hopper, S., Wells, A., Doig, A., Guilliams, T., Latimer, J., McNamee, C., Norris, A., Sanseau, P., Cavalla, D., & Pirmohamed, M. (2019). Drug repurposing: Progress, challenges and recommendations. *Nature Reviews. Drug Discovery*, 18(1), 41–58. doi:10.1038/nrd.2018.168 PMID:30310233

Qin, Y., Xiao, X., Liu, F., de Sa e Silva, F., Shimabukuro, Y., Arai, E., & Fearnside, P. M. (2023, January 02). de Sa e Silva, F., Shimabukuro, Y., Arai, E., & Fearnside, P. M. (2023). Forest conservation in Indigenous territories and protected areas in the Brazilian Amazon. *Nature Sustainability*, 6(3), 295–305. doi:10.1038/s41893-022-01018-z

Qi, Y., Guo, K., Zhang, C., Guo, D., & Zhi, Z. (2018). A VGI-based Foodborn Disease Report and Forecast System. *Proceedings of the 4th ACM SIGSPATIAL International Workshop on Safety and Resilience, EM-GIS 2018*. ACM. 10.1145/3284103.3284124

Qureshi, H. N., Manalastas, M., Zaidi, S. M. A., Imran, A., & Al Kalaa, M. O. (2021). Service Level Agreements for 5G and Beyond: Overview, Challenges and Enablers of 5G-Healthcare Systems. *IEEE Access : Practical Innovations, Open Solutions*, 9, 1044–1061. doi:10.1109/ACCESS.2020.3046927 PMID:35211361

R, J., & B, S. (2022). Social Impacts of Data Science in Food, Housing And Medical Attention Linked to Public Service. *Technoarete Transactions on Advances in Data Science and Analytics*, 1(1). doi:10.36647/TTADSA/01.01.A004

Raghavan, M., Barocas, S., Kleinberg, J., & Levy, K. (2020, January). Mitigating bias in algorithmic hiring: Evaluating claims and practices. In *Proceedings of the 2020 conference on fairness, accountability, and transparency* (pp. 469-481). ACM. 10.1145/3351095.3372828

Rahamathunnisa, U., Sudhakar, K., Murugan, T. K., Thivaharan, S., Rajkumar, M., & Boopathi, S. (2023). Cloud Computing Principles for Optimizing Robot Task Offloading Processes. In *AI-Enabled Social Robotics in Human Care Services* (pp. 188–211). IGI Global. doi:10.4018/978-1-6684-8171-4.ch007

Rai, A. (2020). Explainable AI: From black box to glass box. *Journal of the Academy of Marketing Science*, 48(1), 137–141. doi:10.1007/s11747-019-00710-5

Rajabifard, A., Binns, A., Masser, I., & Williamson, I. (2006). The role of sub-national government and the private sector in future spatial data infrastructures. *International Journal of Geographical Information Science*, 20(7), 727–741. doi:10.1080/13658810500432224

Raji, I. D., Smart, A., White, R. N., Mitchell, M., Gebru, T., Hutchinson, B., ... Barnes, P. (2020, January). Closing the AI accountability gap: Defining an end-to-end framework for internal algorithmic auditing. In *Proceedings of the 2020 conference on fairness, accountability, and transparency* (pp. 33-44). ACM. 10.1145/3351095.3372873

Rama Krishna, S., Rathor, K., Ranga, J., & Soni, A., D, S., & N, A.K. (2023). Artificial Intelligence Integrated with Big Data Analytics for Enhanced Marketing. *2023 International Conference on Inventive Computation Technologies (ICICT),* (pp. 1073-1077). IEEE. 10.1109/ICICT57646.2023.10134043

Ramachandran, K. K., Mary, S. S. C., Painoli, A. K., Satyala, H., Singh, B., & Manoharan, G. (2022). Assessing The Full Impact Of Technological Advances On Business Management Techniques.

Ramalingam, M., Manoharan, G., & Puviarasi, R. (2021). Web-Based Car Workshop Management System—A Review. Next Generation of Internet of Things. *Proceedings of ICNGIoT, 2021,* 321–331.

Ramudu, K., Mohan, V. M., Jyothirmai, D., Prasad, D., Agrawal, R., & Boopathi, S. (2023). Machine Learning and Artificial Intelligence in Disease Prediction: Applications, Challenges, Limitations, Case Studies, and Future Directions. In Contemporary Applications of Data Fusion for Advanced Healthcare Informatics (pp. 297–318). IGI Global.

Ramzy, A. (2007). The Leader's Guide to Storytelling. Mastering the Art and Discipline of Business Narrative. In Corporate Reputation Review, 10(2). doi:10.1057/palgrave.crr.1550044

Rangaswamy, E., Periyasamy, G., & Nawaz, N. (2021). A study on singapore's ageing population in the context of eldercare initiatives using machine learning algorithms. *Big Data and Cognitive Computing, 5*(4), 51. doi:10.3390/bdcc5040051

Rao, K. (2019). *The Path to 5G for Health Care.*

Raschka, S., Patterson, J., & Nolet, C. (2020). Machine learning in python: Main developments and technology trends in data science, machine learning, and artificial intelligence. In Information (Switzerland), 11(4). doi:10.3390/info11040193

Ray, A. M., Pramanik, S., Das, B., Khanna, A. (2023). Hybrid Cryptography and Steganography Method to Provide Safe Data Transmission in IoT. *ICDAM 2023.* Research Gate.

Ray, A., & Bala, P. K. (2020). Social media for improved process management in organizations during disasters. *Knowledge and Process Management, 27*(1), 63–74. doi:10.1002/kpm.1623

Ray, R., Agar, Z., Dutta, P., Ganguly, S., Sah, P., & Roy, D. (2021). MenGO: A Novel Cloud-Based Digital Healthcare Platform for Andrology Powered By Artificial Intelligence, Data Science & Analytics, Bio-Informatics And Blockchain. *Biomedical Sciences Instrumentation, 57*(4), 476–485. doi:10.34107/KSZV7781.10476

Ray, T. R., Kellogg, R. T., Fargen, K. M., Hui, F., & Vargas, J. (2023). The perils and promises of generative artificial intelligence in neurointerventional surgery. *Journal of Neurointerventional Surgery.* PMID:37438101

Razak, A., Nayak, M. P., Manoharan, G., Durai, S., Rajesh, G. A., Rao, C. B., & Ashtikar, S. P. (2023). Reigniting the power of artificial intelligence in education sector for the educators and students competence. In *Artificial Intelligence and Machine Learning in Smart City Planning* (pp. 103–116). Elsevier. doi:10.1016/B978-0-323-99503-0.00009-0

Reim, W., Åström, J., & Eriksson, O. (2020). Implementation of artificial intelligence (AI): A roadmap for business model innovation. *AI, 1*(2), 11. doi:10.3390/ai1020011

Remian, D. (2019). *Augmenting education: ethical considerations for incorporating artificial intelligence in education.* Academic Press.

Ren,, J., Huang,, S., , & Li,, Y. (2006). The application of AR model in the prediction of citrus canker disease . *Journal of Plant Pathology,* (05), 460–465.

Resnik, D. B. (2021). What Is Ethics in Research & Why Is It Important? National Institute of Environmental Health Sciences.

Richardson, B., & Gilbert, J. E. (2021). A framework for fairness: a systematic review of existing fair AI solutions. *arXiv preprint arXiv:2112.05700.*

Rico-Villademoros, F., Hernando, T., Sanz, J.-L., López-Alonso, A., Salamanca, O., Camps, C., & Rosell, R. (2004). The role of the clinical research coordinator—Data manager—In oncology clinical trials. *BMC Medical Research Methodology, 4*(1), 6. doi:10.1186/1471-2288-4-6 PMID:15043760

Rieck, D., Schünemann, B., & Radusch, I. (2015). Advanced Traffic Light Information in OpenStreetMap for Traffic Simulations. In Lecture Notes in Mobility. Springer. doi:10.1007/978-3-319-15024-6_2

Rienties, B., Boroowa, A., Cross, S., Farrington-Flint, L., Herodotou, C., Prescott, L., & Woodthorpe, J. (2016). Reviewing three case-studies of learning analytics interventions at the Open University UK. In S. Dawson, H. Drachsler, & C. P. Rosé (Eds.), *Enhancing impact: Convergence of communities for grounding, implementation, and validation* (pp. 534–535). ACM. doi:10.1145/2883851.2883886

Rizvi, S., Rienties, B., Rogaten, J., & Kizilcec, R. F. (2020). Investigating variation in learning processes in a FutureLearn MOOC. *Journal of Computing in Higher Education, 32*(1), 162–181. doi:10.1007/s12528-019-09231-0

Robert, L. P., Pierce, C., Marquis, L., Kim, S., & Alahmad, R. (2020). Designing fair AI for managing employees in organizations: A review, critique, and design agenda. *Human-Computer Interaction, 35*(5-6), 545–575. doi:10.1080/07370024.2020.1735391

Rodrigues, R. (2020). Legal and human rights issues of AI: Gaps, challenges and vulnerabilities. *Journal of Responsible Technology, 4*, 100005. doi:10.1016/j.jrt.2020.100005

Rodriguez, G., Torres, H., Fajardo, M., & Medina, J. (2021). Covid-19 in Ecuador: Radiography of Hospital Distribution Using Data Science. *ETCM 2021 - 5th Ecuador Technical Chapters Meeting.* IEEE. 10.1109/ETCM53643.2021.9590641

Rogge, N., Agasisti, T., & De Witte, K. (2017). Big data and the measurement of public organizations' performance and efficiency: The state-of-the-art. *Public Policy and Administration, 32*(4), 263–281. doi:10.1177/0952076716687355

Ronmi, A. E., Prasad, R., & Raphael, B. A. (2023). How can artificial intelligence and data science algorithms predict life expectancy - An empirical investigation spanning 193 countries. *International Journal of Information Management Data Insights, 3*(1), 100168. doi:10.1016/j.jjimei.2023.100168

Rosenberg, S. A., Aebersold, P., Cornetta, K., Kasid, A., Morgan, R. A., Moen, R., Karson, E. M., Lotze, M. T., Yang, J. C., Topalian, S. L., Merino, M. J., Culver, K., Miller, A. D., Blaese, R. M., & Anderson, W. F. (1990). Gene Transfer into Humans—Immunotherapy of Patients with Advanced Melanoma, Using Tumor-Infiltrating Lymphocytes Modified by Retroviral Gene Transduction. *The New England Journal of Medicine, 323*(9), 570–578. doi:10.1056/NEJM199008303230904 PMID:2381442

Ross, A., Chen, N., Hang, E. Z., Glassman, E. L., & Doshi-Velez, F. (2021, May). Evaluating the interpretability of generative models by interactive reconstruction. In *Proceedings of the 2021 CHI Conference on Human Factors in Computing Systems* (pp. 1-15). ACM. 10.1145/3411764.3445296

Roy, A., & Pramanik, S. (2023). A Review of the Hydrogen Fuel Path to Emission Reduction in the Surface Transport Industry. *International Journal of Hydrogen Energy.* doi:10.1016/j.ijhydene.2023.07.010

Russell, C. A., Jones, T. C., & Barr, I. G. (2008). The global circulation of seasonal influenza A (H3N2) viruses. *Science, 320*(5874), 340-346.

Russell, S. J. (2010). *Artificial intelligence a modern approach.* Pearson Education, Inc.

Ryan, R. M., & Deci, L. E. (2017). Self-determination theory: basic psychological needs in motivation. In Self-determination theory: Basic psychological needs in motivation, development, and wellness. IEEE.

Ryan, S., & Carr, A. (2010). Chapter 5 - Applying the biopsychosocial model to the management of rheumatic disease. K. Dziedzic & A. Hammond, (eds). Rheumatology. Elsevier. doi:10.1016/B978-0-443-06934-5.00005-X

Saeed, Z. (2021). Interval–valued fuzzy and intuitionistic fuzzy–KNN for imbalanced data classification. *Expert Systems with Applications*, 184.

Sætra, H. S. (2021). A Framework for Evaluating and Disclosing the ESG Related Impacts of AI with the SDGs. *Sustainability (Basel)*, *13*(15), 8503. doi:10.3390/su13158503

Safiullah, M., Pathak, P., Singh, S., & Anshul, A. (2016). Social media in managing political advertising: A study of India. Polish journal of management. *Studies*, 13.

Sagi, T., Lehahn, Y., & Bar, K. (2020). Artificial intelligence for ocean science data integration: Current state, gaps, and way forward. *Elementa*, *8*, 21. doi:10.1525/elementa.418

Saha, M., Patil, S., Cho, E., Cheng, E. Y. Y., Horng, C., Chauhan, D., Kangas, R., McGovern, R., Li, A., Heer, J., & Froehlich, J. E. (2022). Visualizing Urban Accessibility: Investigating Multi-Stakeholder Perspectives through a Map-based Design Probe Study. *Conference on Human Factors in Computing Systems - Proceedings*. IEEE. 10.1145/3491102.3517460

Sakr, S., Wylot, M., Mutharaju, R., Le Phuoc, D., & Fundulaki, I. (2018). *Linked Data_Storing, Querying, Reasoning.* Springer.

Salehi, H., & Burgueño, R. (2018). Emerging artificial intelligence methods in structural engineering. *Engineering Structures*, *171*, 170–189. doi:10.1016/j.engstruct.2018.05.084

Salter, K. L., & Kothari, A. (2014). Using realist evaluation to open the black box of knowledge translation: A state-of-the-art review. In Implementation Science, 9(1). doi:10.1186/s13012-014-0115-y

Saltzer, J. H., & Schroeder, M. D. (1975). The protection of information in computer systems. *Proceedings of the IEEE*, *63*(9), 1278–1308. doi:10.1109/PROC.1975.9939

Sambasivan, N., Arnesen, E., Hutchinson, B., Doshi, T., & Prabhakaran, V. (2021, March). Re-imagining algorithmic fairness in india and beyond. In *Proceedings of the 2021 ACM conference on fairness, accountability, and transparency* (pp. 315-328). ACM. 10.1145/3442188.3445896

Samikannu, R., Koshariya, A. K., Poornima, E., Ramesh, S., Kumar, A., & Boopathi, S. (2022). Sustainable Development in Modern Aquaponics Cultivation Systems Using IoT Technologies. In *Human Agro-Energy Optimization for Business and Industry* (pp. 105–127). IGI Global.

Sampath, B. (2021). *Sustainable Eco-Friendly Wire-Cut Electrical Discharge Machining: Gas Emission Analysis.*

Sampath, B., Sasikumar, C., & Myilsamy, S. (2023). Application of TOPSIS Optimization Technique in the Micro-Machining Process. In IGI:Trends, Paradigms, and Advances in Mechatronics Engineering (pp. 162–187). IGI Global.

Sampath, B., Pandian, M., Deepa, D., & Subbiah, R. (2022). Operating parameters prediction of liquefied petroleum gas refrigerator using simulated annealing algorithm. *AIP Conference Proceedings*, *2460*(1), 070003. doi:10.1063/5.0095601

Sangiambut, S., & Sieber, R. (2016). The V in VGI: Citizens or Civic Data Sources. *Urban Planning*, *1*(2), 141–154. doi:10.17645/up.v1i2.644

Satpathy, A., Samal, A., Gupta, S., Kumar, S., Sharma, S., Manoharan, G., Karthikeyan, M., & Sharma, S. (2024). To Study the Sustainable Development Practices in Business and Food Industry. *Migration Letters: An International Journal of Migration Studies, 21*(S1), 743–747. doi:10.59670/ml.v21iS1.6400

Saura, J. R., Ribeiro-Soriano, D., & Palacios-Marqués, D. (2022). Assessing behavioral data science privacy issues in government artificial intelligence deployment. *Government Information Quarterly, 39*(4), 101679. doi:10.1016/j.giq.2022.101679

Sbailò, L., Fekete, Á., Ghiringhelli, L. M., & Scheffler, M. (2022). The NOMAD Artificial-Intelligence Toolkit: Turning materials-science data into knowledge and understanding. *npj Computational Materials, 8*(1), 250. doi:10.1038/s41524-022-00935-z

Sboui, T., & Aissi, S. (2022). *A Risk-based Approach for Enhancing the Fitness of use of VGI.* IEEE., doi:10.1109/ACCESS.2022.3201022

Schelter, S., Grafberger, S., & Dunning, T. (2021). HedgeCut: Main- tainingRandomised Trees for Low-Latency Machine Unlearning. *VirtualEvent,China,* (June), 20–25.

Schermer, B. W., Custers, B., & van der Hof, S. (2014). The crisis of consent: How stronger legal protection may lead to weaker consent in data protection. *Ethics and Information Technology, 16*(2), 171–182. doi:10.1007/s10676-014-9343-8

Schmitt, C. M. (2002). Clinical research. *Gastrointestinal Endoscopy Clinics of North America, 12*(2), 395–419. doi:10.1016/S1052-5157(01)00018-6 PMID:12180169

Schunk, D. H., Pintrich, P. R., & Meece, J. L. (2008). *Motivation in education: Theory, research and applications* (3rd ed.). Pearson/Merrill Prentice Hall.

Schwaller, P., Vaucher, A. C., Laplaza, R., Bunne, C., Krause, A., Corminboeuf, C., & Laino, T. (2022). Machine intelligence for chemical reaction space. In Wiley Interdisciplinary Reviews: Computational Molecular Science, 12(5). doi:10.1002/wcms.1604

Schwendimann, B. A., Rodriguez-Triana, M. J., Vozniuk, A., Prieto, L. P., Boroujeni, M. S., Holzer, A., Gillet, D., & Dillenbourg, P. (2017). Perceiving learning at a glance: A systematic literature review of learning dashboard research. *IEEE Transactions on Learning Technologies, 10*(1), 30–41. doi:10.1109/TLT.2016.2599522

Seizov, O., & Wulf, A. J. (2020). Artificial Intelligence and Transparency: A Blueprint for Improving the Regulation of AI Applications in the EU. *European Business Law Review, 31*(4).

Seldon, A., Lakhani, P., & Luckin, R. (2021). *The ethical framework for AI in education.* Buckingham. https://www.buckingham.ac.uk/wp-content/uploads/2021/03/The-Institute-for-Ethical-AI-in-Education-The-Ethical-Framework-for-AI-in-Education.pdf

Sen, A., Ryan, P. B., Goldstein, A., Chakrabarti, S., Wang, S., Koski, E., & Weng, C. (2017). Correlating eligibility criteria generalizability and adverse events using Big Data for patients and clinical trials. *Annals of the New York Academy of Sciences, 1387*(1), 34–43. doi:10.1111/nyas.13195 PMID:27598694

Sengeni, D., Padmapriya, G., Imambi, S. S., Suganthi, D., Suri, A., & Boopathi, S. (2023). Biomedical Waste Handling Method Using Artificial Intelligence Techniques. In *Handbook of Research on Safe Disposal Methods of Municipal Solid Wastes for a Sustainable Environment* (pp. 306–323). IGI Global. doi:10.4018/978-1-6684-8117-2.ch022

SEP. (2014, August 19). *Philosophy of Statistics.* Stanford Encyclopedia of Philosophy. https://plato.stanford.edu/entries/statistics/#StaInd

Sestino, A., & De Mauro, A. (2022). Leveraging artificial intelligence in business: Implications, applications, and methods. *Technology Analysis and Strategic Management, 34*(1), 16–29. doi:10.1080/09537325.2021.1883583

Seyed-Ahmad, F. (2016). *Hough-CNN: Deep learning for segmentation of deep brain regions in MRI and ultrasound.* Elsevier.

Seyyed, R. (2020). *Detection of brain tumors from MRI images base on deep learning using hybrid model CNN and NADE.* Elsevier.

Seyyed-Kalantari, L., Zhang, H., McDermott, M. B., Chen, I. Y., & Ghassemi, M. (2021). Underdiagnosis bias of artificial intelligence algorithms applied to chest radiographs in under-served patient populations. *Nature Medicine, 27*(12), 2176–2182. doi:10.1038/s41591-021-01595-0 PMID:34893776

Sezer, A., Deniz, M., & Topuz, M. (2018). Analysis of Accessibility of Schools in Usak City via Geographical Information Systems (GIS). *Tarih Kultur Ve Sanat Arastirmalari Dergisi-Journal Of History Culture And Art Research, 7*(5).

Shaban-Nejad, A., Michalowski, M., & Buckeridge, D. L. (2018). Health intelligence: how artificial intelligence transforms population and personalized health. In npj Digital Medicine. doi:10.1038/s41746-018-0058-9

Shafik, W. (2023a). Cyber Security Perspectives in Public Spaces: Drone Case Study. In Handbook of Research on Cybersecurity Risk in Contemporary Business Systems (pp. 79-97). IGI Global. doi:10.4018/978-1-6684-7207-1.ch004

Shafik, W. (2023b). Making Cities Smarter: IoT and SDN Applications, Challenges, and Future Trends. In Opportunities and Challenges of Industrial IoT in 5G and 6G Networks (pp. 73-94). IGI Global. doi:10.4018/978-1-7998-9266-3.ch004

Shafik, W. (2024b). Predicting Future Cybercrime Trends in the Metaverse Era. In Forecasting Cyber Crimes in the Age of the Metaverse (pp. 78-113). IGI Global. doi:10.4018/979-8-3693-0220-0.ch005

Shafik, W. (2023c). A Comprehensive Cybersecurity Framework for Present and Future Global Information Technology Organizations. In *Effective Cybersecurity Operations for Enterprise-Wide Systems* (pp. 56–79). IGI Global. doi:10.4018/978-1-6684-9018-1.ch002

Shafik, W. (2024a). Introduction to ChatGPT. In *Advanced Applications of Generative AI and Natural Language Processing Models* (pp. 1–25). IGI Global. doi:10.4018/979-8-3693-0502-7.ch001

Shafik, W., Matinkhah, S. M., & Shokoor, F. (2023). Cybersecurity in unmanned aerial vehicles: A review. *International Journal on Smart Sensing and Intelligent Systems, 16*(1), 20230012. doi:10.2478/ijssis-2023-0012

Shafique, K., Khawaja, B. A., Sabir, F., Qazi, S., & Mustaqim, M. (2020). Internet of things (IoT) for next-generation smart systems: A review of current challenges, future trends and prospects for emerging 5G-IoT Scenarios. In *IEEE Access* (Vol. 8, pp. 23022–23040). Institute of Electrical and Electronics Engineers Inc. doi:10.1109/ACCESS.2020.2970118

Shah, P. K., Pandey, R. P., & Kumar, R. (2016, November). Vector quantization with codebook and index compression. In *2016 International Conference System Modeling & Advancement in Research Trends (SMART)* (pp. 49-52). IEEE.

Shaikh, I. A. K., Kumar, C. N. S., Rohini, P., Jafersadhiq, A., Manoharan, G., & Suryanarayana, V. (2023, August). AST-Graph Convolution Network and LSTM Based Employees Behavioral and Emotional Reactions to Corporate Social Irresponsibility. In *2023 Second International Conference on Augmented Intelligence and Sustainable Systems (ICAISS)* (pp. 966-971). IEEE. 10.1109/ICAISS58487.2023.10250754

Shameem, A., Ramachandran, K. K., Sharma, A., Singh, R., Selvaraj, F. J., & Manoharan, G. (2023, May). The rising importance of AI in boosting the efficiency of online advertising in developing countries. In *2023 3rd International Conference on Advance Computing and Innovative Technologies in Engineering (ICACITE)* (pp. 1762-1766). IEEE. 10.1109/ICACITE57410.2023.10182754

Sharma, A. (2023, January). Exploratory data analysis and deception detection in news articles on social media. *Ain Shams Engineering Journal*, 21. doi:10.1016/j.asej.2023.102166

Sharma, A., Agrawal, R., & Khandelwal, U. (2019). Developing ethical leadership for business organizations. *Leadership and Organization Development Journal*, *40*(6), 712–734. doi:10.1108/LODJ-10-2018-0367

Sharma, N., Chakraborty, C., & Kumar, R. (2022). (2022) Optimized multimedia data through computationally intelligent algorithms. *Multimedia Systems*. doi:10.1007/s00530-022-00918-6

Sharma, R. C., Kawachi, P., & Bozkurt, A. (2019). The landscape of artificial intelligence in open, online and distance education: Promises and concerns. *Asian Journal of Distance Education*, *14*, 1–2. http://www.asianjde.com/ojs/index.php/AsianJDE/article/view/432

Shea, K., & Smith, I. (2005). Intelligent structures: A new direction in structural control. *Artificial Intelligence in Structural Engineering: Information Technology for Design, Collaboration, Maintenance, and Monitoring*, 398–410.

Shernoff, D. J., Cukier, K. N., & Anderson, M. (2020). *Learning by doing: A new approach to education*. W. W. Norton & Company.

Shin, D. (2021). The effects of explainability and causability on perception, trust, and acceptance: Implications for explainable AI. *International Journal of Human-Computer Studies*, *146*, 102551. doi:10.1016/j.ijhcs.2020.102551

Shroff, C., Goswami, A., Prabhu, A., Mohapatra, A., Sengupta, A., Sharma, M., Soni, A., Hussain, S., & Tiwari, S. (2023). *Children and Consent under the Data Protection Act: A Study in Evolution*. Cyril Amarchand Blogs.

Shukla, R. K., Prakash, V., & Pandey, S. (2020, December). A Perspective on Internet of Things: Challenges & Applications. In *2020 9th International Conference System Modeling and Advancement in Research Trends (SMART)* (pp. 184-189). IEEE.

Shukla, R. K., Sengar, A. S., Gupta, A., & Chauhar, N. R. (2022, December). Deep Learning Model to Identify Hide Images using CNN Algorithm. In *2022 11th International Conference on System Modeling & Advancement in Research Trends (SMART)* (pp. 44-51). IEEE. 10.1109/SMART55829.2022.10047661

Shukla, R. K., Sengar, A. S., Gupta, A., Jain, A., Kumar, A., & Vishnoi, N. K. (2021, December). Face Recognition using Convolutional Neural Network in Machine Learning. In *2021 10th International Conference on System Modeling & Advancement in Research Trends (SMART)* (pp. 456-461). IEEE. 10.1109/SMART52563.2021.9676308

Shukla, R. K., Tiwari, A. K., & Verma, V. (2021, December). Identification of with Face Mask and without Face Mask using Face Recognition Model. In *2021 10th International Conference on System Modeling & Advancement in Research Trends (SMART)* (pp. 462-467). IEEE. 10.1109/SMART52563.2021.9676204

Shukla, R. K., & Tiwari, A. K. (2020). A Machine Learning Approaches on Face Detection and Recognition. *Solid State Technology*, *63*(5), 7619–7627.

Shukla, R. K., & Tiwari, A. K. (2023). Masked face recognition using mobilenet v2 with transfer learning. *Computer Systems Science and Engineering*, *45*(1), 293–309. doi:10.32604/csse.2023.027986

Shukla, R. K., Tiwari, A. K., & Jha, A. K. (2023). An Efficient Approach of Face Detection and Prediction of Drowsiness Using SVM. *Mathematical Problems in Engineering*, *2023*, 2023. doi:10.1155/2023/2168361

Shute, V. J. (2008). Unfair and unproductive: Grading and reporting practices undermining public confidence. *Educational Researcher*, *37*(4), 19–33.

Siah, K. W., Kelley, N. W., Ballerstedt, S., Holzhauer, B., Lyu, T., Mettler, D., Sun, S., Wandel, S., Zhong, Y., Zhou, B., Pan, S., Zhou, Y., & Lo, A. W. (2021). Predicting drug approvals: The Novartis data science and artificial intelligence challenge. *Patterns (New York, N.Y.)*, 2(8), 100312. doi:10.1016/j.patter.2021.100312 PMID:34430930

Sidik, S. (2021). *How the COVID-19 Pandemic Has Shaped Data Journalism.* GIJN. https://gijn.org/2021/04/13/how-the-covid-19-pandemic-has-shaped-data-journalism/

Sijing, L., & Lan, W. (2018). Artificial intelligence education ethical problems and solutions. *2018 13th International Conference on Computer Science & Education (ICCSE)*, (pp. 1-5). IEEE. 10.1109/ICCSE.2018.8468773

Silva, M., Santos de Oliveira, L., Andreou, A., Vaz de Melo, P. O., Goga, O., & Benevenuto, F. (2020, April). Facebook ads monitor: An independent auditing system for political ads on facebook. In *Proceedings of The Web Conference 2020* (pp. 224-234). ACM. 10.1145/3366423.3380109

Singh, A. (2019). Remote sensing and GIS applications for municipal waste management. *Journal of Environmental Management*, 243, 22–29. doi:10.1016/j.jenvman.2019.05.017 PMID:31077867

Singh, A., & Chouhan, T. (2023). Artificial Intelligence in HRM: Role of Emotional–Social Intelligence and Future Work Skill. In P. Tyagi, N. Chilamkurti, S. Grima, K. Sood, & B. Balusamy (Eds.), *The Adoption and Effect of Artificial Intelligence on Human Resources Management, Part A* (pp. 175–196). Emerald Studies in Finance, Insurance, and Risk Management. doi:10.1108/978-1-80382-027-920231009

Singh, D., Pati, B., Panigrahi, C. R., & Swagatika, S. (2020). Security Issues in IoT and their Countermeasures in Smart City Applications. *Advances in Intelligent Systems and Computing*, 1089, 301–313. doi:10.1007/978-981-15-1483-8_26

Singh, N., Hamid, Y., Juneja, S., Srivastava, G., Dhiman, G., Gadekallu, T. R., & Shah, M. A. (2023). Load balancing and service discovery using Docker Swarm for microservice based big data applications. *Journal of Cloud Computing (Heidelberg, Germany)*, 12(1), 4. doi:10.1186/s13677-022-00358-7

Singh, V., Braddick, D., & Dhar, P. K. (2017). Exploring the potential of genome editing CRISPR-Cas9 technology. *Gene*, 599, 1–18. doi:10.1016/j.gene.2016.11.008 PMID:27836667

Singh, Y. P., Singh, A. K., & Singh, R. P. (2016). *Web GIS based Framework for Citizen Reporting on Collection of Solid Waste and Mapping in GIS for Allahabad City. SAMRIDDHI : A Journal of Physical Sciences.* Engineering and Technology. doi:10.18090/samriddhi.v8i1.11405

Sleeman, D. H., & Brown, J. S. (1979). Editorial: Intelligent tutoring systems. *International Journal of Man-Machine Studies*, 11(1), 1–3. doi:10.1016/S0020-7373(79)80002-4

Slimi, Z., & Villarejo Carballido, B. (2023). Navigating the Ethical Challenges of Artificial Intelligence in Higher Education: An Analysis of Seven Global AI Ethics Policies. TEM Journal, 12(2).

Smaldone, F., Ippolito, A., Lagger, J., & Pellicano, M. (2022). Employability skills: Profiling data scientists in the digital labour market. *European Management Journal*, 40(5), 671–684. doi:10.1016/j.emj.2022.05.005

Smit, K., Zoet, M., & van Meerten, J. (2020). *A review of AI principles in practice.* Academic Press.

Sohail, A. (2023). Genetic Algorithms in the Fields of Artificial Intelligence and Data Sciences. In Annals of Data Science, 10(4). doi:10.1007/s40745-021-00354-9

Soh, C.-K., & Soh, A.-K. (1988). Example of intelligent structural design system. *Journal of Computing in Civil Engineering*, 2(4), 329–345. doi:10.1061/(ASCE)0887-3801(1988)2:4(329)

Solís, P., McCusker, B., Menkiti, N., Cowan, N., & Blevins, C. (2018). Engaging global youth in participatory spatial data creation for the UN sustainable development goals: The case of open mapping for malaria prevention. *Applied Geography (Sevenoaks, England)*, *98*, 143–155. doi:10.1016/j.apgeog.2018.07.013

Sood, K., Pathak, P., Jain, J., & Gupta, S. (2023). How does an investor prioritize ESG factors in India? An assessment based on fuzzy AHP. *Managerial Finance*, *49*(1), 66–87. doi:10.1108/MF-04-2022-0162

Spector-Bagdady, K. (2023). Generative-AI-Generated Challenges for Health Data Research. *The American Journal of Bioethics*, *23*(10), 1–5. doi:10.1080/15265161.2023.2252311 PMID:37831940

Srinivas, B., Maguluri, L. P., Naidu, K. V., Reddy, L. C. S., Deivakani, M., & Boopathi, S. (2023). Architecture and Framework for Interfacing Cloud-Enabled Robots. In *Handbook of Research on Data Science and Cybersecurity Innovations in Industry 4.0 Technologies* (pp. 542–560). IGI Global. doi:10.4018/978-1-6684-8145-5.ch027

Srinivasu, P. N., Ijaz, M. F., Shafi, J., Wozniak, M., & Sujatha, R. (2022). *6G Driven Fast Computational Networking Framework for Healthcare Applications*. IEEE. doi:10.1109/ACCESS.2022.3203061

Srivastava, S., Ahmed, T., & Saxena, A. (2023). An Approach To Secure Iot Applications Of Smart City Using Blockchain Technology. In International Journal of Engineering Sciences & Emerging Technologies, 11(2).

stahlesq.com. (2018). *How the IRS Uses Artificial Intelligence to Detect Tax Evaders*. Stahlesq. https://stahlesq.com/criminal-defense-law-blog/how-the-irs-uses-artificial-intelligence-to-detect-tax-evaders/

Stansbury, N., Barnes, B., Adams, A., Berlien, R., Branco, D., Brown, D., Butler, P., Garson, L., Jendrasek, D., Manasco, G., Ramirez, N., Sanjuan, N., Worman, G., & Adelfio, A. (2022). Risk-Based Monitoring in Clinical Trials: Increased Adoption Throughout 2020. *Therapeutic Innovation & Regulatory Science*, *56*(3), 415–422. doi:10.1007/s43441-022-00387-z PMID:35235192

Steinhubl, S. R., Muse, E. D., & Topol, E. J. (2013). Can Mobile Health Technologies Transform Health Care? *Journal of the American Medical Association*, *310*(22), 2395. doi:10.1001/jama.2013.281078 PMID:24158428

Steiniger, S., Poorazizi, M. E., Scott, D. R., Fuentes, C., & Crespo, R. (2016). Can we use OpenStreetMap POIs for the Evaluation of Urban Accessibility? *International Conference on GIScience Short Paper Proceedings, 1.* ACM. 10.21433/B31167F0678P

Stone, P., Brooks, R., Brynjolfsson, E., Calo, R., Etzioni, O., Hager, G., & Teller, A. (2022). *Artificial intelligence and life in 2030: the one hundred year study on artificial intelligence*. arXiv preprint arXiv:2211.06318.

Stracener, C., Samelson, Q., MacKie, J., Ihaza, M., Laplante, P. A., & Amaba, B. (2019). The Internet of Things Grows Artificial Intelligence and Data Sciences. *IT Professional*, *21*(3), 55–62. doi:10.1109/MITP.2019.2912729

Stuck, M., & Grunes, A. (2016). *Big data and competition policy*. Oxford University Press.

Sturrock, H. J. W., Woolheater, K., Bennett, A. F., Andrade-Pacheco, R., & Midekisa, A. (2018). Predicting residential structures from open source remotely enumerated data using machine learning. *PLoS One*, *13*(9), e0204399. Advance online publication. doi:10.1371/journal.pone.0204399 PMID:30240429

Subrahmanya, S. V. G., Shetty, D. K., Patil, V., Hameed, B. M. Z., Paul, R., Smriti, K., Naik, N., & Somani, B. K. (2022). The role of data science in healthcare advancements: applications, benefits, and future prospects. In Irish Journal of Medical Science, 191(4). doi:10.1007/s11845-021-02730-z

Sun, H., Burton, H. V., & Huang, H. (2021). Machine learning applications for building structural design and performance assessment: State-of-the-art review. *Journal of Building Engineering*, *33*, 101816. doi:10.1016/j.jobe.2020.101816

Sun, L., Shang, Z., Xia, Y., Bhowmick, S., & Nagarajaiah, S. (2020a). Review of bridge structural health monitoring aided by big data and artificial intelligence: From condition assessment to damage detection. *Journal of Structural Engineering, 146*(5), 04020073. doi:10.1061/(ASCE)ST.1943-541X.0002535

Syamala, M., Komala, C., Pramila, P., Dash, S., Meenakshi, S., & Boopathi, S. (2023). Machine Learning-Integrated IoT-Based Smart Home Energy Management System. In *Handbook of Research on Deep Learning Techniques for Cloud-Based Industrial IoT* (pp. 219–235). IGI Global. doi:10.4018/978-1-6684-8098-4.ch013

Tachibana, M., Amato, P., Sparman, M., Woodward, J., Sanchis, D. M., Ma, H., Gutierrez, N. M., Tippner-Hedges, R., Kang, E., Lee, H.-S., Ramsey, C., Masterson, K., Battaglia, D., Lee, D., Wu, D., Jensen, J., Patton, P., Gokhale, S., Stouffer, R., & Mitalipov, S. (2013). Towards germline gene therapy of inherited mitochondrial diseases. *Nature, 493*(7434), 627–631. doi:10.1038/nature11647 PMID:23103867

Tacho, A. (2019). AI in Indian education: Opportunities and challenges. In: *International Journal of Education and Development, 54*(3). https://www.researchgate.net/publication359046086_Education_and_the_Use_of_Artificial_Intelligence

Tamboli, A. (2019). Evaluating Risks of the AI Solution. *Keeping Your AI Under Control*, 31–42. doi:10.1007/978-1-4842-5467-7_4

Tang, A., Li, K. K., Kwok, K. O., Cao, L., Luong, S., & Tam, W. (2023). The importance of transparency: Declaring the use of generative artificial intelligence (AI) in academic writing. *Journal of Nursing Scholarship*, jnu.12938. doi:10.1111/jnu.12938 PMID:37904646

Tarafdar, M., Beath, C. M., & Ross, J. W. (2019). Using AI to enhance business operations. *MIT Sloan Management Review, 60*(4).

Tarmuji, I., Maelah, R., & Tarmuji, N. H. (2016). The impact of environmental, social and governance practices (ESG) on economic performance: Evidence from ESG score. International Journal of Trade. *Economics and Finance, 7*(3), 67–74.

TEDxTalks. (2020, January 16). *Data Privacy and Consent | Fred Cate | TEDxIndianaUniversity* [Video]. YouTube. https://youtu.be/2iPDpV8ojHA

Teo, C. T., Abdollahzadeh, M., & Cheung, N. M. (2023, June). Fair generative models via transfer learning. *Proceedings of the AAAI Conference on Artificial Intelligence, 37*(2), 2429–2437. doi:10.1609/aaai.v37i2.25339

Teoh, T. T., & Goh, Y. J. (2023). *Artificial Intelligence in Business Management*. Artificial Intelligence in Business Management. doi:10.1007/978-981-99-4558-0

Thaker, K., Huang, Y., Brusilovsky, P., & He, D. (2018*). Dynamic knowledge modeling with heterogeneous activities for adaptive textbooks.* Paper presented at the 11th International conference Educational Data Mining (EDM 2018), Buffalo.

Thales. (2021, May 10). *BEYOND GDPR: DATA PROTECTION AROUND THE WORLD*. Thales. https://www.thalesgroup.com/en/markets/digital-identity-and-security/government/magazine/beyond-gdpr-data-protection-around-world

Thavamani, S., Mahesh, D., Sinthuja, U., & Manoharan, G. (2022, May). Crucial attacks in internet of things via artificial intelligence techniques: The security survey. In AIP Conference Proceedings (Vol. 2418, No. 1). AIP Publishing.

Thiagarajan, J. J., Sattigeri, P., Rajan, D., & Venkatesh, B. (2020). Calibrating healthcare ai: Towards reliable and interpretable deep predictive models. *arXiv preprint arXiv:2004.14480*.

Thierer, A. D., Castillo O'Sullivan, A., & Russell, R. (2017). *Artificial intelligence and public policy*. Mercatus Research Paper.

Thomas, B. (2020). *Machine Unlearning: Linear Filtration for Logit-based Classifiers*. arXiv:2002.02730.

Thompson, A. F., Afolayan, A. H., & Ibidunmoye, E. O. (2013). Application of geographic information system to solid waste management. *2013 Pan African International Conference on Information Science, Computing and Telecommunications, PACT 2013*. IEEE. 10.1109/SCAT.2013.7055110

Thuraisingham, B. (2020). Artificial Intelligence and Data Science Governance: Roles and Responsibilities at the C-Level and the Board. *Proceedings - 2020 IEEE 21st International Conference on Information Reuse and Integration for Data Science, IRI 2020*. IEEE. 10.1109/IRI49571.2020.00052

Tian, Y., Yi, C., & Wang, X. (2015). A Method for Identifying Apple Pest Damage Defects and Fruit Stems/Calyx Based on Hyperspectral Imaging. *Journal of Agricultural Engineering*, (4), 325–331.

Tigistu, T. (2021). Classification of rose flowers based on Fourier descriptors and color moments. *Multimedia Tools and Applications*, *80*(30).

Tinkham, S. F., & Weaver-Lariscy, R. A. (1993). A diagnostic approach to assessing the impact of negative political television commercials. *Journal of Broadcasting & Electronic Media*, *37*, 377–399.

To, H., Kim, S. H., & Shahabi, C. (2015). Effectively crowdsourcing the acquisition and analysis of visual data for disaster response. *Proceedings - 2015 IEEE International Conference on Big Data, IEEE Big Data 2015*. IEEE. 10.1109/BigData.2015.7363814

Tong, X., Liu, X., Tan, X., Li, X., Jiang, J., Xiong, Z., Xu, T., Jiang, H., Qiao, N., & Zheng, M. (2021). Generative Models for De Novo Drug Design. *Journal of Medicinal Chemistry*, *64*(19), 14011–14027. doi:10.1021/acs.jmedchem.1c00927 PMID:34533311

Toreini, E., Aitken, M., Coopamootoo, K., Elliott, K., Zelaya, C. G., & Van Moorsel, A. (2020, January). The relationship between trust in AI and trustworthy machine learning technologies. In *Proceedings of the 2020 conference on fairness, accountability, and transparency* (pp. 272-283). ACM. 10.1145/3351095.3372834

Tran, K., Barbeau, S., Hillsman, E., & Labrador, M. A. (2013). GO_Sync - A Framework to Synchronize Crowd-Sourced Mapping Contributors from Online Communities and Transit Agency Bus Stop Inventories. *International Journal of Intelligent Transportation Systems Research*, *11*(2), 54–64. doi:10.1007/s13177-013-0056-x

Tripathi, P. K., Shukla, R. K., Tiwari, N. K., Thakur, B. K., Tripathi, R., & Pal, S. (2022, December). Enhancing Security of PGP with Steganography. In *2022 11th International Conference on System Modeling & Advancement in Research Trends (SMART)* (pp. 1555-1560). IEEE. 10.1109/SMART55829.2022.10046709

Tripathi, M. A., Tripathi, R., Effendy, F., Manoharan, G., Paul, M. J., & Aarif, M. (2023, January). An In-Depth Analysis of the Role That ML and Big Data Play in Driving Digital Marketing's Paradigm Shift. In *2023 International Conference on Computer Communication and Informatics (ICCCI)* (pp. 1-6). IEEE. 10.1109/ICCCI56745.2023.10128357

Truby, J. (2020). Governing artificial intelligence to benefit the UN sustainable development goals. *Sustainable Development (Bradford)*, *28*(4), 946–959. doi:10.1002/sd.2048

Trujillo, J. (2021). The Intelligence of Machines. Filosofija. *Sociologija*, *32*(1), 84–92. doi:10.6001/fil-soc.v32i1.4383

Trujillo-Cabezas, R. (2020). Integrating Foresight, Artificial Intelligence and Data Science to Develop Dynamic Futures Analysis. *Journal of Information Systems Engineering & Management*, *5*(3), em0120. doi:10.29333/jisem/8428

Tsou, M. H., Jung, C. Te, Allen, C., Yang, J. A., Han, S. Y., Spitzberg, B. H., & Dozier, J. (2017). Building a real-time geo-targeted event observation (Geo) viewer for disaster management and situation awareness. *Lecture Notes in Geoinformation and Cartography*. doi:10.1007/978-3-319-57336-6_7

Turing, A. (2012). Computing Machinery and Intelligence (1950). In J. Copeland (Ed.), *A. Turing, The Essential Turing: Seminal Writings in Computing, Logic, Philosophy, Artificial Intelligence, and Artificial Life: Plus The Secrets of Enigma by B* (pp. 433–464).

Turusbekova, N., Broekhuis, M., Emans, B., & Molleman, E. (2007). The role of individual accountability in promoting quality management systems. *Total Quality Management & Business Excellence, 18*(5), 471–482. doi:10.1080/14783360701239917

Tzavella, K., Fekete, A., & Fiedrich, F. (2018). Opportunities provided by geographic information systems and volunteered geographic information for a timely emergency response during flood events in Cologne, Germany. *Natural Hazards, 91.* doi:10.1007/s11069-017-3102-1

Ugandar, R. E., Rahamathunnisa, U., Sajithra, S., Christiana, M. B. V., Palai, B. K., & Boopathi, S. (2023). Hospital Waste Management Using Internet of Things and Deep Learning: Enhanced Efficiency and Sustainability. In M. Arshad (Ed.), (pp. 317–343). Advances in Bioinformatics and Biomedical Engineering. IGI Global. doi:10.4018/978-1-6684-6577-6.ch015

UNESCO Mahatma Gandhi Institute of Education for Peace and Sustainable Development. (2020). *Ethical frameworks for AI in education in India.* UNESCO. https://mgiep.unesco.org/

UNESCO. (2021). *Artificial intelligence in education: Opportunities and challenges for schools.* UNESCO. https://unesdoc.unesco.org/ark:/48223/pf0000366994

UNESCO. (2021). *The Open University of China awarded UNESCO Prize for its use of AI to empower rural learners.* UNESCO. https://en.unesco.org/news/open-university-china-awarded-unesco-prize-its-use-ai-empower-rural-learners

United Nations. (1987). *Report of the World Commission on Environment and Development: Our Common Future ('Brundtland Report').* Oxford University Press.

Urban Davis, J., Anderson, F., Stroetzel, M., Grossman, T., & Fitzmaurice, G. (2021, June). Designing co-creative ai for virtual environments. In Creativity and Cognition (pp. 1-11). doi:10.1145/3450741.3465260

Valcarenghi, L., Pacini, A., Borromeo, J. C., Fichera, S., Gagliardi, M., Amram, D., & Lionetti, V. (2022). Managing Physical Distancing Through 5G and Accelerated Edge Cloud. *IEEE Access : Practical Innovations, Open Solutions, 10*, 104169–104177. doi:10.1109/ACCESS.2022.3210262

Valinsky, J. (2019, April 11). *Amazon reportedly employs thousands of people to listen to your Alexa conversations.* CNN Business. https://edition.cnn.com/2019/04/11/tech/amazon-alexa-listening/index.html

Vamathevan, J., Clark, D., Czodrowski, P., Dunham, I., Ferran, E., Lee, G., Li, B., Madabhushi, A., Shah, P., Spitzer, M., & Zhao, S. (2019). Applications of machine learning in drug discovery and development. *Nature Reviews. Drug Discovery, 18*(6), 463–477. doi:10.1038/s41573-019-0024-5 PMID:30976107

Van Norman, G. A. (2021). Decentralized Clinical Trials. *JACC. Basic to Translational Science, 6*(4), 384–387. doi:10.1016/j.jacbts.2021.01.011 PMID:33997523

Vancauwenberghe, G., Valečkaitė, K., van Loenen, B., & Welle Donker, F. (2018). Assessing the Openness of Spatial Data Infrastructures (SDI): Towards a Map of Open SDI. *International Journal of Spatial Data Infrastructures Research, 13.* doi:10.2902/1725-0463.2018.13.art9

Varian, H. (2018). Artificial intelligence, economics, and industrial organization. In *The economics of artificial intelligence: an agenda* (pp. 399–419). University of Chicago Press.

Varun, G. (2021). Adaptive Machine Unlearning. *35th Conference on Neural Information Processing Systems (NIPS).* IEEE.

Vedaei, S. S., Fotovvat, A., Mohebbian, M. R., Rahman, G. M. E., Wahid, K. A., Babyn, P., Marateb, H. R., Mansourian, M., & Sami, R. (2020). COVID-SAFE: An IoT-based system for automated health monitoring and surveillance in post-pandemic life. *IEEE Access : Practical Innovations, Open Solutions, 8,* 188538–188551. doi:10.1109/ACCESS.2020.3030194 PMID:34812362

Venkateswaran, N., Vidhya, K., Ayyannan, M., Chavan, S. M., Sekar, K., & Boopathi, S. (2023). A Study on Smart Energy Management Framework Using Cloud Computing. In 5G, Artificial Intelligence, and Next Generation Internet of Things: Digital Innovation for Green and Sustainable Economies (pp. 189–212). IGI Global. doi:10.4018/978-1-6684-8634-4.ch009

Venkateswaran, N., Kumar, S. S., Diwakar, G., Gnanasangeetha, D., & Boopathi, S. (2023). Synthetic Biology for Waste Water to Energy Conversion: IoT and AI Approaches. In M. Arshad (Ed.), (pp. 360–384). Advances in Bioinformatics and Biomedical Engineering. IGI Global. doi:10.4018/978-1-6684-6577-6.ch017

Vennila, T., Karuna, M., Srivastava, B. K., Venugopal, J., Surakasi, R., & Sampath, B. (2022). New Strategies in Treatment and Enzymatic Processes: Ethanol Production From Sugarcane Bagasse. In Human Agro-Energy Optimization for Business and Industry (pp. 219–240). IGI Global.

Verbin, I. (2020). *Corporate Responsibility in the Digital Age: A Practitioner's Roadmap for Corporate Responsibility in the Digital Age.* Routledge. doi:10.4324/9781003054795

Verganti, R., Vendraminelli, L., & Iansiti, M. (2020). Innovation and design in the age of artificial intelligence. *Journal of Product Innovation Management, 37*(3), 212–227. doi:10.1111/jpim.12523

Verma, K., Bhardwaj, S., Arya, R., Islam, U. L., Bhushan, M., Kumar, A., & Samant, P. (2019). *Latest tools for data mining and machine learning.*

Verma, S., Sharma, R., Deb, S., & Maitra, D. (2021). Artificial intelligence in marketing: Systematic review and future research direction. *International Journal of Information Management Data Insights, 1*(1), 100002. doi:10.1016/j.jjimei.2020.100002

Vesselinov, V., Alexandrov, B., & O'Malley, D. (2019). Non negative Tensor Factorization for Contaminant Source Identification. *Journal of Contaminant Hydrology, 220,* 66–97. doi:10.1016/j.jconhyd.2018.11.010 PMID:30528243

Villarreal-Torres, H., Ángeles-Morales, J., Cano-Mejía, J., Mejía-Murillo, C., Flores-Reyes, G., Cruz-Cruz, O., Marín-Rodriguez, W., Andrade-Girón, D., Carreño-Cisneros, E., & Boscán-Carroz, M. C. (2023). Development of a Classification Model for Predicting Student Payment Behavior Using Artificial Intelligence and Data Science Techniques. *EAI Endorsed Transactions on Scalable Information Systems, 10*(5). doi:10.4108/eetsis.3489

Vincent-Lancrin, S., & Van der Vlies, R. (2020). *Trustworthy artificial intelligence (AI) in education: Promises and challenges.* OECD.

Vinod, D. N., & Prabaharan, S. R. S. (2020). Data science and the role of Artificial Intelligence in achieving the fast diagnosis of Covid-19. *Chaos, Solitons, and Fractals, 140,* 110182. doi:10.1016/j.chaos.2020.110182 PMID:32834658

Vinuesa, R., Azizpour, H., Leite, I., Balaam, M., Dignum, V., Domisch, S., Felländer, A., Daniela Langhans, S., Tegmark, M., & Fuso Nerini, F. (2020). The role of artificial intelligence in achieving the Sustainable Development Goals. *Nature Communications, 11*(1), 1–10. doi:10.1038/s41467-019-14108-y PMID:31932590

Wagstaff, K. (2013). *Massive Target credit card breach new step in security war with hackers: experts.* NBC News. https://www.nbcnews.com/technology/massive-target-credit-card-breach-new-step-security-war-hackers-2D11778083

Wahdain, E. A., Baharudin, A. S., & Ahmad, M. N. (2019). Big data analytics in the malaysian public sector: The determinants of value creation. *Advances in Intelligent Systems and Computing, 843,* 139–150. doi:10.1007/978-3-319-99007-1_14

Walker, E., Rummel, N., & Koedinger, K. R. (2009). Integrating collaboration and intelligent tutoring data in the evaluation of a reciprocal peer tutoring environment. *Research and Practice in Technology Enhanced Learning, 4*(3), 221–251. doi:10.1142/S179320680900074X

Walker, J., Pekmezovic, A., & Walker, G. (2019). *Sustainable Development Goals: Harnessing Business to Achieve the SDGs through Finance, Technology and Law Reform.* John Wiley & Sons. doi:10.1002/9781119541851

WalkowiakE.MacDonaldT. (2023). Generative AI and the Workforce: What Are the Risks? *Available at* SSRN. doi:10.2139/ssrn.4568684

Walmsley, J. (2021). Artificial intelligence and the value of transparency. *AI & Society, 36*(2), 585–595. https://doi.org/https://doi.org/10.1007/s00146-020-01066-z. doi:10.1007/s00146-020-01066-z

Walters, W. P., & Barzilay, R. (2021). Critical assessment of AI in drug discovery. *Expert Opinion on Drug Discovery, 16*(9), 937–947. doi:10.1080/17460441.2021.1915982 PMID:33870801

Walters, W. P., & Murcko, M. (2020). Assessing the impact of generative AI on medicinal chemistry. *Nature Biotechnology, 38*(2), 143–145. doi:10.1038/s41587-020-0418-2 PMID:32001834

Wamba-Taguimdje, S. L., Fosso Wamba, S., Kala Kamdjoug, J. R., & Tchatchouang Wanko, C. E. (2020). Influence of artificial intelligence (AI) on firm performance: The business value of AI-based transformation projects. *Business Process Management Journal, 26*(7), 1893–1924. doi:10.1108/BPMJ-10-2019-0411

Wan, W. Y., Tsimplis, M., Siau, K. L., Yue, W. T., Nah, F. F. H., & Yu, G. M. (2022). Legal and Regulatory Issues on Artificial Intelligence, Machine Learning, Data Science, and Big Data. Lecture Notes in Computer Science (Including Subseries Lecture Notes in Artificial Intelligence and Lecture Notes in Bioinformatics), 13518 LNCS. Springer. doi:10.1007/978-3-031-21707-4_40

Wang, J. (2012). *Machine learning based prediction of cross species transmission and antigen relationship of influenza A virus.* Huazhong University of Science and Technology.

Wang, T., Zhang, Y., Qi, S., Zhao, R., Xia, Z., & Weng, J. (2023). Security and privacy on generative data in aigc: A survey. *arXiv preprint arXiv:2309.09435.*

Wang, D. Q., Feng, L. Y., Ye, J. G., Zou, J. G., & Zheng, Y. F. (2023). Accelerating the integration of ChatGPT and other large-scale AI models into biomedical research and healthcare. *MedComm - Future Medicine, 2*(2), e43. doi:10.1002/mef2.43

Wang, X. (2018). Overview of the research status on artificial neural networks. *Proceedings of the 2nd International Forum on Management, Education and Information Technology Application (IFMEITA 2017).* Atlantis Press. 10.2991/ifmeita-17.2018.59

Ward, T. M., Mascagni, P., Madani, A., Padoy, N., Perretta, S., & Hashimoto, D. A. (2021). Surgical data science and artificial intelligence for surgical education. In Journal of Surgical Oncology, 124(2). doi:10.1002/jso.26496

Warren, C. (2023). *Not all parts of the rainforest have suffered equally. Satellite images plot a path for Amazon protection.* Anthropocene. https://www.anthropocenemagazine.org/2023/01/not-all-parts-of-the-rainforest-have-suffered-equally-satellite-images-plot-a-path-for-amazon-protection/

Weick, K. E., Sutcliffe, K. M., & Obstfeld, D. (2005). Organizing and the process of sensemaking. In Organization Science, 16(4). doi:10.1287/orsc.1050.0133

Weller, M. (2017). *The machine classroom: Artificial intelligence and education.* John Wiley & Sons.

Weng, Y., Xiao, H., Zhang, J., Liang, X.-J., & Huang, Y. (2019). RNAi therapeutic and its innovative biotechnological evolution. *Biotechnology Advances, 37*(5), 801–825. doi:10.1016/j.biotechadv.2019.04.012 PMID:31034960

Westermayr, J., Gastegger, M., Schütt, K. T., & Maurer, R. J. (2021). Perspective on integrating machine learning into computational chemistry and materials science. *The Journal of Chemical Physics, 154*(23), 230903. doi:10.1063/5.0047760 PMID:34241249

Whittaker, M., Crawford, K., Dobbe, R., Fried, G., Kaziunas, E., Mathur, V., & Schwartz, O. (2018). *AI now report 2018.* AI Now Institute at New York University.

Widayanti, R., & Meria, L. (2023). *Business Modeling Innovation Using Artificial Intelligence Technology. International Transactions on Education Technology.* ITEE.

Will, D. H. (2020). Predictive policing algorithms are racist. They need to be dismantled. *MIT Technology Review.* https://www.technologyreview.com/2020/07/17/1005396/predictive-policing-algorithms-racist-dismantled-machine-learning-bias-criminal-justice/

Williamson, B. (2015). Governing methods: Policy innovation labs, design and data science in the digital governance of education. *Journal of Educational Administration and History, 47*(3), 251–271. doi:10.1080/00220620.2015.1038693

Williamson, B. (2018). Silicon startup schools: Technocracy, algorithmic imaginaries and venture philanthropy in corporate education reform. *Critical Studies in Education, 59*(2), 218–236. doi:10.1080/17508487.2016.1186710

Wing, J. (2019). The Data Life Cycle. *Harvard Data Science Review, 1*(1), 6. doi:10.1162/99608f92.e26845b4

Wirtz, B. W., Weyerer, J. C., & Sturm, B. J. (2020). The dark sides of artificial intelligence: An integrated AI governance framework for public administration. *International Journal of Public Administration, 43*(9), 818–829. doi:10.1080/01900692.2020.1749851

Wong, W., & Hinnant, C. C. (2022). Competing perspectives on the Big Data revolution: A typology of applications in public policy. *Journal of Economic Policy Reform.* doi:10.1080/17487870.2022.2103701

Woodman, R. J., & Mangoni, A. A. (2023). A comprehensive review of machine learning algorithms and their application in geriatric medicine: Present and future. *Aging Clinical and Experimental Research, 35*(11), 2363–2397. doi:10.1007/s40520-023-02552-2 PMID:37682491

World Innovation Summit for Education. (2015). *The Pratham Education Foundation: Transforming learning in India.* WISE. https://www.wise-qatar.org/2015-summiteducation-invest-impact/

Wu, W., Doan, A., & Yu, C. (2009). Modeling and Extracting Deep-Web Query Interfaces. Berlin, Germany: Springer.

Wu, Y., Zhang, Z., Kou, G., Zhang, H., Chao, X., Li, C. C., Dong, Y., & Herrera, F. (2021). Distributed linguistic representations in decision making: Taxonomy, key elements and applications, and challenges in data science and explainable artificial intelligence. In Information Fusion (Vol. 65). doi:10.1016/j.inffus.2020.08.018

Wu, F., Lu, C., Zhu, M., Chen, H., Zhu, J., Yu, K., Li, L., Li, M., Chen, Q., Li, X., Cao, X., Wang, Z., Zha, Z., Zhuang, Y., & Pan, Y. (2020). Towards a new generation of artificial intelligence in China. *Nature Machine Intelligence, 2*(6), 312–316. doi:10.1038/s42256-020-0183-4

Xie, J., Nozawa, W., Yagi, M., Fujii, H., & Managi, S. (2019). Do environmental, social, and governance activities improve corporate financial performance? *Business Strategy and the Environment, 28*(2), 286–300. doi:10.1002/bse.2224

Xie, X., Zhou, Y., Xu, Y., Hu, Y., & Wu, C. (2019). OpenStreetMap Data Quality Assessment via Deep Learning and Remote Sensing Imagery. *IEEE Access : Practical Innovations, Open Solutions, 7*, 176884–176895. doi:10.1109/AC-CESS.2019.2957825

Xiuli, L., Zhaohao, H., & Yongqiang, B. (2021). Research on facial recognition algorithms based on improved LBP and DBN. *Industrial Instruments and Automation Devices*, (5), 80-2.

Yan, M. (2020). Research on the Application of Internet of Things Technology in the Development of Modern Agriculture. *Small and Medium sized Enterprise Management and Technology (Second Edition)*, (11), 191-3.

Yang, D., Liu, H., Goga, A., Kim, S., Yuneva, M., & Bishop, J. M. (2010). Therapeutic potential of a synthetic lethal interaction between the *MYC* proto-oncogene and inhibition of aurora-B kinase. *Proceedings of the National Academy of Sciences of the United States of America, 107*(31), 13836–13841. doi:10.1073/pnas.1008366107 PMID:20643922

Yang, S. J., Ogata, H., Matsui, T., & Chen, N. S. (2021). Human-centered artificial intelligence in education: Seeing the invisible through the visible. *Computers and Education: Artificial Intelligence, 2*, 100008. doi:10.1016/j.caeai.2021.100008

Yarger, L., Cobb Payton, F., & Neupane, B. (2020). Algorithmic equity in the hiring of underrepresented IT job candidates. *Online Information Review, 44*(2), 383–395. doi:10.1108/OIR-10-2018-0334

Yingwei, Y., Dawei, M., & Hongchao, F. (2020). A research framework for the application of volunteered geographic information in post-disaster recovery monitoring. *Tropical Geography, 40*(2). doi:10.13284/j.cnki.rddl.003239

Yogarajan, V., Dobbie, G., Leitch, S., Keegan, T. T., Bensemann, J., Witbrock, M., Asrani, V. M., & Reith, D. M. (2022). Data and model bias in artificial intelligence for healthcare applications in New Zealand. *Frontiers of Computer Science, 4*, 1070493. doi:10.3389/fcomp.2022.1070493

Yoon, B., Lee, J. H., & Byun, R. (2018). Does ESG performance enhance firm value? Evidence from Korea. [DOI]. *Sustainability (Basel), 10*(10), 3635. doi:10.3390/su10103635

Young, E., Wajcman, J., & Sprejer, L. (2023). Mind the gender gap: Inequalities in the emergent professions of artificial intelligence (AI) and data science. *New Technology, Work and Employment, 38*(3), 391–414. doi:10.1111/ntwe.12278

Yudianto, M. R. AAgustin, TJames, R M. (2021). Rainfall Forecasting to Recommend Crops Varieties Using Moving Average and Naive Bayes Methods. *International Journal of Modern Education and Computer Science, 13*(3).

Yu, S., Chai, Y., Samtani, S., Liu, H., & Chen, H. (2023). Motion Sensor–Based Fall Prevention for Senior Care: A Hidden Markov Model with Generative Adversarial Network Approach. *Information Systems Research*, isre.2023.1203. doi:10.1287/isre.2023.1203

ZamanB. U. (2023). Transforming Education Through AI, Benefits, Risks, and Ethical Considerations. Authorea Preprints.

Zandbergen, P. A. (2014). Ensuring Confidentiality of Geocoded Health Data: Assessing Geographic Masking Strategies for Individual-Level Data. *Advances in Medicine, 2014*, 1–14. doi:10.1155/2014/567049 PMID:26556417

Završnik, A. (2021). Algorithmic justice: Algorithms and big data in criminal justice settings. *European Journal of Criminology, 18*(5), 623–642. doi:10.1177/1477370819876762

Zawacki-Richter, O., Marín, V. I., Bond, M., & Gouverneur, F. (2019). Systematic review of research on artificial intelligence applications in higher education – Where are the educators? *International Journal of Educational Technology in Higher Education, 16*(1), 39. doi:10.1186/s41239-019-0171-0

Zeide, E. (2019). Artificial intelligence in higher education: Applications, promise and perils, and ethical questions. *EDUCAUSE Review*, 31–39. https://er.educause.edu/-/media/files/articles/2019/8/er193104.pdf

Zeng, L. (2023). Generative AI in Public Opinion Guidance during Emergency Public Events: Challenges, Opportunities, and Ethical Considerations. *Public Opinion Guidance during Emergency Public Events: Challenges, Opportunities, and Ethical Considerations (March 30, 2023).*

Zeng, X., Wang, F., Luo, Y., Kang, S. G., Tang, J., Lightstone, F. C., Fang, E. F., Cornell, W., Nussinov, R., & Cheng, F. (2022). Deep generative molecular design reshapes drug discovery. *Cell Reports Medicine, 3*(12), 100794. doi:10.1016/j.xcrm.2022.100794 PMID:36306797

Zhang, B., Korolj, A., Lai, B. F. L., & Radisic, M. (2018). Advances in organ-on-a-chip engineering. *Nature Reviews. Materials, 3*(8), 257–278. doi:10.1038/s41578-018-0034-7

Zhang, P., & Kamel Boulos, M. N. (2023). Generative AI in Medicine and Healthcare: Promises, Opportunities and Challenges. *Future Internet, 15*(9), 286. doi:10.3390/fi15090286

Zhang, X., Rane, K., Kakaravada, I., & Shabaz, M. (2021). Research on vibration monitoring and fault diagnosis of rotating machinery based on internet of things technology. *Nonlinear Engineering, 10*(1), 245–254. doi:10.1515/nleng-2021-0019

Zhao, J., & Fariñas, B. G. (2023). Artificial Intelligence and Sustainable Decisions. *European Business Organization Law Review, 24*(1), 1–39. doi:10.1007/s40804-022-00262-2

Zhong, H., Chang, J., Yang, Z., Wu, T., Mahawaga Arachchige, P. C., Pathmabandu, C., & Xue, M. (2023, April). Copyright protection and accountability of generative AI: Attack, watermarking and attribution. In *Companion Proceedings of the ACM Web Conference 2023* (pp. 94-98). ACM.

Zhou, Z. H. (2021). *Machine learning.* Springer Nature. doi:10.1007/978-981-15-1967-3

Zilske, M., Neumann, A., & Nagel, K. (2011). OpenStreetMap For Traffic Simulation. M. Schmidt, G. Gartner *(Eds.), Proceedings of the 1st European State of the Map – OpenStreetMap Conference, No. 11-10.* Springer.

About the Contributors

Rajeev Kumar is a proficient academician and academic administrator with more than 14 years of experience in developing the strategy towards the excellence in professional education. Prof. Kumar is currently serving as Professor of Computer Science and Engineering Department, Moradabad Institute of Technology, Moradabad, Uttar Pradesh, India. Prof. Kumar, earned the intellect in distinction as Ph.D. (Computer Science), D.Sc. (Post-Doctoral Degree) in Computer Science, Postdoctoral Fellowship (Malaysia). He has done certification in Data Science and Machine Learning using python and R Programming from IIM Raipur and certification from IBM, Google, etc, Senior member of IEEE and core team member of IEEE young professional committee and he is having membership of Computer Society of India, and AIEEE. He has participated many training programs in leadership, and also delivers the expert talks on how to develop or enrich the curriculum, development of PO, PSO, PEOs, and how to design the vision and mission and the implementation. His academic areas of interest and specialization include Artificial Intelligence, Cloud Computing, e-governance and Networking.

Ankush Joshi, a seasoned professional with 13 years of experience, holds an impressive academic background. His pursuit of knowledge continued with postgraduate qualifications, including an MCA and M.Tech, culminating in a Ph.D. Currently serving as an Assistant Professor at COER University, Roorkee, Dr. Joshi specializes in the dynamic fields of Artificial Intelligence, Machine Learning, and Data Science. His extensive expertise and dedication make him a valuable asset in shaping the academic landscape at COER. He is authored and co-authored more than 10 papers in refereed international journals and IEEE conferences, Served as a reviewer and chaired a session in IEEE conferences.

Hari Om Sharan, serving as Dean – Academic Affairs & FET at Rama University Uttar Pradesh, Kanpur (India), he is having more than 15 Years of experience in academic as well in research, his research area is AI, Security and HPC. Dr. Sharan received his Ph.D in DNA Computing: A Novel Approach towards the solution of NP-Complete Problems, and his UG (B.Tech) & PG (M.Tech) in Computer Science and Engineering, He has done International certification in Data Science and Artificial Intelligence Machine Learning Deep Learning and its Application. Dr. Sharan also published three (03) books on Mobile Network Technology. Dr. Sharan developed short term course on Artificial Intelligence Machine Learning Deep Learning and its Application, and published 14 patents (National/International) & copyrights mostly in Computer Science and Engineering to serve the nation in the field of research. Dr. Sharan authored and coauthored more than 60 papers in refereed international journal & many international Conferences and National Conferences & serve as editor/Reviewer of different international journals, Springer International Conferences.

Sheng-Lung Peng is a Professor and the director (head) of the Department of Creative Technologies and Product Design, National Taipei University of Business, Taiwan. He received the PhD degree in Computer Science from the National Tsing Hua University, Taiwan. He is an honorary Professor of Beijing Information Science and Technology University, China, and a visiting Professor of Ningxia Institute of Science and Technology, China. He is also an adjunct Professor of Mandsaur University, India. Dr. Peng has edited several special issues at journals, such as Soft Computing, Journal of Internet Technology, Journal of Real-Time Image Processing, International Journal of Knowledge and System Science, MDPI Algorithms, and so on. His research interests are in designing and analyzing algorithms for Bioinformatics, Combinatorics, Data Mining, and Networks areas in which he has published over 100 research papers.

Anshika, a proactive BTech CSE student specializing in Business Systems at Chandigarh University. With a knack for frontend development, Anshika is a skilled React and Tailwind CSS enthusiast, dedicated to crafting seamless user interfaces. Beyond her coding prowess, she harbors a deep interest in UI/UX, aiming to create designs that seamlessly merge functionality and aesthetics. Anshika's academic journey is not just about mastering technology; it's a mission to bridge the gap between tech and business, driven by innovation and a commitment to transformative change. In her free time, she stays abreast of industry trends, contributing to open-source projects and attending tech meetups, poised to make a significant impact in the ever-evolving tech landscape.

Anurag A. S. obtained his BE degree in Electrical and Electronics from Anna University in 2013, followed by an MBA degree in Operations from Indira Gandhi National Open University in 2017. Accumulating nearly a decade of practical working experience in both the Engineering and Management fields, Anurag has garnered invaluable insights within various organizations across two culturally diverse countries. His extensive professional journey has deepened his understanding of the global significance of management functions. Currently pursuing a Ph.D. at the Central University of Kerala within the Management Studies Department, Anurag's research revolves around the intersection of Management and Artificial Intelligence. Fueled by a unique blend of engineering and business acumen, Anurag is wholeheartedly committed to unraveling the intricacies of how technology, specifically Artificial Intelligence, can enhance and streamline various facets of management in the contemporary business landscape.

Munir Ahmad, Ph.D. in Computer Science, brings over 24 years of invaluable expertise in the realm of spatial data development, management, processing, visualization, and quality assurance. His unwavering commitment to open data, crowdsourced data, volunteered geographic information, and spatial data infrastructure has solidified him as a seasoned professional and a trusted trainer in cutting-edge spatial technologies. With a profound passion for research, Munir has authored more than 30 publications in his field, culminating in the award of his Ph.D. in Computer Science from Preston University Pakistan in 2022. His dedication to propelling the industry forward and sharing his extensive knowledge defines his mission. Connect with Munir to delve into the world of Spatial Data, GIS, and GeoTech. #SpatialData #GIS #GeoTech

Abhay Bhatia is Working as Assistant Professor in Department of Computer Science and Engineering at Roorkee Institute of Technology, Roorkee, Haridwar Uttarakhand. He is having 12+ years of academic experience and worked with various reputed engineering institutions. He has completed his B.Tech in Computer Science and Engineering from AKTU (formely UPTU), M.Tech in Computer Science and

Engineering from Rajasthan and Ph.D. in Wireless Sensor Networks. He is currently an active member of IEEE as well as a reviewer for several journals too. He is having distinguish record of research papers with 17+ Indexed, Scopus, IEEE and SCI papers. He also visited many institutes for guest lecture on various upcoming research. Moreover 4 patents with 5 book chapters are also in his bucket, he is also author to a book on IoT, as a researcher with research area of Artificial Intelligence, Machine Learning, Image Processing and Wireless Sensor Network his work is up heading to great research.

Pankhuri Bhatia is working as assistant professor in department of management of GRD IMT Dehradun Uttarakhand India. She did her Bcom from Shree Dev Suman University after which completed her MBA in HR and International Business. Currently she is pursuing PhD in Management. Moreover she had been member of several committees, and currently working on topic related to the progression growth of Uttarakhand. She is not only a good academician as well as having keen interest in mandla art.

Rohit Dhalwal is an MBA student at Lovely Professional University, India.

Dwijendra Nath Dwivedi is a professional with 20+ years of subject matter expertise creating right value propositions for analytics and AI. He currently heads the EMEA+AP AI and IoT team at SAS, a worldwide frontrunner in AI technology. He is a post-Graduate in Economics from Indira Gandhi Institute of Development and Research andis PHD from crackow university of economics Poland. He has presented his research in more than 20 international conference and published several Scopus indexed paper on AI adoption in many areas. As an author he has contributed to more than 8 books and has more than 25 publications in high impact journals. He conducts AI Value seminars and workshops for the executive audience and for power users.

Ashok Singh Gaur is a Assistant Professor, School of Computer Application, Noida Institute of Engineering and Technology Greater Noida. He is having 12+ years of teaching and research experience in various institutions. He is pursuing his Ph.D. degree in Computer Science and Engineering from Rama University, Kanpur, Uttar Pradesh. He has done his M.C.A. in year 2010. His research interests are in Artificial Intelligence, Blockchain Technology, Machine Learning, etc. So far, he has published 02 research papers in reputed international/national journals and conferences. In addition, he has also participated in many reputed national conferences/ FDP/ Workshops etc.

Ankur Gupta has received the B.Tech and M.Tech in Computer Science and Engineering from Ganga Institute of Technology and Management, Kablana affiliated with Maharshi Dayanand University, Rohtak in 2015 and 2017. He is an Assistant Professor in the Department of Computer Science and Engineering at Vaish College of Engineering, Rohtak, and has been working there since January 2019. He has many publications in various reputed national/ international conferences, journals, and online book chapter contributions (Indexed by SCIE, Scopus, ESCI, ACM, DBLP, etc). He is doing research in the field of cloud computing, data security & machine learning. His research work in M.Tech was based on biometric security in cloud computing.

Sanjiv Kumar Jain is working as an A.P. (S.G.) at the Department of Electrical Engineering, Medi-Caps University, Indore, India. He completed his PhD from Maulana Azad National Institute of Technology, Bhopal, India. He received best teaching award and best project mentor award at university level. His research interests include power system control and planning, evolutionary algorithms and neural networks.

Achukutla Kumar is dedicated, motivated, and detail-oriented clinical researcher having experience in Drug & Vaccine development, Epidemiological and various clinical research studies. Also worked in Infectious disease surveillance and Pandemic management. Well versed in data management as well. Skilled in scientific literature review, writing manuscripts and publications. Having an overall experience in the field of Clinical, biomedical and implementation research.

Dinesh Kumar (DJ) is an assistant professor at Mittal School of Business, Lovely Professional University. He is the founder of Mission Dost-E-Jahan and Pomento. He can be contacted at linktr.ee/realdrdj.

Anil Kumar is Working as Assistant Professor in Department of Information Technology at Ajay Kumar Garg Engineering College, Ghaziabad India. He is having more than 16 years of academic experience and worked with various reputed engineering institutions. He has completed his B.Tech in Computer Science and Engineering from AKTU (formely UPTU), M.Tech in Computer Science and Engineering from Rajasthan and Ph.D. Pursuing. He had reviewed several journals articles too. He is having distinguish record of research papers with more than 15 International, Scopus and SCI papers. He is also in his bucket as a researcher with research area of Artificial Intelligence, Machine Learning, Image Processing and Computer Network.

Pradeep Kumar is working in the Dept. of College of computing Sciences & IT, Teerthanker Mahaveer University, Moradabad. He has 12 years of experience in the reputed organization of the computer application department. Currently he is pursuing Ph.D from Babasaheb Bhimrao Ambedkar Bihar University, India in computer application. He has completed his M.Tech(CSE) from Uttarakhand Technical University, Dehradun. His research area is Blockchain Technology that he has working on the web security process.

Shivani Malhan is working as an Assistant Professor in Chitkara University. She completed her PhD in July,2020. Also, she has a corporate experience of two years in Tata Motors where she worked as a Territory Sales Manager and Regional Accessories manager. Moreover, she has worked as an Assistant Professor in DAV University in Marketing Management for five years. She has published many research papers in UGC Care listed journals and scopus indexed journals and has attended many national and international conferences and seminars. She has been awarded the "Best research paper presentation award" by IIT Roorkee.

Geetha Manoharan is currently working in Telangana as an assistant professor at SR University. She is the university-level PhD program coordinator and has also been given the additional responsibility of In Charge Director of Publications and Patents under the Research Division at SR University. Under her tutelage, students are inspired to reach their full potential in all areas of their education and beyond through experiential learning. It creates an atmosphere conducive to the growth of students into independent thinkers and avid readers. She has more than ten years of experience across the board in the business world, academia, and the academy. She has a keen interest in the study of organizational behavior and management. More than forty articles and books have been published in scholarly venues such as UGC-refereed, SCOPUS, Web of Science, and Springer. Over the past six-plus years, she has participated in varied research and student exchange programs at both the national and international levels. A total of five of her collaborative innovations in this area have already been published and patented.

Emotional intelligence, self-efficacy, and work-life balance are among her specialties. She organizes programs for academic organizations. She belongs to several professional organizations, including the CMA and the CPC. The TIPSGLOBAL Institute of Coimbatore has recognized her twice (in 2017 and 2018) for her outstanding academic performance.

Sabyasachi Pramanik is a professional IEEE member. He obtained a PhD in Computer Science and Engineering from Sri Satya Sai University of Technology and Medical Sciences, Bhopal, India. Presently, he is an Associate Professor, Department of Computer Science and Engineering, Haldia Institute of Technology, India. He has many publications in various reputed international conferences, journals, and book chapters (Indexed by SCIE, Scopus, ESCI, etc). He is doing research in the fields of Artificial Intelligence, Data Privacy, Cybersecurity, Network Security, and Machine Learning. He also serves on the editorial boards of several international journals. He is a reviewer of journal articles from IEEE, Springer, Elsevier, Inderscience, IET and IGI Global. He has reviewed many conference papers, has been a keynote speaker, session chair, and technical program committee member at many international conferences. He has authored a book on Wireless Sensor Network. He has edited 8 books from IGI Global, CRC Press, Springer and Wiley Publications.

Chakunta Rao is serving as Professor & Dean, School of CS & AI as well as Director of Evaluation at SR University.

Wasswa Shafik is an IEEE member, P.Eng received a bachelor of science in Information Technology Engineering with a minor in Mathematics in 2016 from Ndejje University, Kampala, Uganda, a PhD and master of engineering in Information Technology Engineering (MIT) in 2020, from the Computer Engineering Department, Yazd University, Islamic Republic of Iran. He is an associate researcher at the Computer Science department, Network interconnectivity Lab at Yazd University, Islamic Republic of Iran, and at Information Sciences, Prince Sultan University, Saudi Arabia. His areas of interest are Computer Vision, Anomaly Detection, Drones (UAVs), Machine/Deep Learning, AI-enabled IoT/ IoMTs, IoT/IIoT/OT Security, Cyber Security and Privacy. Shafik is the chair/co-chair/program chair of some Scopus/EI conferences. Also, academic editor/ associate editor for set of indexed journals (Scopus journals' quartile ranking). He is the founder and lead investigator of Digital Connectivity Research Laboratory (DCR-Lab) since 2019.

Anu Sharma, working as an Assistant Professor in Moradabad Institute of Technology, Moradabad, UP, India. She is pursuing Ph.D in CSE from Uttrakhand Technical University, Dehradaun. She has done M.Tech (CSE) from MMU, Mullana, Ambala. Her current research interest includes data Mining, Neural Network, Machine Learning. She is having experience of more than 15 years. She has published more than 35 research papers in International Journals/ Conferences including SCI/Scopus

Gurwinder Singh, is currently working as an Associate Professor at the Department of AIT-CSE, Chandigarh University, Punjab, India. With a strong focus on optimization techniques, particularly heuristic and meta-heuristic algorithms, his research work revolves around their application to various combinatorial optimization problems, including different variants of the Transportation Problem. He has made significant contributions to the academic community with a notable publication record. He has authored five research papers published in SCI journals and ten papers in IEEE/Scopus-indexed conferences. As a passionate researcher, he serves as a peer reviewer for prestigious journals, including the International Journal of Systems Assurance Engineering and Management and the International Journal of Soft Computing.

Saadia Ureeb is a PhD scholar working on collaborative pan Industrial Studies and exploring the Information technology advancement impact on various financial industries. The aim is to minimize the operational losses while working on data analytics. She has a diverse experience in Banking and Risk Management. A graduate from International Islamic University researched in MS(CS) on computer networking and security. During her studies, she worked on Artificial Intelligence, Congestion issues and Migrating inputs. Currently her scope is exploring the operational risk and financial losses due to inadequate processes and Knowledge Management.

Index

A

accountability 23, 35, 39, 57, 62-64, 100, 123, 156-158, 160, 163, 168-169, 173, 175-176, 178, 184, 187, 189-190, 195, 200, 206, 233, 236, 238-239, 244, 295, 298, 303-305, 307, 310, 325-326, 337, 339, 364-368, 371-372, 376-377, 380

AI Accountability 178

AI Algorithms 3, 7, 19, 23, 25, 47, 73, 87-89, 158-161, 163-165, 174, 181-184, 189, 208, 250, 273, 299, 301, 363, 370

AI Bias and Fairness 187, 295

AI Ethics 37-38, 157, 177, 188, 196-197, 199, 201, 208, 248, 255, 303-304, 362, 365, 372, 377

AI Ethics KPIs 196

AI Frameworks 27, 62-63, 65, 70, 367

AI Implementations 18, 196, 200

AI Integration 20, 34-36, 77-78, 185, 291

AI Solutions 64, 70, 75, 77, 80, 178, 203, 306, 380

AI Transparency 196, 295

AI Winter 297, 309, 311-312

Alan Turing 295, 297

Algorithmic Governance 196

architectural framework 343, 352, 354-355, 358

Artificial Intelligence (AI) 1, 3, 13, 18, 20, 31, 33-36, 38-39, 55-58, 60, 64, 68, 70-71, 75, 78-79, 88-89, 93, 97-98, 100-101, 107, 156, 159, 163, 169-170, 194, 196-203, 245-248, 250-251, 272-274, 278, 295, 309, 311, 320, 325, 335, 362, 383-384, 387

Artificial intelligence in education 36-39, 50-52, 94, 97, 174, 177, 179, 308, 379

B

Bias 4, 18-20, 23, 25-27, 35, 39, 45-46, 50, 52, 66, 68, 85, 93, 100, 103, 107, 157-169, 171-172, 174-178, 184, 187, 189, 192, 198-199, 201, 205, 208-210, 225-226, 228-229, 233, 236, 245, 287, 295, 298-300, 303-304, 307-311, 362, 365, 368-369, 375

B

Big Data 3, 15-16, 53, 67, 69, 72, 76, 92, 94-95, 97-98, 101, 136, 155, 177, 179, 209, 240, 242, 245-246, 249-250, 253-254, 256, 293, 295-296, 298, 300-301, 308-311, 315, 320-321, 325-326, 335, 337-342, 364, 383-384, 387

Blockchain 4, 14, 27, 67, 95, 270, 291, 313, 316, 319-320, 325, 335, 360, 385

Brain tumor 110, 115-117

Business Intelligence 72, 76-77, 83, 94, 253, 339

C

Chatbot 79, 81, 199, 245

Clinical Compliance 1

CNN 110-113, 115-117, 141, 271, 300, 311

Collaboration Between Industry Stakeholders and Regulators 70

Consent 2, 4, 103, 118-119, 123-134, 136-137, 139, 157, 181, 184-185, 187-189, 228, 233, 257, 269, 298, 300-302, 308, 310, 364, 367, 372

Cookie 125, 301-302, 311

Cyber security 252, 257-258, 265, 268-269, 313, 315-316, 386

Cyber-crime 258, 263-264

D

Dashboards 39, 42, 50

Data 1, 3-7, 9-13, 15-16, 18-20, 22-23, 25, 27, 29-31, 33-36, 39-40, 42-45, 47-54, 56-83, 85, 87-95, 97-98, 100-101, 103-107, 109-113, 116-130, 132, 135-137, 139-151, 154-155, 157-160, 162-167, 170-172, 174, 176-177, 179-187, 189-192, 194, 196-204, 208-210, 212-216, 218-223, 225-226, 228-233, 236, 239-264, 266-271, 273-274, 276-278, 285-287, 292-293, 296-299, 301-318, 320-326, 329-355, 358-370, 372-379, 381-387

Data Ethics 295, 304, 306, 309, 377

Data Integration 70, 343-345, 352, 358, 360, 385

Data Privacy 4, 18-20, 23, 33-36, 39, 46, 56, 61, 71, 93, 118-119, 122, 124, 136, 139, 154, 157, 186, 189, 201, 203, 226, 245, 295, 300-302, 304, 306-308, 310, 363, 370, 375-376

Data Protection 7, 30-31, 45, 53, 119, 121-123, 128-129, 137, 184, 186-187, 201, 203, 302, 304-305, 310, 372, 376

Data Quality and Availability 70

Data Science (DS) 245-246, 248, 362

Data Scientist 253, 304-305, 309, 312

Data Security 58, 65, 70, 85, 100, 181, 184, 189, 269, 298, 309, 316, 333, 367, 376

Data Solutions 62-63, 70

DDoS 257, 259, 317

Decentralized Trials 1, 13, 15

Decsion Making 225

deep learning 2, 12, 77, 91-92, 110, 116-117, 163, 168, 177, 191, 204, 206, 214, 245-246, 248, 270-271, 291, 293, 342, 359-360

Difficulties and Barriers 70

E

E Data 118

Economies 63, 93, 294

Education 18-26, 29-54, 61, 63, 71, 75, 92, 94-95, 97-98, 108, 122, 156-157, 169, 174, 177, 179, 188-189, 191, 204-205, 211, 224, 226, 236, 238-240, 242-243, 245, 250-251, 254-255, 268, 304, 306-308, 311, 315, 320, 326, 330, 333, 335, 342, 374-375, 379, 381-382, 387

Educational Technology 18, 44-45, 48, 50-51, 54

Efficient Data Removal 138

E-Health Care 313

Environmental, Social, and Governance (ESG) 55, 57-60, 63-65, 68, 70-71

Ethical AI 25, 29, 34-35, 47, 62-66, 70-71, 156, 158, 162, 169-170, 175, 185, 188, 198, 204, 303-307, 365, 367, 371, 380-381

Ethical and Privacy Concerns 71

Ethical Challenges 18-20, 22, 35-36, 38-39, 50, 61, 157-159, 163, 180-181, 184, 188-189, 208, 227, 295, 304, 306-307, 336, 362-365, 367, 371

Ethical Frameworks 38, 163, 180-181, 185, 187, 189, 204, 287, 371, 378, 380

Ethical Governance 197

Ethics 33, 36-39, 49-50, 53, 57, 61, 64, 66-68, 72, 119, 136-137, 156-157, 169, 175-177, 181, 188, 191, 196-199, 201-202, 204-205, 207-209, 225-228, 236-241, 243-245, 248, 255, 288, 295, 298, 303-

304, 306, 309-310, 312, 336, 362, 365, 372-373, 375, 377-378, 380

Explainable AI 156, 165, 167, 173-174, 176, 178, 368

F

Forecasting Future Trends and Anticipating Challenges 71

G

Gene Editing 1, 7-8, 13, 16

Generative AI 156, 180-182, 184, 190-195, 386

Generative Artificial Intelligence 180, 192-194

GIS 225, 239, 241, 340-341

H

Healthcare Disparities 180-181, 184, 190

Healthcare Ethics 180-181

High Risk AI 196

I

Inadequate Knowledge and Understanding 71

incorporation of ESG 55, 58

Incremental Learning 138, 141, 143, 145, 148

Influence of a Datapoint 138

Information Management 67, 69, 109, 118, 174, 193, 385

Information security 122-123, 128, 132, 257, 373

Informed Consent 2, 118-119, 123-124, 127, 131, 133-134, 136, 181, 184-185, 188, 269, 367

Integrating Artificial Intelligence (AI) With Environmental, Social, and Governance (ESG) Factors 71

Intelligent Structural Engineering 272, 282

IoT 67, 69, 75-76, 79, 96, 101, 209, 211-214, 222-223, 246, 253, 257, 259, 269, 285, 287, 291-293, 313-320, 325, 335, 379, 386

J

John McCarthy 246, 296-297, 307

John Wilder Tukey 297, 307

Justice 46, 53, 56, 64, 157, 159-160, 162, 164-165, 170, 175, 179, 188-189, 191-192, 200, 202, 229, 233, 243, 304-306, 324, 362-363, 365-366, 368, 372, 375, 377

L

Learner-facing 39-43, 45, 49

Limitations on Expenses and Available Resources 61, 71

M

Machine Learning (ML) 1, 3, 13, 59, 245-246, 248, 272-277, 286, 296, 298

machine learning algorithms 6, 82, 107, 116, 157, 180, 194, 198, 211, 253, 273, 276-277, 285, 287, 340, 363, 375-376

Malware and Viruses 257

ML Drug Development 1

moral issues 55, 100

MRI 110-111, 115-117

N

Natural Language Processing 6, 39, 56, 59, 80-81, 83, 89-91, 167, 177, 248, 295, 297-298, 379, 386

Network Security 251, 257, 260, 266-268, 314

Neural Networks 59, 110, 143, 145, 157, 180, 223, 246, 248, 250, 276, 297, 302-303, 311

O

OpenStreetMap 320-322, 324, 329-331, 336-337, 340-342

operational framework design 343, 359

Opportunities 2-3, 15, 18, 20, 30, 32, 35-36, 38-39, 42, 47-48, 50, 55, 65, 69, 76-77, 94, 101, 130-131, 164, 170, 175, 177, 181, 190, 192-193, 195, 207, 229, 239, 257, 264, 269, 291, 298, 307-308, 310, 329, 333, 338-339, 341, 383, 385-386

Optimization Strategies 272, 274, 278

P

Patient Privacy 181-182, 184, 186-187

Patient Research 1

Performance contrast 272

Phishing 257-259, 263, 267

placement of nodes 211

Predictive Analytics 71, 76, 83, 89, 91, 95, 181, 186, 249, 253, 325-326, 375

Privacy 4, 18-20, 23, 25, 27-29, 33-36, 39, 42, 46, 49-51, 56, 58, 61, 66, 70-71, 85, 93, 100, 103, 107, 118-126, 129, 136, 138-140, 142, 154, 157, 161, 180-182, 184-187, 189, 192, 194, 197, 200-203, 205, 225-

226, 228-229, 232-233, 245, 257, 269, 287, 295, 297-298, 300-302, 304-310, 314-318, 333, 339, 362-363, 365, 367-368, 370, 372, 375-376, 381, 386

Public Sector 57, 118, 320-321, 325-326, 332-337, 339-340, 342, 385

Q

Quantum Computing 252, 288-289

R

responsible AI 20, 31, 34-36, 163, 169, 178, 188, 192, 198, 202, 306, 308, 366, 378, 381

S

social responsibility 55-58, 61, 63, 66, 70, 209, 304

socioeconomic Groups 164, 362

Spatial Data 225-233, 239-244, 321, 332, 335, 341

Statistics 70, 155, 213, 246, 248, 254-255, 296-298, 304, 309-310, 312, 337

Sustainability 33, 55-61, 63-69, 71, 95, 209, 223, 243, 273-275, 277, 279, 281-289, 291-293, 324, 330, 335-336, 338, 376, 379, 382

sustainable solutions 55-56, 232

T

Teacher-facing 39-41, 43, 45

technological uses 55

TPB 118, 130-131, 135

Transparency 4, 23, 26-27, 29, 33, 35, 49-50, 56-57, 62-63, 68, 71, 76, 123, 157-158, 160, 162, 165, 167-169, 173, 175-176, 178, 184-187, 189-190, 194, 196, 201-203, 209, 225-226, 228, 232-233, 236, 238-239, 295, 298, 302-305, 309, 311, 326, 335, 337, 362-365, 368, 370-372, 374, 376-377, 380-381

U

unstructured data source 343, 359

V

View Dependent Data Integration System 343, 352, 358

Volunteered Geographic Information 320-325, 327-335, 337-339, 341-342

W

web-query-interface 343, 359

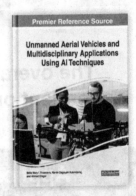

Ensure Quality Research is Introduced to the Academic Community

Become an Reviewer for IGI Global Authored Book Projects

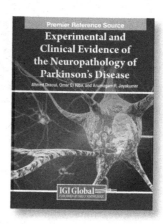

The overall success of an authored book project is dependent on quality and timely manuscript evaluations.

Applications and Inquiries may be sent to:
development@igi-global.com

Applicants must have a doctorate (or equivalent degree) as well as publishing, research, and reviewing experience. Authored Book Evaluators are appointed for one-year terms and are expected to complete at least three evaluations per term. Upon successful completion of this term, evaluators can be considered for an additional term.

If you have a colleague that may be interested in this opportunity, we encourage you to share this information with them.

Submit an Open Access Book Proposal

Have Your Work Fully & Freely Available Worldwide After Publication

Seeking the Following Book Classification Types:

Authored & Edited Monographs • Casebooks • Encyclopedias • Handbooks of Research

Gold, Platinum, & Retrospective OA Opportunities to Choose From

Easily Track Your Work in Our Advanced Manuscript Submission System With **Rapid Turnaround Times**

Double-Blind Peer Review by Notable Editorial Boards (*Committee on Publication Ethics* (COPE) Certified

Publications Adhere to All **Current OA Mandates & Compliances**

Affordable APCs *(Often 50% Lower Than the Industry Average)* Including Robust Editorial Service Provisions

Direct Connections with **Prominent Research Funders** & OA Regulatory Groups

Institution Level OA Agreements Available (Recommend or Contact Your Librarian for Details)

Join a **Diverse Community** of 150,000+ Researchers **Worldwide** Publishing With IGI Global

Content Spread Widely to Leading Repositories (AGOSR, ResearchGate, CORE, & More)

Retrospective Open Access Publishing

You Can Unlock Your Recently Published Work, Including Full Book & Individual Chapter Content to Enjoy All the Benefits of Open Access Publishing

Learn More

Printed in the United States
by Baker & Taylor Publisher Services

Printed in the United States
by Baker & Taylor Publisher Services